GW00541887

ISBN 978-1-332-95662-3
PIBN 10442707

er's ionary:

OR,

Gentleman and Architect's COMPANION.

Explaining not only the

TERMS of ART

In all the ſeveral

PARTS of ARCHITECTURE,

But, alſo containing the

THEORY and PRACTICE

Of the

Various BRANCHES thereof, requiſite to be known by

MASONS,	PLAISTERERS,	TURNERS,
CARPENTERS,	PAINTERS,	CARVERS,
JOINERS,	GLAZIERS,	STATUARIES,
BRICKLAYERS,	SMITHS,	PLUMBERS, &c.

Alſo Neceſſary Problems in

ARITHMETIC, GEOMETRY, MECHANICS, PERSPECTIVE, HYDRAULICS, and other MATHEMATICAL SCIENCES.

Together with

The Quantities, Proportions, and Prices of all Kinds of MATERIALS uſed in BUILDING; with DIRECTIONS for Chuſing, Preparing, and Uſing them: The ſeveral Proportions of the FIVE ORDERS of ARCHITECTURE, and all their Members, according to VITRUVIUS, PALLADIO, SCAMOZZI, VIGNOLA, M. LE CLERC, &c.

With RULES for the Valuation of HOUSES, and the EXPENCE calculated of Erecting any FABRICK, Great or Small.

The Whole Illuſtrated with more than Two Hundred FIGURES, many of them curiouſly Engraven on COPPER-PLATES: Being a Work of great Uſe, not only to ARTIFICERS, but likewiſe to GENTLEMEN, and others, concerned in BUILDING, &c.

Faithfully Digeſted from the moſt Approved Writers on theſe Subjects.

In TWO VOLUMES.

LONDON:

Printed for A. BETTESWORTH and C. HITCH, at the *Red-Lion* in *Pater-noſter-Row*; and S. AUSTEN, at the *Angel* and *Bible* in *St. Paul's Church-Yard.*

M.DCC.XXXIV.

January 11. 173¾.

W*E have perused these Two Volumes of the* Builder's Dictionary, *and do think they contain a great deal of useful Knowledge in the Building Business.*

Nicholas Hawksmoor,
John James,
James Gibbs.

21
1–

THE PREFACE.

ARCHITECTURE is one of thoſe Arts which Neceſſity has made univerſal: From the Time that Men firſt felt the Inclemencies of the Seaſons, it had its Beginning; and accordingly it has ſpread whereſoever the Severities of the Climate demanded Shelter or Shade: It is to be traced in the *Indian*'s Hut and the *Icelander*'s Cave; and ſtill ſhews, in thoſe barbarous Parts of the Globe, from what mean Original it roſe to its preſent Glory.

As Diſtreſs was the Parent of it, ſo Convenience was the firſt Object it regarded: Magnificence and Decoration were the Reſult of long Refinement, and deſigned to flatter the Oſtentation of the Owners: Politeneſs is but a more delicate Term for Luxury; and was it not natural for Men to grow wanton with Eaſe and Affluence, all the Sciences in general had laid inactive, nor ever ſtarted into Being.

'Tis eaſy to conclude from hence, That *Convenience* ſhould ſtill be the Builder's firſt View: Every Structure is raiſed to anſwer ſome particular End; and the moſt obvious and ſimple Means are always the beſt to obtain it. When ſuch a Plan as this is uniformly and conſiſtently laid; when all its Uſes may be comprehended at a ſingle Glance, and all appear undeniably reaſonable and perfect; then the Artiſt is at Liberty to add Grandeur and Elegancy to Strength and Propriety,

priety, and finish the Whole with the full Splendor of Beauty and Grace.

By this Division of *Architecture* into Beauty and Use, it will be demonstrable to every Reader, that 'tis partly an Art, and partly a Science; that the first is mechanical, and the last the Result of Genius and superior Understanding: One calls in all the Aid of Fancy and Imagination, grows poetical in Design, and picturesque in Decoration; the other lays down fix'd and stated Rules, proceeds in the same invariable Track of Reasoning, and comes always to the same Conclusions. Hence it happens, that many an excellent Workman has proved himself a mere Mechanick, and many a surprizing Genius, that he was ignorant of the very Principles of the Art he made it his Profession to understand. To make a thorough Master, both must be united; for the Propriety of a Plan is seldom attended to, and seldomer understood; and a glaring Pile of Beauty, without Use, but mocks the Possessor with a Dream of Grandeur he can never enjoy.

The Design of this DICTIONARY is chiefly for the Assistance of such, who study the Mechanical Part of Building, and will be of the greatest Service to all Professions that have any Relation to it: The Elements of the Art will be fully explained, and in so regular a Method too, that it can hardly be in the Power even of a Novice to mistake. Neither is it impossible that the most finish'd Artist, or most perfect Critick, should stand in need of its Help: It will serve, at least, as a kind of Remembrancer, or Common Place-Book, where all their Knowledge lies regularly digested, and may be referred to with Ease and Pleasure.

To do this more effectually, all the valuable Authors which have wrote on the Subject have been

examin'd,

examin'd, consulted, and reduced into Method and Consistency with each other: We may quote a great Variety of eminent Names; but as *Le Clerc* has been referred to the most, we shall content ourselves with his Authority only, and recommend the Steps he, in particular, has pointed out, as the surest Methods to attain to any Degree of Perfection in this Art.

ARITHMETICK is the first, as being the Ground-Work of Mensuration, either as to Extent or Solidity, as being the Medium of all Calculations, and the only Road to any Degree of practical Knowledge in the Mathematicks: For these Reasons, we have made no Scruple to add all the necessary common Rules, and some brief Examples for the Extraction of the Square and Cube Roots; as likewise the Use of several Instruments; such as Measuring by Scale and Compass, the Use of *Gunter*'s Line, Sliding Rule, *&c.* Under this Head too we have added various Tables for calculating the Value of various Kinds of Work belonging to Building, according to their Dimensions, and at several different Prices: Which, though no Way of a Piece with the Theory of this Art, are of no small Use in the Practice.

GEOMETRY follows in the next Place, and is indeed the Foundation that all Students must build upon, since 'tis impossible to attain to any Perfection in *Architecture* without it: 'Tis *Geometry* that lays down all the first Principles of Building, that adjusts all Bearings and Proportions, and measures Points, Angles, and Solidities. In short, there is no being a Master of Architecture, without being perfect in all the Parts of *Geometry*; and he that is so, though he may err in Decoration, can never do the same, either in Strength or Proportion. To answer this important Purpose, we have not only inserted all such Articles of *Geometry* as are necessary to be known by the Architect, but even such as may be of Use to Carpenters, Joiners, Masons, *&c.* containing, at once, the Definitions, Theorems, Problems, and their Demonstrations; and likewise the engrav'd Representations of the Figures so defined. And hence, we flatter ourselves, the young Practitioner will be much better able to form his Models, by making himself Master of these Rules, than by the most reasonable Notions which may result from his own uncertain Fancies and Conjectures.

MASONRY, or the Mechanical Means of raising Perpendiculars, turning Arches, erecting Bridges, and forming Stair-Cases, is another Branch of this Art, and must be understood with great Accuracy and Readiness; as being the Execution of the Whole which the Student desires to learn. On this Head, therefore, we have collected the best Instructions to be found, from heaving the Stones out of the Quarry, to their Arrangement in the Structure; and in regard little is to be found on *Bridges* in *Vitruvius*, or, beside *Palladio*, in almost any of the modern Architects, we have given an Extract from a *French Treatise of Bridges*, published by *M. Gautier*, Architect and Ingineer to *Lewis the Fourteenth*.

LEVELLING, and HYDRAULICKS, are likewise of great Importance to the Builder: The first at once enabling him, to understand good Situations, or amend them if they are otherwise: And the last, of course, directing the Conveyance of Water, the Draining of low Grounds, and teaching the whole Secrets of collecting Reservoirs, or afterwards employing them to the best Advantage. In short, on these depend both the necessary Use of Water for Family-Supply, and also all the beautiful Effects that can result from it in Gardens, by Basons, Fountains, Cascades, *&c.* On this Head, we have added a Description of the most useful Instruments for these Purposes, as

likewise

likewise, the most approved Methods of employing them to Advantage.

MECHANICKS is another Essential in this noble Art. 'Tis by understanding their Power and Effect, that such Machines are contriv'd, as alone are able to raise up the heavy Materials to Buildings of any considerable Height, or empty Waters from a Bottom, or drain a Level, or force them upwards, as Art would direct, or Necessity require. And to answer these Purposes the better, we have not only annexed Definitions, Theorems, and Problems, which are the Fundamentals of this Study, but have likewise added Plates, with explanatory Figures, for the greater Ease and Facility of learning the Mechanical Powers, the Balance, Lever, Pully, Screw, Wheel, and Wedge.

These, with the Art of Sketching and Drawing, are all the different Branches of Study which are necessary to form a Compleat Mechanical Architect. But when he is thoroughly initiated in them all, so as not to err, even in Principles or Practice, if he cannot add as much Knowledge more of his own, in their Use and Application, he will be fit for nothing more than the Overseer of a Work, or a Judge of the mere Methods to carry on and finish the Whole.

The Science of DESIGNING is still wanting to form a Great Master, or produce such Plans as would vie with the antient Beauties of *Greece* and *Rome*. But if this is not in the Genius, it is never to be learn'd: To be able to enter into this Secret, the Student must have great natural Parts; a noble and fruitful Imagination, a thorough Insight and Acquaintance with Beauty, and Judgment sedate and cool enough to form a just and delicate Taste. Without Taste, even Genius itself wanders blindfold, and spends itself in vain. Genius is, indeed, the first Quality of the Soul; but Taste must be added, or we shall censure the Wildness, instead of

admiring

admiring the Beauty; we shall be dissatisfy'd with the Irregularity, instead of being pleased with the Magnificence.

But though Genius cannot be learn'd, it may be improv'd: And though the Gift of *Designing* is born with a Man, it may be methodized by Study and Observation.

The principal Points, therefore, that the *Designer* should have in view, are first Convenience, as has been hinted already, and then Beauty and Magnificence. With regard to Convenience, few Directions can be given, since it means no more than contriving all the Requisites belonging to your Plan, in the most clear and elegant Manner, and then laying out the Space they are to be ranged in with the most perfect Order and Oeconomy. As to Beauty and Magnificence, they are Themes never to be exhausted; and though many Volumes have been written on them already, as many more might still be added.

SIMPLICITY is generally understood to be the Groundwork of Beauty, and Decoration of Magnificence. 'Tis certain, the fewer Parts a Building is composed of, if they are harmoniz'd with Elegance and Proportion, the more beautiful it appears: The Eye is best satisfy'd with seeing the Whole at once, not in travelling from Object to Object; for then the Whole is comprehended with Pain and Difficulty, the Attention is broken, and we forget one Moment what we had observed another.

But a *Contrast of Figure* must be preserved even in the Midst of this Simplicity. 'Tis in Building, as in Musick; the Parts are various and disagreeing in themselves, 'till reconcil'd by the Skill and Judgment of the Master. A Sameness of Form betrays a Poverty of Imagination, and is the same in Architecture, as Dulness is in Writing: The Mind is glutted

with

with it instantly, and turns away dissatisfy'd. It is therefore a principal Thing to be regarded by the Student, to design simply and variously at the same Time, and Beauty will infallibly be the Result of the Whole.

PERSPECTIVE is another grand Part of Designing; which demands the Master's most critical Regard, in as much as nothing contributes more to Grandeur and Beauty, if well understood; and nothing is understood with more Difficulty or Study. By Perspective, is commonly meant the thorough Inside Prospect of a Building: But if it cannot be applied with Propriety to the Art, we would take the Liberty of substituting the Painters Word *Keeping* in the Stead of it. For in all Buildings, as in Pictures, there must be one principal Figure, to which all the others must be subordinate; and from whence you must set out to examine the Parts, and to which you must return to determine of the Whole.

DECORATION, or Choice and Disposition of Ornaments, is the last grand Requisite to make a compleat Architect: And this depends partly on Genius, and partly on Fancy; but both must be under the Conduct of the severest Judgment and exactest Taste. In short, all Ornaments are ill-placed, that may be spar'd without being missed; as all empty Spaces are absurd, where Nakedness hurts the Eye, and Propriety would admit of Decoration.

We can't sufficiently recommend to all Persons, who build sumptuously, to calculate their Buildings according to the Point of Light from whence they are to be view'd: If they may, or should be seen from far, their Parts should be simple, great, and noble; if the Prospect is near, the Workmanship should be neat and little, that it may be seen and understood, as the Nature of its Situation will give Leave.

Upon

The PREFACE.

Upon the Whole, nothing but Nature, and a long Study of the antient and modern Structures, will enrich the Mind ſufficiently to excel in this noble Art; and this *Dictionary* will be found a proper Key, to explain their Beauties, as well as a needful Caution to avoid their Defects.

To conclude; We have nothing more to add, but our grateful Acknowledgments to thoſe Gentlemen and Artiſts, who have favoured us with their Aſſiſtance in this uſeful Undertaking; and that we hope our Labours will lie ſecure from Cenſure at leaſt, if they may not be judg'd altogether worthy of Applauſe.

Directions to the Binder, *for placing the* Plates.

Vol. I.

Plate I. II. at the End of Sheet K, facing it.
III. in Sheet P, facing Chissel.
IV. in Sheet Y, facing Doors.
V. in Sheet Z, facing E A.
VI. in Sheet Aa, facing Fluids.
VII. in Sheet Bb, facing *Vulgar* Fractions.
VIII. in Sheet Ee, facing Graticulation.
IX. in Sheet Ee, facing Gravity
X. in Sheet Ff, facing Groin.
XI. in Sheet Gg, facing Hip-Roof
XII. in Sheet Gg, following Plate XI.
XIII. in Sheet Ii, at the End.
XIV. in Sheet Ii, at the End.
XV. in Sheet Ii, at the End.

Vol. II.

XVI. in Sheet H, facing Orle.
XVII. in Sheet I, facing Parabolick.
XVIII. in Sheet K, facing Persian Order.
XIX. in Sheet M, facing Point.
XX. in Sheet N, facing Pump.
XXI. in Sheet N, fac. Quadrant.
XXII. in Sheet O, facing Rides.
XXIII. XXIV. XXV. XXVI. in Sheet P, facing Rudenture.
XXVII. in Sheet S, facing Steps.
XXVIII. in Sheet Y, facing Triglyphs.
XXIX. in Sheet Bb, facing Vagina.
XXX. in Sh. Cc, fac. Wainscot
XXXI. in Sheet Dd, fac. Water
XXXII. in Sheet Gg, fac. Wedge
XXXIII. in Sheet Gg, f. Windmil
* Plate facing Triangle.

THE

THE NEW BUILDER'S *Dictionary*: OR, Gentleman's *and* Architect's COMPANION.

ABACUS [is *Latin* of 'Αβαξ, Gr. which ſignifies ſeveral Things; as a ſquare Trencher, and ſometimes a Cup-board, &c.] But in Architecture, *Abacus* is the upper Member of the Capital of a Column, ſerving as a kind of Crowning, both to the Capital and the whole Column.

Others define it to be ſquare Table, Liſt, or Plinth, in the upper Part of the Chapiters of the Columns, eſpecially of thoſe of the *Corinthian* Order, ſerving inſtead of a *Drip* or *Corona* to the *Capital*, ſupporting the nether Face of the Architrave and whole Trabeation.

In Columns of the *Corinthian* Order, it repreſents a kind of ſquare Tile covering a Basket, ſuppos'd to be encompaſs'd with Leaves.

In the *Tuſcan*, *Doric*, and antient *Ionic*, it is a flat, ſquare Member, well enough reſembling the original Title; whence it is called by the *French Tailloir*, *i. e.* a Trencher, and by the *Italians Credenza*.

In the richer Orders, it loſes its native Form, the four Sides or Faces of it being arch'd or cut inwards with ſome Ornament, as a Roſe, or other Flower, a Fiſh's Tail, &c. in the Middle of each Arch. Others ſay, that in the *Corinthian* and *Compoſite*, it is compoſed of an *Uvolo*, a Fillet, and a *Cavetto*.

But ſome Architects take other Liberties, both as to the Name, Place, and Office of the *Abacus*.

Thus in the *Tuscan* Order, where it is the largest and most massive, as taking up one Third of the whole Capital, and is sometimes call'd the Dye of the Capital.

In the *Doric*, it is not always the uppermost Member of the Capital, it having a Cymatium frequently placed over it.

In the *Ionic*, some make it a perfect Ogee, and crown it with a Fillet; or 'tis compos'd of a *Cima reversa*, and Fillet only.

Andrea Palladio, in the *Tuscan* Order, calls the Plinth above the (Echinus) or Boultin, *Abacus*; which, as he says, is commonly called a *Dado*, or *Dye*, from its Form, and is one Third of the whole Height of the Capital.

In the *Ionic* Order, he defines it to be an Ogee, with a Fillet over it; which is one Third of the whole Height of the Capital, and *M. Mauclerc* after him does the same.

He also calls the *Plinth* above the *Boultin* of the Capital of the *Doric* Order, *Abacus*, and places a *Cymatium* above it for the uppermost Member of the Capital.

He also describes the *Abacus* of the *Corinthian* Order to be one seventh Part of the whole Capital, divided into three Parts, the uppermost of which is a Boultin, and one Third of the next Third below is the Fillet, and the remaining Part below, which is one, and two Thirds is the *Plinth* of the *Abacus*.

Besides this, the *Abacus* is not constantly restrain'd to the Capital of the Column, *Scamozzi* using the Name *Abacus* for a concave Moulding in the Capital of the *Tuscan* Pedestal.

Vitruvius, and others after him, who give the History of the Orders, inform us, that the *Abacus* was originally intended to represent a square Tile laid over an Urn, or rather a Basket. See *Acanthus*.

ABBREUVOIR } [*Abbreuvoir*, in
ABREVOIR } *French*, signifies a Watering-place.] In Masonry it signifies the Joint or Juncture of two Stones, or the Interstice or Space left between to be filled up with Mortar or Cement.

ABBUTTALS } the Buttings and
ABUTTALS } Boundings of a Piece of Land, expressing on what other Lands, Streets, Highways, &c, the several Extremes thereof abutt or terminate.

ABSCISSA } [in *Conicks*] is
ASSCISSE } the Part of the Diameter of a Curve, or transverse Axis of a Conic Section, intercepted between the Vertex, or some other fix'd Point, and a Semi-ordinate.

ABUNDANT NUMBERS are such, whose Quota-Parts, added together, make more than the whole Number; as of those of 12, make more than the whole Number they are Parts of, whose Quota-Parts being 1, 2, 3, 4, and 6, being added together, make 16; and so the Parts of 20 make 22, &c.

ACANTHUS [of Ἄκανθος, *Gr*. a Thorn, being so called, as being prickly, or of the Thistle Kind. It is called in *Latin Brancha Ursina*, in *English* Bear's Breech, on account of some supposed Resemblance it bears thereto: And also *Branc Hircina*, i. e. Goat's

Goat's Horn, becauſe its Leaves bend and twiſt ſomewhat like a Goat's Horn.]

Acanthus in Architecture, is an Ornament in the *Corinthian* and *Compoſit* Orders, being the Repreſentation of the Leaves of the Plant in the Capitals of them.

Callimachus, an ingenious Statuary of *Athens*, is ſaid to have been the Inventor of this Ornament, on the Occaſion following: An *Athenian* Old Woman happening to place a Basket, covered with a little Tile, over the Root of an *Achanthus*, which grew on the Grave of a young *Corinthian* Lady, the Plant ſhooting up the following Spring, encompaſs'd the Basket all around, till meeting with the Tile, it curled back in a kind of Scrolls: He paſſing by, and obſerving it, immediately executed a Capital on this Plan, repreſenting the *Tile* by the *Abacus*, and the Leaves of the *Acanthus* by the Volutes or Scrolls; and the Baſket (*Tambour*, as the *French* call it) by the Vaſe or Body of the Capital.

There are two Kinds of the Plant *Acanthus*; the one wild, and full of Prickles; and the other of the Garden, without Prickles. The *Greek* Maſons adorn'd their Works with the Garden *Acanthus*, and the *Gothic* Maſons with that of the wild; repreſenting it not only in their Capitals, but alſo in other Ornaments.

Garden *Acanthus* is more indented than the wild, and carries a near Reſemblance to Parſley, or Smallage, as it is found repreſented in the *Compoſit* Capitals of *Titus*, and *Septimius Severus* at *Rome*.

Theſe Leaves make the principal Characters in diſtinguiſhing the two richeſt Orders from the reſt; and theſe two Orders are diſtinguiſhed from each other by their different Number and Arrangement.

ACCELERATED *Motion* [in *Mechanicks*] is the Increaſe of Velocity in a moving Body.

The Motion of falling Bodies is an *accelerated Motion*; and ſuppoſing the Medium they fall through, *i.e.* the Air void of Reſiſtance, the ſame Motion may be conſidered as uniformly accelerated.

ACCELERATION is the ſame as *accelerated Motion*, and is chiefly uſed of heavy Bodies, tending to the Center of the Earth by the Force of Gravity.

It is evident from various Conſiderations, that natural Bodies are accelerated in their Deſcent: And it is actually found, that the greater Height a Body falls from, the greater Impreſſion it makes, and with the greater Violence does it ſtrike the Thing it falls upon.

ACCESSIBLE HEIGHT [in *Geometry*, &c.] is either that which may be mechanically meaſured, by applying a Meaſure to it, or it is a Height whoſe Baſe and Foot may be approached to, and a Diſtance meaſured thence on the Ground.

ACCIDENTAL POINT [in *Perſpective*] is a Point in the horizontal Line, where Lines parallel to one another, though not perpendicular to the Repreſentation, meet.

ACCLIVITY, the Steepneſs or Slope of a Line or Plane inclining to the Horizon, reckon'd upwards; as the *Aſcent* of an Hill is the *Acclivity*, and, on the contrary, the *Deſcent* is the *Declivity*.

ACCOMPANIMENT, ſomething attending on, or added to another, by way of Ornament, or for the ſake of Symmetry, or the like.

ACERRA, among the *Romans*, a kind of Altar erected near the Gate of a Perſon deceaſed, on which his Friends and Acquaintance daily offered Incenſe till the Time of his Interment.

ACOUSTICKS ['Ακȣςικὰ of ἀκȣ́ω, Gr. to hear] the Doctrine or Theory of Sounds.

Dr. *Hook* ſays, it is not impoſſible to hear the loweſt Whiſper that can be made to the Diſtance of a Furlong; and that he knows a Method to hear any Perſon ſpeak through a Brick Wall of three Foot thick. See *Whiſpering-Place*.

ACROTERIA } [Ακροτήρια of ACROTERES } ἄκρος, Gr. the Extremity of any Thing] ſmall Pedeſtals, uſually without Baſes, placed on Pediments, and ſerving to ſupport Statues.

Thoſe at the Extremes ought to be half the Height of the *Tympanum*, and that in the Middle, according to *Vitruvius*, one eighth Part, or more.

ACROTERIA alſo are ſometimes uſed to ſignify Figures, either of Stone or Metal, placed as Ornaments or Crownings on the Tops of Temples, or other Buildings.

Sometimes the Name is uſed to ſignify thoſe ſharp Pinacles or ſpity Battlements, which ſtand in Ranges about flat Buildings, with Rails and Balluſters.

ACUTE Angled Triangle, is one whoſe three Angles are all

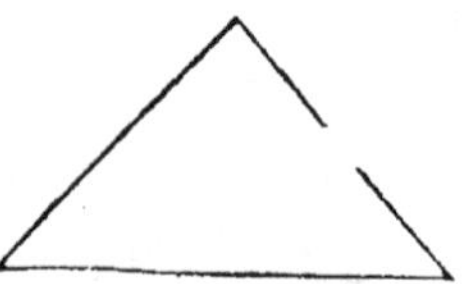

alike; and is alſo called an Oxigonous Angle, as in the Figure.

ACUTE Angular Section of a Cone, a Name by which Antient Geometricians called the *Ellipſis*.

ADJACENT Angles are made by continuing out one Side

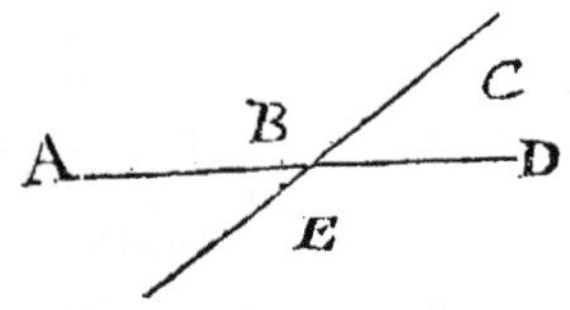

of an Angle; whence adjacent Angles are contiguous, but not on the contrary.

ADIT, the Shaft or Entrance into a Mine.

ADYTUM ['Αδυτον of a Priv. and δύω, Gr. to enter] a ſecret or retir'd Place in the Pagan Temples, where Oracles were given; and into which, none but the Prieſts were permitted to enter; in Imitation of the *Sanctum Sanctorum* of the *Jewiſh* Temple.

AERIAL, conſiſting of the Air, or ſomething that has a Reſemblance to it.

AERIAL *Perſpective*, is that which is repreſented both weak and diminiſh'd, in Proportion to the Diſtance from the Eye.

It is founded on this, That the longer a Column of Air an Object

ject is ſeen through, the weaker do the viſual Rays emitted from it affect the Eye.

The Object of Aerial Perſpective is chiefly Colours of Objects, whoſe Force and Luſtre it takes off more or leſs, to make them appear as if more or leſs remote.

AGEOMETRICAL, Ungeometrical, or defective in Point of Geometry.

AJUTAGE [in Hydraulicks] Part of the Apparatus of an artificial Fountain; being a ſort of *Jet d'eau*, or kind of Tube fitted to the Mouth or Aperture of a Veſſel, through which the Water is to be play'd, and by it determin'd into this or that Form or Figure.

ALABASTER, a Kind of Stone ſofter than Marble; but harder than Plaiſter of *Paris*; if it be ſo ſoft, as that it can be cut, it is called *Gypſum*.

It is found or digged in the *Indies*, *Ægypt*, *Syria*, &c. There is alſo ſome found in *Lincolnſhire* and *Staffordſhire*.

It is found of ſeveral Colours; ſome extreamly white and ſhining, which is the moſt common; ſome red, like Coral; and ſome of the Colour of the Onyx; which thence is called Onyx, though it differs very much from it in Nature or Quality.

Its Uſe is chiefly in making Monuments in Churches, &c. where there are many Figures in Relief, or Baſs-Relief, &c. carved: It is alſo uſed for carving Coats of Arms cut in Relief, to be ſet in Brick or Stone in the Fronts of Houſes.

Alabaſter cuts very ſmooth and eaſy, and is much uſed by Sculptors in making little Statues Vaſes, Columns, &c.

It is alſo uſed like Plaiſter of *Paris*, being firſt burnt and calcin'd; then mix'd up with Water to a thin conſiſtence; which being afterwards caſt into a Mould, it very readily coagulates into a ſolid Body.

ALCOVE [of *Alcoba*, *Spaniſh*, of *Elkans*, *Arabick*, a Cabinet; or of *Elcobat*, a Tent, or Place to ſleep in] is a Receſs or Part of a Chamber, ſeparated by an Eſtrade or Partition of a Column and other correſpondent Ornaments; in which is placed a Bed of State, and ſometimes Seats to entertain Company. The Bed is frequently raiſed up two or three Aſcents, with a Rail at the Feet. Theſe Alcoves are frequent in noble Houſes in *Spain*, and other Places.

ALDER is an Aquatick Tree, too well known to need any Deſcription.

In former Times, large *Alders* were uſed for building Boats; and now they are very much eſteem'd for ſuch Parts of Works which lie continually under Water; where it will become as hard as Stone; but if ſuffer'd to lie ſome Times expos'd to the Weather, and at others to lie under the Ground in watry Places, it will decay in a little Time.

We are inform'd by *Vitruvius*, that the *Moraſſes* about *Ravenna* in *Italy*, were piled with *Alder* Timber, in order to build upon; for which Uſe he highly commends it.

And the *Rialto*, that famous Bridge at *Venice*, which paſſes over the Grand Canal, and bears

 a vaſt

a vaſt Weight, is built upon Piles of this Wood.

Trunks of Trees or Poles of this Wood, are extraordinary uſeful in making Pumps, Water-Pipes, *&c.*

They are uſed (in the Country) for Water-Pipes for the Conveyance of Water through Bays and Dams; and alſo for Water-Pipes for conveying Water from any Spring, to ſupply a Houſe with it; and large Poles or Trees of this Wood are uſed for Ground Guts, for conveying Water out of Stews. Theſe Poles are about eight or ten Inches diameter, and the Cavity in them about four, or four and a half; for boring and fitting up of which Size, they give about 3*s.* 6*d.* *per* Rod for Workmanſhip.

But for Water-Pipes, the Poles need not be above four or five Inches diameter, and the Cavity about an Inch and quarter, or an Inch and half diameter.

As to the Method of boring Alder *Poles,* Theſe Poles being laid on Horſes or Treſſels of a fit Heighth, to reſt the Augur upon while they are boring, they ſet up a Lath, to turn the leaſt End of the Poles, to fit them to the Cavities of the great End of the others: The Lath being ſet up, and the Poles cut to the Lengths, they will conveniently hold, *viz.* eight, ten, or twelve Foot. They turn the ſmall Ends of the Poles about five or ſix Inches in Length, to the Size they intend to bore the bigger Ends, about the ſame Depths, *viz.* five or ſix Inches (this is deſigned to make the Joint to ſhut each Pair of Poles together, the concave Part being the Female Part, and the other Part, the Male of the Joint.) In turning of the Male Part, they turn a Channel in it, or a ſmall Groove at a certain Diſtance from the End; and in the Female Part, they bore a ſmall Hole to fit over this Channel.

This being done, they bore the Poles through; and to prevent them from boring out at the Sides, they ſtick great Nails at each End, to be a Guide to them in boring ſtrait through; though they uſually bore them at both Ends; ſo that if a Pole be crooked one Way, they can bore it through, and not ſpoil ir.

The Poles being bored, they form them into Pipes in the Ground; in order to which, they dig a Trench, and prepare it with Clay, to ram them in the Female Part, which is firſt bound with an Iron Ring round it, to prevent its Splitting; afterwards they drive in the Male Part till the Groove in it is juſt under the Hole; and pour melted Pitch hot into the Hole, in the Female Part, which will flow round in the Groove which was turned in the Male Part: By this Means, the Junctures are render'd very ſtaunch and cloſe; and in this Manner they proceed till they have laid all the Poles or Pipes in their Order.

As to the Charge of preparing theſe Pipes: For the Workmanſhip only, they uſually require about 2*s.* 6*d.* or 3*s.* *per* Rod, 1*s.* for boring and fitting them; but the Charge of all the Work and Materials, *Boring Digging* the Trench, *Laying* and *Ramming* in the Clay, *&c.* and alſo

alſo the Charge of the *Poles*, *Clay*, *Pitch*, and *Iron Rings*, will amount from 4*s*. to 6*s*. *per* Rod, according as the Materials can be procur'd.

ALGEBRA is a Method of reſolving Problems by Means of Equations.

ALIQUANT *Part* [*in Arithmetick*] is that which cannot meaſure or divide any Number exactly; but that there will be at laſt ſome Remainder; as 5 is an Aliquant Part of 12; for being taken twice, it falls ſhort, and if taken three Times, it exceeds 12.

ALIQUOT *Part* [*in Arithmetick*] is ſuch a Part of a Number as will meaſure it exactly without any Remainder, as 3 is an Aliquot Part of 9, and 4 of 12.

ALLEY [in *Perſpective*] is that which is larger at the Entrance, than at the Iſſue; to give it a greater Appearance of Length.

ALTERNATE *Angles* are the internal Angles made by a Line cutting two Parallels, and lying on the oppoſite Sides of

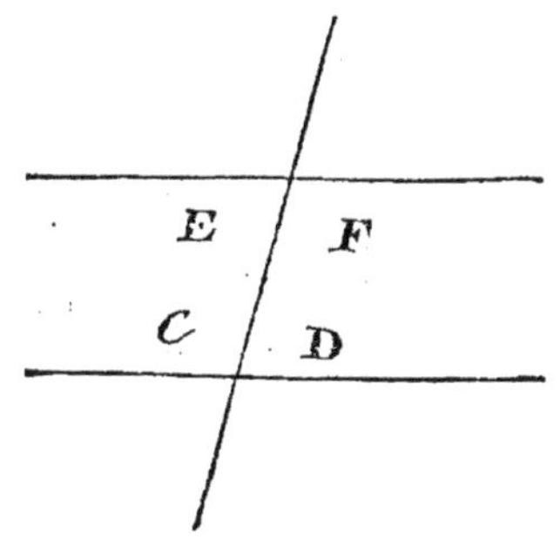

the Cutting Line; the one below the firſt Parallel, and the other above the ſecond, as the Angles E and D and F and C.

ALTERNATE *Proportion*, or *Ratio*, [in *Arithmetick*,] is the aſſuming an Antecedent to an Antecedent, as the Conſequent to the Conſequent; as if A B, C D, then by alternate Proportion, will A C, B D.

ALTERNATION is uſed by ſome for the different Changes or Alterations of Order in any Number of Things propoſed.

This Alternation is eaſily found by only multiplying continually all the Numbers, begining at Unity: As ſuppoſe it be required to know how many Changes can be rung upon five Bells; you need only write down 1 2 3 4 5, and then multiply all thoſe Numbers continually one into another, and the laſt Product will be 120, the Number of Changes.

ALTIMETRY [of *alta* high Things, and *metiri* to meaſure] the Art of taking or meaſuring of Altitudes or Heights, whether acceſſible, or inacceſſible.

ALTITUDE, the third Dimenſion of Body; called alſo Heighth, or Depth.

The ALTITUDE, or HEIGHTH *of Figures, is the parallel Diſtance between the Top of a Figure and the Baſe.* So the

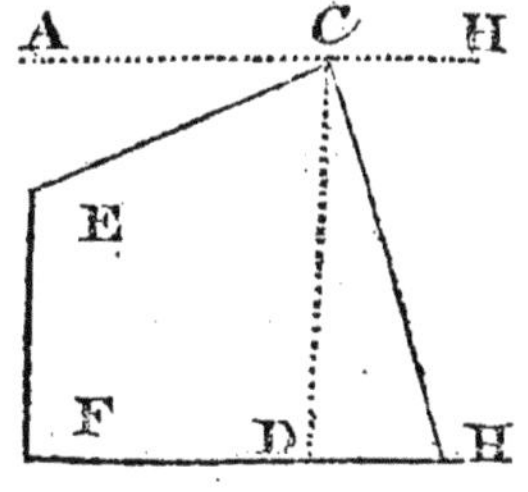

Heighth of the *Trapezium* C E F H is the Perpendicular C D; because

becauſe it is in the neareſt Diſtance between the parallel Lines AH, the Top, and FDH, the Baſe. And it is the ſame of other Figures; as a Triangle, Hexagon, &c.

AMBLYGONIAL [in *Geometry*] Obtuſe-Angular.

AMBLYGONOUS *Angle*, is an Obtuſe Angle, or Angle conſiſting of more than 90 Degrees.

AMPHIPROSTYLE [in *Antient Architecture*] a Kind of Temple which had four Columns in the Front, and the ſame Number in the Face behind.

AMPHITHEATRE [of Ἀμφιθέατρον of ἀμφί, on both Sides, and θεάομαι, I behold. *Gr.*] and is a ſpacious Edifice or Building in either a circular or oval Form, having its Area encompaſſed with Rows of Seats ariſing gradually one above another; where Spectators might ſit to behold Spectacles, as Stage-Plays, Combats of Gladiators, and thoſe of wild Beaſts, &c.

The Theatres of the Antients were built in the Form of a Semicircle, only exceeding a juſt Semicircle by one fourth Part of the Diameter; and the *Amphitheatre* is nothing elſe but a double Theatre, or two Theatres join'd together; ſo that the longeſt Diameter of the *Amphitheatre*, was to the ſhorteſt, as One and a half to One.

The Amphitheatre of *Veſpaſian*, call'd the *Coliſæum*, that at *Verona* in *Italy*, and that at *Niſmes* in *Languedoc*, are the moſt celebrated that we have now remaining of Antiquity.

Pliny makes mention of an Amphitheatre built by *Curio*, which turned on large Iron Pivots; ſo that of the ſame Amphitheatre, two ſeveral Theatres were made occaſionally, on which different Entertainments were exhibited at one and the ſame Time.

ANABATHRUM [Ἀνάβαθρον of ἀναβαίνω to aſcend, *Gr.*] a Place that is aſcended to by Steps.

ANALOGY [in *Geometry*, &c.] is the Compariſon of ſeveral *Ratio's* together; and is the ſame as Proportion.

ANAMORPHOSIS, or *Monſtrous Projection of an Image*, [in *Perſpective*,] is the Deformation of an Image on a Plane, or the Superficies of ſome Body, which ſeen at a certain Diſtance, will appear formous.

If it be required to delineate a *monſtrous Projection* on a Plane, proceed thus:

Firſt, Make a Square ABCD, (called the *craticular Protolype*,) of a Bigneſs at Pleaſure, and divide the Side AB into a Number of equal Parts, that ſo the ſaid Square may be divided into a Number of Areola's, or leſſer Squares.

Secondly,

Secondly, In this Square let the Image, to be repreſented deform'd, be drawn.

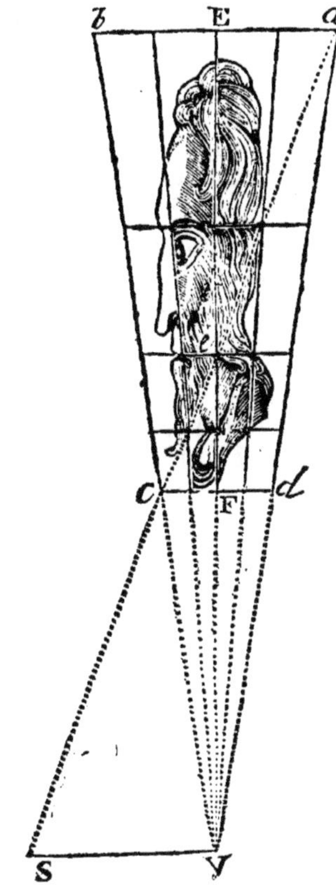

Thirdly, Draw the Line *ab*= AB, and divide it into the ſame Number of equal Parts as the Side of the *Protolype* AB is divided.

Fourthly, In E, the Middle thereof, erect the Perpendicular EV, ſo much the longer, as the Deformation of the Image is to be greater.

Fifthly, Draw VS perpendicular to EV, ſo much leſs in Length.

Sixthly, From each Point of Diviſion, draw ſtrait Lines to V, and join the Points *a* and S, as alſo the Right Line *a* S.

Seventhly, Through the Points *d*, *e*, *f*, *g*, draw Right Lines parallel to *ab*; then will *a*, *b*, *c*, *d*, be the Space that the *monſtrous Projection* is to be delineated in, called the *craticular Ectype*.

Eighthly, In every Areola, or ſmall Trapezium of this Space *abcd*, let there be drawn what appears delineated in the correſpondent Areola of the Square ABCD; and by this Means you will obtain a deform'd Image, which will appear formous to an Eye diſtant from it by the Length FV, and raiſed above it the Height VS.

Ninthly, It will be very diverting to manage it ſo, that the deform'd Image does not repreſent a mere Chaos; but ſome other Image different from it, which by this Contrivance ſhall be deform'd.

As there has been ſeen a River with Soldiers, Waggons, &c. marching along the Side of it; ſo drawn, that being look'd at by an Eye in the Point S, appear'd to be the ſatyrical Face of a Man.

Tenthly, An Image may be deform'd mechanically; if you place the Image, having little Holes here and there made in it with a Needle or Pin, againſt a Candle or Lamp, and obſerve where the Rays, going through theſe little Holes, fall on a Plane or Curve Superficies; for they will give the correſpondent Points of the Image deform'd, by which Means the Deformation of the Image may be compleated.

ANCHORS [in *Architecture*] a Sort of carving something resembling an Anchor, or Arrow-Head. They are commonly placed as Part of the Enrichments of the Boultins of Capitals of the *Tuscan*, *Doric*, and *Ionic* Orders; and also of the Boultins of Bed-Mouldings of the *Doric*, *Ionic*, and *Corinthian* Cornishes; these Anchors and Eggs being carved alternately throughout the whole Buildings.

ANCONES, are the Corners or Coins of Walls, Cross-Beams, or Rafters. *Vitruvius* calls the Consoles (a Sort of Brackets, and Shouldering-Pieces) *Ancones*.

An ANGLE is an indefinite Space, terminated by two Right

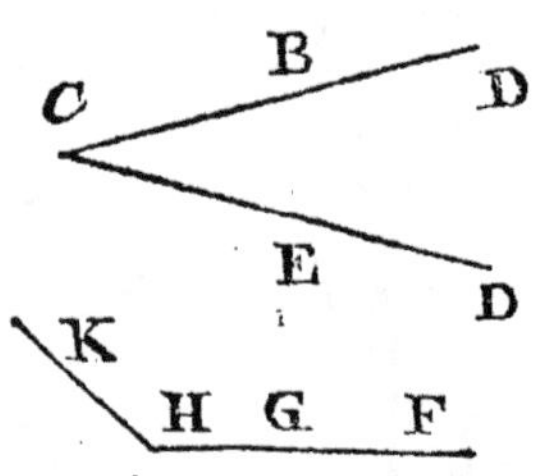

Inclining Lines which meet together in one Point, as the Right Lines DB, and DE; which being continu'd in their own Positions, will meet at C, and by that generate an Angle.

So likewise the Right Lines F G, and K, being continu'd, will meet at H, and form an Angle also.

But if two Lines meet in such a Manner, as to have no Inclination the one to the other, they will generate a Right Line; and a Right Line equal to both their Lengths, without forming any Angle. And on the contrary, if any Right Line meet another Right Line in any different Position, they will constitute an Angle at their Point of Meeting.

An ANGLE is signified by three Letters, of which the middlemost always denotes the Angle; so in the Case of the two Right Angles in the Figure RIGHT ANGLE, the one is denoted by the Letters ECD, and the other by the Letters ECB.

A RIGHT ANGLE is constituted by the Meeting of two Right Lines, with an equal Inclination; that is, if a Line, as EC, meet another Line, as DB, and inclines no more towards

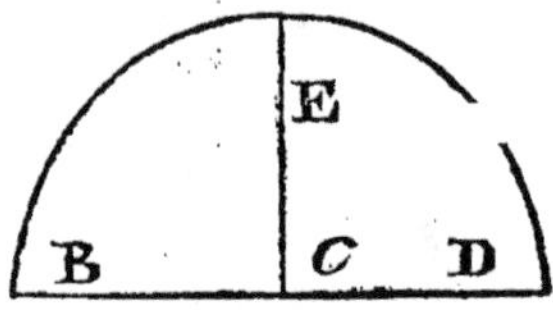

D, than it does towards B, but stands directly square between both, then the Angle is called a Right Angle, and the Line EC is therefore called a Perpendicular Line to the Line DB.

The Line EC standing upon the Line B D, at C, is perpendicular thereto; because if you describe a Semicircle on C, with any Radius, as BED, the Arch BE will be equal to the Arch DE; and since that both Arches are equal to each other, and to a Semicircle also, being taken together, it therefore follows, that both the Angles on either Side are equal, and are therefore called *Rect* or *Right Angles*.

Now

Now Since the Semicircle BED contains 180 Degrees, being just the Half of 360 Degrees, contain'd in every whole Circle, and is equally divided in C, by the Perpendicular Line EC; it therefore follows, that the Angles BEC, and DEC, are equal to each other, and must each consist of 90 Degrees; therefore a Right-angled Triangle is that whose Arch contains 90 Degrees precisely.

ACUTE ANGLE is an Angle whose Inclination is nearer than a Right Angle; so that when any two Lines incline nearer to one another than DC

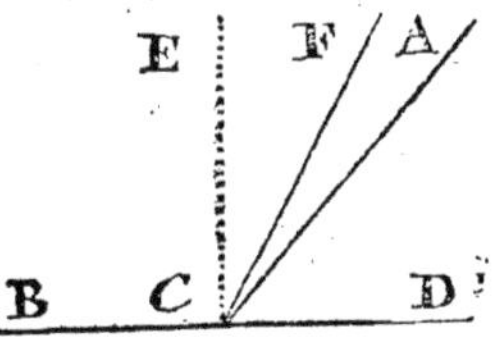

doth to EC, as the Lines FC and DC, or AC and DC, then by their Meeting they form sharper Angles than the Right Angle ECD, and are therefore all called Acute Angles.

OBTUSE ANGLE is an

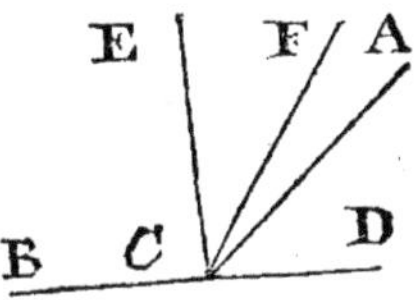

Angle constituted by the Meeting of two Right Lines, whose Inclination is greater from one another, than the Lines of the Right Angle, as the Angles made by the Meeting of the Lines FC and CB, or AC and EB, by which they form Angles that are more blunt than the Right Angle, and therefore are called *Obtuse Angles*.

To make the *Angle* FHK equal to the Angle CAE.

First, On the given Angle A, with any Opening of the Compasses, describe an Arch as BD; and then having drawn a Right Line at pleasure, as HK, on any of its Ends, as H, with the Opening AD describe the Arch IG.

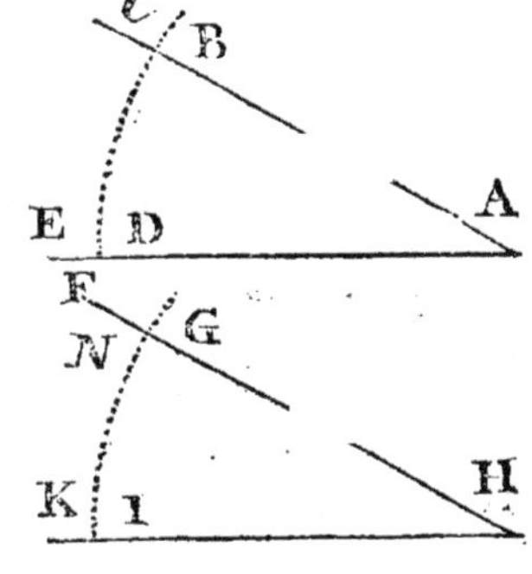

Secondly, Make IN equal to DB, and then from H, through N, draw the Right Line HF, which completes the Angle FHK = the given Angle CAE, as required.

To divide the *Angle* BAE into two equal Parts by the Right Line AH.

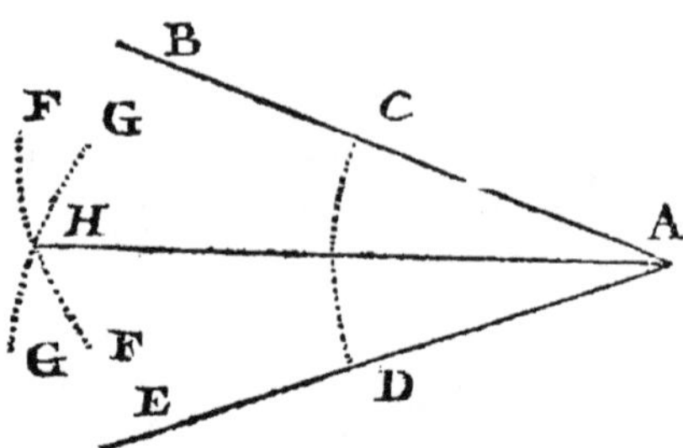

First, On the Point A describe an Arch of a Circle of any Radius, as CD; and with the same Opening of the Compasses, on the Points C and D, describe the Arches GG and FF, intersecting each other in H.

Secondly, From the angular Point A, draw to H the Right Line AH; which will divide the Angle into two equal Parts, as required.

The *Quantity* or *Measure* of an *Angle*, is the Arch of a Circle described on the angular Point, intercepted between the two Sides of that Angle.

To make an *Angle* of any given Magnitude, suppose 50 Degrees.

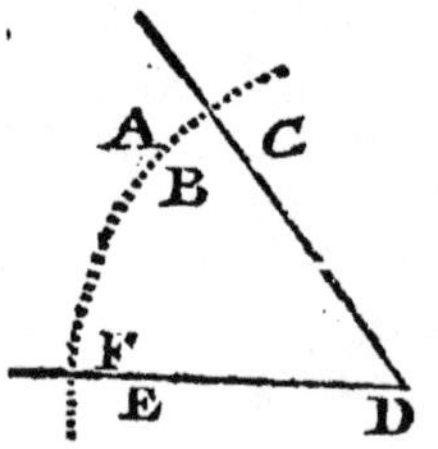

First, Draw a Line at pleasure, as FD; then take 60 Degrees from your Line of Chords, and on one End thereof, as at D, describe an Arch, as EC.

Secondly, Take from your Line of Chords 50 Degrees, the Quantity of the given Angle, and set it on the Arch from EC to EB; then drawing the Right Line DA from A through E, it will complete the Angle required.

The *Complements* of *Angles* are the same as the Complements of Arches; because their Quantities are measured by Arches of Circles.

The *Angle* BAE being given, to find its Quantity,

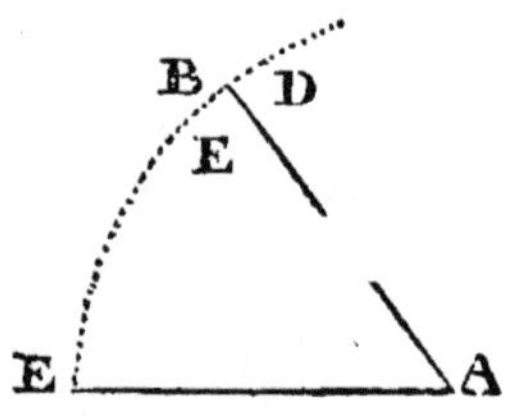

First, Take 60 Degrees from your Line of Chords, and with that Distance, on the angular Point A, describe the Arch FED.

Secondly, Take the Arch EF in the Compasses, and applying that Extent upon your Line of Chords, from the Beginning of it, the extended Point of the Compasses will fall upon the Number of Degrees and Minutes, which the

the *Angle* contains; *viz.* 60 Degrees, 00 Minutes.

ANNELETS } [of *Annulus*
ANNULETS } a Ring, *L.*] are small square Members in the *Doric* Capital, under the Quarter-Round, *&c.*

Annulets are used in Architecture to signify a narrow, flat Moulding, (of which see *Capital*;) which is common to divers Places of the Columns, as in the Bases, Capitals, *&c.*

It is the same Member which *M. Mauclerc*, from *Vitruvius*, calls a *Fillet*; and *Palladio*, a *Listel*, or *Cincture*; and *M. Brown*, from *Scamozzi*, a *Supercilium*, a *List*, *Tinea*, *Eye-brow*, *Square*, and *Rabit.*

ANTA [in *Architecture*] is used by *M. Le Clerc* for a kind of Shaft of a Pilaster, without Base or Capital, and even without any Moulding.

ANTÆ, Pillars adjoining to a Wall. Vide *Parastatæ.*

ANTE-CHAMBER } An
ANTI-CHAMBER } outer Chamber, before the principal Chamber of an Apartment, where the Servants wait, and Strangers stay till the Person to be spoken withal is at Leisure, *&c.*

2. *As to its Proportions.*] A well-proportion'd *Ante-Chamber* ought to be in Length the Diagonal Line of the Square of the Breadth, and not to exceed the Bread and half at most.

3. *As to their Height.*] They are made either arched or flat; if they are flat, their Height ought to be, from the Floor to the Joists, two third Parts of their Breadth. But if you have a mind to have it higher, divide the Breadth into seven Parts, and take five of them for the Height. Or else divide the Breadth into four Parts, and take three of them for the Height.

In large Buildings, the *Ante-Chamber*, *Hall*, and other Rooms of the first Story, may be arched, which will render them handsome, and less subject to Fire: And in such Buildings, the Height may be five Sixths of the Breadth, from the Floor to the Bottom of the Key of the Arch.

But if this Height shall be thought too low, you may make the Height seven Eighths of the Breadth. Or it may be eleven Twelfths, which will render it yet more stately.

4. *Of their Situation.*] *Ante-Chambers*, &c. ought to be so situated, that they may be on each Side of the Entry, and of the *Hall*: And this likewise ought to be observed, that those on the Right Hand may answer, and be equal (or nearly so) to those on the Left; to the End, that the Buildings may, on all Sides, bear equally on the Roof. See *Halls.*

ANTERIOR, before another, in respect to Place; in which Sense the Term stands opposite to *Posterior.*

ANTICK [in *Sculpture* and *Painting*] is used to signify a confused Composure of Figures, of different Natures and Sexes, *&c.* as of Men, Beasts, Birds, Flowers, Fishes, *&c.* And also such-like Fancies as are not to be found in Nature.

It would be tedious to enumerate all the *Antick* Forms and Fancies by which the Heathens represented their several Gods, and their Poets, Painters, and Sculptors described them.

They

They had alſo ſtrange and monſtrous Figures of human Creatures, (and ſo repreſented them in Sculpture, *&c.*) as of Centaurs, half Men and half Horſes; Sagitaurs, half Men and half Bulls; Syrens, or Mermaids, half Women and half Fiſh; Harpies, half Women and half Birds; Griffins, half Beaſts and half Birds; Dragons, Part Serpents and Part Birds; the Spread-Eagle with two Heads; and many other of the like Nature. They alſo repreſented divers ſorts of Flowers and Fruits growing on the ſame Plant, *&c.* and many ſuch like Fictions, which we have reaſon to believe are not to be found in Nature; though the Belief of their Exiſtences hath been propagated by Poets, *&c.* upon account of their Fitneſs to be made uſe of in the Way of Similitude.

Theſe Sorts of Repreſentations the *Italians* call *Groteſca*, and the *French Groteſque*; which ſignifies comical, pleaſant, apt to make one laugh; alſo ridiculous; and their Word *Groteſques* ſignifies idle, fooliſh Fancies.

ANTICUM, a Porch before a Door, a Fore-door, a *Hatch*, *L.*

ANTIPAGMENTS, Ornaments or Garniſhings in Carv'd Work, ſet on the Architrave, (Jaumbs, Poſts, or Puncheons of Doors,) whether of Wood or Stone, after the *Latin* antique Word *Antipagmenta*.

ANTIQUE, ſomething that is ancient.

The Term is chiefly uſed by Architects, Sculptors, Painters, *&c.* who apply it to ſuch Pieces of Architecture, Sculpture, Painting, *&c.* as were made at the Time when their Arts were in their greateſt Perfection among the antient *Greeks* and *Romans*, *viz.* between the Time of *Alexander the Great*, and that of the Emperor *Phocas*, when *Italy* became over-run by the *Goths* and *Vandals*, about the Year 600, about which Time the noble Arts were extinguiſh'd.

Thus we ſay an *Antique Building*, or a Building after the *Antique*; an *Antique Buſt*, or *Baſs Relievo*, *Antique Manner*, *Taſte*, &c.

ANTIQUE, is ſometimes even contradiſtinguiſhed from *Antient*, which denotes a leſſer Degree of Antiquity, when the Art was not in its utmoſt Purity.

Thus *Antique Architecture* is frequently diſtinguiſhed from *Antient Architecture*.

Alſo ſome Writers uſe the compound Word

ANTIQUO MODERN, in ſpeaking of old *Gothic* Churches, to diſtinguiſh them from thoſe of the *Greeks* and *Romans*.

APERTIONS } [from the APERTURES } *Latin aperio*, to open, ſignifies Openings.] In Architecture the Words are uſed to ſignify Doors, Windows, Stair-caſes, Chimneys, Outlets and Inlets for Smoke, Light, *&c.* which ought to be as few in Number, and moderate in Dimenſions, as poſſible, it being a Rule in Architecture that all Openings are Weakenings; nor muſt they be made too near the Angles of the Walls; for indeed it would be a great Solecism to weaken that Part which ought to ſtrengthen all the reſt.

APERTURE, the Opening of any Thing, or a Hole, Cleft, or vacant Place, in ſome otherwiſe

wiſe ſolid or continuous Subject.

APERTURE [in *Geometry*] is uſed for the Space left between two Lines, which mutually incline towards each other, to form an Angle.

APHORISM [ἀφορισμός of ἀφορίζω, Gr. to ſelect] is a Maxim, general Rule, or Principle of a Science.

APOPHYGE [of Ἀποφυγὴ, Gr. Flight or Eſcape; whence the *French* call it *Eſcape*, *Congee*, *&c.*] In *Architecture* it ſignifies that Part of a Column where it begins to ſpring out of its Baſe, and ſhoot upwards.

The *Apophyge*, in its Original, was no more than the Ring or Fertil heretofore faſtened at the Extremities of wooden Pillars, to keep them from ſplitting, which afterwards was imitated in Stone-Work.

APPEARANCE [in *Perſpective*] is the Repreſentation or Projection of a Figure, Body, or the like Object, upon the Perſpective Plane.

APPROXIMATION [in *Arithmetick*] a continual Approach nearer ſtill, and nearer to a Root or Quantity ſought, without a Poſſibility of ever arriving at it exactly.

AQUEDUCT [*Aquæductus*, q. d. *Ductus Aquæ*, L.] a Conveyance made for carrying of Water from one Place to another.

It is a Conſtruction of Stone or Timber, made on uneven Ground to preſerve the Level of the Water, and convey it by a Canal from one Place to another.

Some *Aqueducts* are under Ground, and others rais'd above it, ſupported by Arches.

The *Romans* were extraordinary ſumptuous and magnificent in their *Aqueducts*, ſome of which extended 100 Miles.

Frontinius, who had the Direction of them, informs us of nine which emptied themſelves through 13,514 Pipes, of a Inch Diameter. And

Blaſius has computed that the City of *Rome* received from theſe *Aqueducts* no leſs than 500,000 Hogſheads of Water in twenty-four Hours Time.

The *Aqueduct* built near *Maintenon*, for carrying the River *Bure* to *Verſailles*, is the greateſt in the World. It is in Length 7000 Fathoms, and its Elevation 2560 Fathoms, containing 242 Arcades.

AQUATICK, living, breeding, or growing about the Water; as Animals, Plants, Trees; as *aquatick Trees*, are ſuch as grow on the Banks of Rivers, Marſhes, Ditches, *&c.*

ARABESQUE } ſomething
ARABESK } done after the Manner of the *Arabians*.

Arabeſque, *Groteſque*, and *Moreſque*, are Terms applied to ſuch Paintings, Ornaments of Freezes, *&c.* on which there are no human or animal Figures; but which conſiſt wholly of imaginary Foliages, Plants, Stalks, *&c.*

The Terms are deriv'd from the *Arabs*, *Moors*, and other *Mahometans*, who uſe theſe kinds of Ornaments, becauſe their Religion forbids them to make any Images or Figures of Men, or other Animals.

ARÆOSTYLE

ARÆOSTYLE [of ἀραιὸς thin ſet, and ςύλος, *Gr.* a Column] a Term uſed by *Vitruvius* to ſignify the greateſt Interval or Diſtance which can be made between Columns; which conſiſts of eight Modules, or four Diameters.

ARC, the ſame as *Arch.*

ARC BOUTANT [of *arc* and *bouter*, Fr. *to abut*] is a flat Arch, or Part of an Arch abutting againſt the Reins of a Vault, to ſupport and prevent its giving Way.

Arcs boutants are only *Arch-Buttreſſes.*

ARCH [of *Arcus*, L.] is a Part of any Curve Line, *e.gr.* as of a Circle, an Ellipſis, and the like.

ARCH *of a Circle*, is a Part of the Circumference of it, leſs than half a Semicircle.

The Baſe or Line that joins the two Extremes of the *Arch*, is called the *Chord*; and the Perpendicular raiſed in the Middle of that Line, is the Sine of the *Arch*, as A and B in the Figure.

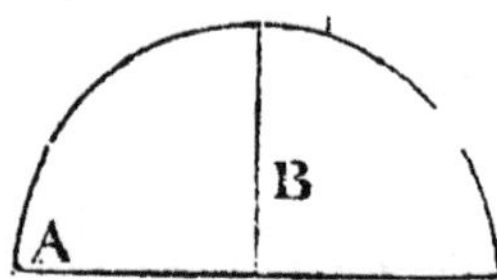

Every Circle is ſuppoſed to be divided into 360 Degrees, and an *Arch* is eſtimated according to the Number of theſe Degrees it takes up:—Thus an *Arch* is ſaid to be 20, 30, 50, 80, 100 Degrees.

Equal ARCHES are ſuch *Arches* of the ſame or equal Circles, as contain the ſame Number of Degrees.

Similar ARCHES are ſuch as contain the ſame Number of Degrees of unequal Circles, as the Figures annexed; which, though

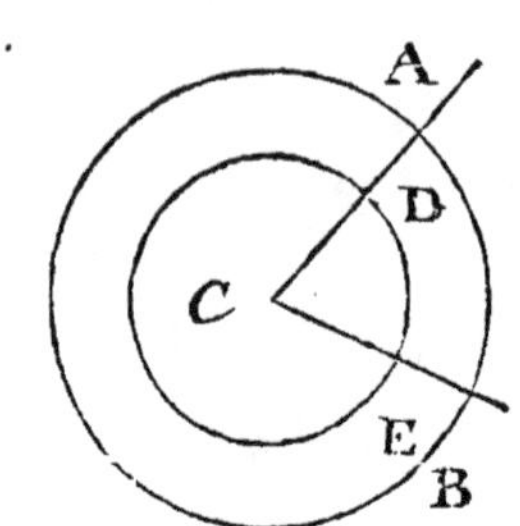

they are of Circles of different Magnitudes, yet are ſimilar, both containing the ſame Number of Degrees, as ſuppoſe 45.

ARCH [in *Architecture*] is a concave Building, rais'd with a Mould bent in the Form of the Arch of a Curve, and ſerving as an inward Support of any Structure.

Sir *Henry Wotton* ſays, an *Arch* is nothing but a narrow contracted Vault; and a Vault a dilated *Arch.*

Arches are uſed in large Intercolumniations of ſpacious Edifices; in Portico's, both within and without Temples; in publick Halls, as Cielings; the Courts of Palaces, Cloiſters, Theatres, and Anti-Theatres.

Arches are alſo uſed as Buttreſſes and Counter-Forts, to ſupport large Walls, and deep in the Earth; for Foundations of Bridges and Aqueducts; for Triumphal Arches, Gates, Windows, *&c.*

Arches are either *circular*, *elliptical*, or *ſtrait.*

Circular Arches are of three Kinds; *ſemicircular*, *ſcheme* or *ſheen*; or of the third and fourth Point,

Point, as ſome Workmen call them; though the *Italians* call them *Di Terzo* and *Quarto Acuto*, becauſe they always meet in an Acute Angle at the Top.

Semicircular Arches, are thoſe *Arches* which are an exact Semicircle; and have their Center in the Middle of the Diameter, (or Chord of the *Arch*,) or the Right Line that may be drawn betwixt the Feet of the *Arch*.

Of this Form the *Arches* of Bridges, Windows of Churches, and great Gates, are ſometimes made in modern Buildings.

Scheme or Skeen Arches, are thoſe which are leſs than a Semicircle, and conſequently are flatter *Arches*, containing ſome 90, ſome 70, and others 60 Degrees

Semicircular, are eaſily diſtinguiſhed from *Scheme Arches* by this; That the *Chord* (or Right Line) drawn between the Feet of a *Semicircular Arch*, is juſt double to its Height, (being meaſured from the Middle of the *Chord* to the Key-Piece or Top of the *Arch*;) whereas the *Chord* of a *Scheme Arch* of 96 Degrees will be more than four Times its Height, and the *Chord* of a *Scheme Arch* of 60 Degrees will be more than ſix Times its Height.

The famous *Alberti*, in his *Architectura*, ſays as follows: In all Openings in which we make *Arches*, we ought to contrive to have the *Arch* never leſs than a Semicircle, with an Addition of the ſeventh Part of half its Diameter; the moſt experienced Workmen having found that *Arch* to be by much the beſt adapted for enduring, in a manner to Perpetuity; all other *Arches* being thought leſs ſtrong for ſupporting the Weight, and more liable to Ruin.

It is alſo thought that the half Circle is the only *Arch* that has no Occaſion either for Chain, or any other Fortification; whereas all others are found either to burſt out, or fall to ruin by their own Weight, if they are not either chain'd, or ſome Weight be placed againſt them for a Counterpoiſe.

I will not here omit (ſays he) what I have obſerv'd among the Antients, a Contrivance certainly very excellent and praiſe-worthy: Their beſt Architects plac'd theſe Apertures, and the Arches of the Roofs of Temples, in ſuch a Manner, that even though you took away every Column from under them, yet they would ſtill ſtand firm, and not fall down, the Arches on which the Roof was placed being drawn quite down to the Foundation with wonderful Art, known but to a few: So that the Work upheld itſelf by being only ſet upon *Arches*; for thoſe *Arches* having the ſolid Earth for their Chain, no wonder they ſtood firm without any Support.

Arches of the third and fourth Point. Theſe conſiſt of two *Arches* of a Circle meeting in an Angle at the Top, and are drawn from the Diviſion of a Chord into three, or four, or more Parts, at Pleaſure.

Of this Kind are many of the *Arches* in old *Gothick* Buildings; but on account both of their Weakneſs, and Unſightlineſs, they ought, in the Opinion of Sir *Henry Wotton*, to be for ever excluded out of all Buildings.

Elliptical Arches. These Arches consist of a Semi-Ellipsis, and were formerly much us'd instead of Mantle-Trees in Chimneys. They are commonly described on three Centers; but they may be drawn otherwise: These consist of three Parts, *viz.* two *Hanches*, and a *Scheme*. Each End of these *Arches* are called *Hanches* by Workmen; and these *Hanches* are always the Arches of Circles, smaller than the *Scheme*, which is the middle Part of these Arches, and consists of a Part of a larger Circle, which is drawn betwixt the two *Hanches*, to conjoin them all together, in order to make, as it were, one *Heliacal Line*, and consequently, an *Elliptical Arch*.

To these Arches there are commonly a *Key-Stone* and *Chaptrels*: The *Key-Stone* is that which is the very Summity or Top of the Arch, and is equally distant from both Ends; and the Breadth of this Key-Stone at the Top, ought to be equal to the Height of the Arch (which is usually about fourteen Inches, when made of Brick;) and *Summer*, (or Point with two Edges, to the Center of the Scheme.) The Key-Stone ought to be so much without the Arch, as the *Chaptrels* project over the *Jaumbs*.

The *Chaptrels*, I suppose to be the same that most Architects call Imposts; and 'tis those on which the Feet of the Arches stand, the Height or Thickness of which, ought to be equal to the Breadth of the lower Part of the Key-Stone.

N. B. That each other Course in these Arches, consists of two *Stretchers*, which are seven Inches long each, (when the Arch is fourteen Inches deep,) and the other Courses betwixt these of three *Headers*, and two *Closers*; the Length of the *Headers* ought to be three Inches and a half, and the *Closers* one Inch and three quartets: Thus one Course of the Arch will be divided into two *Stretchers*; and the other alternately into three *Headers*, and two *Closers* throughout the whole Arch.

How to describe an *Elliptical Arch* to any Rise or Width, by the Intersection of Right Lines.

First draw the Line AB, then draw BC perpendicular to AB,

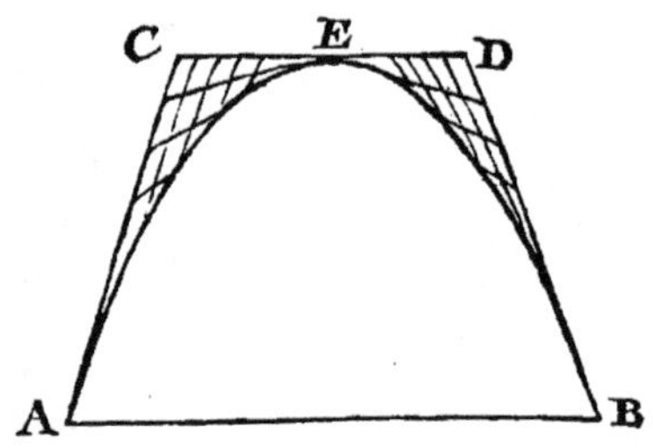

and as high as you design the Arch shall rise, and draw the Line CD parallel to AB, which divide into two equal Parts at E; then divide AC and BD into any Number of equal Parts; also CE and ED into the same Number of equal Parts, and draw Right Lines to each correspondent Division, as from 1 to 1, from 2 to 2, and so on; and then will the Intersections of those Lines create the *Arch* AEB.

How

How to draw and *Elliptical Arch* reverſe.

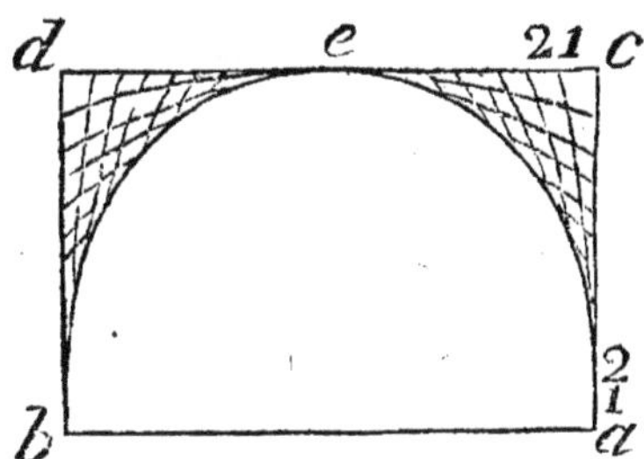

Firſt draw the Baſe Line AB then draw the Line CD parallel and equal to the Line AB, and ſo far diſtant as you deſign the *Arch* ſhall riſe, and draw the Lines CA and DB; then divide CA and DB into any Number of equal Parts, alſo CE and ED into the ſame Number of equal Parts, and draw Right Lines to each correſpondent Diviſion, as from 1 to 1, from 2 to 2, and ſo on, till you have deſcribed the *Arch* AEB. Which was to be done.

To ſtrike and find the Moulds of an *Elliptical Arch*, either in Brick or Stone.

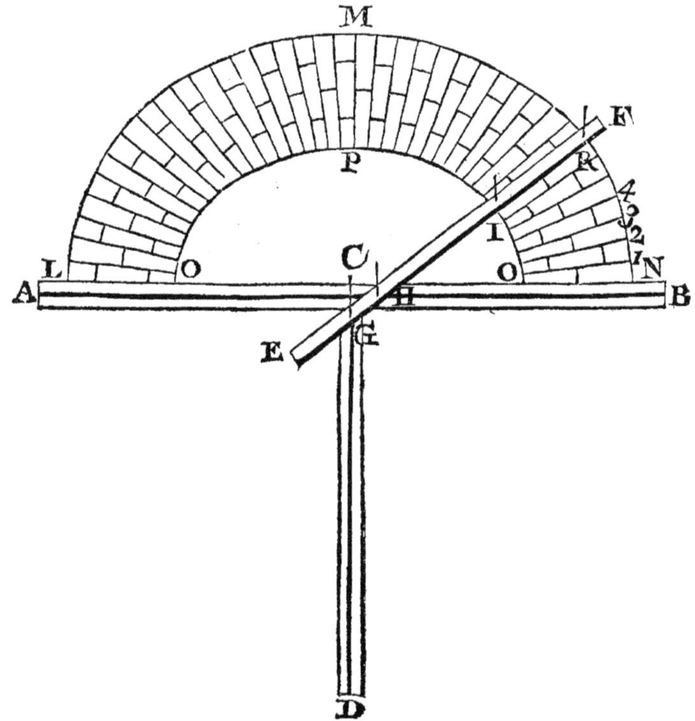

Firſt place the Trammel ABCD, (on which is a Groove, as AB and CD;) then propoſe the Widths OQ, OL, and QN, prepare a ſtraight Lath EF, ſomewhat longer than half the Baſe

CL or CN; then put in a Pencil or Marker at K; alſo at I, ſo that IK is equal to QN, or OL, or PM; then put in a wooden Pin at H and G, letting IH be equal to CP, alſo GK to CN; then fix one Hand at K, and the other at G, and keep the two Pins G and H in the Grooves AB and CD, and turn about the Lath FE; then will the Markers marked I and K, create the two *Arches* LMN and OPQ.

To give the *Bricks* or *Stones* the true Summering, divide the *Arch* LMN into ſo many equal Parts as the Thickneſs of the Brick will allow, as 1, 2, 3, 4, &c.

Bring down the Sliding-Lath to 1, and on its Edge draw the Summering or Joint of the firſt Brick, then move it to 2; in the ſame manner draw the Summering of the next Brick, and ſo on. The Croſs-Joints are drawn by the ſame Rule, as the *Arches* LMN or OPQ

To draw the *Elliptical Arch* ramping.

Firſt draw the Level Line AF, and divide it in the Middle

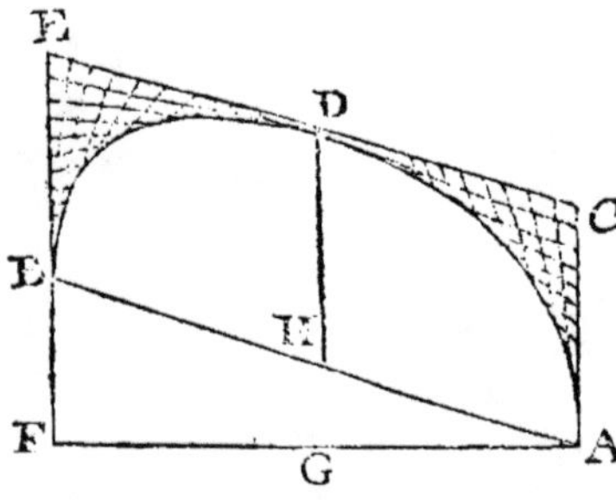

at G; then erect a Perpendicular at Pleaſure from F to E, alſo from G towards D, and from A towards C; then draw the Raking Line AB, and ſet up the Height of the *Arch* from A to C, and from B to E, and draw the Line CE; then divide the Lines AC and CD into any Number of equal Parts; alſo the Lines BE and ED, and draw the Right Lines, as in the foregoing Examples, which will create the *Arch* ADB.

Straight ARCHES are thoſe whoſe upper and under Edges are ſtraight; whereas in the others, they are curved; and thoſe two Edges alſo parallel, and the Ends and Joints all pointing towards one certain Centre. Theſe are principally uſed over Windows, Doors, &c.

And it is a general Rule among Workmen, that according to the Breadth of the Peers between the Windows, ſo ought the Skew-back, or Summering of the Arch, to be; for if the *Peers* be of a good Breadth, as ſuppoſe three or four Bricks in Length, then the *Straight Arch* may be deſcribed from the *Oxi*, (as it is vulgarly called by Workmen,) being a Contraction of the Word *Oxygonium*, which is the Name of an Equilateral Triangle: But if the Peers are ſmall, as they are ſometimes, being but of the Length of two Bricks, and ſometimes, again, but of one Brick and a Half, then the Breadth of the Window, or more, may be the Perpendicular (to the Middle of the Under-ſide of the Arch,) at the End of which, below, ſhall be the Centre, for the *Skew-back* or *Summering* to point to.

Theſe *Straight Arches* are uſually about a Brick and half, which when rubb'd, makes about twelve Inches high, although ſometimes they are but eleven

eleven Inches, or thereabouts, which answers to four Courses of Bricks; but notwithstanding, they may be made either more or less in Height, according as Occasion requires.

N. B. By the Term *Skewback*, is to be understood the Levelling-End of the Arch; and by *Summering*, the Level-Joints betwixt the Courses of Bricks in an Arch. These Arches usually consist of a *Stretcher* and a *Header* in Height; the *Stretchers* being a whole Brick's Length, and the *Headers*, a Brick's Breadth.

The Doctrine and Use of Arches is well deliver'd by Sir *Henry Wotton*, in the five following *Theorems*.

Theorem I. All Matter, unless impeded, tends to the Centre of the Earth in a perpendicular Line, or descends perpendicularly downwards; because Ponderosity is a natural Inclination to the Centre of the Earth, and Nature performs her Motions by the shortest Lines.

Theorem II. All solid Materials, as Bricks, Stones, &c. moulded in their ordinary Rectangular Form, if laid in Numbers, one by the Side of another, in a level Row, and their Extream Ends sustain'd between two Supporters, all the Pieces between will necessarily sink even by their own natural Gravity; and must much more, if they are pressed down, or suffer any Pressure by a super-incumbent Weight; because their Sides being parallel, they have Room to descend perpendicularly, without Impediment, according to the former *Theorem*; therefore, to make them stand, either their Figure or their Position must be altered.

Theorem III. Stones, Bricks, or other Materials, being figur'd *cuneatim*, i. e. wedge-wise, somewhat broader above than below, and laid in a level Row, with their two Extreams supported, as in the preceding Theorem, and pointing all to the same Centre; none of the Pieces between can sink, till the Supporters or Butments give way; because they want Room in that Situation, to descend perpendicularly.

But this is but a weak Structure; because the Supporters are subject to too much Impulsion, especially where the Line is long; For which Reason, this Form of Straight Arches is seldom used, but over Doors and Windows, where the Line is short.—— Therefore, in order to fortify the Work, the Figure of the Materials must not only be chang'd, but the Position of them too; as will appear in the following Theorem.

Theorem IV. If the Materials be shap'd wedge-wise, and disposed in Form of a Circular *Arch*, and pointing to some Centre: In this Case, neither the Pieces of the said Arch can sink downwards for want of Room to descend perpendicularly, nor can the Supporters or Butments of this Arch suffer so much Violence, as in the preceding flat Form; for the Roundness or rather Convexity, will always make the incumbent Weight rather rest upon the Supporters, than heave or shove them outwards; whence this Corollary may be fairly deduced, That the safest or most secure of all the Arches above mention'd, is the Semicircular; and of all Vaults, the Hemispherical, although not absolutely exempted from some natural Imbecility, (which is the sole Prerogative of Perpendicular Lines and Right Angles,) as has been observ'd by *Bernardino Baldi*, Abbot of *Guastalla*, in his Commentary upon *Aristotle's Mechanicks*; where, by the way, it is to be noted, that when any Thing is demonstrated mathematically to be weak, it is much more so mechanically; Errors always occurring more easily in the Management of gross Materials, than in Lineal Designs.

Theorem V. As Semicircular Arches, or Hemispherical Vaults, rais'd on the whole Diameter, are the strongest and securest by the precedent Theorem, so they are also the most beautiful; which keeping precisely to the same Height, are yet distended one Fourteenth Part longer than the said Diameter; which Addition of Width will contribute greatly to their Beauty, without diminishing any Thing considerable of their Strength.

However, it is to be observ'd, that according to Geometrical Strictness, in order to have the strongest *Arches*, they must not be Portions of Circles, but of another Curve, called the *Catenaria*; the Nature of which is such, that a Number of Spheres dispos'd in this Form, will sustain each other, and form an *Arch*. See *Catenaria*.

Dr. *Gregory*, *Philosoph. Transactions*, N° 231. has shewn, that *Arches* constructed in other Curves, only stand or sustain themselves by Virtue of the *Catenaria* contain'd in their Thickness; so that if they were made infinitely slender or thin, they must of course tumble; whereas the *Catenaria*, though infinitely slender, must stand, by reason that no one Point of it tends downwards more than any other. Vide *Vault*.

Of Measuring Arches. Whether the *Arches* be straight or circular, they must be measured in the Middle, *i.e.* if a *straight Arch* be ten Inches in Height or Depth, the Length must be measured in the Middle of the ten Inches; which Length will not be any longer, than if it were measured at the Under side next to the Head of the Window, by so much as one Side of the *Springing Arch* is skew'd-back from the Upright of the Jaumbs, Peers, or Coins of the Windows.

And also in *Circular Arches*, it is to be observed, that the upper Part of the *Arch* is longer (if girt about) than the under Part, by reason that it is the Segment of a greater Circle, cut off by the same Right Line that the lesser is, and for that Reason must be girt in the Middle.

As to the Price. As for the Workmanship of *Straight Arches*, (of Brick,) handsomely set, and well rubb'd, in *London*, about eight Pence or nine Pence a Foot; but if the Workman finds Materials, he will have ten Pence, or one Shilling *per* Foot. But in some Parts of *Sussex* and *Kent*, they will require one Shilling *per* Foot; nor will they do it under running Measure.

Scheme

Scheme or *Sheen Arches*, and *Elliptical* ones of *rubb'd Brick*, are uſually much about the ſame Price as *Straight Arches*. But if the *Scheme Arches* are of *unrubb'd Bricks*, they are uſually included in the Plain Work, unleſs the Plain Work be done at a low Price: But you muſt take notice, that the Owner or Maſter of the Building muſt be at the Charge of the Centres to turn the *Arches* on, and not the Workman, unleſs an Allowance be made him for it in the Price of the Work.

How to deſcribe a *Scheme Arch*, when the Baſe and Perpendicular are given.

Firſt draw the Line AB, then draw a Line at Right Angles

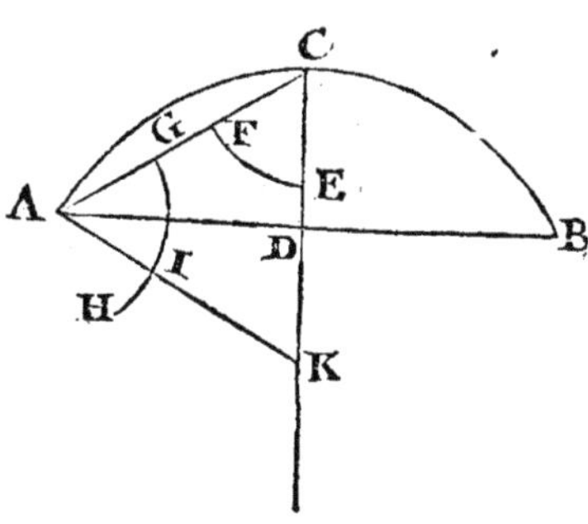

with it, through the Middle D, at Pleaſure, and ſet up the Height you deſire to riſe from D to C, and draw the Line CA; then open your Compaſſes to any convenient Diſtance, ſet one Foot in C, and ſtrike the *Arch* FE; with the ſame Opening of the Compaſſes, ſet one Foot in A, and ſtrike the *Arch* from G to H, at Pleaſure; then take the Radius EF in your Compaſſes, and ſet it on the *Arch* GH, as at I, and draw a Right Line from A through I, to cut the Perpendicular, as at K; then is K the Centre to ſtrike the *Arch* ACB. Which was to be done.

The Baſe and Perpendicular of a *Scheme Arch* being given, how to deſcribe it by an Interſection of Lines.

Firſt draw the Baſe AB, and Middle at E, from whence ſet

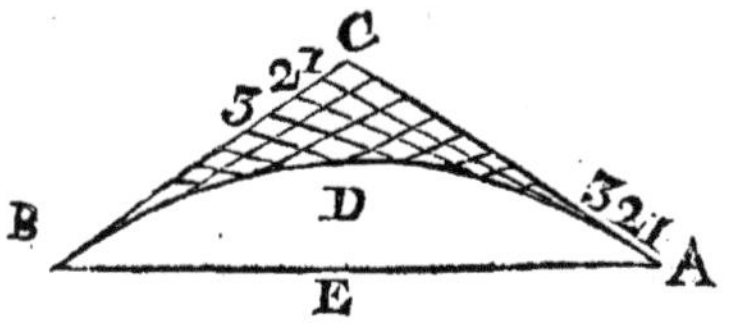

up perpendicularly to C, twice as much as you would have the *Arch* to riſe, and draw the Lines CB and CA, and divide each Line into any Number of equal Parts, and draw Right Lines to every correſpondent Diviſion, as from 1 to 1, from 2 to 2, from 3 to 3, and ſo on; and then will the Interſections of thoſe Lines create the *Arch* ADB. Which was to be done.

It is the ordinary-Proportion of *Arches*, that the Height be made double the Width; but this may be varied, and made a little more or a little leſs, as Occaſion requires. *Le Clerc.*

When *Arches* are to be at ſome Diſtance from each other, for the Conveniency of any Appartments, either above or underneath, the Columns which ſeparate them ought to be in Couples; but when they are in Couples, they ſhould have but one Pedeſtal, if they have any Pedeſtal at all.

Arch is particularly uſed for the Space between the two Peers of a Bridge. The Chief or Maſter *Arch*, is that in the Middle, which is the wideſt, and commonly higheſt, and the Water that runs under it the deepeſt, being deſign'd for the Paſſage of Boats, or other Veſſels. Some Relations mention Bridges in the *Eaſt* having 300 *Arches*. See *Bridge*.

A *Triumphal Arch* is a Gate or Paſſage into a City, magnificently adorn'd with Architecture, Sculpture, Inſcriptions, &c. which, being erected either of Stone or Marble, are uſed not only as Decorations in Triumphs, on account of ſome Victory, but alſo to tranſmit the Memory of the Conqueror to Poſterity.

The moſt celebrated *Triumphal Arches*, that are now remaining, are thoſe of *Titus*, of *Septimus Severus*, and of *Conſtantine* at *Rome*.

To find the Length of an *Arch-Line* geometrically;

Divide the Chord Line AB into four equal Parts, and ſet one

of thoſe Parts from B to C, and draw a Line from C to three of thoſe Parts at D; ſo ſhall CD be equal to half the Arch Line ACB.

To find the Length of an *Arch-Line* Arithmetically; Multiply the Chord of half the Segment AC or CB by 8, and from the Product ſubtract the Chord of the whole Segment AB, and divide the Remainder by 3, the Quotient will be the *Arch-Line* ACB ſought.

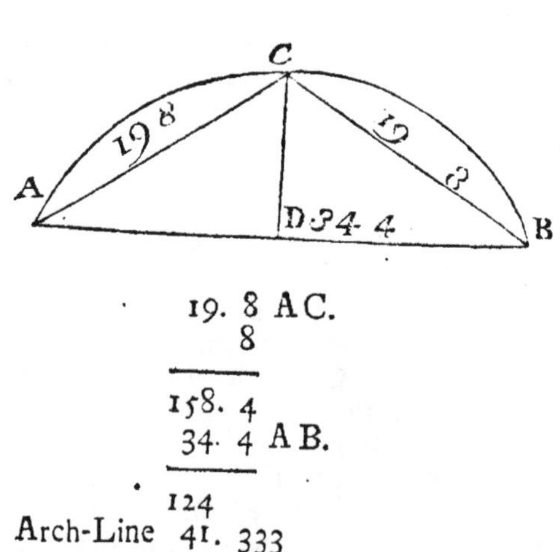

19. 8 AC.
8

158. 4
34. 4 AB.

124
Arch-Line 41. 333

Another Way.

From the double Chord of half the Segment's Arch ſubtract the Chord of the Segment, one third Part of the Difference added to the double Chord of half the Segment's Arch, the Sum is the

the Arch Line of the whole Segment. Thus, if AC 19. 8 be doubled, it makes 39. 6; from which, if you fubtract 34. 4, the Remainer is 52; which divided by 3, the Quotient is 1. 733: This being added to 39. 6, (the double Chord of the Half Segment) the Sum will be 41. 333. So if the *Arch-Line* ACB were ftretch'd out ftrait, it would then contain 41. 333 fuch Parts as the Chord AB contains, 34. 4 of the like Parts.

ARCHITECT ['Αρχιτεκτων, of ἀρχὸς Chief, and τεκτων an Artificer or Builder, *Gr.*] a Mafter-Workman in a Building, he who defigns the Model, or draws the Plot, Plan, or Draught of the whole Fabrick; whofe Bufinefs it is to confider the whole Manner and Method of the Building; and alfo to compute the Charge and Expence. In the managing of which, he ought to have regard to its due *Situation*, *Contrivance*, *Receipt*, *Strength*, *Beauty*, *Form*, and *Materials*.

The Name *Architect* is alfo ufed for the *Surveyor*, or *Superintendant* of an Edifice, the Management being wholely committed to his Circumfpection; wherefore he ought to manage the whole Affair prudently and advifedly, with the utmoft Caution, that all Matters may be ordered and difpofed, (in all Circumftances,) fo as to anfwer the Owner's Defign, and be confentaucous to Reafon.

But notwithftanding the Care of the whole Fabrick be incumbent on this *Surveyor*, or *Superintendant*, yet Sir *Henry Wotton* advifes the having a fecond *Superintendant*, (or *Officinator*, as he is called by *Vitruvius*,) whofe Bufinefs is to chufe, (or examine,) and fort all the Materials for every particular Part of the Building.

Vitruvius enumerates 12 Qualifications requifite for a compleate *Architect*; that he be docile and ingenious, literate, skill'd in Defigning, in Geometry, Opticks, Arithmetick, Hiftory, Philofophy, Mufick, Medicine, Law, and Aftrology.

The moft celebrated ancient *Architects* are *Vitruvius*, *Palladio*, *Scamozzi*, *Serlio*, *Vignola*, *Barbaro*, *Cataneo*, *Alberti*, *Vida*, *Bullant*, *De Lorme*, and many others.

ARCHITECTONICK, that which builds a Thing up regularly, according to the Nature and Intentions of it. The Term is ufually apply'd to that plaftick Power, *Spirit*, or whatever elfe it be, which hatches the Ova of Females into living Creatures, which is called the *Architectonick Spirit*; yet it is alfo apply'd to the chief Overfeer of Buildings, or an Architect.

ARCHITECTURE, the Art of Building, or a Mathematical Science, which teaches the Art of crecting *Edifices proper either for Habitations or Defence*; being a Skill obtain'd by the Precepts of *Geometry*; by which it gives the Rules for defigning and railing all forts of Structures, according to the Rules of *Geometry* and *Proportion*, and contains under it all thofe Arts which conduce any Thing to the framing Houfes, Temples, &c.

The *Scheme* or *Projection* of a Building is ufually laid down in three feveral Defigns or Draughts.

The *firſt* is a *Plan*, which exhibits the Extent, Diviſion, and Diſtribution of the Ground into Apartments and other Conveniences.

The *ſecond* ſhews the Stories, their Heights, and the outward Appearances of the whole Building: And this is uſually called the *Deſign* or *Elevation*.

The *third* is commonly called the *Section*, and ſhews the Inſide of the Fabrick.

From theſe three Deſigns, the Undertaker frames a Computation of the Charges of the whole Building, and the Time requiſite to complete it.

As to the Antiquity of Architecture: *Architecture* is ſcarce inferior, in Point of Antiquity, to any other Arts. Nature and Neceſſity taught the firſt Inhabitants of the Earth to build or ſet up Huts, Tents, and Cottages; from which, in Proceſs of Time, they gradually advanced to raiſing more regular and ſtately Dwellings, ſet off with Variety of Ornaments, Proportions, &c.

Antient Writers aſcribe the carrying of *Architecture* to a tolerable Height to the *Tyrians*, who were therefore ſent for by *Solomon* for the Building of his Temple.

But *Villapandus* will not allow thoſe who were ſent for from *Tyre* to be any more than under Workmen, ſuch as Artificers in Gold, Silver, Braſs, &c. and ſuppoſes that the Rules of *Architecture* were delivered by God himſelf to *Solomon*.

So that the *Tyrians* rather learnt *Architecture* from *Solomon*, than he from them; which they afterwards communicated to the *E-*

common Paſſion for *Architecture*; but Luxury and Diſſoluteneſs had a greater Share in it, than true Magnificence. In the Time of *Trajan*, *Apollodorus* excell'd in the Art; by which he merited the Favour of that Prince, and erected that famous Column called *Trajan*'s, which is remaining to this Day.

But after his Time *Architecture* began to decline; though it was for ſome Time ſupported by the Care and Magnificence of *Alexander Severus*, yet it fell with the Weſtern Empire, and ſunk into Corruption; from whence it was not recovered for the Space of 1200 Years.

All the moſt beautiful Monuments of Antiquity were deſtroyed by the Ravages of the *Viſigoths*; and from that Time *Architecture* became ſo coarſe and artleſs, that their profeſſed Architects knew nothing at all of juſt Deſigning, wherein the whole Beauty of *Architecture* conſiſts: Hence a new Manner of *Architecture*, called *Gothic*, took its Riſe.

Charlemagne ſet himſelf induſtriouſly about the Reſtoration of *Architecture*; and the *French* applied themſelves to it with Succeſs, under the Encouragement of *Hugh Capet*. His Son *Robert* proſecuting the ſame Deſign, the modern *Architecture*, by Degrees, ran into as great an Exceſs of Delicacy, as the *Gothic* had before done of Maſſiveneſs.

We may add to theſe the *Arabesk*, *Moresk*, or *Mooriſh Architecture*, which were much of the ſame Kind with the *Gothic*; except, that as the former was brought from the North by the *Goths* and *Vandals*, the latter was brought from the South by the *Moors* and *Saracens*.

The Architects of the 13th, 14th, and 15th Centuries, who had ſome Knowledge of Sculpture, ſeem'd to make Perfection conſiſt altogether in the Delicacy and Multitude of Ornaments, which they beſtow'd on their Buildings with abundance of Care; but often without Conduct, or Taſte.

In the two laſt Centuries, the Architects of *Italy* and *France* were induſtriouſly bent upon retrieving the primitive Simplicity and Beauty of antient *Architecture*; nor did they fail of Succeſs: Inſomuch, that now our Churches, Palaces, &c. are wholly built after the *Antique*.

Civil Architecture may be diſtinguiſhed, with reſpect to the ſeveral Periods or States of it, into *Antique*, *Antient*, *Gothic*, *Modern*, &c.

Another Diviſion of *Civil Architecture* ariſes from the different Proportions, which the different Kinds of Buildings rendered neceſſary, that there might be ſome proper for every Purpoſe, according to the Bulk, Strength, Delicacy, Richneſs, or Simplicity required.

From hence proceeded the five Orders or Manners of Building, all invented by the Antients, at different Times, and on different Occaſions, *viz.* *Tuſcan*, *Doric*, *Ionic*, *Corinthian*, and *Compoſite*.

That which forms an Order, is the Column, with its Baſe and Capital, ſurmounted by an Entablature conſiſting of Architrave, Freeze, and Cornice, ſuſtain'd by a *Pedeſtal*.

We

We have no *Greek* Authors now extant on Architecture: The firſt who wrote was *Agathereus* the *Athenian*. He was ſeconded by *Democritus* and *Theophraſtus*.

Of all the Ancients, *Vitruvius* is the only Author we have entire, notwithſtanding that he relates that there was 700 Architects in *Rome* in his Time.

Vitruvius, in the Time of *Auguſtus*, wrote a complete Syſtem of Architecture, in ten Books, which he dedicated to that Prince.

The Moderns cenſure two Things in this excellent Work, *viz.* want of Method and Obſcurity.

The Mixture of *Latin* and *Greek* in *Vitruvius* is ſuch, that *Leon. Baptiſt. Alberti* has obſerved, that he wrote *Latin* to the *Greeks*, and *Greek* to the *Latins*: He alſo ſays, that there are abundance of Things ſuperfluous and foreign to the Purpoſe contain'd in that Work.

For this Reaſon, M. *Perrault* has extracted all the Rules out of the prolix Work of *Vitruvius*, methodiz'd and publiſh'd them in a ſmall Abridgment.

Several Authors have attempted to explain the Text of *Vitruvius*, as particularly *Philander*, *Barbaro*, and *Salmaſius*, in Notes added to their ſeveral Editions in *Latin*; *Rivius* and *Perrault*, in the Notes to their *German* and *French* Verſions; and *Baldus*, in his *Lexicon Vitruvianus*..

M. *Perrault* alſo has compos'd an excellent Treatiſe *of the Five Orders*, which may be look'd upon as a Supplement to *Vitruvius*, he having left the Doctrine of the Orders imperfect.

The Authors who have written on Architecture ſince *Vitruvius*, are *Leon. Baptiſt. Alberti*, who publiſhed in *Latin* ten Books of the Art of Building, deſigning to outvie *Vitruvius*; but however has not ſucceeded in his Deſign, although his Books contain a great Number of good Things, but is deficient in the Doctrine of the Orders.

Sebaſt. Serlio alſo wrote ſeven Books of Architecture, five of which were concerning the Five Orders, were publiſhed in 1602, through the Whole of which he ſtrictly keeps to *Vitruvius*'s Rule; the ſeventh was publiſhed afterwards in the Year 1675.

Philip de Lorme publiſhed nine Books of Architecture in *French*, in the Year 1667.

J. Barozzio de Vignola publiſhed his Rules of the Five Orders in *Italian*, in the Year 1631; which have been ſince tranſlated, with large Additions, by *Daviler*, under the Title of *Cours d'Architecture*.

Alſo *Vincent Scamozzi*'s Idea of *Univerſal Achitecture* was publiſhed in *Italian*, in the Year 1615; and *Car. Phil. Dieuſſart*'s *Theatre of Civil Architecture*, was publiſhed in *High Dutch*, in the Year 1697; in which he not only delivers the Rules of *Architecture*, but alſo explains and compares the Five Orders, as laid down by *Palladio*, *Vignola*, *Scamozzi*, &c.

R. Freart de Cambray alſo purſued the ſame Deſign in *French*, in a Parallel of the Ancient Architecture with the Modern; which

which was publiſhed in 1650, and tranſlated into *Engliſh* by Mr. *Evelyn*, with Additions.

Fr. Blondel, Director of the Royal Academy of Painting, &c. in 1698, gave a *Courſe of Architecture* in *French*, which was a Collection from all the celebrated Writers upon the Subject of the Orders, &c.

Nich. Goldman has alſo done good Service, by reducing the Rules and Orders of *Architecture* to a farther Degree of Perfection, and ſhewing how they may be eaſily delineated, by Means of certain Inſtruments invented by him. This Treatiſe was publiſhed in *Latin* and *High Dutch*, in the Year 1661.

Alſo Sir *Henry Wotton* has laid down *the Elements of Architecture*, which have been reduced by *Sturmius* and *Wolfius* to certain Rules and Demonſtrations. And by theſe Gradations has Architecture been brought to a Mathematical Art; by the firſt in his *Matheſis Juvenilis*; and by the ſecond, in his *Elementa Matheſeos, Tom.* II. *Anno* 1715.

Military Architecture is the Art of ſtrengthening and fortifying Places, to ſcreen and defend them from the Inſults of Enemies, and the Violence of Arms. This is more commonly called Fortification, and conſiſts in the erecting Forts, Caſtles, Fortreſſes, with Ramparts, Baſtions, &c.

Naval Architecture is the Art of Building, or that which teaches the Conſtruction of Ships, Gallies, and other floating Veſſels for the Water; alſo Ports, Moles, Docks, &c. on the Shore.

Architecture in Perſpective, is a ſort of Building, wherein the Members are of different Meaſure and Modules, and diminiſh proportionably to their Diſtance, in order to make the Work appear longer and larger to the View, than really it is.

Of this Kind is the celebrated Pontifical Staircaſe of the *Vatican*, built in the Time of Pope *Alexander* VII. by the Cavalier *Bernini*.

Counterfeit Architecture, is that which conſiſts of Projectures, painted either in Black or White, or Colours, after the Manner of Marble; as is to be ſeen perform'd in the Facades and Palaces in *Italy*, and in the Pavilions of *Marli*.

This Painting is done in Freſco upon plaiſtered Walls, and in Oil on ſtone Walls.

Alſo, under the Title of *Counterfeit Architecture*, is to be comprehended, that which may be alſo called *Scene Work*, i. e. that painted on ſlight Boards, or wooden Planks, on which Columns, Pilaſters, and other Parts of Building, ſeem to ſtand out with a *Relievo*; the Whole being coloured in Imitation of various Marbles, Metal, &c. ſerving for the Decorations of Theatres, Triumphal Arches, &c.

ARCHITRAVE [of ἀρχὴ chief, and *trabs*, L. a Beam] is that Part of a Column, or Order of Columns, which lies immediately upon the Capital; the *Architrave* is the loweſt Member of the Frize, and even of the whole Entablature.

The *Architrave* is ſuppoſed to repreſent the principal Beam in Timber Buildings; from whence it takes its Name, as above.

But

But to this it is objected by ſome, that they do not well underſtand what is meant by the principal Beam of a Building; becauſe they do not ſuppoſe that it can properly be applied to all Buildings, but only to ſome peculiar Kinds, ſuch as are called *Portico's*, *Piazza's*, or *Cloiſters*, (by which are uſually underſtood long Kinds of Galleries, or walking Places, whoſe Roofs are borne or ſupported by Columns or Pillars, at leaſt on one Side, which have not Arches ariſing from them, to ſupport the ſuper-incumbent Part of the Fabrick,) but have a Beam reſting or lying upon the Tops of the Columns, by which the ſuperior Part of the Edifice is ſupported; upon which Account, it is probable, it is called the principal Beam.

'Tis true, according as Mr. *Perrault* defines it, it is the firſt Member of the *Entablement*, being that which bears upon the Column, and is made ſometimes of a ſingle *Summer*, as appears in the moſt antient Buildings; and ſometimes of ſeveral Haunches, as is uſually ſeen in the Works of the Moderns.

Architrave is alſo ſometimes called the *Reaſon-Piece*, or *Maſter-Beam*, in Timber Buildings; as Portico's, Cloiſters, &c. In Chimneys it is called the *Mantle-Piece*; and over Jaumbs of Doors, and Lintels of Windows, *Hyperthyron*.

Architrave Doors, are thoſe which have an Architrave on the Jaumbs, and over the Door, upon the Cap-Piece, if ſtraight, or on the Arch, if the Top be curv'd.

Architrave Windows, of Timber, are commonly an Ogee, raiſed out of the ſolid Timber, with a Liſt over it; though ſometimes the Mouldings are ſtruck, and laid on, and ſometimes are cut in Brick.

The upper *Fatio* is called the *Header*, or Heading *Architrave*.

Architects take a deal of Latitude as to *Architraves*, ſome uſing more Members than others, and many of them having two or three Forms of *Architraves*.

Sometimes they are according to one of the Five Orders of Architecture; but at others, they are according to the Fancy of the Workman.

As ſome, for an *Architrave* round a Door, have put firſt a ſmall Bead next the Door, then a broad Plinth, or *Fatio*, and above that an Ogee and Liſt.

There are *Architraves* of Stone and Brick, as well as of Timber.

Brick Architraves are uſually cut in the Length of a Brick, and ſometimes in the Length of a Brick and half, and then every other Courſe alternately conſiſts of the Breadth of two Bricks; the upper one, on which the Ogee is cut, and Part of the upper *Fatio*, they call *Header*, or *Heading Architrave*; and the Breadth or Head of Bricks, on which the lower *Fatio*, and Part of the upper one is cut, they call a *Jak Architrave* of Stone.

The Kinds: *Architraves* are diſtinguiſhed into five Kinds, *viz.*

Tuſcan,

Tuſcan, *Doric*, *Ionic*, *Corinthian*, and *Compoſite*, according to the five Orders of Columns.

Of the Parts, *or Members:* Theſe are more numerous than the Kinds, becauſe there are two different Sorts of *Architraves* to ſome of the Orders; and that which yet more increaſes the Number is, that ſome Authors differ from others in the Form of the ſame Oders.

The *Tuſcan Architrave*, according to *Vitruvius*, ought to be half a Mode, or M. in Height. This general Member he has deſcribed in two Forms: The firſt conſiſts of three Parts or Members, *viz.* two *Fatio's*, and a *Cymatium*, and is thus divided: The whole Height is divided into ſix Parts; which are divided after this Manner, *viz.* the uppermoſt ſixth Part is the *Cymatium*; which being ſubdivided into three, the upper Part is to be the *Fillet*, and the two lower Parts the *Ogee*.

The five grand Diviſions which remain, are to be divided into nine Parts, five of which are to be the upper *Fatio*, and the other four the lower one.

His ſecond Form is as follows, conſiſting of but three Members, or Parts, *viz.* a large *Plinth* or *Planchier*, a *Caſement*, and a large *Fillet*, and is ſubdivided as follows: The whole Height is divided into ſix; and the upper Part is for the *Fillet*, (which projects in Square beyond the *Plinth*;) the fifth Part is for the *Caſement*, (which riſes from the Plain of the *Plinth*, and ends in a Quadrant as the lower Corner of the *Fillet*;) the other four Parts remaining are for the *Plinth*, or *Planchier*, or *Fatio*.

Palladio has alſo two diſtinct Forms for the *Tuſcan Architrave:* The firſt conſiſts of two *Fatio's* and a *Liſt*: The lower *Fatio* is twelve and a half M. high, the upper *Fatio* is ſeventeen and a half M. ending with a quadrantal *Caſement* tiling with its Plain, and ending with the loweſt Corner of the *Liſt*; the *Liſt* is five M. high; and ſo the whole Height of the *Architrave* is thirty-five M.

His ſecond *Architrave* is only a plain *Fatio* of thirty-five M. high.

Scamozzi, according to his Delineations, makes the *Tuſcan Architrave* thirty-one and a half M. high, which he divides into four Parts, or Members, *viz.* two *Fatio's*, a *Liſt*, and a *Plinth*. He makes his firſt *Fatio* ten M. his ſecond ſixteen and a half M. his *Liſt* one and a half M. and his *Plinth* three and a half M. all which make Thirty-one one Third M. Though, according to this verbal Account, he ſays, it muſt be thirty-two and a half M. except it be a typographical Error.

Vignola deſcribes the *Architrave* with the ſame Parts, Height, and Form with *Vitruvius*'s ſecond.

The *Doric*. This *Architrave*, according to *Vitruvius*, is half a M. in Altitude, which he delineates in two Forms: The *firſt* he divides into ſeven Parts; the uppermoſt of which is the *Tænia*, the other ſix Parts which remain he makes a *Faſcia* under the *Tænia*, and

and places Drops, whose Height are, one Seventh of the Architrave: A Fourth of this Seventh is the *Fillet*, to which the *Drops* hang; the Drops are in Number six, placed under, and of the same Breadth with the *Triglyphs*.

His *Second Figure* of his *Architrave* consist of the same Members with the first, and the whole Heighth is equal to the first. But he divides the Altitude only into six Parts; the upper one of which, is his *Tænia*, and the other Five, the *Fascia*; the uppermost of which, is the Altitude of his Drops, which have a *List*, which is one Quarter of their Height.

Palladio makes this *Architrave* of the same Altitude with *Vitruvius*, but of a different Form; for he makes it to consist of three Parts or Members, *viz.* two *Fascia's* and *Tænia*: He divides the whole Altitude into six Parts, one of which being five M. he assigns for the *Guttæ*, *Drops*, or Bells, and the *Listella* of the *Drops* is one Fifth of the whole Height, and one Third M. and the Drops Two and two Thirds M. The *Tænia* above the Drops, (or rather of the *Architrave*,) he also makes Four one half M. and the *Prima* (or Upper) *Fascia*, Fourteen one half M. and the *Secunda* (or Lower) eleven M. in all thirty M. which is the whole Altitude.

Scamozzi (according to the Portraiture of this *Architrave*) makes it thirty-five M. in Altitude; and he makes this grand Member to comprehend three petty Members, *viz.* two *Fascia's*, and a List; the Dimensions of which are, beginning at the Top, and so descending; the List he makes five M. the upper *Fascia* eighteen M. and the lower one twelve M. in all thirty-five M. Divides the Drops or Bells thus: He designs the List above to be One one half M. and the Bells or Drops Four one half M. so that the whole Height is six M.

Vignola makes this *Architrave* thirty M. in Altitude, the same as *Vitruvius* and *Palladio*; both which he also imitates in the lesser Member: For he has two distinct Forms; one like to that of *Vitruvius*, which contains two Members or Parts, the one a *List*, the other a *Fascia*; his other Form is like that of *Palladio*, comprehending three petty Members, *viz.* one *Tænia*, and two *Fatio's*.

The *Ionick*, according to *Vitruvius*'s Order, this grand Member ought to be half a M. in Heighth. He describes two Forms of *Architraves* in the *Ionick* Order, *viz.* one for the *Ionick* Column, without a Pedestal; and the other, with a Pedestal.

He composes that without a Pedestal, of four minuter Parts, *viz.* three *Fascia's*, and a *Cymatium*; which is divided, as follows: He divides the whole Altitude into seven Parts, the uppermost of which, he allots to the *Cymatium*, which he subdivides into three Parts; the uppermost of which, is for the *List*, and the two remaining for the Ogee.

The other six remaining Parts are divided into twelve; five of which he makes the upper *Fascia*, four the middle one, and three the lowest.

The

The other for the *Ionick* Column with a Pedeſtal, he proportions, as follows, *viz.* he reckons the whole Altitude of the *Architrave*, *Freeze*, and *Corniſh*, to be two Mod. which are divided into ten Parts; three of which, are for the *Architrave*, (which is thirty-ſix M.) which he diſtinguiſhes into ſix minuter Parts, or Members; which he names as follows, (beginning at the Top, and ſo deſcending,) *viz.* a *Fillet*, a *Cima*, a *Thorus*, and three *Faſcia's*; all which ſmaller Members he thus finds, *viz.*

Firſt, he divides the whole Altitude into ſix equal Parts; the uppermoſt of which Parts, he ſubdivides into four Parts: The higheſt of theſe four is for the *Fillet*, the two next of the four are allotted to the *Cima*, and the Fourth, that remains, is for the *Thorus*.

The five grand Diviſions, which remain, are ſubdivided into twelve, which are diſtributed as follows, *viz.* five for the upper, four for the middle, and three for the lower *Faſcia*.

Palladio aſſigns thirty-four M. for the Height of this *Architrave*. According to his Scheme of this Member, it is compoſed of ſeven Parts, *viz.* a *Liſt*, a *Cima*, three *Faſcias*, and two *Aſtragals*; which are proportion'd as follows: To the *Liſt* (which is above the *Cima*,) he allots two three Tenths M. to the *Cima*, four three Fifths M. to the upper *Faſcia* he allows ten one Eighth M. to the Aſtragal, at its Foot, one Third M. the middle *Faſcia* is to contain ſeven fifty-two Sixtieths; and the Aſtragal at its Foot one Third M. to the lower *Faſcia* he aſſigns ſix nine Tenths M. All which being added together, amount to thirty-four one half M.

Scamozzi makes the *Ionick Architrave* thirty-five M. high, and of the ſame Form with that of the Second of *Vitruvius*, conſiſting of ſix Parts, *viz.* a *Liſt*, *Cima*, *Aſtragal*, (or *Thorus*,) and three *Faſcia's*; which he proportions, as follows: He allots two one half M. to the *Liſt*, to the *Cima* four, to the *Thorus* two to the upper *Faſcia* eleven one half, to the middle one eight one half, and to the lower one ſix one half.

Vignola allows thirty-ſeven one half to the *Ionic Architrave* in Altitude; and as to the Form, it is much the ſame with that of *Vitruvius*'s firſt Order.

The *Corinthian Architrave*, according to *Vitruvius*, ought to be half a Mod. in Height; but it is to be obſerv'd, that this is for the *Corinthian* Column without a Pedeſtal. This Member he divides into ſeven Parts, of which the uppermoſt is the *Cymatium*; the ſix remaining Parts he divides into twelve, of which he allots five to the upper *Faſcia*; alſo allowing one Eighth of this *Faſcia* for a *Bead* at its Foot, he allows four of the twelve Parts to the middle *Faſcia*, and one Eighth of this *Faſcia* for the Bead at its Foot, and makes the lower *Faſcia* of the three remaining Parts.

The *Architrave* for the *Corinthian* Order with a Pedeſtal, according to *Vitruvius*, is allow'd a greater Altitude than that without, conſiſting of the ſame

Members, both as to Number and Form with the former Architrave, but differing in Dimensions.

The whole Altitude of the *Architrave* ought to be one Quatter of the Altitude of the Column, nearly to two Eighths of the Body of the Column below, which is —— to forty one half M. This Altitude he divides into ſeven equal Parts, and makes a *Cymatium* at the uppermoſt of theſe ſeven, and divides the ſix that remain into twelve equal Diviſions ; of which, five he allots to the upper *Faſcia*, four to the middle, and three to the lower one. He ſubdivides the upper and middle *Faſcia*, each into eight Parts, and allows one of theſe Eighths for a Bead at the Foot of each of theſe *Faſcia's*.

Palladio makes this *Architrave* to contain eight Parts, *viz.* one *Liſt*, one *Cima*, three *Beads*, and three *Faſcia's* ; the Height of all which, he allows to be thirty-ſix M. Which are thus ſubdivided, *viz.* to the *Liſt* (or upper Member) he allows two three quarters M. the next in Order is a *Cima*, and the next in Order is of two M. high, at the Foot of which is a Bead ; after this is the upper *Faſcia*, its Bead at its Foot, both which contain about thirteen one half M. After this is the middle *Faſcia*, and its Bead at its Foor, which contain eight one Eighth M. and laſt of all, the lower *Faſcia*, which is ſix one quarter M. in Height.

Scamozzi makes the whole Altitude of this Architrave to be forty M. Which he ſubdivides in-

Palladio makes this *Architrave* forty-five M. in Height, and distributes them into seven particular minuter Members (beginning at the Top, and so descending:) First, he allows two one Eighth M. to the *List*, four one Eighth to the *Casement*, to the *Ogee* nine one quarter, to the *Bead* one one quarter, to the upper *Fascia* fifteen M. two one quarter to the *Ogee* at its Foot, and eleven M. to the lower *Fascia*.

Scamozzi makes this *Architrave* forty M. in Height, which he distributes among these eight following Members, *viz.* (descending:) First, a *List* of two M. Secondly, an *Ogee* of four one half M. Thirdly, an *Astragal* of two M. Fourthly, the upper *Fascia* of eleven three quarters M. Fifthly, a *Bead* at its Foot of two one quarter M. Sixthly, the middle *Fascia* of eight one half M. Seventhly, at its Foot, one one half M. Eighthly, the lower *Fascia* of six one half.

Vignola makes this *Architrave* forty-five M. in Height, which he divides into seven Members, a *List*, a *Casement*, a *Boultin*, a *Fillet*, a *Fascia*, a *Bead*, and a *Fascia*.

Measuring of Architraves. Architraves in Buildings (either of Brick or Stone) are usually done by the Foot lineal; and therefore, having taken the Length in Feet, you have also the Content at the same Time.

The Price of Architraves. That is different, according to the Breadth or Width of them. *Stone Architraves*, about Doors or Windows (according to Mr. *Wing*) are usually reckon'd at a Penny an Inch in Breadth at one Foot *e g.* if it be nine Inches broad, it's worth nine Pence a Foot, ten Inches ten Pence a Foot, *&c.*

The Faces of an *Architrave*, says M. *L Clerc*, ought not to have Ornaments, but to be left plain; and particularly when the Frieze is inrich'd.

The Proportion of *Architraves* by equal Parts.

The *Ionic Architrave* is divided into nine, giving one and three Fourths to the first Face, two and a half to the second, and three to the third; one and one Fourth to the Ogee, and one half apart to the Fillet: The Projection of the second and third Faces have a quarter of a Part each, and the Whole two of these Parts.

The *Corinthian Architrave* is divided into nine, giving one and a half to the first Face, one Fourth to the small Bead, two to the second Face, three Fourths to the small Ogee, two and a half to the third Face, half apart to the Bead, one to the Ogee, and half apart to the Fillet: The Projection of the second Face hath one Fourth of a Part, the third Face one of these Parts, and the Whole two.

The Height of the *Composite Architrave* into nine, giving two and a half to the first Face, one half apart to the Ogee, three and one half to the second Face; one Fourth to the Astragal, three Fourths to the Ovolo, one to the Hollow, and half apart to the Fillet: The Projection of the second Face hath one half apart, the Ovolo one and one Fourth, and the Whole two.

AREA properly denotes any plain Surface whereon we walk. The Word is *Latin.*

Area, in Architecture, signifies the Extent of a Floor, *&c.*

Area, in Geometry, denotes the Compass, or superficial Content of any Figure; thus, *exempli gratiâ*, if a Plot of Ground be exactly square, and its Side be 30 Feet, the *Area* will be 30 multiply'd by 30, i. e. 900 Feet.

ARITHMETICK [Ἀριθμετικὴ of ἀριθμὸς, Gr. Number] is the Art of Numbering, or that Part of the Mathematicks which considers the Powers and Properties of Numbers, and teaches how to compute or calculate truly, and with Expedition and Ease.

Vulgar Arithmetick is that which is conversant about Integers and vulgar Fractions.

Decimal Arithmetick is the Doctrine of decimal Fractions.

Sexagesimal Arithmetick is the Doctrine of sexagesimal Fractions, or that which proceeds by Sixties.

Binary Arithmetick, *Diadick Arithmetick*, } is an *Arithmetick* or Way of Numbering, wherein but two Figures, *viz.* Unity, or 1 and 0, are used.

Decadal Arithmetick is that which is wrought by a Series of ten Characters; so that the Progression is from 10 to 10, such as is the common Arithmetick which we use, wherein the ten *Arabick* Figures are used, as 0, 1, 2, 3, 4, 5, 6, 7, 8, 9, after which we begin 11, 12, *&c.*

Tetradick Arithmetick is that where only the Figures 1, 2, 3, and 0, are used.

Instrumental Arithmetick, is that where the common Rules are perform'd by Means of Instruments, contrived for Ease and Dispatch, as by the Lines on a common Carpenter's Rule, a Sector, Napier's Bones, *&c.*

Numerous Arithmetick is that which gives the *Calculus* of Numbers, or indeterminate Quantities, and is perform'd by the common Numeral, or *Arabick* Characters.

Specious Arithmetick, is that which gives the *Calculus* of Quantities, using Letters of the Alphabet, instead of Figures, to denote the Quantities, commonly called ALGEBRA.

Theoretical Arithmetick, is the Science of the Properties, Relations, *&c.* of Numbers consider'd abstractedly, with the Reasons and Demonstrations of the several Rules of Arithmetick.

Practical Arithmetick, is the Art of Computing, that is, from certain Numbers given, to find others whose Relation to the former is not known.

Political Arithmetick is the Application of Arithmetick to Political Subjects, as the Revenues of Princes, Number of Subjects, Births, Burials, *&c.*

Arithmetick of Infinites, is the Method of summing up a Series of Numbers, consisting of infinite Terms, or of finding the Ratio's of them.

ARRANGEMENT, the Disposition of the Parts of a Whole in a certain Order.

ARSENAL, a Royal or Publick Building, or Magazine, for the making and keeping of Arms, necessary

neceſſary either for Defence, or Aſſault.

ARTICLE [in Arithmetick] ſignifies the Number 10, or any Number juſtly diviſible into ten Parts, as 20, 30, 40, &c. which are ſometimes called *round Numbers*, and ſometimes *Decades*.

ASH, next to the Oak itſelf, is reckon'd one of the moſt uſeful Sorts of Timber we have, ſerving for ſo many Uſes for the Carpenter, the Cooper, &c. and like the Elm, is good for Mortoiſes, Tenons, &c.

The Price of ſawing Aſh. In ſome Places, they have 3*s*. *per* Hundred, in others, 3*s*. 6*d*. and ſometimes 4*s*. the Price varying according to the Cuſtom of the Place; but it is certain, it is worth 6*d*. *per* Hundred at leaſt, more than 'tis to ſaw Oak.

ASHLAR, a Term uſed by Builders, by which they mean common Freeſtones, as they come out of the Quarry, of different Lengths and Thickneſſes. Nine Inches is the common Thickneſs.

As to the Price of Aſhlars, Mr. *Wing* ſays, that they commonly value them in *Rutland* at 3*d*. the Foot in the Quarry.

In *Suſſex* and *Kent*, they commonly ſell them by the Load, being a common or ordinary Sort of Stone. About eighteen or twenty Foot makes a Load; which, if they come rough from the Quarry, coſt about 3*d*. a Foot; laid down at the Place where they are to be uſed; but if they are ready ſcapted, they are reckon'd at 4*d*. the Foot.

But if they are bought rough at the Quarry, they may be had at 2*d*. a Foot; but if ſcapted, 3*d*. And in ſome Places of *Kent* and *Suſſex*, they may be bought rough at the Quarry for three Halfpence a Foot, and ſcapted for two Pence Half-penny; but if they are laid down at the Place rough, then they are uſually reckon'd at two Pence Halfpenny a Foot, and if ready ſcapted, at three Pence Halfpenny a Foot.

But Places differing as to the Price of *Aſhlar*, it is impoſſible to give a certain Rule for the Price; which is different, Firſt, according to the different Cuſtoms of the Places; Secondly, the Circumſtances of the Quarry. And, Thirdly, the Goodneſs of the *Aſhlar*.

1. As to the Cuſtom of the Place, in reſpect to Carriage, Stones, in one Place, have been carry'd above a Mile for 1*s*. 8*d*. a Load; and at another Place, have coſt 2*s*. a Load, for carrying them half a Mile.

2. As to the Circumſtances of the Quarry. As, Firſt, whether the Stones be drawn on Incloſed Land, or the Lord's Waſte, *viz*. on the Highways, or on Commons, &c.

For if they are drawn within Land, (as they commonly call it,) the Perſon who is the Proprietor of the Land, will be paid well for the Damage done to his Ground, both by drawing and carrying the Stones out of his Land. But if they are drawn on the Lord's Waſte, the Lord has uſually no more than a ſmall Acknowledgment, either by Load, or otherwiſe, for treſpaſſing upon his Waſte.

3. As to the Goodneſs of the Stones, either for their Durableneſs, or Largeneſs. As for their Durableneſs, that only is to be known by Experience: For at the firſt opening of a new Quarry, no body can tell how the Stones may prove. For ſome Stones, when firſt taken out of the Quarry, are very ſoft and friable, and will moulder to Sand, by being expos'd to the Weather but a few Years: Whereas others of thoſe ſoft Stones will be indurated, or hardened, by being expoſed to the open Air.

Firſt, Thoſe Stones that come hard out of the Quarry, are generally durable, being of a more firm and ſolid Conſiſtence.

Secondly, as to their Largeneſs, I need not ſay much, all knowing that large Stones muſt needs be better, and make firmer Work than ſmall ones; which are only fit for filling Work in thick Walls; or to be uſed in ſuch Places where the Country affords no better.

There is a great Difference in Quarries, in Reſpect to the Poſition of the *Stones* in the Ground; which may be conſider'd under two Heads, *viz.* Firſt, as to their Depth in the Ground: For when they lie a conſiderable Depth in the Ground, it requires a great deal of Labour to uncope them, (as they term it) i. e. to remove the Earth. Secondly, if they do lie almoſt even with the Surface of the Ground, then it will require the leſs Labour to uncover them.

And beſides, an Allowance is to be made, as to the Manner of of their lying in the Ground: For of the Quarry is a Rock.

In this Caſe, it requires the more Labour to raiſe the Stones, and break them fit for Uſe, than if they lay ſeparate and diſunited: So that theſe Circumſtances ſo vary the Price, that ſome have been drawn for 9*d.* the Load, when others have coſt 3*s.*

ASHLERING [with *Builders*] Quartering to tack to in Garrets, about two Foot and a half or three Foot high, perpendicular to the Floor, up to the Underſide of the Rafters.

The Workmanſhip is from 4*d.* to 6*d.* a Square.

ASSEMBLAGE, the Joining or Uniting of ſeveral Things together; alſo the Things themſelves ſo joined or united: Of which Aſſemblages, there are divers Kinds and Forms uſed by Joiners, as with Mortoiſes, Tenons, Dove-tails, *&c.*

Aſſemblage of Orders. *M. Le Clerc* ſays, when two Columns are placed one over another, they muſt be of different Orders, the Stronger always to ſupport the Weaker.

For Inſtance, 1. The *Doric* may be placed over the *Tuſcan*, the *Ionic* over the *Doric*, the *Roman* over the *Ionic*, the *Spaniſh* over the *Roman*, and the *Corinthian* over the *Spaniſh*.

2. That the upper Order muſt always be leſs maſſive than the Under, agreeable to the Maxim, *That the Strong ought to ſupport the Weak.*

3. That the Columns ought to ſtand exactly over each other; ſo that their two Axis's may be both found in the ſame Perpendicular.

4. The

4. The Diſtances between the lower Columns, muſt be determin'd by the Intercolumniations of the Order, that is, without Pedeſtals; and the Diſtances of the upper Columns, by the Intercolumniations of the Order, with Pedeſtals, taking Care, by the Way, that the firſt Order be mounted on a pretty high Zocle, or an Aſcent of ſeveral Steps, to ſerve inſtead of a continu'd Pedeſtal, or Foot.

He gives a Pedeſtal to the Upper Order; becauſe being confin'd to the Breadth of the Intercolumniation of the Lower Order, its Columns, by this Means, are render'd ſmaller, inſomuch, that the Diameter of their Baſe does not exceed that of the Top of the Under Column; which is a Rule (in his Opinion) not to be diſpenſed withal.

He Remarks that *Vitruvius* will not allow the Upper Order more than three quarters of the Height of the Under.

But if this Reduction were follow'd, the Columns would be too ſmall, and conſequently too far aſunder, with Reſpect to their Height, if placed one over another.

In order to find the Mod. of an Order that is to be placed over another, he propoſes, for Inſtance, to place the *Ionic* over the *Doric*; and adviſes,

To conſider, firſt, that in the *Doric* Order, without a Pedeſtal, which is to give the Meaſures of that firſt Order, that the Columns are placed at the Diſtance of eleven M. from each other, in Portico's.

That in the *Ionic* Order with a Pedeſtal, the Columns are fifteen M. a-part; and that to place this Order upon the *Doric*, you muſt divide the Intercolumn, or its Equal, into fifteen equal Parts; one of which Fifteen will be the M. for raiſing the *Ionic* Order, with its Pedeſtal.

He likewiſe obſerves, that when two Portico's are placed over each other, the Higher ought to be regulated by the Lower: He means, that the Width of the Upper Arch ſhould be made equal to that of the Under; it being but juſt, that the two Arches ſhould have the ſame Width.

On ſuch an Occaſion, the lower Arch may be made ten or twelve Minutes narrower than uſual, that the Width of the Upper Arch may be better proportion'd.

When Columns are to be without Portico's, he ſays, there needs only be four Triglyphs made between the *Doric* Columns, that is, an Interval of eight Mod. four Minutes, which are equivalent to twelve M. in the *Ionic*, as appears by the Rule of Proportion; and that the ſame Thing may be obſerv'd of coupled Columns.

The *Roman* Order, he ſays, does not match perfectly well with the *Ionic*; becauſe its Capital is higher, with reſpect to its Column, than the *Ionic* Capital, with reſpect to the *Ionic* Column; and becauſe the Denticles of the *Ionic* appear ſomewhat weak underneath the Modillions of the *Roman*.

However, the *Roman* Order being in this Place leſs than the the *Ionic*, the Diſproportion between their Capitals, becomes less sensible, as well as that between the Denticles of the one, and the Modillions of the other.

To find the M. for raising a Corinthian *Column over a* Spanish *Order*, he says,

It is evident, that the Modillions of the Upper Order must be the same in Number with those of the Under, in order to have them exactly one over another.

Now the Inter-Modillions of of the *Corinthian* Order containing just 40 Minutes, where the Column has no Pedestal, these 40 Minutes must be multiplied by the Number of Modillions; which being 11, the Product will be 440; which being divided by 30, the Mod. the Quotient will be 14 M. 20 Minutes; which is the Division of the Scale for raising the *Corinthian* Order.

He observes, that there is a Difficulty in placing three Orders over each other; which consists in this, That the Second Order having a Pedestal, the Columns of the Third become a little to big at the Bottom; though 'tis so very little, that the Eye can hardly perceive it. But this Inconveniency, however, may be remedied, by taking the Excess away imperceptibly, wholly from the Base of the Column. It is true, this will occasion a little Swelling; but that won't do any Harm.

Again, he is of Opinion, it would not be proper to undertake the placing of more than three Orders of Columns over one another:

For, besides that in the fourth Order, the Columns would be too far asunder, in respect to their Height, it ought likewise to be consider'd, that four Columns raised over one another, can't well be very strong: Indeed, the first may have a Rustick Order, whereon it is raised, and which may serve it as a Foot.

Assemblage of Pilasters. See PILASTER.

ASTRAGAL ['Αςράγαλος, *Gr.* which signifies the Ankle, or Ankle-Bone] is, in Architecture, a little round Member, in the Form of a Ring, or Bracelet, serving as an Ornament at the Tops and Bottoms of Columns.

The *Astragal* is also sometimes us'd to separate the *Fasciæ* of of the Architrave. In which Case, it is wrought in Chaplets, or Beads and Berries.

It is also used both above and below the Lists adjoining immediately to the Dye or Square of the Pedestal.

The *Astragal* of a Column, *M. Le Clerc* says, ought always to be plain, excepting in the *Ionic* Order, where the Astragal of the Shaft is converted into a Chaplet of Pearls and Olives, for the Capital.

Astragal, or Baguette, has the Figure of a Staff, when it is join'd to a Fillet; the Height of which Fillet, *M. Le Clerc* divides into three Parts; two of which, he gives to the Astragal. And this Rule, he says, he observes on all Occasions.

This

This *Astragal* is frequently carv'd with Pearls and Olives, which the *French* call *Paternosters*.

ASYMMETRY, a Want of Symmetry, or Proportion.

ASYMPTOTES, are properly Straight Lines, which approach nearer and nearer to the Curve they are said to be Asymptotes of; but if they, and their Curve, are indefinitely continu'd, they will never meet.

Or *Asymptotes*, are Tangents to their Curves, at an infinite Distance.

And two Curves are said to be asymptotical, when they continually approach to one another; and if indefinitely continu'd, do not meet.

As two Parabola's, which have their Axis placed in the same Straight Line, are asymptotical to one another.

Of Curves of the second Kind, that is the Conic Sections only the Hyperbola has Asymptotes, being two in Number.

All Curves of the third Kind, have at least, one Asymptote; but they may have three.

And all Curves of the fourth Kind, may have four Asymptotes.

The *Conchoid*, *Cissoid*, and Logarithmick Curve, have each one Asymptote.

The Nature of an *Asymptote* will be very easily conceiv'd from that of the *Conchoid*: For if CDE be a Part of the Curve of the Conchoid, and A its Pole, and the Right Line MN be so drawn, that the Parts BCGD FE, &c. of Right Lines drawn from the Pole A, be equal to each other, then the Line MN will be the Asymptote of the Curve, because the Perpendicular Dp is shorter than BC, and Ep than

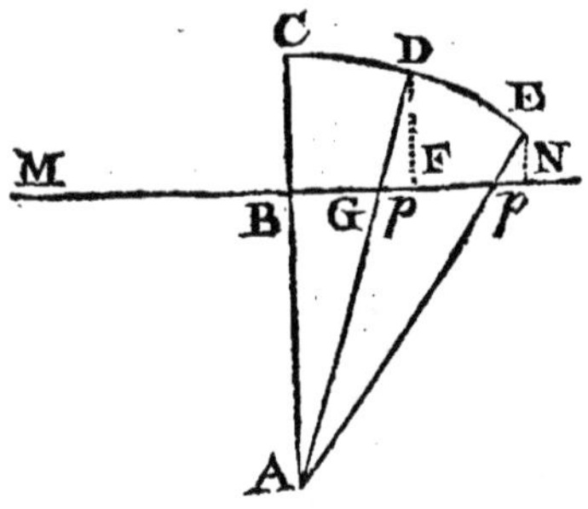

Dp, and so on; and the Points E, &c. and p, can never coincide.

ATLASSES [in *Architecture*] a Name given to those Figures, or half Figures of Men, so commonly used instead of Columns or Pilasters, to support any Member in Architecture, as a Balcony, or the like. These are otherwise called *Telamones*.

ATTIC, signifies something relating to *Attica*, or the City of *Athens*.

Attic is also used in Architecture for a kind of Building, wherein there is no Roof or Covering to be seen; thus called, because usual at *Athens*.

Attic, *or Attic Order*, is a sort of little Order raised upon a larger one, by Way of crowning, or to finish the Building.

It is also sometimes used for the Conveniency of having a Wardrobe, or the like, and instead of regular Columns, has only Pilasters of a particular Form.

Attic Order, according to M. *Le Clerc*, is a kind of rich Pedeſtal. Some Architects give it the ſeveral Capitals of all the Orders of Columns; but he ſays, the *Ionic*, *Roman*, and *Corinthian*, do not at all become it. The beſt Way, in his Opinion, is only to diſtinguiſh the Capitals by a Difference in their Mouldings; which may be made more or leſs ſimple, and more or leſs delicate, according to the Relation they are to bear to the Architecture underneath.

The Name *Attic*, is alſo given to a whole Story, into which, this Order enters; this little Order being always found over another that is greater.

This Pedeſtal, or Falſe Pilaſter, he ſays, ought always to have the ſame Breadth with the Column, or Pilaſter underneath; and its Height may be equal to a Third, or even a Half of the ſame Column or Pilaſter, by which it is ſupported.

Attic of a Roof, is a kind of a Parapet to a Terras, Platform, and the like.

Attic Continu'd, is that which encompaſſes the whole Pourtour of a Building, without any Interruption, following all the Jets, the Returns of the Pavilions, *&c.*

Attic Interpos'd is one ſituate between two tall Stones, ſometimes adorn'd with Columns or Pilaſters.

Attic Baſe, is a peculiar Kind of Baſe uſed in the *Ionic* Order by the ancient Architects; and alſo by *Palladio*, and other Moderns, in the *Doric*. It is the moſt beautiful of all *Baſes*.

ATTITUDE [in *Sculpture* and *Painting*] is the Poſture of a Statue or Figure, or the Diſpoſition of its Parts; by which, we diſcover the Action it is engaged in, and the very Sentiments ſuppoſed to be in its Mind.

The repreſenting theſe in a ſtrong and lively Manner, makes what they call a good Expreſſion.

The Word comes from the *Italian Attitudo*, which ſignifies the ſame Thing.

ATTRIBUTES [in *Sculpture*, *&c.*] are Symbols added to ſeveral Figures, to denote their particular Office and Character; as a Club is the Attribute of *Hercules*; a Trident, of *Neptune*; a Palm, of *Victory*; the Eagle, of *Jupiter*; a Peacock, of *Juno*, *&c.*

AVIARY, a Houſe or Apartment for the keeping, feeding, and breeding of Birds.

AUREOLA, a kind of Crown of Glory, given by Statuaries, *&c.* to Saints, Martyrs, *&c.* as a Mark of the Victory they have obtain'd.

AXIS properly ſignifies a Line or long Piece of Iron, or Wood, paſſing through the Centre of a Sphere, which is moveable upon the ſame.

Spiral Axis, in Architecture, is the Axis of a twiſted Column drawn ſpirally, in order to trace the Circumvolutions without.

Axis of the *Ionic* Capital, is a Line paſſing perpendicularly through the Middle of the Eye of the Volute.

Axis in *Mechanicks*, as the *Axis*, of a *Balance*, is the Line upon which is moves or turns.

Axis of *Rotation*, or *Circumvolution* in *Geometry*, is an imaginary Right Line, about which any plain Figure is conceived to revolve, in order to generate a Solid.

Thus a Sphere is conceived to be form'd by the Rotation of a Semicircle about its Diameter, or Axis, and a Right Cone by that of Right-angled Triangle about its Perpendicular Leg; which is here, its Axis.

Axis of a *Circle* or *Sphere*, is a Right Line passing through the Circle, or Sphere, and terminating at each End in the Circumference of it.

Axis is yet more generally used for a Right Line proceeding from the Vertex of a Figure, to the Base of it.

Axis of a Cylinder, is properly that quiescent Right Line about which the Parallelogram turns, by the Revolution of which the Cylinder is form'd.

Though both in Right and Oblique Cylinders, the Right Line joining the Centres of the opposite Bases, is also call'd the Axis of the Cylinder.

Axis of a Cone, is a Right Line or Side upon which the Right-angled Triangle forming the *Cone*, makes its Motion.

Axis of a Vessel, is that quiescent Right Line passing through the Middle of it perpendicularly to its Base, and equally distant from its Sides.

Axis of a Conick Section, is a Quiescent Right Line passing through the Middle of the Figure, and cutting all the Ordinates at Right Angles.

Axis in *Opticks*, is a Ray passing through the Centre of the Eye; or it is that Ray, which proceeding out of the Middle of the luminous Cone, falls perpendicularly on the Chrystalline Humour, and consequently passes through the Centre of the Eye.

Axis of *Oscillation*, is a Right Line parallel to the Horizon passing through the Centre, about which a Pendulum vibrates.

Common or *Mean Axis*, is a Right Line drawn from the Point of Concourse of the two Optick Nerves, through +, which joins the Extremity of the same Optick Nerves.

Axis of a Lens, or Glass, is a Right Line passing along the Axis of that Solid, of which that Lens is a Segment,

Thus a Spherical Convex Lens being a Segment of some Sphere, the *Axis* of the *Lens* is the *Axis* of the *Sphere*; or it is a Right Line passing through the Centre of it.

Axis of Incidence, in *Dioptricks*, is a Right Line drawn through the Point of Incidence perpendicularly to the Refracting Surface.

Axis of Refraction, is a Right Line, continu'd from the Point of Incidence, or Refraction, perpendicularly to the Refracting Surface along the further Medium.

Spiral Axis, in *Architecture*, is the Axis of a twisted Column drawn spirally, in order to trace the Circumvolutions without.

Axis of the Ionic Capital, is a Line passing perpendicularly through the Middle of the Volute.

The *Axis in Peritrochio*, consists of a Circle concentrick with the Base of the Cylinder, and is moveable together with it about its Axis.

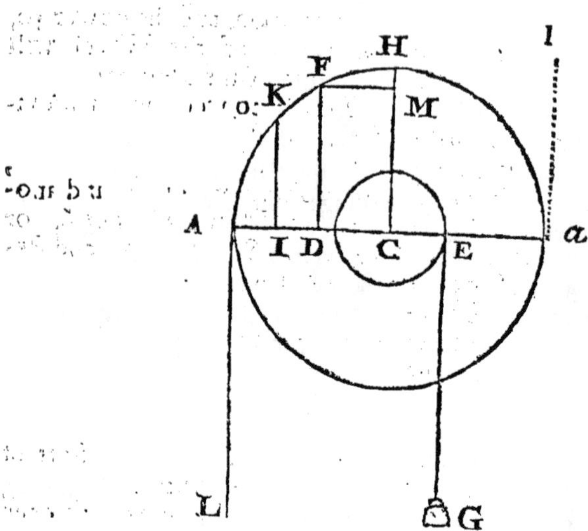

The Cylinder is called the *Axis*; the Circle the *Peritrochium*; and the *Radii*, or Spokes, which are sometimes fitted immediately into the Cylinder, without any Circle, the *Scytalæ*.

Round the Axis winds a Rope, whereby the Weight, &c. is to be rais'd.

The *Axis in Peritrochio* takes Place in the Motion of every Machine, where a Circle may be conceiv'd, describ'd, about a fix'd Axis, concentrick to the Plane of a Cylinder, about which it is placed; as in Crane-Wheels, Mill-Wheels, Capstans, &c. See Wheel.

The Doctrine of the *Axis in Peritrochio*, is as follows:

1. If the Power apply'd to an *Axis in Peritrochio*, in the Direction A L perpendicular to the Periphery of the Wheel, or to the Spoke, be a Weight G, as the Radius of the Axis G C is to the Radius of the Wheel C A, or the Length of the Spoke, the Power will just sustain the Weight, i. e. the Weight and the Power will be *in æquilibrio*.

2. If a Power be apply'd to the Wheel in F, according to the Line of the Direction F D, which is oblique to the Radius of the Wheel, though parallel to the Perpendicular Direction, it will have the same Proportion to a Power; which acts according to the perpendicular Direction A L, which the whole Sine has to the Sine of the Angle of the Direction D F C.

Hence, Since the Distance of the Power in A, is the Radius C A the Angle of Direction being given, the Distance D E C, is easily found.

3. Powers

3. Powers apply'd to the Wheel, in several Points F and K, according to the Directions F D, and K I, and I parallel to the Perpendicular on A L, are to each other, as the Distances from the Centre of Motion C D, and D I reciprocally.

Hence, as the Distance from the Centre of Motion increases, the Power decreases, and *vice versâ.*

Hence also, since the Radius A C is the greatest Distance, and agrees to the Power, acting according to the Line of Direction, the perpendicular Power will be the smallest of all those able to sustain the Weight G, according to the several Lines of Direction.

4. If a Power, acting according to the Perpendicular A L, lift the Weight G, the Space of the Power will be to the Space of the Weight, as the Weight to the Power; for in each Revolution of the Wheel, the Power passes through its whole Periphery; and in the same Time, the Weight is rais'd equal to the Periphery of the *Axis:* The Space of the Power therefore, is to the Space of the Weight, as the Periphery of the Wheel to that of the *Axis:* But the Power is to the Weights, as the Radius of the *Axis* is to that of the Wheel; therefore, &c.

5. *A Power and Weight being given to construct an* Axis in Peritrochio, *whereby it shall be sustain'd.*

Let the Radius of the *Axis* be big enough to support the Weight without breaking. Then, as the Power is to the Weight, so make the Radius of the Wheel, or the Length of the Spoke, to the Radius of the *Axis.*

Hence, if the Power be but a small Part of the Weight, the Radius of the Wheel must be vastly great, *ex. gr.* suppose the Weight 3000, and the Power 50, the Radius of the Wheel will be to that of the *Axis*, as 60 to 1.

This Inconvenience is provided against by increasing the Number of Wheels, and *Axes*, and making one turn round another, by Means of Teeth, or Pinions. See WHEEL, and PERITROCHIO.

B A

BACK, See BAGUETTE.

BACK-NAILS, a Sort of Nails made with flat Shanks, so as to hold fast, and not to open the Grain of the Wood, used in nailing Guts together, for saving Water under the Eaves of a House; or by Back-Makers, in nailing of Boards together for Coolers, or any Vessels made of Planks or Boards for containing Liquors.

BACULOMETRY, the Art of measuring accessible or inaccessible Lines, by the Help of one or more Staves.

BAGNIO, an *Italian* Term, signifying a Bath. Thence *Bagnio* is become a general Name in *Turky*, for the Prisons in which the Slaves are confin'd, it being usual to have Baths in those Prisons.

BAKE-HOUSE, is a Room of Office, or an Appartment belonging to noble Buildings, and other private Buildings, in which an Oven is built.

As to the Position, it ought (according to the Rules laid down by Sir *Henry Wotton*) to be placed

ced on the South-Side of any Building.

BAGUETTE [in *Architecture*, a little round Moulding, less than an *Astragal*; sometimes carv'd and enrich'd with Foliages, Pearls, Ribands, Lawrels, &c. Though, according to M. *Le Clerc*, when a *Baguette* is inrich'd with Ornaments, its Name is chang'd, and it is called a *Chaplet*. Or *Baguette*, is a Term used by Carpenters, for a kind of *Astragal*, or Hip-Moulding; by which is meant the outward Angle, or the Hips or Corners of a Roof; which in square Frames, where the Roof is three quarters pitch, contains an Angle of oue hundred and sixteen Degrees and twelve Minutes.

BALANCE, in *Mechanicks*, is one of the six simple Powers principally used for determining the Equality or Difference of Weights in heavy Bodies, and consequently other Masses and Quantities of Matter.

The *Balance* is of two Kinds, *viz.* the *Antient* and *Modern*.

The *Antient*, or *Roman*, called the *Statera Romana*, or *Steel-Yard*, consists of a Lever, or Beam, moveable on a Centre, and suspended near one of its Extremes: On one Side the Centre are applied the Bodies to be weigh'd, and their Weight measured by the Divisions mark'd on the Beam, in a Place where a Weight moveable along the Beam, being fix'd, keeps the *Balance in æquilibrio*. This is still in use in Markets, &c. where large Bodies are to be weigh'd.

The *modern Balance* now ordinarily in use, consists of a Lever, or Beam, suspended exactly by the Middle, to the Extremes whereof are hung Scales.

In each Case the Beam is called the *Brachia*; the Line on which the Beam turns, or which divides its *Brachia*, is called the *Axis*; and when considered with respect to the Length of the *Brachia*, is esteem'd but a Point, and called the *Centre of the Balance*; and the Places where the Weights are applied, *the Points of Suspension, or Application*.

In the *Roman Balance*, therefore, the Weight used for a Counter Balance is the same; but the Points of Application various: In the common Balance, the Counter-Poise is various, and the Points of Application the same. The Principle on which each is found is the same, and may be conceiv'd from what follows.

The Doctrine of the Balance.

The Beam A B, the principal Part of the Balance, is a Lever

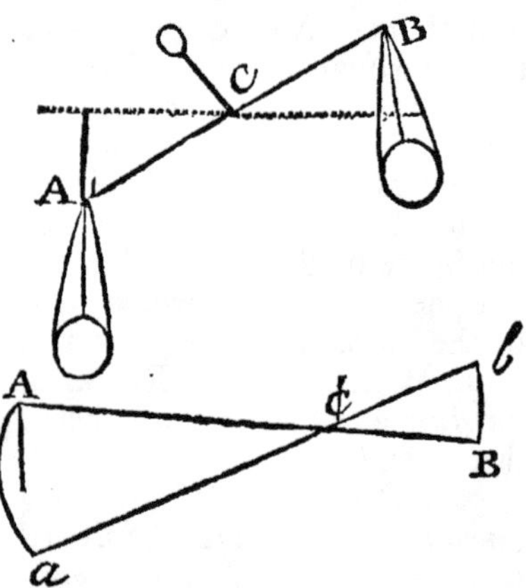

of the first Kind, which (instead of resting on a *Fulcrum* at C, the Centre of its Motion,) is suspended by something fastened to C its Centre

tre of Motion. Hence the Mechaniſm of the Balance depends on the ſame Theorem as that of the Lever. See the *Figure.*

Wherefore as the known Weight is to the unknown, ſo is the Diſtance of the unknown Weight from the Centre of Motion to the Diſtance of that of the known, where the two Weights will counterpoiſe each other; conſequently the known Weight ſhews the Quantity of the unknown Weight: Or thus, the Action of a Weight to move a *Balance* is by ſo much greater, as the Point preſs'd by the Weight is more preſs'd diſtant from the Centre of the *Balance*, and that Action follows the Proportion of the Diſtance of the ſaid Point from that Centre.

When the *Balance* moves about its Centre, the Point B deſcribes the Arch B*b*, whilſt the Point A deſcribes the Arch A*a*, which is the biggeſt of the two; therefore in that Motion of the Balance, the Action of the ſame Weight is different, according to the Point to which it is applied: Hence it follows, that the Proportion of the Space gone through by that Point at A, is A*a*, and at B as B*b*; but thoſe Arches are to one another as CB, CA.

Varieties in the Application of a Balance.

If the *Brachia* of a *Balance* be divided into equal Parts, one Ounce apply'd to the ninth Diviſion from the Centre, will equiponderate with three Ounces at the third; and two Ounces at the ſixth Diviſion, act as ſtrongly as three at the fourth, &c.

Hence it follows, that the Action of a Power to move a *Balance* is in a Ratio compounded of the Power itſelf, and its Diſtance from the Centre; for that Diſtance is as the Space gone through in the Motion of the *Balance.*

It may be here obſerved, that the Weight equally preſſes the Point of Suſpenſion at whatever Height it hangs from it, and in the ſame manner as if it was fixed at that very Point; for the Weight at all Heights equally ſtretches the Cord by which it hangs.

A *Balance* is ſaid to be *in æquilibrio*, when the Actions of the Weights upon each *Brachium* to move the Balance are equal, ſo as mutually to deſtroy each other.

When a *Balance* is *in æquilibrio*, the Weights on each Side are ſaid to equiponderate. Unequal Weights may alſo equiponderate; but then the Diſtances from the Centre muſt be reciprocally as the Weight. In which Caſe, if each Weight be multiplied by its Diſtance, the Product will be equal; which is the Foundation of the Steel-yard.

Thus in a *Balance*, whoſe *Brachia* are very unequal, a Scale hanging at the ſhorteſt, and the longeſt divided into equal Parts; if ſuch a Weight be applied to it, as at the firſt Diviſion ſhall equiponderate with one Ounce in the Scale, and the Body to be weigh'd put into the Scale, and the above mentioned Weight be moved along the longeſt *Brachium*, till the *æquilibrium* be found, the Number of Diviſions between the Body and the Centre, ſhews the Number of Ounces that the Body weighs; and the

the Subdivisions the Parts of an Ounce.

On the same Principle also is founded the *deceitful Balance*, which cheats by the Inequality of the *Brachia*: For instance; take two Scales of unequal Weights, in the Proportion of nine to ten, and hang one of them at the tenth Division of the *Balance* above described; and the other at the ninth Division, so that there may be an *Æquilibrium*; if then you take any Weights, which are to one another as nine to ten, and put the first in the first Scale, and the second in the other Scale, they will equiponderate.

Several Weights hanging at several Distances on one Side, may equiponderate with a single Weight on the other Side: To do this, it is required, that the Product of that Weight, by its Distance from the Centre, be equal to the Sum of the Products of all the other Weights, each being multiplied by its Distance from the Centre. To demonstrate which, hang three Weights of an Ounce each at the second, third, and fifth Divisions from the Centre, and they will equiponderate with one single Ounce apply'd at the tenth Division of the other *Brachium*; and the Weight of one Ounce at the sixth Division, and another of three Ounces at the fourth Division, will equiponderate with the Weight of two Ounces on the other Side at the ninth Division.

Several Weights unequal in Number on either Side, may equiponderate: In this Case, if each of them be multiplied by its Distance from the Centre, the Sums of the Product on either Side will be equal; and if those Sums are equal, there will be an *Æquilibrium*.

To prove which, hang on a Weight of two Ounces at the fifth Division, and two others, each of one Ounce, at the second and seventh; on the other Side hang two Weights, each also of one Ounce, at the ninth and tenth Divisions; and these two will equiponderate with those three.

To the Perfection of a *Balance* 'tis required, that the Points of Suspension be exactly in the same Line as the Centre of the *Balance*, that they be precisely equidistant from that Point on either Side; that the *Brachia* be as long as conveniently they may; that there be as little Friction as possible in the Motion of the Beam and Scales; and lastly, that the Centre of Gravity of the Beam be placed a little below the Centre of Motion.

First, Note that AC *taken together, are called the* Beam of the Balance.

The Point B, *the fix'd Point on which it moves, and equally divides* AB *and* BC, *which are called* two Brachia's.

THEOREM.

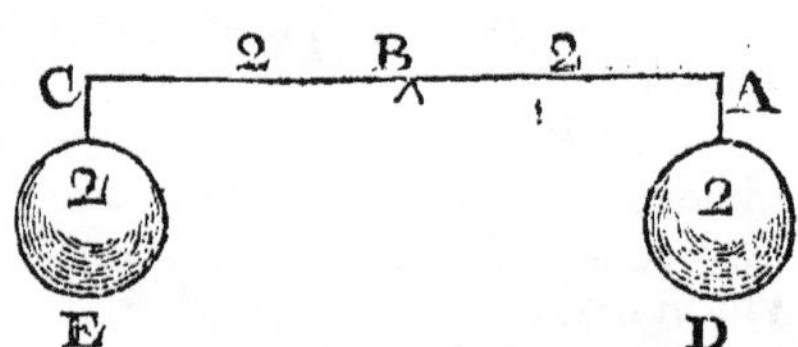

If two Weights tied to the End of an horizontal *Balance*, are to one another reciprocally, as their Distance from the fix'd Point, they will hang in *æquilibrio*.

Of

Of Equal Brachia's.

DEMONSTRATION I.

Let A B = B C, and the Weight D = E, then I say, as D 2 : E 2 : : A B : B C.

Or, as D 2 : B C : : E 2 : A B, which was to be demonstrated.

DEMONSTRATION II.

Of Unequal Brachia's.

1. Let B be the the fix'd Point.
2. Let A C be = 24 Feet.
3. Let A B = 8 and B C = 16.
4. Let the Weight D = 12 Pound, and E = 6 Pound. Then I say,

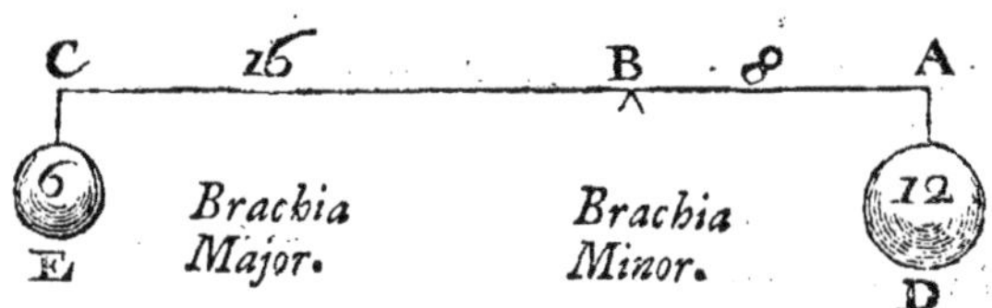

As the Lesser Brachia A B 8 is to the lesser Weight E 6, so is the Greater Brachia B C 16 to the greater Weight D 12, which is requir'd to equipoize E.

Or, as the Lesser Brachia A B 8 is to the Greater Brachia B C 16, so is the Power apply'd at C, *viz.* E 6, to the Weight it will equipoize at A, *viz.* D 12. Q. E. D.

PROPOSITION II.

The Weight of two heavy Bodies being known, apply'd to the Ends of a Balance of a known Length, to find upon that Balance the common fix'd Point, or Centre of Motion, whereon the two given Bodies will hang in *Equilibrio.*

DEMONSTRATION.

Let the Balance A B be = 24 Foot,

The Weight D = to 12, and E = to 6: Then I say

As the Sum of both Weights, 18 is to the lesser Weight D 6, so is the whole Balance A C 24 to the lesser Brachia A B 8; or as the Sum of D + E is to A C

So is the greater Weight D 12 to the Greater Brachia B C 16: Therefore it is evident, that the Point B is is the fixed Point or Centre of Motion required Q. E. D. See the preceding Figure.

N. B. It is here suppos'd, that the Balance A C is without Weight in itself, as a Line, &c. —— as before noted.

PROPOSITION III.

The Length and Weight of a Balance being given, which has at one of its Ends, a Body of known Weight, to find the fix'd Point, about which, the Weight of the Balance and the Weight of the Body shall remain in *Equilibrio.*

E 1. Let

1. Let the Balance weigh 16 Pound, and its Length be 12 Feet.

2. The Body E 8 Pounds.

Then I say, as 24, the Sum of the Weights of the Balance 16 Pounds, and the Body 8 Pounds taken together, is to the whole Length of the Balance 12 Feet; So is 8 the Weight of the Body E to the Lesser Brachia AB 4.

Or thus,

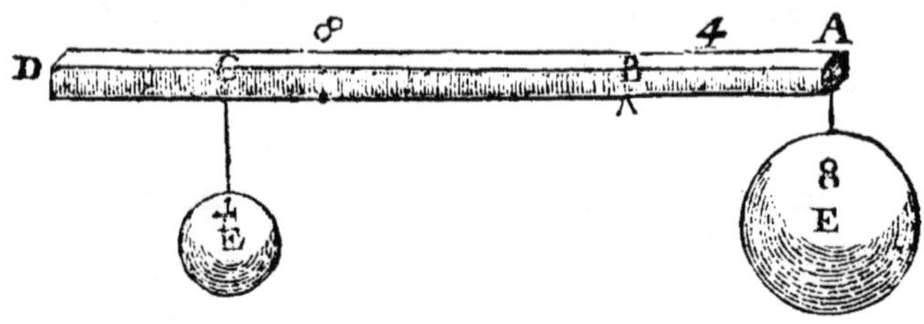

As 24, the Sum of the Weights of AD + E is to 12, the Length of the Balance, so is 16 the Weight of the Balance, to BD 8 the Greater Brachia.

Therefore, it is evident that the Point B is the Point requir'd.

PROPOSITION IV.

Two Bodies being given, the heaviest of which, hangs at one of the Ends of a Balance of known Length and Weight, and given fix'd Point (as Steel-Yards,) to hang that of least Weight in such Manner, that being assisted by the Weight of the Balance, it may keep the heaviest Body in *Equilibrio.*

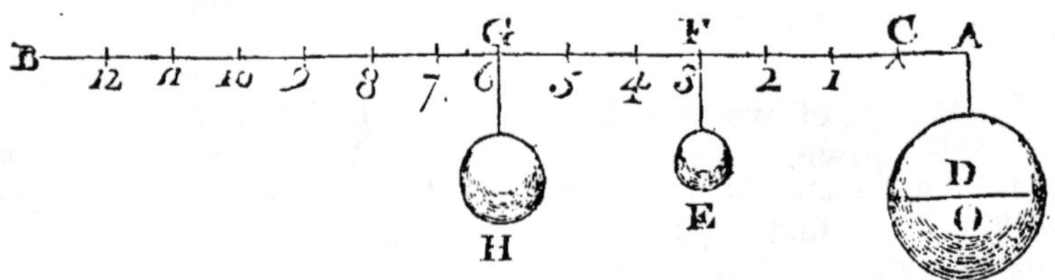

1. Let the Balance AB weigh 2 Ounces, and 14 Inches long.

2. Let G be the fix'd Point, an Inch from the End A, and let the Part D of Body DO, weigh 15 Ounces.

3. Let E be a Body of an Ounce, and moveable at Pleasure, to find the Point F, where the Body E, with the Assistance of the Gravity of the Brachia CB of the Balance AB, shall keep the Weight DO in *Equilibrio* about the Centre of Motion G.

4. Divide the Balance AB into two equal Parts at G, which (if made of equal Matter) will be its Centre of Gravity.

5. Suppose

5. Suppose the Body H to be in Weight equal to the Balance AB, *viz.* = 2 Ounces, to hang from the the Point G.

Then as the Distance AC is to the Distance CG, so is the Weight H, = the Weight of the Balance, to a fourth Proportional = one Part of the Weight DO, *viz.* = 12 Ounces; wherefore, the other Part of it remaining, is = 3 Ounces.

Now again,

As 1 Ounce, the Weight of the Body E, is to the Part D, 12 Ounces, the last Number found, so the Distance of AC to 3, the Distance of F from C, whereon the Body E being hung, will keep the Body DO in *Equilibrio*.

BALCONY, a Projecture beyond the Naked of a Wall or Building, supported by Pillars or Consoles, and encompassed with a Ballustrade. Or it is a kind of open Gallery for People to stand in, to behold any publick Shew, as Pageants, Cavalcades, publick Entries of Ambassadors, &c. in Cities; or for taking the Air, &c.

This jutty or projective Building is usually placed in the Middle of a Front of a House, or publick Hall, &c. if there be but one; and is usually level with the first Floor, up one Pair of Stairs.

Some of these are made with Wood, and others with Iron; Wooden Balconies consist of Rails and Ballusters; and so sometimes do those of Iron; but at other Times, are made of Cast-Iron, of various Figures in Semi-Relief; and some again, of wrought Iron, in Crail'd Work, or Flourishes, in different Forms, according to the Fancy of the Workman, &c.

As to the Price, Wooden Balconies are commonly paid for by the Yard, from 3*s.* to 5*s. per* Yard, according to the Workmanship the Carpenter bestows upon it.

Those of Iron, are commonly paid for by the Pound or Hundred Weight, from 4*d.* to 8*d. per* Pound, according to the Curiosity of the Workmanship.

It may be proper here, to take Notice of what Sir *Henry Wotton* says, concerning all In-lets and Out-lets, such as Balconies, Windows, &c. that they ought not to approach too near the Corner of Walls; it being an essential Error, to weaken that Part which strengthens all the rest.

This, says he, is a Precept well recorded, but ill practis'd, even by the *Italians* themselves, particularly at *Venice*; where he had observ'd divers *Pergoli*, or *Maucina*, (as they seem to be called by *Vitruvius*,) which are certain Balustraded Out-Standings, made for standing in, to satisfy the Curiosity of the Sight, very dangerously set forth upon the very Point itself of the Mural Angle.

M. *Le Clerc* says, the Parts of a Balcony are the Terrass, the Balustrade that incloses it, and the Consoles which support it: Or, to explain himself the more accurately, a Balcony is a Piece of Architecture raised in the Air, inclosed with a Balustrade, and supported by a little Entablature,

 whereof

whereof the Cornice, or uppermost Part, makes a Terras; the Frieze and Architrave being only continu'd at the Bottom and Sides; and the whole Balcony further supported by Consoles.

The Frieze is made with a little Sweep, that the Zocle of the Pedestal above, may not appear ill supported; and that the Console coming to contract, or straighten itself, at the Bottom, may do it the more gracefully; without which, it would appear too heavy.

The Height of the Consoles may be equal to their Projecture; but it will be an Addition both to the Beauty and Strength of the Work, if they be made higher.

A Balcony may be continu'd quite through the *Facade* of a Building, by adding Consoles from Space to Space; to be dispos'd between the Windows, which will be underneath.

He is of Opinion, that Iron Balconies will do much better than those of Stone, as being lighter, and less subject to Decay; which, if they be gilt, they will be exceedingly magnificent, and a very proper Ornament for a Palace.

BALDACHIN [of *Baldachino, Ital. Baldagum, Fr*] a Piece of Architecture, in Form of a Canopy, supported with Columns, and serving as a Crown or Covering to an Altar. It properly signifies a Canopy carry'd over the Host in *Roman* Catholick Countries. Some also give the Name *Baldachin*, to a Shell over the Front-Door of an House.

BALKS, Poles or Rafters, over Out-Houses or Barns; and among Bricklayers, great Beams, such as are used in making Scaffolds; also, so some call great Pieces of Timber, coming from beyond Seas by Floats.

BALLON, in *Architecture*, a *French* Term, used to signify the round Globe on the Top of a Peer, or Pillar.

BALUSTRADE, an Assemblage of one or more Rows of little turn'd Pillars, call'd Balusters, made of Marble, Iron, Wood, or Stone, of a Height fit for a Man to rest his Elbows upon, fixed upon a Terras, or the Top of a Building, or else to make a Separation between one Part of it and another.

BALUSTER, which is usually corruptly pronounced *Banister*, is a small Column, or Pilaster of different Sizes, *viz.* from an Inch and three quarters, to four Inches square or Diameter: their Dimensions and Forms are various, according to the Fancy of the Workman. They are frequently adorn'd with Mouldings.

Du Cange derives the Word from *Balaustrum*, or *Balaustrium*, a Place, among the Antients, where their Baths were rail'd in. others derive the Name from *Balaustrum* in *Latin*, from Βαλαυστιν, in the *Greek*, which signifies the Flower of a Wild Pomegranate; which it is supposed to resemble.

As to their Use, they are placed with Rails on Stairs, in the Fronts of Galleries in Churches, &c. Also round Altar-Pieces in Churches, on Terras-Walks, and in Balconies and Platforms, &c.

As to their Price, with Rails, *&c.* of Wood, on Stairs, Balconies, Platforms, *&c.* according to the Work, abour 3 d. *per* Yard, running Meaſure.

For turning them only, the uſual Rate is 1 *d. per* Inch for Workmanſhip.

The Charge for painting them: Balluſtrades, with their Appurtenances, are uſually painted by the Yard. They are cuſtomarily meaſur'd thus: Borh Sides of the Baluſters are meaſur'd as though they were flat Meaſure, including the Spaces between the Baluſters; which being caſt up in Feet and Parts, are reduced into Yards, like other flat Painting is.

Mr. *Leyburn* tells us, that he has ſeen the Experiment try'd by girting the Baiuſters, to find the Difference betwixt that Way, and meaſuring them, and the Spaces on both Sides, as though upon a Flat, and found that the Difference would not countervail the Trouble of Girting. But it is reaſonably to be ſuppoſed, that they ſhould be nearly the ſame, it being the common Practice to ſet them no more than their Square or Diameter aſunder; and then the Flanks make good the Spaces.

BALUSTERS } are a kind of
BALISTERS } little Pillars join'd by a Rail, at a convenient Height, for the Elbows to reſt upon.

Of theſe, M. *Le Clerc* propoſes various Forms, accommodated to the various Orders of Architecture, where Baluſtrades may be uſed.

By BALUSTRADE, he means a Series or Row of Baluſters, with their Rail, ſerving as a Tablette, or Reſt to the Elbows; and at the ſame Time, as a Fence and Incloſure to Balconies, Altars, Terraſſes, Stair-Caſes, Water-Works, large Windows, *&c.*

Baluſtrades conſiſt of one or more Ranges or Rows of Baluſters, terminated by Pedeſtals of the ſame Height.

If in a Stone or Marble Baluſtrade, the Diſtance from one Pedeſtal to another, be too great for a Tablette or Rail of a ſingle Stone, it muſt be made of two: In which Caſe, it will be proper to have the Juncture or Aſſemblage ſupported with a Dye, if a Baluſter be judged too weak to ſuſtain it.

He is of Opinion, that the Ranges ought to terminate in half Baluſters, join'd to the Pedeſtals. Though ſome Architects have other Sentiments.

Though there are Baluſters of various Figures, he would have the round and the ſquare have the Preference.

He alſo would have every Baluſtrade have its Zocle; though *Palladio*, indeed, gives an Inſtance of the contrary, in his *Ægyptian Hall*, p. 110. But this, he ſays, is not to be imitated.

Round *Baluſters* are not ſo heavy as the ſquare ones: They are frequently made of very hard Stone, as that of *Lions*; which works better than any other kind of Stone, except Marble. When it is deſired to have a Baluſtrade richer, and more delicate than ordinary, ſuch as are ſometimes ſeen before Altars, it may be caſt of Braſs, or Silver; unleſs, to ſave Expences, it be thought better to have it of Wood gilded; for such Balustrades may be made as rich in Ornaments as you please.

All Balustrades, being intended to be Breast-high, none should ever exceed three Foot and a quarter at most, nor be lower than two Foot and a quarter at least. The Measures for this, to be taken from the Royal *Paris* Foot, which is to the *English* Foot, as 1068 to 1000.

He observes, that in the Balustrades of Stair-Cases, the Zocle should always be the Height of the Steps; and that the Balustrade terminates much better with a Pedestal on the Ground, than a Pedestal on the Descent.

And also, that whether the Pedestal be on the Descent, or not, it must always have a Buttress, in manner of a Console, to sustain and bear up against the Pressure of the Balustrade.

And that in Balustrades which are between Pedestals, without either Bases or Cornices, the Tablette should only consist of a Plat-Band, sustain'd by a Fillet, or little Talon underneath; and the Zocle may have a little Cavetto over it.

Again, when a *Balustrade* is independant, the Pedestals may be proportioned to the Balusters; of which M. *Le Clerc* gives several Instances: But when it is used in Orders of Columns, whereon it has some Dependance, the Pedestals, in that Case, can't be managed at Pleasure.

If the Pedestals, which terminate a *Balustrade*, be compleat and well-proportion'd to the Pillars which they support, their Cornices will be found too weak to be continu'd alone, and so to serve for a Tablette to the *Balustrade:* Therefore it will be necessary, to add a Plat-Band underneath, which will make a Symmetry with that round the Table or Pannel of the Pedestal.

Instead of Balusters there is sometimes an Entrelas of Crailed Work, which are not inferior to the others in Beauty. Of these M. *Le Clerc* proposes various Designs in his Book.

These also may be further inrich'd by making the Wreaths, Roses, and Foliages, of Brass; which would still do infinitely better if they were gilt.

These *Entrela's* should be made more or less delicate, according to the Places where they are to be used: As for Instance, those which are placed on a Top of a Building, and which can only be view'd from afar, should be less delicate, than those that are to be view'd near at Hand.

BAND, in Architecture, is a general Name for any flat low Member, or on that is broad, and not very deep; which is also called *Face*, from the *Latin Fascia*, which *Vitruvius* uses for the same Thing. And sometimes *Fillet*, *Plinth*, &c.

BANDELET, [is deriv'd from the *French Bandelette*, a little Fillet or Band] is any little Band or flat Moulding, as that which crowns the *Doric* Architrave. 'Tis also called *Tenia*, from the *Latin Tænia*, which, *Vitruvius* uses for the same Thing. It is also used by Architects, to signify the three Parts that compose an *Architrave*.

BANISTER;

BANISTER, See BALUSTER.

BARACK, } [of *Baraccas*, BARAQUE, } *Spanish*, little Cabins which the Fishermen make on the Sea-Shore] a Hut, or little Lodge, for Soldiers in a Camp: Those for the Horse, were formerly called *Baracks*; and those for the Foot *Huts*: But now *Barack* is used indifferently for both. They are generally made, by fixing four forked Poles in the Ground, and laying four others across them; afterwards the Walls are built up with Sods, Wattles, or what proper Materials the Place affords; the Top is either planked, thatched, or cover'd with Turf, according as the Country supplies them.

When an Army is in Winter Quarters, the Soldiers usually build Baracks; but in Summer, only make Shift with Tents.

BARBACAN } [Some derive BARBICAN, } it from the *French*; but others, of *Barbacane*, *Italian*] a Canal, or Opening left in the Wall, for Water to come in and go out at, when Buildings are erected in Places liable to be overflow'd; or to drain off the Water from a Terras. It is also used to signify an Outwork in a Building.

BARGE-COURSE [with *Bricklayers*] a Term used for Part of the Tiling which projects over without the principal Rafters, in all Sorts of Buildings, where there is either a Gable, or a *Kirkin-Head*.

BARN. Mr. *Worlidge* advises, as to the Situation of Barns, that it will be very inconvenient to build Barns, or Stables, or Places for the like Uses, too near to the Dwelling-House; because Cattle, Poultry, *&c.* require to be kept near to Barns, *&c.* which would be an Annoyance to the House.

As to the Prices of Framing, &c. The Carcass of a Barn has been built for 3*s.* 6*d.* *per* Square, for Carpenters Work alone; and 8*s.* *per* Square has been given for Carpenters Work, the Felling, Hewing, and Sawing of his Timber-Boards, and finding Nails.

Some Workmen say, that the Charge of a Square of building of the Timber-Work of a Timber Barn may be computed in the Manner following; *viz.* 4*s.* a Square for the sawing the Boards (considering that they lap over one another) and the staving of the Logs, 2*s* a Square; for sawing the Timber-Members, 3*s.* 6*d.* a Square; for framing the Carcass, from 4*s.* to 7*s.* a Square for the Value of the Timber, reckoning the Price of the Timber from 12*s.* to 21*s.* *per* Ton; and one Ton to make 3*s.* square of Frame in Barn-Work.

Rough Timber, is that unhew'd or unsquar'd; and a Ton of rough Timber, has been reckon'd equal to a Load of hew'd. From these Computations, we may compute the whole Value of a Square of such Timber-Work to be worth from 3*s.* 6*d.* to 16*s.* 6*d.* *per* Square.

BARS of Iron, upright ones for Windows; their usual Price is three Pence Halfpenny, or four Pence a Pound in *London*.

BAR-POSTS, a Sort of Posts, two of which, and five Rails or Bars, serve instead of a Gate, for an Inlet into Fields and other Inclosures: These Posts consist each of five Mortoises; and the Posts are usually six Foot, or six Foot and a half long, four of which stand above the Ground.

These Posts are, in some Places, made by the Piece, *viz.* a Penny or three Halfpence *per* Post hewing, and a Halfpenny *per* Hole, for mortising.

BASE [of Βάσις, *Gr.* Rest, Support, or Foundation] is used to signify any Body which bears another; but particularly, for the lower Parts of a Column, and a Pedestal.

The *Base* is also sometimes called *Spira*, from *Spiræ*, the Folds of a Serpent laid at Rest, which form a Figure not much unlike it.

The Base of a Column, is that Part between the Shaft and the Pedestal, if there be any Pedestal; or if there be none, between the Shaft and the Plinth, or Zocle.

The Base is supposed to be the Foot of the Column; or, as some define it, it is that to a Column, that a Shoe is to a Man.

The Members or Ornaments whereof a Base is compos'd, are supposed by others, to have been originally design'd to represent the Iron Circles with which the Feet of Trees and Posts were girded, which supported the Houses of the Antients, in order to strengthen them.

The Base is different in the different Orders.

The *Tuscan Base* is the most simple of all Orders; consisting, according to some, only of a single *Tore*, besides the *Plinth*.

The *Doric Base* has an *Astragal* more than the *Tuscan*, although that was introduced by the Moderns.

The *Ionic Base* has a large *Tore* over two slender *Scotia*, separated by two *Astragals*: But there are no Bases at all in the most antient Monuments of this Order; which Architects are at a Loss to account for.

The *Corinthian Base* has two *Tores*, two *Scotia's*, and two *Astragals*.

The *Composite Base* has an *Astragal* less than the *Corinthian*.

The *Attic Base*, is so denominated, because it was first used by the *Athenians*. It has two *Tores*, and a *Scotia*, and is very proper for *Ionic* and *Composite* Columns.

The Parts or Members exceed the Number of the Kinds; because some Authors differ from others in their Form, according to the Account following:

The *Tuscan Base*, according to *Vitruvius*, is to be one half Mod. in Height. This cross Member consists of three smaller Members, or Parts, *viz.* a *Plinth*, a *Torus*, and a *Fillet*; and is divided, and subdivided as follows: The whole Height being 30, is divided into two equal Parts; the lower of which Parts is for the *Plinth*; and the upper Part of the two is to be subdivided into three equal Parts; the upper of which is for the *Fillet*, and the lower, for the *Torus*.

Palladio also allows the Altitude of this Base to be thirty Minutes; which he distributes among three smaller Members, *viz.* a *Plinth*, or *Orlo*, a *Torus*, and a *Listella*, or *Cincture*; the *Plinth* is fifteen Minutes the *Torus* twelve and a half, and the *Listella*, two and a half high.

Scammozzi also allows thirty Minutes to the Altitude of this *Base*; but then he allows it but two

two Members, or Parts; which are, a *Plinth* of eighteen Minutes, and a *Thorus* of twelve Minutes; altho' at the ſame Time he places a *Liſt* of three Minutes above the *Thorus*.

Vignola alſo makes this Baſe to conſiſt of three Parts, a *Plinth*, *Thorus*, and *Fillet*: All which he reckons thirty Minutes—half a Module.

The *Doric Baſe*. This Baſe *Vitruvius* makes to conſiſt of ſix Parts, *viz.* a *Plinth*, two *Thorus's*, one *Scotia*, and two *Liſts*: To the whole Height of all theſe, he allows thirty Minutes, which he thus divides, *viz*: Firſt, into three Parts; the lower one of which is for the *Plinth*, and ſubdivides the two remaining Parts into four; the upper one of which he allots to the upper *Thorus*, and the three lower Parts of theſe four he divides into two; the lower of which two is for the lower *Thorus*: After which he ſubdivides the upper Part of theſe two into ſeven equal Parts; the upper and lower of theſe ſeven Parts, are for the two *Liſts*, and the five which are betwixt them, are for the *Scotia*.

But among all theſe ſix Membets, or Parts of the Baſe, there is one large Fillet, which is one twelfth Part of the Mod. but he does not reckon this *Fillet* to be any Part of the Baſe, but a Part of the Body of the Column.

Palladio aſſigns thirty Minutes to the Height of this Baſe: According to his Scheme of this Member, it is compoſed of ſeven Parts, *viz.* a *Plinth*, two *Thorus's*, three *Annulets*, and a *Scotia*, or *Cavetto*, which he proportions as follows: To the *Plinth*, which is wrought hollow, (and might perhaps be more properly called a *Scotia*, or *Caſement*,) he allows ten Minutes; to the lower *Thorus* ſeven one Third Minutes, and to the lower *Annulet*, One one Fourth Minute, and to the *Cavetto* Four two Thirds Minutes; to the middle *Annulet* One one Fourth Minute, to the upper *Thorus* four one Fourth, and to the upper *Annulet* One one Fourth Minute.

Scamozzi makes the *Doric Baſe* thirty Minutes in Altitude, which are ſubdivided among ſix ſmaller Members, *viz.* 1*ſt*, a *Plinth*, (beginning below, and ſo aſcending,) allowing to it ten Minutes one Sixth. 2*dly*, A *Thorus* of eight Minutes. 3*dly*, A *Liſt* of one Minute. 4*thly*, A *Scotia* of four Minutes. 5*thly*, A *Liſt* of one Minute. And 6*thly*, A *Thorus* of five Minutes and a half. Above all theſe he places a Liſt of two Minutes, which he does not reckon into the Baſe, but to the Part of the Body of the *Column*.

Vignola, in like manner, allows the Altitude of the Baſe to be one half of the Diameter of the Column below; but he makes it to conſiſt of but four Parts, *viz.* a *Plinth*, a large and a ſmall *Thorus*, and a *Liſt*.

The *Ionic Baſe*, according to *Vitruvius*, is half a Mod. in Height, and in this Order deſcribes two ſorts of Baſes; the one for the *Ionic Column*, without a Pedeſtal, and the other for that with. Each of which Baſes conſiſts of ſmaller Members: But the Baſes are different in the Dimenſions of their Parts.

The Members of which they conſiſt are theſe that follow, *viz.* a *Plinth*, four *Fillets*, two *Scotia's*, *tia's*, two *Astragals*, and a *Thorus*.

First, as to the *Ionic Base* without a Pedestal. He divides and subdivides this Base as follows: He divides the whole Altitude of the Base into three equal Parts; the lower one of which is the Height of the *Plinth*, the two upper and remaining Parts he subdivides into seven equal Parts, the upper three of which make the *Thorus*; and the four Sevenths remaining, he subdivides into eight equal Parts; Half of the lower Eighth makes the lower *Fillet*, the other Half, and the second Eighth, and half the third Eighth, make the first *Scotia*, and the upper half of the third Eighth makes the second *Fillet*; the fourth and fifth makes the two *Astragals*, half the sixth Eighth makes the third *Fillet*, the upper half of the sixth Eighth, and all the seventh, and one Third of the last, or uppermost Eighth, makes the second *Scotia*; the two Thirds of the last Eighth which remains, makes the upper *Fillet* which subjoins to the *Thorus*. He also places another Fillet above the *Thorus*, which he does not account any Part of the Base, but a Part of the Body of the *Column*; which Fillet is one Twelfth of the Body of the Column, = five M.

The *Ionic Base*, with a Pedestal, according to *Vitruvius*, is divided into Parts, as follows: First, into three equal Parts, the lower of which is the Altitude of the *Plinth*; the two Thirds remaining, he divides into three equal Parts, the uppermost of which he assigns for the *Thorus*; and the two Thirds remaining, he subdivides into twelve equal Parts; half the lower one Twelfth he assigns for the *Fillet* above the *Plinth*; the remaining Half of one Twelfth, and the three next Twelfths, make the first *Scotia*; the fifth Twelfth makes the second *Fillet*, the sixth and seventh make the two *Astragals*, and Half the eighth makes the next *Fillet*; the other Half of the eighth and ninth, tenth and eleventh, make the second *Scotia*; and the twelfth and last Part makes the upper *Fillet* which is under the *Thorus*.

There is also a *Fillet* above the *Thorus*, which is of the same Height with that without the Pedestal.

Palladio assigns thirty Minutes for the Height of this Base; and according to his Scheme of this Member, he divides it into six smaller Members, *viz.* 1*st*, A *Plinth*, or rather, as he delineates it, a *Casement* of ten Minutes. 2*dly*, A *Thorus* of seven Minutes and a half. 3*dly*, A *List* of one Minute one Fourth. 4*thly*, A *Scotia* of four Minutes three Fourths. 5*thly*, Another *List*, or *Cincture*, of one Minute one Fourth. 6*thly*, A *Thorus* of five Minutes one Fourth. All which, being added together, make up thirty Minutes, which compleats the Base.

Above the *Base*, on the Foot of the Body of the *Column*, he places an *Astragal* of two Minutes one Fourth; and above that a *Ceincture* of one Minute one Fourth: All which together make thirty-three Minutes and a half.

Scamozzi also makes the *Ionic Base* thirty Minutes in Height, and

d confifting of the fame Num-
r of Parts and Form with that
' *Palladio*, viz. 1*ft*, A *Plinth*
'hich is concave) of ten Mi-
ıtes and a half. 2*dly*, A *Tho-*
ıs of eight Minutes. 3*dly*, A
ift of one Minute. 4*thly*, A
cotia of four Minutes and a
ılf. 5*thly*, A *Lift* of one Min.
nd, 6*thly*, another *Thorus* of five
in. All which, added together,
ıake the Bafe thirty Minutes.

Above which, on the Column, re two fmall Members more, *iz.* an *Aftragal* of two Minutes and a half, and a Lift of one Minute and a half: All which together make the Height thirty-four Minutes.

Vignola compofes his *Ionic Bafe* of the fame Number of fmall Members, and of the fame Form with that of *Vitruvius*.

The *Corinthian Bafe*, according to *Vitruvius*, is half a Mod. in Height, both in the *Corinthian Column* with a Pedeftal, and that without a Pedeftal; both of which he makes to confift of four fmaller Members, *viz.* 1*ft*, A *Plinth*. 2*dly*, Two *Thorus's*. 3*dly*, Four *Fillets*. 4*thly*, Two *Scotia's*, and two *Aftragals*.

This Bafe he divides thus: *Firft*, he divides the whole Height into four equal Parts; the lower of which Divifions he affigns for the *Plinth*; the three remaining Parts are again fubdivided into five equal Parts, the upper of which five is affigned to the upper *Thorus*, (which is the higheft Member of the Bafe;) the lower *Thorus* is made to contain five Quarters of one of thefe fifth Parts, *viz.* all the firft, or lower fifth Part, and half of the fecond; fo that one Fifth being taken form the upper *Thorus*, and the one Fifth and one Fourth of the one Fifth below for the lower *Thorus*, there remains but two of thefe Fifths, three of one Fifth, of which he fubdivides into twelve equal Parts.

Of half the lower Twelfth, he makes the 1*ft*, (or loweft *Fillet*;) then of the other half, all the 2*d*, 3*d*, 4*th*; and half the 5*th*, he makes the lower *Scotia*; of the remaining half of the fifth twelfth Part, he makes the fecond *Fillet*; of the 6*th* and 7*th* Parts, he makes the two *Aftragals*; of half the 8*th*, he makes the third *Fillet*; of the other half of the 8*th*, and all the 9*th*, 10*th*, and 11*th*, and half the 12*th*, he makes the fecond *Scotia*; and of the laft half of the twelfth Part, he makes the fourth or laft *Fillet* which fubjoins the under Side of the upper *Thorus*.

He alfo adds a *Fillet* above the Bafe, which is one Twenty-fourth of the Diameter of the *Column* in Altitude, which is 24 Minutes and a half.

The *Bafe* of the *Corinthian Column* with its Pedeftal, is of the fame Altitude and Number of Parts; and each Part has the fame Dimenfions with that which has no Pedeftal.

Palladio makes this Bafe to contain eight fmaller Members, *viz.* one *Orlo*, two *Thorus's*, two *Aftragals*, two *Cinctures*, and one *Scotia*; (though probably the Author, or Engraver, has made fome Miftake in the Divifion and Subdivifion of the Bafe; but I fhall give it as it is there found.)

He makes the *Orlo* nine Minutes two Thirds; the lower *Thorus* feven Minutes; the lower *Aftragal* three Fourths of a Minute,

Minute, (which ſeems to be too little;) the lower *Cincture* one Fourth of a Minute; the *Scotia* three Minutes three Fourths; the next *Cincture* has not any Number ſet to it, but it appears of the ſame Size with the other *Cincture*: Then comes the next *Aſtragal* of half a Minute, and then the upper *Thorus* of five Minutes; and he places another *Aſtragal* of two Minutes and a half above theſe eight Members of the *Baſe*, and another *Aſtragal* of two Minutes and a half, and a *Cincture* above that.

This, it is true, is but a lame Account; but the Fault lies either in the Author, or Engraver, or both.

Scamozzi portraits this Baſe of thirty Minutes in Altitude, and divides this grand Member into eight petty ones, of the ſame Form with thoſe of *Palladio*, *viz.* 1*ſt*, An *Orlo* of nine Minutes and a half. 2*dly*, A *Thorus* of ſeven Minutes. 3*dly*, An *Aſtragal* of two Minutes. 4*thly*, A *Liſt* of one Minute. 5*thly*, A *Scotia* of three Minutes and a half. 6*thly*, A *Liſt* of one Minute. 7*thly*, Another *Aſtragal* of one Minute and a half. And 8*thly*, and laſt of all, a *Thorus* of four Minutes and a half: All which, added together, make up thirty Minutes.

He alſo places above the Baſe two other Members on the Foot of the Column, *viz.* an *Aſtragal* of two Minutes and a half, and a *Liſt* of one Minute.

Vignola allows thirty Minutes to the Altitude of this Baſe; and as to the Form, he makes it the ſame with that of *Vitruvius*.

The *Compoſite* or *Roman Baſe*, according to *Vitruvius*, contai[ns] thirty Minutes in Height.

This grand Member he divid[es] into ten ſmaller, *viz.* a *Plint[h]* two *Thorus's*, (one of which [is] in the Middle, where the tw[o] *Aſtragals* are in the *Corinthia[n]* Order,) four *Fillets*, and tw[o] *Scotia's*.

He firſt divides this Memb[er] into four Parts, the lower [o]f which is for the Altitude of th[e] *Plinth*; and then he ſubdivide[s] the other three Parts into five. Of the uppermoſt of the five, h[e] makes the upper *Thorus*; of th[e] lower Fifth, and one Fourth o[f] the ſecond, he makes the lower *Thorus*; (ſo that the lower *Thorus* is four Fifths high:) The ſecond fifth Part and three Fourths that remain, he ſubdivides into twelve equal Parts; of half the lower twelfth, he makes the firſt *Fillet*; of the other Half, and all the 2*d*, 3*d*, 4*th*, and half the 5*th*, he makes the firſt *Scotia*; of the remaining Half of the 5*th*, he makes the ſecond *Fillet*; of the 6*th* and 7*th*, he makes the middle *Thorus*; of half the 8*th*, he makes the third *Fillet*; of the remaining Half of the 8*th*, and all the 9*th*, 10*th*, and 11*th*, and half the 12*th*, he makes the ſecond *Scotia*; of the remaining Half of the 12*th*, he makes the laſt *Fillet*, which is juſt under the upper *Thorus*.

Above the *Baſe*, on the Foot of the *Column*, he makes a *Fillet*, which is one twenty-fourth of the Diameter of the *Column* below.

Palladio makes this Baſe thirty Minutes in Altitude, and divides it into eleven ſmall Members, *viz.* an *Orlo*, two *Thorus's*, four *Liſts*, two *Scotia's*, and two *Aſtragals*.

To

rus may be inrich'd; of which we have an Inſtance in the new Chapel at *Verſailles*, where 'tis done with a great deal of Prudence.

For as nothing ſhould be expoſed to the Eyes of a great Prince, but what is ſome Ways diſtinguiſh'd by its Richneſs; and as the King here has in ſight the Baſes of the Columns of his Seat, 'tis but juſt they ſhould be inrich'd like the reſt of the Chapel, which is extremely pompous: But, ſetting aſide ſuch Occaſions, it would be a Fault to adorn the Baſes of Columns; though *Scamozzi* is of another Opinion.

A *Baſe of any ſolid Figure*, is its lowermoſt plain Side, or that on which it ſtands; and if the Solid has two oppoſite parallel plain Sides, one of them is the *Baſe*; and then the other is alſo called its *Baſe*.

BASIL, with *Joiners*, &c. the Angle to which the Edge of an iron Tool is ground. To work on ſoft Wood, they uſually make their *Baſil* twelve Degrees; for hard eighteen Degrees; it being obſerv'd, that the more acute or thin the *Baſil* is, the better and ſmoother it cuts; and the more obtuſe, the ſtronger and fitter for Service.

BASILIC } [of Βασιλικη, Gr.
BASILICA } a Royal Houſe or Palace] a Term antiently uſed for a large Hall, or publick Place, with Iſles, Portico's, Galleries, Tribunals, &c. where Princes ſat and adminiſtred Juſtice in Perſon.

But the Name has ſince been transferred, and is now applied to ſuch Churches, Temples, &c. which, by their Grandeur, as far ſurpaſs other Churches, as Princes Palaces do private Houſes. As alſo, to certain ſpacious Halls in Princes Courts, where the People hold their Aſſemblies. And alſo ſuch ſtately Buildings where Merchants meet and converſe together: As for Inſtance, that of the Palace at *Paris*, the *Royal Exchange* in *London*, *&c.*

BASON, a Reſervatory of Water, as the Baſon of a *Jet d'Eau*, or Fountain; the *Baſon* of a Port, *Bath*, *&c.* which *Vitruvius* calls *Labrum*.

BASSO-RELIEVO, } is a
BASS-RELIEF, } Piece of Sculpture, the Figures of which do not project far, or ſtand out from the Ground in their full Proportion. M. *Felibien* diſtinguiſhes three Kinds of *Baſſo-Relievo's*: In the firſt, the front Figures appear almoſt with the full *Relievo*; in the ſecond, they ſtand out no more than one Half; and in the third, much leſs, as in Coins, Vaſes, *&c.*

BATEMENT, a Term in Carpentry, ſignifying an Abatement or Waſte of a Piece of Stuff, by forming it to a deſign'd Purpoſe or Uſe: Thus inſtead of asking how much was cut off from ſuch a Board or Piece of Stuff, they ſay, what Batement had that Piece of Stuff?

BATTEN, is a Name that Workmen give to a Scantling of Wooden Stuff, from two to four Inches broad, and about one Inch thick; the Length is pretty conſiderable, but undetermin'd.

This Term is chiefly uſed in ſpeaking of Doors, and Windows of Shops, *&c.* which are not fram'd of Whole Deal, or one quarter Inch-Oak, with *Stiles*, *Rails*, and *Pannels*, (as Wainſcot is fram'd;) and yet they are made to appear as if they were, by Means of theſe Pieces or *Battens*, bradded on upon the plain Boards which are joined together for the Door, or Window, all round, and ſometimes croſs them, and up and down, *&c.* according to the Number of the Pannels, the Workman deſigns the Door, or Window ſhall appear to have.

Theſe Pieces, which are thus bradded on, to repreſent *Stiles*, *Rails*, and *Montans*, and are of different Breadths, according to the Deſign of the Workman, as from two to ſix or ſeven Inches; and there is uſually ſome Moulding ſtruck, as a *Bead*, an *Ogee*, or the like, on one Edge of thoſe that repreſent the Stiles, and the upper and lower Rails, and on both the Edges of thoſe which are deſign'd to appear like *Montans*, and middle *Rails*.

BATTEN-DOORS are ſuch as ſeem to be Wainſcot ones, but are not; for in Wainſcot ones, the *Pannels* are groov'd into the Framing; but in theſe, they firſt joint and glew the Boards, which are cut to the full Length and Breadth of the Door-Caſe; which Gluing being dry, they traverſe them over with a long Plane; and being ſmooth'd, the *Battens* are fitted on, on the Front-Side. And theſe are called ſingle *Batten-Doors*; for there are others, call'd double *Batten-Doors*, *viz.* ſuch as are battened on both; though this is but rarely done.

But there are batten'd Doors, which are called double Doors, ſuch as Front or Outer-Doors; which are uſually made of whole

Deal,

Deal, and afterwards batten'd on the Outside, and Pieces, four or five Inches broad, mitred round the Edges on the Inside of the Door; and then it is lin'd cross the Door betwixt these Pieces, with thin Slit-Deal, which renders it level with the mitred Pieces.

Some Doors have been lin'd with Pieces laid Bevelling, and not at Right Angles, but near Mitre to the Sides of the Door; and when all has been plain'd off level, it has been divided into Rhombus's, and struck with a Pencil, and round-headed Nails driven in at the Angles of the Rhombus's, which added something of Beauty to the Work.

This Way of Lining upon the Doors, *viz.* pointing from the lower Corner behind, towards the upper Corner before, seems to be a good Way to prevent a Door from sagging or sinking at the Fore-Corner, when ever the Joints shall happen to unglue.

As to the Price of Batten-Doors. For the Workmanship of making Batten-Doors of slit Deal, about an Inch in Thickness (or of thin whole Deals,) glu'd and batten'd on one Side, 4*s.* *per* Door, is a moderate Price between Master and Workman: But for such as have been mention'd above (which are for Front and other Outer Doors, *viz.* both batten'd and lin'd are worth 7*s.* *per* Door, for Workmanship.

BATTER, a Term used by Bricklayers, Carpenters, *&c.* to signify that a Wall, Piece of Timber, or the like, doth not stand upright, but leans from you-ward, when you stand before it; but when, on the contrary, it leans toward you, they say it *over-hangs*, or *hangs over.*

BATTLEMENTS, are Indentures or Notches in the Top of a Wall, or other Building, in the Form of Embrazures, for the sake of looking through them, *&c.*

BAY, a Term used to signify the Magnitude of a *Barn*; as if a Barn consists of a Floor and two Heads, where they lay Corn, the call it a Barn of two *Bays.*

These *Bays* are from fourteen to twenty Foot long, and Floors, from ten (which is the smallest Size) to twelve broad, and usually twenty Foot long, which is the Breadth of the Barn.

If a *Bay* be twenty Foot in Length, then there is usually a Pair of *Prick-Posts* in the Middle, and a *Beam* to hold in the Rod from bending the *Raisons*; but if the *Bays* are not more than sixteen Foor, and the Timber stout, then there are no Posts; but at the End of each Bay, where there are always hanging Braces, fram'd into the *Beam* and *Posts*; and also a Cross-Cell, to hold in the Side-Cells from flying out when the Barn is full: And also, 'tis common for large Barns to consist of divers *Bays.*

BAY-WINDOW, one that is composed of an Arch of a Circle; and consequently such a one will stand without the Stress of the Building. By which Means Spectators may better see what is done in the Street.

BEACON, a Signal for the better Security of the Kingdom from Foreign Invasions. These are long Poles or Posts set up on certain Eminencies, on which are fasten'd Pitch-Barrels, to be fir'd by Night, and Smoke made by Day, to give Notice to the whole

whole Kingdom, in a few Hours of an approaching Invaſion.

BEAD [in *Architecture*] a round Moulding, commonly made upon the Edge of a Piece of Stuff, in the *Corinthian* and *Roman* Orders, cut or carv'd, in ſhort Emboſſments, like Beads in Necklaces, in Semi-Relief. See BAGUETTE.

A *Bead* is uſually one quarter of a Circle, and only differs from a *Boultin* in Size: For when they are large, Workmen commouly call them *Boultins*. Sometimes a Bead-plain is ſet on the Edge of each *Faſcia* of an Architrave and ſometimes, likewiſe an Aſtragal is thus carv'd: In both which theſe Carvings are called *Beads*.

A *Bead* is often placed on the Lining-Board of a Door-Caſe, and on the upper Edges of Skirting-Boards.

BEAK [in *Architecture*) a little Fillet, left on the Edge of a *Larmier*, which forms a Canal, and makes a kind of Pendant.

Chin-Beak, a Moulding, the ſame as the Quarter-round, except that its Situation is inverted. This is very frequent in modern Buildings, though few Examples of it are found in the Antient.

BEAM, in a Building, is the largeſt Piece of Wood in a Building, which always lies croſs the Building or the Walls, and ſetving to ſupport the principal Rafters of the Roof, and into which the Feet of the the principal Rafters are fram'd.

No Building has leſs than two of theſe Beams, *viz.* one at each Head. Into theſe, the Girders of the Garret-Floor are alſo fram'd; and if the Building be of Timber, the *Teazle-Tenons* of the Poſts are fram'd.

The *Teazle-Tenons* are made at Right Angles to thoſe which are made on the Poſts to go into the *Raiſons*; and the *Reliſh* or *Cheats* of theſe *Teazle-Tenons*, ſtand up within an Inch and half of the Top of the *Raiſon*, and the Beam is *cauked* down (which is the ſame Thing as Dove-tailing acroſs) till the Cheeks of the Mortiſes in the Beam conjoin with thoſe of the *Teazle-Tenons* on the Poſts.

As to the Size of Beams. The Proportions of Beams in or near *London*, are fix'd by a Statute or Act of Parliament for the rebuilding of the City of *London*, after the Fire in 1666, and were appointed to be of the following Scantlings.

A Beam fifteen Foot long, muſt be ſeven Inches on one Side its Square, and five on the other; if it be ſixteen Foot long, one Side muſt be eight Inches, the other ſix: If ſeventeen Foot long, one Side muſt be ten Inches, and the other ſix; and ſo proportionable to their Lengths. In the Country, where Wood is more plenty, they uſually make their Beams ſtronger.

Sir *Henry Wotton* adviſes, That all *Beams*, *Summers*, and *Girders*, be made of the ſtrongeſt and moſt durable Timber.

Herrera informs us, That in *Ferdinand Cortez*'s Palace in *Mexico*, there were ſeven thouſand Beams of Cedar: But then he muſt be underſtood to uſe the Word *Beam* in a greater Latitude than it is uſed with us. The

French,

French, under the Word *Poutre*, which ſignifies a Beam, take in not only the Pieces which bear the Rafters, but alſo all thoſe which ſuſtain the Joiſts for the Ceilings.

Some *French* Authors have conſider'd the Force of *Beams*, and brought their Reſiſtance to a preciſe Calculation; as particularly, M. *Varignon*, and M. *Parent*: The Syſtem of the latter of which, is as follows:

When two Plans of Fibres, which were contiguous before, are ſeparated in a *Beam*, which breaks parallel to its Baſe, (which is ſuppos'd to be a Parallelogram,) there is nothing to be conſider'd in theſe Fibres, but their Number, Bigneſs, and Tenſion, before they are broken, and the Lever, by which they act: All theſe together making the Reſiſtance of the Beam remaining to be broke.

Then ſuppoſe another *Beam* of the ſame Wood, where the Baſe is likewiſe a Parallelogram, and of any Bigneſs, with regard to the other, at Pleaſure. The Height of each of theſe, when laid Horizontal, being divided into an indefinite Number of equal Parts, and their Breadth into the ſame Number, in each of their Baſes will be found an equal Number of ſmall quadrangular Cells, proportional to the Baſes of which they are Parts; then theſe will repreſent little Baſes; or, which is the ſame Thing, the Thickneſſes of the Fibres to be ſeparated for the Fracture of each Beam, and the Number of Cells being equal in each *Beam*, the Ratio of the Baſes of both *Beams* will be that of the Reſiſtance of their Fibres, both as to Number and Thickneſs.

Now the two *Beams* being ſuppoſed to be of the ſame Wood, the Fibres moſt remote from the Points of Support, which are thoſe which break the firſt, muſt be equally ſtretch'd when they break.

Thus the Fibres, *v. g.* of the tenth Diviſion, are equally ſtretched in each Caſe, when the firſt breaks; and in whatever Proportion the Tenſion be ſuppoſed, it will be ſtill the ſame in both Caſes; ſo that the Doctrine be entirely free, and unembaraſs'd with any Syſtem of Phyſicks.

Laſtly, It is evident, that the *Levers*, by which the Fibres of the two *Beams* act, are repreſented by the Height of their Baſes; and conſequently the whole Reſiſtance of each *Beam* is the Product of its Baſe by its Height, or, which is the ſame Thing, the Square of the Height being multiply'd by the Breadth, which holds not only of parallelogrammick, but alſo of Elliptical Baſes.

Hence, if the Baſis of two *Beams* be equal, though both their Heights be unequal, their Reſiſtance will be as their Heights alone; and conſequently one and the ſame *Beam* laid on the ſmalleſt Side of its Baſe, will reſiſt more than when laid flat, in Proportion, as the firſt Situation gives it a greater Height than the ſecond. And thus an Elliptical Baſe will reſiſt more, when laid on its greateſt Axis; than on its ſmalleſt.

Since in *Beams* equal in Length, it is the Baſes which determine the Proportion of their Weights or Solidities; and since their Bases being equal, their Heights may be different; two *Beams* of the same Weight, may have Resistance different to Infinity. Thus if in the one, the Height of the Base be conceiv'd infinitely great, and the Breadth infinitely small; while in the other, the Dimensions of the Base are finite, the Resistance of the first will be infinitely greater than that of the second, though their Solidity and Weight be the same.

If therefore, all requir'd in Architecture were to have *Beams* capable of supporting vast Loads, and at the same Time have the least Weights possible, 'tis plain they must be cut thin as *Laths*, and laid edge-wise.

If the Bases of the two *Beams* are suppos'd to be unequal, but the Sum of the Sides of the two Bases equal, *v. g.* if they be either 12 and 12, or 11 and 13, or 10 and 14, *&c.* so that they always make 24; and further, if they are supposed to be laid edge-wise, persuing the Series, it will appear, that in the *Beam* of 12 and 12, the Resistance will be 1728, and the Solidity or Weight 144, or that in the last, or 1 and 23, the Resistance will be 529, and the Weight 23: Therefore the first, which is square, will half the Strength of the last with regard to its Weight.

Hence, M. *Parent* remarks, that the common Practice of cutting the Beams out of Trees as square as possible, is ill Husbandry: And thence he takes Occasion to determine Geometrically, what Dimensions the Base of a Beam to be cut out of any Tree propos'd, should have, in order to its having the greatest Resistance possible; or which is the same Thing, a Circular Base being given, he determines the Rectangle of the greatest Resistance that can be inscrib'd, and finds that the Sides must be nearly as 7 to 5; which agrees with Observation.

Hitherto we have supposed the Length of the *Beams* to be equal; if it be unequal, the Bases will resist so much the less, as the *Beams* are the longer.

To this it may be added, That a *Beam* sustain'd at each End, breaking by a Weight suspended from its Middle, does not only break at the Middle, but at each Extreme; or if it does not actually break there, at least, immediately before the Moment of the Fracture, which is that of the *Equilibrium* between the Resistance and the Weight, its Fibres are as much stretch'd at the Extremes, as in the Middle; so that of the Weight sustain'd by the Middle, there is but one third Part that acts at the Middle to make the Fracture; the other two only acting to induce a Fracture in the two Extreams.

A *Beam* may be supposed to be either loaden only with its own Weight, or with other foreign Weights, apply'd at any Distance, or only with those foreign Weights. Since, according to M. *Parent*, the Weight of a *Beam* is not ordinarily above one seventieth Part of the Load given to sustain it, it is evident, that in considering several Weights they

they muſt be all reduced by the common Rules, to one common Centre of Gravity.

M. *Parent* has alſo calculated Tables of the Weights, which will be ſuſtain'd by the Middle in *Beams* of various Baſes and Lengths, fitted at each End, into Walls, on a Suppoſition, that a Piece of Oak of an Inch ſquare, and a Foot long, retain'd Horizontally by the two Extreams, will ſuſtain three hundred and fifteen Pounds in its Middle before it breaks; which, it has been found by Experiments, that it will. See *The Memoirs of the* French *Academy*, Anno 1708.

BEAM-COMPASSES, an Inſtrument made either of Wood or Braſs, with ſliding Sockets or Curſors, which ſerve to carry ſeveral ſhifting Points for drawing and dividing Circles with very long *Radii.*

They are of Uſe in large Projections, for drawing the Furnitures on Wall-Dials, &c.

BEAM-FILLING [in Building] is Plaiſterer's Work; and is the Filling up the vacant Space between the Raiſon and Roof, whether Tiling, Thatching, or any other Roof, with Stones or Bricks laid between the Rafters on the Raiſon, and plaiſtered on with Loom, frequent where the Garrets are not pargeted or plaiſter'd; or ſometimes they ſet ſome *Tiles* with one Edge upon the Raiſon, and the other leaning againſt the Roof; and then theſe Tiles are plaiſtered over with Loam. This Sort of Work is very common in the Country, where they do not parget or plaiſter their Garrets.

The Price. The uſual Price for Workmanſhip only, in the Country, is a Halfpenny *per* Foot, or three Halfpence *per* Yard, lineal Meaſure.

To BEAR *Timber*, is ſaid to *bear* at its whole Length, when neither a Brick-Wall, or Poſts, &c. ſtand between the Ends of it; but if either a Brick-Wall, or Poſts be trimm'd up to the Timber, then it is ſaid to bear only at the Diſtance between the Brick-Wall, or Poſt, and either End of the Timber.

Thus Carpenters uſually ask, What *Bearing* ſuch a Piece of Timber has? The Anſwer to ſuch a Queſtion, is ten, fifteen, or twenty Foot, &c. according to the Length of the whole Timber; or elſe, according to the Diſtance between either End of the Timber.

BEARER [*in* Architecture] *a Poſt or Brick-Wall trimmed up between the two Ends of a Piece of Timber, to ſhorten its* Bearing; *or to prevent its* bearing *with the whole Weight at the Ends only.*

BEARING *of a Piece of Timber*, with Carpenters, the Space, either between the two fix'd Extreams thereof, when it has no other Support, which they call *Bearing at Length*; or between one Extream and a Poſt, Brick-Wall, &c. trimm'd up between the Ends, to ſhorten its *Bearing.*

BED *of* STONE [in *Maſonry*] a Courſe or Range of Stones; and the Joint of the *Bed* is the Mortar between two Stones, placed over each other.

BED-MOULDING, } BEDDING-MOULDING, } a Term uſed by Workmen to ſignify thoſe Members in a Cornice which are placed below the

Coronet: And now, a *Bed-Moulding*, with Joiners, usually consists of these four Members, an *Ogee*, a *List*, a large *Boultin*, and another List under the Coronet.

BEVEL, } [in *Masonry* and BEVIL, } *Joinery*] a kind of Square, one Leg of which is frequently crooked, according to the Sweep of an Arch or Vault. It is moveable on a Point or Centre, and so may be set to any Angle. The Make and Use of it are pretty much the same, as those of the common Square and Mitre, except that those are fix'd, the first at an Angle of ninety Degrees, and the second, at forty-five: Whereas the Bevel being moveable, it may in some Measure supply the Office of both, and yet supply the Deficiency of both, which it is chiefly intended for, serving to set off or transfer Angles, either greater or less than ninety or forty-five Degrees. Hence,

Any Angle that is not square, is called a Bevel-Angle, whether it be more obtuse, or more acute, than a Right Angle: But if it be one half as much as a Right Angle, *viz.* forty-five Degrees, then Workmen call it *Miter*. They have also a Term *Half-Miter*, which is an Angle that is one quarter of a Quadrant or Square, *viz.* an Angle of twenty-two Degrees and a half. This they call a *Half-Miter*.

Green BICE, is a Colour of a sandy Nature, and therefore not much used. But it is to be wash'd, before it be used. See Washing *of Colours*.

Blue Bice bears the best Body of all bright *Blues* us'd in common Work, but 'tis the palest in Colour; it works indifferently well, but inclines a little to be sandy; therefore it requires good Grinding, and that on a very hard Stone. It is a Blue that lies best near the Eye of any now in Use, except *Ultramarine*, a Colour produced from the Tincture of *Lapis Lazuli*; but this is so very dear, that it is not to be used but in Pieces of great Price. This *Bice* is also to be wash'd. See Washing *of Colours*.

BILL, an Edg'd-Tool fitted to a Handle, used in lopping Trees. When it is long, it is called a *Hedging-Bill*; when short, a *Hand-Bill*.

BINDING-JOISTS, are those Joists in any Floor, into which the Trimmers of Stair-Cases (or Well-Holes for the Stairs) and Chimney-Ways are fram'd. These Joists ought to be stronger than common Joists.

As to the Scantling and Size of these, as well as all other Timber Members, it was settled by an Act of Parliament before the Rebuilding of *London*; according to which Act, *Binding-Joists* which contain in Length

Feet.			*Inches*	*Inches.*
7	}	must be	} 6	}
9	}	in their	} 7	} and 5.
11, or 12	}	Squares,	} 8	}

Se

So large they were ordered to be, and not leſs; but probably, they might be as much bigger as they pleas'd.

BISSECTION [in *Geometry*] the Diviſion of any Quantity into two equal Parts; and is the ſame with Bipartition. Thus to biſſect any Line, is to divide it into two equal Parts.

BLACKS, uſed in Painting, as LAMP-BLACK: This Colour is nothing but a Soot rais'd from the Roſiny and fat Parts of Fir-Trees. It comes to us moſtly from the Northern Countries, as *Sweden* and *Norway*. This *Black* is more generally us'd in Painting than any other, becauſe of its Plenty and Cheapneſs; and and proves a very good *Black* for moſt Uſes: 'Tis of ſo fine a Body, that if only temper'd with Lin-Seed Oil, it will ſerve to work with on moſt common Occaſions, without grinding.

But being thus uſed, it will require a long Time to dry, unleſs there be a good deal of drying Oil mix'd with it; or, which is better, ſome Verdegreaſe finely ground: This, and the drying Oil together, will make it dry quickly. Alſo ſome add Oil of Turpentine, which gives it alſo a drying Quality; but without ſome of theſe, it will be a long Time a drying: For in the Subſtance of the Colour is contained a greaſy Fatneſs, that is an Enemy to Drying. To remedy which, if it be burnt in the Fire, 'till it be red-hot, and ceaſe to ſmoak, the Fatneſs will be conſum'd, and then it will dry much ſooner. But when it is burnt, it muſt then of Neceſſity be ground with Oil, the Fire being of that Nature, that it is apt to harden moſt Bodies that paſs through it.

This Colour is uſually made up in ſmall Barrels, &c. of Deal of ſeveral Sizes, and is ſo brought over to us.

Lamp-Black is burnt, or rather dry'd in the following Manner:

It is put into an Iron Ladle, or a Crucible, and ſet over a clear Fire, and there let it remain till it be red-hot, or ſo near it, that there is no manner of Smoke ariſes from it.

Beſides this *Black*, there is another Sort of *Black*, which is the Soot of a Lamp; which ſome commend as a much better *Black* for any Uſe, than the former, it being of a finer Body, and brighter Colour. But this not being to be procur'd in very great Quantities, is therefore only uſed in very fine Work.

Ivory-Black, is made of Comb-Makers Raſpings, and other waſte Fragments of Ivory; being burnt or charr'd to a black Coal in a Crucible cloſe ſtopped up, which proves a very delicate *Black*, when ground very fine. It is ſold at Colour-Shops, well prepar'd and levigated, or ground very fine with Water on a Marble-Stone, and then dry'd in ſmall Lumps: Being thus prepar'd, 'tis the more eaſily ground in Oil, with which it will lie with as ſmooth a Body as moſt Colours do; but being pretty dear, is therefore not uſed in any common Work.

Some uſe Willow-Charcoal. This, if ground very fine in Oil, makes a very good *Black*; but is not ſo much uſed as the Lamp-Black, not being ſo eaſily to be gotten.

Ivory must be burnt also to make a *Black*, thus: Fill two Crucibles with Ivory Shavings, then clap their two Mouths together, and bind them fast with an Iron Wire, and lute the Joints close with Clay, Salt, and Horse-Dung, well beaten together; then set it in a Fire, covering it all over with Coals, and let it remain therein, till you are sure the Matter inclosed in the Crucibles be thoroughly red-hot; then take it from the Fire, but open not the Crucible, till they are perfectly cold; for if you should open them while hot, the Matter would turn to Ashes. The same will be done, if the Joints are not lured close; for 'tis only the Exclusion of all Air, that prevents any Matter whatever that's burnt to a Coal, from turning to a white Ash, and preserves the Blackness.

BLOCK *of Marble*, a Piece of Marble, as it comes out of the Quarry, before it has assum'd any Form from the Hand of a Workman.

BOARD-MEASURE. To measure a Board, is nothing else but the measuring a long Square.

EXAMPLE.

If a *Board* be sixteen Inches broad, and thirteen Feet long, how many Feet are contain'd therein?

Multiply sixteen by thirteen, and the Product will be two hundred and eight; which being divided by twelve, gives seventeen Feet, and four remaining, which is a third Part of a Foot, thus:

12 : 13 : : 16
13
——
48
16
——
12) 208 (17 $\frac{4}{12}$
12
——
88
84
——
4

Or, you may multiply one hundred and fifty-six (the Length in Inches) by sixteen, and the Product will be two thousand four hundred and ninety-six; which being divided by one hundred and forty-four (the Number of Inches in a Foot square,) the Quotient will be seventeen Feet, and forty-eight remaining, which is a third Part of one hundred and forty-four, as before, thus:

144 : 156 : : 16
16
——
936
156
——
144) 2496 (17 $\frac{48}{144}$
144
——
1056
1008
——
48

By Scale and Compasses.

Extend the Compasses from twelve to thirteen; the same Extent will reach from sixteen to seventeen

ſeventeen Feet and one Third, the Content. Or,

Extend them from one hundred forty-four to one hundred fifty-ſix, (the Length in Inches;) and the ſame Extent will reach from ſixteen to ſeventeen Feet one Third, the Content.

Example II. If a Board be nineteen Inches broad, how many Inches in Length will make a Foot?

Divide one hundred and forty-four by nineteen, and the Quotient will be ſeven very near; and ſo many in Length, if a Board be nineteen Inches broad, will make a Foor.

Inc.	*Inc.*	*Inc.*	*Inch.*
19 :	144 ::	1 :	7. 11 or 58 fete.
	133		——
	——		19
	11		

Again; Extend the Compaſſes from nineteen to one hundred forty-four; that Extent will reach from one to ſeven, fifty-eight, *i. e.* ſeven Inches and ſomewhat more than a half; ſo that if a Board be nineteen Inches broad, if you take ſeven Inches, and a little more than a half in your Compaſſes, from a Scale of Inches, and run that Extent along the Board from End to End, you may find how many Feet that Board contains, or you may cut off from that Board any Number of Feet deſired.

For this Purpoſe there is a Line upon moſt ordinary Joint-Rules, with a little Table placed upon the End, of all ſuch Numbers as exceed the Length of the Rule, as in this little Table annexed.

0	0	0	0	5	0	8	6
12	6	4	3	2	2	$1\frac{1}{2}$	1
1	2	3	4	5	6	7	8

Here you ſee if the Breadth be one Inch, the Length muſt be twelve Feet; if two Inches, the Length is ſix Feet; if five Inches broad, the Length is two Feet five Inches, *&c.*

The reſt of the Lengths are expreſs'd in the Line thus: If the Breadth be nine Inches, you will find it againſt ſixteen Inches, counted from the other End of the Rule; if the Breadth be eleven Inches, then a little above thirteen Inches will be the Length of a Foot, *&c.*

BOARDING of Walls. See *Weather Boarding.*

BOAT NAILS, a certain Sort of *Nails.*

BODY [in *Geometry*] is that which has three Dimenſions, Length, Breadth, and Thickneſs. As a Line is form'd by the Motion of a Point, and a Superficies by the Motion of a Line, ſo a *Body* is general by the Motion of a Superficies.

To BEAR A BODY [with *Painters.*] A Colour is ſaid to *bear a Body*, when it is of ſuch a Nature, as is capable of being ground ſo fine, and mixing with the Oil ſo intirely, as to ſeem only a very thick Oil of the ſame Colour; and of this Nature are *White-Lead* and *Ceruſe*, *Lamp-Black*, *Vermillion*, *Lake*, *Pink*, *Yellow Oker*, *Verdegreaſe*, *Indigo*, *Umber* and *Spaniſh Brown*; *Blue Bice* and *Red-Lead* are not ſo fine; but yet ſo fine, that they may be ſaid *to bear a very good Body.* All theſe may be ground ſo fine, as to be even like Oil itſelf;

ſelf; and then they alſo may be ſaid to work well, ſpreading ſo ſmooth, and covering the Body of what you lay upon it ſo intirely, as that no Part will remain viſible where the Pencil hath gone, if the Colour be work'd ſtiff enough.

Whereas, on the contrary, Verditers and Smalts, with all the grinding imaginable, will never be well imbodied with the Oil, nor work well: Indeed Bice and Red-Lead will hardly grind to an oily Firmneſs, nor lie intirely ſmooth in the working; yet may be ſaid to *bear an indifferent Body*, becauſe they will cover ſuch Work very well, that they are laid upon; but ſuch Colours as are ſaid not to *bear a Body*, will readily part with the Oil, when laid on the Work; ſo that when the Colour ſhall be laid on a Piece of Work, there will be a Separation, the Colour in ſome Parts, and the clear Oil in others, except they are temper'd extreme thick.

BOLTS of Iron, for Houſe-Building, are diſtinguiſh'd by Iron-Mongers into three Kinds, *viz. Plate*, *Round*, and *Spring Bolts*: *Plate* and *Spring Bolts* are uſed for the faſtening Doors and Windows, and theſe are of different Sizes and Prices: Small *Spring Bolts* have been ſold for three Pence Half-penny *per* Piece, others at nine Pence, others at fourteen Pence; and ſo likewiſe *Plate Bolts* ſome nine Pence, and ſome ten Pence a Piece.

There are alſo *Braſs-knob'd Bolts*, ſhort and long; the ſhort are ſold for about ten Pence *per* Piece; and long for Folding-Doors, at eighteen Pence *per* Piece; and Iron *Balcony Bolts* at about ten Pence *per* Piece; and *Round Bolts*, (or long Iron Pins,) with a Head at one End, and a Key-Hole at the other, which are uſually ſold by the Pound, *viz.* three Pence Half-penny or four Pence *per* Pound.

BOND, a Term among Workmen, as make *good Bond*, means that they ſhould faſten the two or more Pieces together, either by tenanting, mortiſing or dovetailing, &c.

BOSSAGE } in *Architecture*,
BOSCAGE } is a Term uſed for any Stone that has a Projecture, and is laid in a Place in a Building Lineal, to be afterwards carved into Mouldings, Capitals, Coats of Arms, &c.

Boſſage is alſo that which is otherwiſe called *Ruſtick Work*; which conſiſts of Stones which ſeem to advance beyond the Naked of a Building, by reaſon of Indentures or Channels left in the Joinings: Theſe are chiefly uſed in the Corners of Edifices, and thence called *Ruſtick Quoins*.

The Cavities or Indentures are ſometimes round, and ſometimes chain-fram'd, or bevell'd; ſometimes in a Diamond Form, and ſometimes it is inclos'd with a Cavetto, and ſometimes with a Liſtel.

BOTHAM NAILS, a ſort of *Nails* ſo called by Ironmongers.

BOULDER WALLS, a kind of *Walls* built of round Flints or Pebbles, laid in a ſtrong Mortar; uſed where the ſea has a Beach caſt up, or where there are plenty of Flints.

As to the Manner of building theſe *Walls*, a Bricklayer which has been uſed to this kind of

Work.

BRADS, a kind of Nails uſed in Building, which have no ſpreading Heads, as other Nails have. Theſe are diſtinguiſhed by Ironmongers by ſix Names, as *Joiners Brads*, *Flooring Brads*, *Batten Brads*, *Bill Brads*, or *Quarter Heads*, &c.

Joiners Brads, for hard Wood-Wainſcot, from one Inch to two and a quarter in Length.

Batten Brads, for ſoft Wood-Wainſcot the ſorts are One-penny, Two-penny, Three-penny; *ditto*, large Four-penny; *ditto*, large Five-penny, Six-penny.

For *Flooring*, plain or ſoft Wood Joiſts, the ſorts are fourteen, fifteen, eighteen, nineteen, twenty, twenty-one, twenty-two, twenty-three, twenty-eight, thirty-two, and thirty-ſix Pound *per* M.

Ditto ſtrong, fit for hard Joiſts, the ſorts are fifteen, eighteen, nineteen, twenty-four, and thirty-two Pound *per* M.

Quarter Heads, for ſoft Wood, the Sorts are ten, thirteen, fifteen, eighteen, nineteen, twenty, twenty-two, twenty-three, twenty-eight, and thirty-two Pound *per* M.

Ditto ſtrong, for hard Wood Joiſts, the Sorts are fourteen, twenty, thirty-four, forty-four, and fifty-four Pound *per* M.

N. B. All *Bill Brads*, alias *Quarter Heads*, are very fit for ſhallow Joints that are ſubject to warp; or for Floors which are laid in haſte, or by unskilful Perſons, becauſe the *Bill* with the Head will hinder the Boards from ſtarting from the Joiſts; but do not make ſo ſmooth Work as the plain *Brads*.

As to the *Prices of Brads*, of which I ſhall ſet down a few, as follows: As *Joiners Brads*, the uſual Price

			d.
Of a M. of	2	Inch is	20
	$1\frac{1}{2}$		15
	1		11

Quarter-Heads, or Bill-Brads, for ſoft Wood Floors, the uſual Price

	lb.		s.	d.
Of a M. of	15	is	4	9
	18		5	6

BRANCHES [in Architecture] are the Arches of *Gothic* Vaults. Theſe Arches traverſing from one Angle to another Diagonal-wiſe, form a Croſs between the other Arches, which make the Sides of the Square of which the Arches are Diagonals.

BRAZING, the Soldering or Joining two Pieces of Iron, by Means of thin Plates of Braſs melted between the two Pieces to be join'd.

If the Work be very fine, as when the two Leaves of broken Saws are to be join'd, it is cover'd with beaten Borax, moiſtened with Water, that it may be incorporated with the Braſs Duſt, which is here added; and the Piece is expoſed to the Fire without touching the Coals, 'till the Braſs be obſerv'd to run.

Laſtly, to braze with a ſtill greater Degree of Delicacy, they uſe a Solder made of Braſs, with a tenth Part of Tin; or another, one Third Braſs, and two Thirds Silver; or Borax and Roſin; obſerving, in all theſe Manners of *Brazing*, that the Pieces be join'd cloſe throughout; the Solder only holding in thoſe Places that touch.

To BREAK IN [in Architecture] is a Term uſed by Carpenters, when they cut, or rather *break* a Hole in Brick-Walls, with a ripping Chizzel.

BREST, a Term in Architecture, uſed, by ſome, to ſignify the ſame Member in a Column, that others call a *Thorus*.

BREST-SUMMERS, in Timber Buildings, are Pieces in the outward Parts of a Building, into which the Girders are framed in all the Floors but the Ground-Floor, then they call it a Cell; and Garret-Floor, then it is call'd a Beam.

As to their Size and Square, it is the ſame according to the Act of Parliament, with that of Girders; which ſee.

It is here to be obſerved, that it is not here meant, all the Pieces which have Girders in them, (and are not in the Garret, or Ground-Floor;) but all ſuch as are in the exterior Part of the Building; whether in the Front, Flanks, or exterior Part of the Building; for the Pieces in the Internal Part of the Building, into which the Girders are fram'd are call'd *Summers*.

Mr. *Leybourn* ſays that the *Breſt-Summers*, in *London*, are meaſur'd by the Foot, running Meaſure; but it is uncertain whether he means only for the Work or Timber, or both.

Com. Comer ſays, that *Breſt-Summers*, in *London*, are valued by the ſolid Foot; if of Oak, 3*s.* *per* Foot; and if of Fir, 2*s.*

BREW-HOUSE. Sir *Henry Wotton*, in his *Elements of Architecture*, ſays, that all Offices which require Heat, as *Brew-Houſes*, Bake-Houſes, Waſh-Houſes, Kitchens, and the like, ought to be placed in the South Part of the Building, if the Poſition of the Houſe, in reſpect to the High-Street, or the like, will admit of it; for it would be but an odd Contrivance, if a Houſe ſtood on the North-Side of a High-Street, to place all the Offices in the Front of it; and it would be very ridiculous, to paſs through a Bake-Houſe, *Brew-Houſe*, or Waſh-Houſe, into Rooms of Entertainment in a Nobleman's or Gentleman's Houſe.

BRICKS, are a fat, reddiſh Earth, form'd into long Squares, four Inches broad, and eight or nine long, by Means of a Wooden Mould, and then bak'd or burnt in a Kiln, to ſerve for the Uſes of Building.

Bricks are of a very antient Standing, as appears from Sacred Hiſtory, the Tower of *Babel* being built with them; and, as ſome ſay, the Remains thereof, are ſtill in Being.

In the Times of the firſt Kings of *Rome*, they built with maſſive ſquare Stones; which they learn'd from the *Tuſcans* Towards the latter End of the Republick, they began to uſe Brick, having borrow'd the Practice from the *Greeks*. And the greateſt, as well as moſt durable Edifices, of the ſucceeding Emperors, as the *Pantheon*, &c. were built with *Brick*.

In the Time of *Gallienus*, the Buildings were compoſed alternately, of an Order of *Brick*, and an Order of *Tofus*, a ſort of ſoft gritty Stone.

After his Time, they laid aſide the Uſe of *Bricks*, and reſum'd Flints.

In the Eaſt, they bak'd their *Bricks* in the Sun. The *Romans* us'd them unburnt, only leaving them to dry for three, four, or five Years in the Air.

The *Greeks* principally uſed three Kinds of *Bricks*. The firſt were call'd Didoron, Διδωρον, i. e. of two Palms. The ſecond, Tetradoron, Τετραδωρον, i. e. of four Palms. And the third, Pentadoron, Πεντάδωρον, i. e. of five Palms. They had alſo other *Bricks* juſt half of theſe, to render their Works more ſolid, and alſo more agreeable to the Sight, by the Diverſities of the Figures and Sizes of the *Bricks*.

Of the Matter whereof Bricks *are made*. *Pliny* ſays, if you would have good *Bricks*, they muſt not be made of any Earth that is full of Sand or Gravel, nor of ſuch as is gritty or ſtony, but of a greyiſh Marl, or whitiſh Chalky Clay, or at leaſt, of a reddiſh Earth: But in caſe there is a Neceſſity to uſe that which is ſandy, that is to be made choice of, which is tough and ſtrong.

He alſo adds that the beſt Seaſon for making *Bricks*, is in the Spring; becauſe they will be ſubject to crack and be full of Chinks, if made in the Summer. He directs, that the Loam of which the Bricks are made, be well ſteep'd or ſoak'd, and wrought with Water.

Mr.

Mr. *Inco* in *Mech. Exer.* ſays, that the Bricks made of the whitiſh Chalky Sort of Earth, and the reddiſh are the beſt.

At *Lunenburgh* in *Saxony*, their *Bricks* are made of a fat Earth, full of Allum.

At *Patana*, in *Aſia*, they make good *Bricks* of a Pumice Sort of Earth, which being dry'd, will ſwim in Water, and not ſink.

But here in *England*, they are for the moſt part made of a yellowiſh-colour'd fat Earth, ſomewhat reddiſh, vulgarly called Loam.

As for thoſe *Bricks* made in *England*, they ſhould not be of ſandy Earth, which will make them both heavy and brittle; nor muſt the Loam be too fat, which will make them crack in drying; they ſhould alſo be made either in the Spring or Autumn.

Mr. *Leyburn* ſays, that the Earth for *Bricks* ought to be digged before Winter; but not made into *Bricks* 'till the Spring Seaſon.

When *Bricks* have been made, they ſhould be ſheltered from the Sun, if it be too hot, but yet muſt be expoſed to the Air to dry. If they be made in froſty Weather, they muſt be covered with Sand, and in hot Weather, with wet Straw.

Of their Kinds and Appellations. *Bricks*, among us, are various, acquiring various Forms, Dimenſions, Uſes, Method of making, Place where, *&c.* Thoſe from their Form, are *Compaſs-Bricks*, of a circular Form, uſed in ſteyning of Walls. 2. *Concave*, or *Hollow Bricks*, on one Side flat, like a common Brick, on the other, hollow'd. They are the Work at the Pilaſters of Fence-Walls which are built of *Great Bricks*. *Coping-Bricks* are form'd on Purpoſe for Coping-Walls; *Paving-Bricks*, or Tiles, are of ſeveral Sizes in ſeveral Counties and Places.

7. Thoſe from Accident, are *Clinkers*, *Samel* or *Sandal*. *Clinkers* are ſuch *Bricks*, as are glazed by the Heat of the Fire, in making. *Samel* or *Sandal-Bricks*, are ſuch as lie outmoſt in a Kiln or Clamp, and conſequently are ſoft and uſeleſs, as not being thoroughly burnt.

Of all which, I ſhall treat in their Order.

1. *Compaſs-Bricks*, are, as has been ſaid, of a circular Form; and their Uſe is for ſteyning of Walls; which is perform'd in the Manner following:

A good Bed of Clay is firſt laid for the Bottom, and then it is pav'd with *common* or *Statute-Bricks*, only laid down, and well ſettled on it; which being done, the Compaſs-Work is begun with *Compaſs-Bricks*, and as the Courſes are carried up, they ram Clay in behind them, (Room being left behind for that Purpoſe,) which cauſes all the *Bricks* to pen tight and cloſe together.

An experienced Wotkman ſays, he has done this Sort of Work, where the Walls have gone but a little Depth in the Ground, and in a looſe and open Mould, where the Water has been brought in by *Concave-Bricks*; which held very well for thirty Years.

As to the Price of theſe Bricks, it is not certain; but they are not much dearer than *common* or *Statute Bricks*; but the Perſon who has them made for his Uſe, is uſually at the Charge of a Mold, made according to the Circumference of his Wall.

2. *Concave* or *Hollow-Bricks*. Theſe are like *common* or *Statute-Bricks* on one Side, but have a Concavity or Hollowneſs, on ther, which is ſemicircular.

This Hollowneſs is about three Quarters of an Inch deep, and an Inch and half broad, ſo that two of theſe Bricks being placed with their Hollows together, they are like a Pipe of an Inch and half Bore. Theſe *Bricks* are uſually about twelve Inches in Length, four and a half in Breadth, and two and a half in Thickneſs.

As to the Manner of laying them in the Ground, it is uſually done in Clay: But it ſhould be carefully minded, that no Trees, Buſhes, or Brambles, be ſuffer'd to grow over where theſe *Bricks* are laid for the Conveyance of Water, nor yet very near them; becauſe their Roots are apt to get in betwixt the Joints of the Bricks, and there to ſpread themſelves with Fibrous Roots, which meeting together like a Ball of Hair in the Concavity, will in Time ſtop up the Paſſage, and hinder the Currency of the Water. Which Inconvenience could it infallibly be prevented, it would be the cheapeſt Way of conveying Water to a Houſe; for *Bricks* to the Value of ſeven or eight Shillings, will do about ſix Rods; and ſuppoſing the Workmanſhip in digging the Trench, laying the *Bricks*, Charge of the Clay, and ramming up again, to be as much

much more, one Rod would cost but two Shillings and two Pence, or two Shillings and eight Pence; and would not be one sixth Part of the Charge of Leaden Pipes, and altogether as serviceable, if not more; because they would last (as may be said) for ever: And preferrable to Lead, if (as we may suppose) the Frost would not hurt these; whereas it frequently bursts Leaden Pipes. For though the Water should be frozen up in them, and we may reasonably suppose that the Ice would then, by expanding itself, open the Joints of the *Bricks*; yet it is as reasonable to suppose, that after the Frost is gone, they will come together in their due Places, by the natural Gravity of the Earth; for then there will be no solid Body betwixt the Joints to hinder the *Bricks* from closing again.

And as to *Alder Pipes*, altho' they be much cheaper than *Lead*, these of *Bricks* will not come to much above (if any thing at all) half the Price of Alder Pipes.

As to the Price of these concave Bricks: They have been sold in *Kent* for four Shillings *per* Hundred, and in *Sussex* for three Shillings; two Hundred of these *Bricks*, being a Foot in Length, will lay six Rods.

3. *Cogging Bricks*, are a sort of *Bricks* in Use in some Parts of *Sussex*, for making their Toothing or Indented Work under the Coping of Walls, which are built of great *Bricks*.

They are in Length about ten Inches, in Breadth four, and in Thickness two and a half, and are usually sold at the Price of common *Bricks*.

They muſt be laid in Sand.

They are alſo uſed in making Fats and Ciſterns for Soap-Boilers.

Of theſe *Bricks*, which are ſix Inches and a quarter long, and two and a half broad, allowing a quarter of an Inch for the Joints, 72 will pave a Yard ſquare; but if they be ſet on Edge, it will take up about 113 to pave a Yard ſquare.

But of the other Size, of ſix Inches in Length, three in Breadth, and one in Thickneſs, ſixty-three being laid flat, will pave a Yard ſquare; but being ſet edge-ways, it will require one Hundred and ſixty-five to do the ſame.

Theſe *Bricks* are commonly ſold at *London* at two Shillings *per* Hundred.

6. *Clinkers*, are ſuch *Bricks* as have much Nitre or Salt-Petre in them; which, by reaſon of the Violence of the Fire, runs and glazes them.

7. *Didoron*, a ſort of *Bricks* uſed by the Ancients; in Length two Spans, or a Foot and half, [the Word δωρον, *Doron*, being *Greek* for a Span, or the Space between the Top of the Thumb and little Finger extended,] and a Foot in Breadth.

Theſe were the ſmalleſt ſort of *Bricks* uſed by the *Greeks* in their private Buildings; but for their publick Buildings, they had two larger Sizes, as you will find hereafter, called *Tetradoron*, and *Pentadoron*.

8. *Feather-edg'd Bricks*, are a ſort of *Bricks* formerly uſed in *Kent* and *Suſſex*; they are of the ſame Size as common *Bricks*, but are made thinner at one Edge than they are at the other, on purpoſe to pen up their Brick Pannels (as they call them) in Timber Buildings, and were commonly ſold amongſt the Statute *Bricks* for that Purpoſe.

9. *Great Bricks*, are a ſort of *Bricks* that are twelve Inches long, ſix Inches broad, and three Inches thick.

The Weight of one of theſe *Bricks* has been found to be fifteen Pound; ſo that one Hundred will weigh one thouſand five hundred Pounds, and one Thouſand 15000 Pounds; which amounts to ſix Tons, thirteen Hundred, three Quarters, and twenty Pounds; and ſo one Hundred and fifty will be a Ton Weight.

Theſe *Bricks* are uſed in building Fence-Walls, together with *Pilaſter* or *Buttreſs Bricks*, and *Coping Bricks*. Theſe Walls are no more than ſix Inches thick, except at the *Pilaſters*, where they are twelve; and it is uſual to ſet a Pilaſter at every ten Feet.

There is a Wall, about nine Feet high, built with theſe ſort of *Bricks*, that ſtands very well, that has been built near thirty Years.

Theſe Walls are reckon'd by ſome to be much cheaper than thoſe of a *Brick* and half, or fourteen Inches of *Statute Bricks*. See *Walls*.

Theſe *Bricks* are ſold by the Thouſand at forty Shillings the Thouſand, or four Shillings the Hundred.

10. *Paving Bricks:* Of theſe there are various Sizes, according to the Fancy of the Workman, and the Cuſtom of the

Places.

Places. Mr. *Leybourn* ſays, they are ſix, eight, ten, and twelve Inches ſquare, and are ſold from ſix to twenty Shillings *per* Hundred. Some call them *Tiling-Bricks*.

Note, That		*Bricks* of		Inches ſquare, will pave a Yard ſquare.
	9		12	
	13		10	
	21		8	
	36		6	

Paving-Bricks, are made in *Surry*, and ſeveral Counties in *England*, of three ſeveral Largeneſſes, *viz.* twelve Inches ſquare, and an Inch and half thick; ten Inches ſquare, and an Inch and quarter thick; and eight Inches ſquare, and one Inch thick: Either of which Sorts being poliſhed or rubbed with ſharp Sand on the Surface, and well-joined, and the Sides rendred equal, by hacking them with a Brick-Ax, and rubbing them with a Rubbing-Stone with ſharp Sand, makes an excellent Pavement, very handſome to the Sight, eſpecially when laid Arras-ways.

There have been made in *Suſſex Paving-Bricks* ſix Inches and a half ſquare, and an Inch and ſeven Eighths thick; two of which have weighed 11 *lb.* ſo that a hundred of them would weigh 550 *lb.* and a thouſand, 5500 *lb.* and conſequently, four hundred and ſeven of them would weigh a Ton.

Some have been made in *Suſſex* nine Inches ſquare, which were uſually ſold for 8 *s.* *per* Hundred.

An experienced Brick-Maker ſays, that he had made *Paving-Bricks* of Clay, fifteen Inches ſquare, but found much Trouble to prevent their warping.

Theſe *Bricks*, when burnt, were of a pale red Colour; as were alſo ſome which he made ſix Inches ſquare of another ſort of Clay, ſome Miles diſtant from the former.

He likewiſe ſays, that *Paving Bricks* made of Loam, are reddeſt in Colour, when burnt; but ought to be made of better Earth than *common Bricks*, though they ſeldom are by thoſe who make them for Sale.

He adds, that beſide the Goodneſs of the Earth, in *Paving-Bricks*, there ought to be a great deal of Care taken in the drying them, to hinder them from warping, and that when they have been dry'd, in dreſſing them ſmooth and ſtraight, eſpecially on the uppermoſt Surface, and alſo in paring the Edges ſtraight, and a little under, making an acute Angle with the upper Side, and in ſeeing that they be exactly ſquare before they are put into the Kiln to be burnt.

The common Price of nine or ten Inch *Paving-Bricks* is from 8 *s.* to 12 *s.* a Hundred in the Country.

Thoſe of ten Inches have been brought by Water from *Surrey*, to Seaport Towns in *Kent* and *Suſſex*, and ſold for 10 *s.* a Hundred.

11. *Pentadoron-Bricks*, are a kind of *Bricks* anciently in Use among the *Greeks*, being three Foot nine Inches long, and one Foot broad, with which they built their publick Edifices.

12. *Place-Bricks*, are all Sorts of *Bricks*, made after the following Method, from whence they derive their Name:

Workmen, say they, are forced to use more than one Method of making *Bricks*; not purely for the sake of Fancy, but out of pure Necessity: The Reason of which proceeds from certain different Qualities inherent in different *Earths*.

Place-Bricks and *Stock-Bricks*, are the two Kinds that take their Names from the Method of their making.

Place-Bricks are generally made in the Eastern Part of *Sussex*, so called, because there is a Place hard by, where they strike (or mold) their *Bricks*, which is a level smooth Piece of Ground, prepar'd for the Bearer-off, (i. e. him who carries the *Bricks* from the Striker,) to lay them down singly in Rows (which are by them called Ricks,) as soon as they are molded; where they are let lie 'till they are a little dry'd, *viz.* 'till they are stiff enough to be turn'd on their Edges, and dress'd, (i. e. 'till their Inequalities are cut off,) and when they are dry'd, they carry them to the Hacks, (or Places where they are row'd up like a Wall of two *Bricks* thick, with some small Intervals betwixt them, to let in the Wind and Air to dry them.) When the Hack is filled, they cover them with Straw on the Top, till they are dry enough to be carried to the Kiln, to be burnt.

13. *Pilaster*, or *Buttress-Bricks*, are made of the same Length, Breadth, and Thickness, with the *Great Bricks*, six and nine. The only Thing they differ from them, is this; they leave a Notch at one End, which is half the Breadth of the *Brick*, and made of the same Mold with the *Great Bricks*, only in making *Pilaster-Bricks*, they put a Cube of Wood of three Inches square into one Corner of the Mold, which Piece makes the Notch in the *Bricks* in the Molding. These *Bricks* are used to bond the Work at the *Pilaster* of *Fence-Walls*, built of *Great Bricks*.

These Pilasters are made of a Foot square, *viz.* a *Brick* in Length, or two *Bricks* in Breadth alternately, throughout the whole Height of the Pilaster: So that the Pilaster stands out three Inches beyond the Surface of the Wall, on each Side.

14. *Samel*, or *Sandal-Bricks*, are such as lie outmost in a Kiln or Clamp, where the Salt-Petre not being digested for Want of Heat, they are very soft, and will soon moulder to Dust.

15. *Stock-Bricks*, differ not from *Place-Bricks* in Form; but their Difference lies in the Quality of the Earth. They are made upon a Stock, *viz.* the Mold is put on a Stock, after the same Manner of molding or striking of Tiles; and when they have molded one *Brick*, they lay it upon a little Piece of Board, somewhat longer than the *Brick*, laying another Piece of Board on that *Brick* like the first, and ano-

G ther

ther *Brick* on that; after this Manner, laying three *Bricks* one upon another, continuing so to strike, and lay them on the Stage, as they do Tiles, till the Stage is full; and then they are carried away by Three, and Three, successively, to the Hacks, and there turn'd down on their Edges; so that there will be the Thickness of a thin Piece of Board, between every *Brick*.

When the Hack has been filled with one Height of *Bricks* from End to End, then they begin to set them upon those which were first laid on the Hack, by which Time, they will be a little dry'd, and will bear the others, being molded of very stiff Earth. By that Time they come to set a Second, or Third, &c. At which Time, they cater them a little, (as they call it,) to prevent their reeling: And when the Hack is of a proper Height, they cover them with Straw, after the same Manner, as they do *Place-Bricks*, till they are dry enough for Burning.

This, they say, being more Trouble than the other Way, *viz.* of making *Place-Bricks*, for making and burning (besides the digging of the Earth,) they have 6*s.* a thousand, which is 1*s.* more than they usually have for making *Place-Bricks*; but they are under a Necessity to make them after this Manner; or else if they were laid abroad in a Place to dry, as the *Place-Bricks* are, the Quality of the Earth is such, that they would burst to Pieces.

There is an Instance of this, which is related by an experienced Brick and Tile-Maker, (who was used to work in *Kent* and *Sussex*,) who being sent for to *Rumford* in *Essex*, to make a hundred thousand of *Bricks*, he unadvisedly, not knowing the Quality of the Earth, having struck about a thousand, they being set down to dry, after the Method of *Place-Bricks*, and lain till about Ten a-Clock, the Sun beginning to shine very hot, the whole thousand of *Bricks* burst to Pieces, so that he was forced to throw them away, and go to work afresh; and thaching them (i. e. covering them) with Straw till the next Morning, and then raking it off, the *Bricks* did very well, when they came to be set on the Hack; and after they had been burnt, were curious red *Bricks*, which would ring, being struck with any hard Thing.

At this Place, they made none but *Stock-Bricks* before he found out the Way of making *Place-Bricks* of this Earth.

16. *Statute*, *Small*, or *Common Bricks*. Their Dimensions, *viz.* of the Mold, according to the Statute, ought to be as follows, *viz.* nine Inches in Length, four and a half in Breadth, and two and a quarter in Thickness within.

Bricks made in a Mold of these Dimensions (the Earth being first well-temper'd) being dry'd and burnt, will be lesser and lighter; yet they shrink but little in Thickness, less in Breadth, and scarce any Thing discernable in Length. As to the Weight of *Bricks*, that is uncertain, there being a great Difference in the Gravities of Earths; yet in common, a single

gle *Brick* will weigh about five Pounds, and contain ninety Cubick Inches; and from ſome Molds, a hundred.

Four *Bricks* being meaſured, and weighed, each being nine Inches long, four and a quarter broad, and two and one Third thick, weigh'd twenty-two Pounds; ſo that a ſingle *Brick* weigh'd five Pounds and a half, a hundred of which, at that Rate, would weigh five hundred and fifty Pounds; and a thouſand, five thouſand five hundred Pounds; and about four hundred and ſeven would be a Ton weight.

Theſe were *Suſſex Bricks*, of which they uſually reckon five hundred to the Load; which Number of *Bricks*, according to this Proportion, will weigh about twenty-four hundred and a half.

Theſe *Bricks* are often uſed in paving Cellars, Waſh-Houſes, Sinks, and Fire-Hearths, and the like; thirty of which made, according to the Statute, will pave a Yard ſquare, and three hundred and thirty of them, a Square of an hundred Foot, being laid the flat Way, and not ſet edge-ways; for then it will require near as many more.

It has been found by Obſervation, that thirty-two *Bricks* laid flat, will pave a Yard ſquare; and ſixty-four ſet on Edge will do the ſame.

It is alſo found by Experience, that four thouſand ſix hundred *Statute-Bricks* will be required to make a ſuperficial Statute-Rod of Brick-Work of a *Brick* and half thick, and conſequently ſeventeen hundred to the Square, and an hundred and fifty-five to the ſuperficial Yard, on a Wall of a Brick and half thick.

As to the Price of theſe Statute or Common Bricks, it is various, according to the different Places; for they have different Prices in different Parts of the Kingdom; which is not all neither, for *Bricks* in the ſame Kiln, ſhall have a different Price, according to the Diſtance the Maſter of the Brick-Work is to ſend them; and alſo ſome Conſideration is to be had of the Price of Fuel, and Workmen's Wages.

Mr. *Leybourn* ſays, he never knew them cheaper than 9*s.* nor dearer than 18*s.* delivered in any Part of *London*.

Statute or *Common Bricks*, have been ſold in ſome Parts of *Suſſex* and *Kent*, for 16*s.* a Thouſand, laid down within two Miles diſtant from the Kiln; and at other Times, they have been ſold at 20*s.* a Thouſand.

At another Place in *Suſſex*, they are ſold at 25*s.* a Thouſand, if laid down within two or three Miles diſtant from the Kiln: Whereas within twelve or fifteen Years they have been ſold for 20*s.* a Thouſand. But ſince the Beginning of the late Wars, the Iron-Works in that Part, have conſum'd ſo great a Quantity of their Wood, that Fuel of late Years is grown a fourth Part, or more, dearer than it uſed to be; for which Reaſon, they have ſince raiſed their *Bricks* to 25*s.* a Thouſand.

Mr. *Wing* ſays, that in *Rutland*, *Bricks* are but 12*s.* a Thouſand at the Kiln.

[As to the Price of making *Statute-Bricks*,] the common Price in in the Country, is 6*d* a Thousand for the Molder, 4*d.* for the Bearer-off, and 4*d.* for the Digger and Temperer of the Earth, fit for Use; and the Digger of the Earth for making it ready, after it is digged, the Digging not being reckon'd into the Making, Molding, Bearing off, *&c.* and Burning, the usual Price is 5*s.* a Thousand.

Mr. *Leyburn* informs us, that about *London* they allow the Molder 4*d.* 5*d.* or 6 *d.* a Thousand; and that *Bricks* made at Home, will stand the Maker of them in (besides the Value of the Earth,) betwixt 5 and 6*s. per* Thousand.

But it will be more in some Parts of *Kent* and *Sussex*.

17. *Tetradoron*, a sort of ancient *Grecian Bricks*, which were three Feet or four Spans in Length, and one Foot in Breadth, being one of their larger Size, with which they built their publick Buildings.

Triangular Bricks. Daniel Barbaro, Patriarch of *Aquilia*, in his Comment on *Vitruvius*, recommends another Form of *Bricks*, *viz.* Triangular ones, every Side a Foot long, and some an Inch and half thick: These, he observes, would have many Conveniences above the rest. As, First, being more commodious in the Management. Secondly, of less Expence. And Thirdly, of fairer Shew; adding much Beauty and Strength to the Mural Angles, when they fall gracefully into an Indented Work.

Sir *Henry Wotton* wonders they have never been brought into Use, being recommended by so great an Authority.

[The Method of burning *Bricks.*] The *Bricks* or Kiln being set, and cover'd with Pieces of *Bricks*, they first put in some Cord or great Wood to dry them with a gentle Heat or Fire; and this they continue till the *Bricks* are pretty dry, which is known by the Smoak's turning from a whitish, darkish Colour, to a black transparent Smoke; they then leave off putting in Wood, but proceed to make ready for burning, which is perform'd by putting in Bush, Furze, Spray, Heath, Brake, or Fern Faggots; but before they put in any Faggots, they dam up the Mouth, or Mouths, of the Kiln with Pieces of *Bricks*, (which they call Shinlog,) piled up one upon another, and close it up with wet Brick-Earth instead of Mortar.

This Shinlog they make so high, that there is but just Room above it, to thrust in a Faggot betwixt one Foor and a half, and two Foot; for the whole Height of the Mouth is but three Foot.

The Mouth being thus shinlogg'd, they then proceed to put in more Faggots, till the Kiln and its Arches look white, and the Fire appear on the Top of the Kiln, and the Kiln and Ashes below begin to change from a white to a greyish Colour; then they slacken the Fire for some Time, *viz.* for half an Hour or an Hour, that the Fire or Heat may ascend to the Top of the Kiln, by the Motion of the Air in at the Mouth; and also that the lower Ware may settle and cool, and not be burnt more than that above it.

Thus

Thus they continue to do, heating and ſlackening alternately, till all the Ware is thoroughly burnt; which will be commonly in about forty-eight Hours.

As to the cooling of Kilns of Ware, ſome unskilful Burners do, as ſoon as the Ware is burnt, immediately ſtop up the reſt of the Mouth of the Kiln, which was left open above the Mouth of the Shinlog, which cauſes it to be long a cooling; by which Means a Kiln will be ordinarily a Fortnight or three Weeks in ſetting, burning, cooling, and drawing: Whereas an experienced Burner has affirmed, he has ſet, burnt, cool'd, and drawn a Kiln a Week, for ſeveral Weeks ſucceſſively one after another; but then he never ſtopped up the reſt of the Kiln's Mouths above the Shinlog, but left it open, for the Air to go in and cool the Kiln of *Bricks*.

He adds alſo, that ſix hundred of Faggots will burn a Kiln of ten or eleven thouſand *Statute-Bricks*. And Mr. *Wing* ſays, that a Chaldron of Coals will burn four thouſand two hundred *Bricks*.

By the foregoing Method, a Kiln of *Bricks* may be burnt ſo equally, that thoſe on the Top ſhall be burnt as hard as thoſe at the Bottom: So that an expert Burner affirms, he has burnt ſeveral Kilns of Tiles and *Bricks* together, about three thouſand *Bricks*, and ten or eleven thouſand of Tiles, and has not had above fifty waſte, broken and Sandal Tiles in all: Whereas ſuch Brick-Burners as continue their Fire without any Intermiſſion, render their lower *Bricks* extreme hard, and thoſe on the Top, *Samel-Bricks*, or Tiles: Nay, what is worſe, they cauſe the lower ones to run ſo, by exceſſive Heat, that they are almoſt united in one entire Body; ſo that they are forced to get them out with Wringers, (or Iron Bars,) and each Bolt of Tiles ſhall be one entire Maſs.

About *London*, they burn their *Bricks* in Clamps built of the *Bricks* themſelves, after the Manner of Arches in Kilns, with a Vacancy between each *Brick's* Breadth, for the Fire to play through; but with this Difference, that inſtead of arching, they ſpan it over, by making the *Brick's* project one over another on both Sides the Place for the Wood and Coals to lie in, till they meet, and are bounded by the *Bricks* at the Top; which cloſe all up, projecting over inwards, till they meet in the Middle; which they will do in about three or four Courſes of *Bricks* in Height.

The Place for the Fuel is carried up ſtraight on both Sides, or, which is the ſame Thing, upright on both Sides, till it is about three Foot high; then they fill it almoſt with Wood, and lay a Covering of Sea-Coal over that; and when that is done, they over-ſpan the Arch: But they ſtrew Sea-Coal alſo over the Clamp, betwixt all the Rows of *Bricks*; for they are not laid contingent in their vertical Rows; and one Courſe of *Bricks* is laid one one Way, and another, another; ſo that there are ſmall Interſtices through all the

the *Bricks*, for the Coals to be ſtrewed into. When this is done, they fire the Wood, and that fires the Coals, and when all is burnt, they conclude the *Bricks* are burnt enough.

Mr. *Goldman* obſerves, that *Bricks* will have double the Strength, if, after one Burning they be ſteep'd in Water, and burnt afreſh.

If the Earth be to fat, it muſt be temper'd with Sand, and that trod out again, firſt by Cattle, afterwards by Men.

Bricks made of common Earth melt, nay, often vitrify, by too much Heat: For which Reaſon, the Kilns are made of Stones, which will themſelves calcine, that the Vehemence of the Fire may be broken: Beſides which, they uſually place other *Bricks* made of an Argillous Earth, which would melt next the Fire.

What Quantity of Earth will make a thouſand *Bricks*?

Some ſay, that a Load of Loam (a Load being twelve Buſhels) will make about two hundred *Statute-Bricks*; and if ſo, conſequently five Loads will make a thouſand: Alſo that nineteen Load will make ſixteen hundred of *Great Bricks*; and twelve will be ſufficient for a thouſand of the ſame.

Of the Choice of Bricks. Pliny adviſes in chuſing of *Bricks* for building, to procure (if it can be) ſuch *Bricks* as are two Years old at leaſt.

There are generally in all Kilns, or Clamps, *Bricks* of three Degrees in Goodneſs.

The firſt and beſt Sort are ſuch as lie next the Fire, (*viz.* thoſe are beſt for laſting,) which have as it were a Gloſs on them, which proceeds from the Saltpetre which is inherent in them, and which runs, and glazes them, by Means of the Violence of the Fire. Theſe are called Clinkers.

The ſecond, and moſt general Sort for building, are thoſe which lie next in the Kiln, or Clamp, to thoſe Clinkers before mentioned.

The third, and worſt Sort, are thoſe which lie on the Outſides of the Kilns and Clamps, where the Saltpetre in them is not digeſted for want of due Heat. And theſe, when they come to be expoſed to the Weather for ſome Time, will moulder away into Duſt. Theſe are called by Bricklayers, *Samel* or *Sandal-Bricks*.

It is an Obſervation, that whilſt *Bricks* are burning, thoſe on the Inſide of a Clamp are the worſt of all.

Mr. *Worlidge*, in his *Syſtema Agriculturæ*, is for exciting the Brick-Makers to try their Skill in making a Compoſition of Clay and Sand, to form in Molds, Window-Frames for Houſes of different Forms and Sizes; and alſo Chimney-Pieces, and Frames for Doors, &c. in ſeveral Pieces made in Molds, that when they have been burnt, they may be ſet together with a fine Cement, and ſeem to be but one entire Piece; by which Means all manner of Stone-Work now uſed in Building, may be imitated: Which would very well ſupply the Deficiency of Stones, where they are either wanting, or ſcarce and dear; and at the ſame Time, ſave

ſave a great Deal of Timber now uſed in Brick-Buildings, and appear much more compleat and beautiful, and be of greater Strength, and more durable for laſting, than Timber or ordinary *Brick*.

And one would imagine it very practicable, as may be perceiv'd by the Earthen Pipes made fine, thin, and durable, for the Conveyance of Water under Ground at *Portſmouth* in *Hampſhire*, and by the Earthen Backs for Grates and Chimneys, formerly made by Sir *John Winter*, at *Charing-Croſs*. Which are evident and ſufficient Demonſtrations of the Poſſibility of making Work fine, thin, and light, for Tiles of a large Size and Thickneſs, either plain, or curved, and of making larger Work in Molds, and by burning them for Doors, Windows, and Chimney-Frames.

This, ſays he, is one of the moſt feaſible and beneficial Operations that I know to be neglected in *England*.

Another Author ſays, he really thinks much might be done as to making of Chimney-Pieces, Stone-Moldings, and Architrave-Doors and Windows, and Architraves, or Faſcia's for Fronts of Buildings. If Men of this Profeſſion would but apply their Minds to find out ſome good Compoſition of Earth, and a Method of managing it well in molding, burning, &c.

It may be queſtionable, whether a Compoſition of Earth, ſomething like to common Crockers Earth, would not in ſome Meaſure anſwer the Purpoſe, ſince it appears plainly, that what Form ſoever they put their Earth into, the ſame it retains after drying and burning, altho' Crocks, and ſuch like Things, are form'd very thin.

Now ſuppoſe that Chimney-Pieces, or the like, were made in Molds, and afterwards dry'd and burnt, if they were not thought ſmooth enough when they were ſet up, they might be poliſh'd with ſharp Sand and Water, or a Piece of ſharp Stone and Water.

Or if Care were taken of ſuch Things as theſe (which are for Ornament, as well as Uſe,) when they were half dry, or more, in the Air, they might be poliſh'd over with an Inſtrument for the Purpoſe of either Copper, Iron, or ſome hard Body, and then left to dry till they were dry enough for burning. If ſo, 'tis probable, they would not need much poliſhing afterwards.

It is likewiſe as probable, that Ingenious Workmen might make very handſome and beautiful Chimney-Pieces, Stone-Mouldings for Doors, &c. fit for Noblemens Houſes, and all others who would be at the Charge.

Theſe might be glaz'd, as Potters do their fine Earthern Ware, or elſe vein'd in Imitation of Marble, or be painted and anneal'd with Figures of various Colours, either Hiſtory, Perſpective, or the like; which would be much cheaper, if not alſo as durable as Marble itſelf.

It is not, ſays a certain Author, the Want of Materials, but Want of Skill, and Diligence in managing them, that makes our *Engliſh* Buildings in the leaſt Meaſure inferior to any of thoſe in foreign Countries.

A certain *Engliſh* Ambaſſador made this Obſervation, That we ought not to be diſcouraged with our ignoble Materials for Building, which we uſe in *England*, in Compariſon with the Marbles of *Aſia* and *Numidia:* For, ſays he, I have often view'd with much Pleaſure, at *Venice*, an Anti-Porch after the *Greek* Manner, erected by *Andreas Palladio*, upon eight Columns of the *Roman* Order, the Backs of Stone without Pedeſtals, the Shafts or Bodies of mere *Brick* three Foot and a half in Diameter below, and of Conſequence, thirty-five Foot high, than which, he ſaith, his Eyes never beheld any Columns more ſtately of Stone, or Marble. The *Bricks* being firſt form'd in a circular Mold were cut before they were burnt into four Quarters, or more Parts, and afterwards, jointed ſo cloſely and nicely in laying, and the Points ſo exactly concentred, that the Pillars appeared to be of one entire Piece.

[Things worthy to be obſerved in buying and laying of *Bricks.*]

I. As to buying. The ſeventh Number will be a ſufficient Direction to any Workman, (who does not underſtand it.) to chuſe good *Bricks*. And in the 16th Section of *Bricks*, *viz.* under the Head of *Statute-Bricks*, there are Directions as to the Number of *Bricks*, that will make a Square or Rod of Work; though 'tis impoſſible to be exactly certain to a very few; becauſe, Firſt, the Workman's Hand may vary in laying the Mortar. Secondly, many *Bricks* may warp in burning; and the Seller will bring you ſome ſuch. Thirdly, Some will be broken and ſpoiled in the Carriage. Fourthly, you will often find the Tale deficient, if you be not extraordinary careful.

And beſides this, when *Bricks* are dear, and Lime cheap, and you put your Work out by the Great, or by Meaſure, and the Workman is to find Materials, he, except he be well looked after, will uſe the more Mortar, and the fewer *Bricks*, making large Joints, which is a Defect in any Building.

II. As to laying *Bricks*, which is a Thing of no ſmall Conſequence in any Building, in order to the Well-working and Bonding of Brick-Work (or as it is called by ſome Workmen, Breaking of Joint,) conduces very much to its Strength. It will not be therefore improper to add ſome particular Directions concerning it, which have been recommended by experienced Workmen.

1. Take Care to procure good ſtrong Mortar. See MORTAR.

2. If your *Bricks* are laid in Winter, let them be kept and laid as dry as poſſible. If they are laid in Summer-Time, it will quit Coſt, to employ Boys to wet them; becauſe being wetted, they will unite much better with the Mortar, than if they were laid dry, and will render the Work much ſtronger.

But

But if it ſhall be objected, that if the Building be large, it will be a great deal of Trouble to wet all the *Bricks*, by dipping them in Water; and alſo that it will make the Workmens Fingers ſore in laying them.

To prevent theſe Inconveniences, Water may be thrown on each Courſe of *Bricks* after they have been laid; as is ſaid to have been done by the Order of the Ingenious Mr. *Robert Hook*, the Surveyor at the Building of the *Phyſician's-College* in *Warwick-Lane*.

3. If *Bricks* are laid in the Summer-Time, don't fail to cover them, to prevent their drying too faſt; for if the Mortar dry too haſtily, it doth not cement ſo firmly to the *Bricks* as when it dries leiſurely.

4. If the *Bricks* are laid in the Winter-Time, take care to cover them well, to defend them from Rain, Snow and Froſt; the laſt of which is a mortal Enemy to all Mortar, eſpecially to all ſuch as has taken Wet juſt before the Froſt ſeizes it.

5. Take Care that *Bricks* be not laid Joint on Joint on the Middle of Walls, but as ſeldom as may be; but let there be good Bond made there, as well as on the Outſides: For ſome Workmen, in working a Brick and half Wall, lay the Header on one Side of the Wall, perpendicular on the other Side of the Wall; and ſo all along through; which, indeed, neceſſarily follows, from the unadviſed Setting up of the Quoin at a Toothing; for it is common to tooth in the Stretching-Courſe two Inches with the Stretcher only; and the Header on the other Side to be perpendicular over the Header on this Side; which cauſes the Headers to lie Joint in Joint in the Middle of the Work.

Whereas if the Header on one Side of the Wall, were tooth'd as much as the Stretcher on the other Side, it would be a ſtronger Toothing, and the Joints of the Headers on one Side, would be in the Middle of the Headers of the Courſe they lie upon on the other Side.

All that can be pretended in Excuſe of this ill Practice in working thus, is this: That the Header will not hang two Inches over the *Bricks* underneath it.

This, indeed, is an Objection: But yet the Inconveniency may be avoided without much Difficulty, *viz.* as follows: By having a Piece of Wood of the Thickneſs of a Courſe of *Bricks*, and two Inches broad, and laying it on the laſt Toothing-Courſe, to bear it, or a Brick-Bat, put upon the laſt Toothing, will bear it till the next Quoin is ſet upon it, and then the Bat may be taken away.

6. The ſame Inconvenience, at an upright Quoin in a Brick and half Wall; where it is uſual to lay a Cloſer next the Header, on both Sides of the Walls; and in ſo doing, 'tis Joint in Joint all the Length of the Wall, except by Chance, a Three-quarter Bat happen to be laid.

In order to avoid this Inconveniency, and by that Means to make the Wall much firmer, lay a Cloſer on one Side; but lay a Three-quarter Bat on the Stretching-Courſe, and join a Header

next

next to the Header, at the Quoin in the Heading-Courſe.

7. Alſo in two Brick-Walls, it will be the beſt Way in Stretching-Courſes, in which Stretching is laid on both Sides the Walls next the Line, to lay alſo Stretching in the Middle of the Wall, and Cloſers next to each Stretching-Courſe which lies next the Line.

[What Number of *Bricks* may be laid in a Day.] A Bricklayer and his Labourer (all their Materials being ready) will lay in Day about a thouſand *Bricks*, in whole Work, on a ſolid Plain; and ſome dextrous Bricklayers will lay twelve, and ſome fifteen hundred.

[Of Facing Timber-Buildings with *Bricks*.] This may be more properly called Caſing; it being covered all over on the Outſide with *Bricks*, ſo that no Timber is to be ſeen. The Manner of performing it, is as follows; *viz.* All betwixt the Timber and the Wall is a *Brick's* Length thick, (or Nine-Inch Brick-Wall,) but againſt the Timber, the Wall is but four Inches and a half, or a half *Brick* thick, beſide the Timber.

But experienced Workmen do not approve of this Method; becauſe the Mortar does ſo much corrode and decay the Timber.

An experienced Bricklayer ſays, that in pulling down Work at *Eridge-Place*, (which is one of the Lord *Abergavenny's* Country Seats,) the Timber was extreamly corroded and eaten by the Mortar.

BRICKLAYERS WORK. Whar.] In the City of *London*, *&c.* it conſiſts of ſeveral Kinds, *viz.* Walling, Tiling, Chimney-Work, and Paving with Bricks and Tiles.

But in the Country, tis common for the *Bricklayers* Trade to comprehend thoſe of the Maſon and Plaiſterer alſo: But I ſhall here conſider it only as to the particular Branches of Walling, Tiling, Chimney-Work, Paving, *&c.*

[As to writing a *Bricklayer's* Bill.] A *Bricklayer's* Bill may be made after the Method following:

Mr.

Mr. William Blakewey's *Bill of Materials had of, and Work done, by* Thomas Halling, *Bricklayer,* June 5. 1732.

	l.	*s.*	*d.*
For eight thousand of Bricks, at 12 *s. per* M.	4	16	0
For four thousand of Tiles, at 20 *s. per* M.	4	00	0
For fifteen Hundred of Lime, at 12 *s. per* C.	9	00	0
For Fourteen Load of Sand, at 2 *s.* 6 *d. per* L.	1	15	0
For five Hundred of Nine-Inch Paving-Tiles, at 11 *s. per* Hundred	2	15	0
For thirty Ridge-Tiles, at 1 *d.* ¾ *per* Piece	0	04	4 ½
For three Weeks and two Days Work, for myself, at 3 *s. per Diem*	3	00	0
For twenty-five Days Work and a half for my Man, 2 *s.* 6 *d. per Diem*	3	03	9
For a Labourer, twenty-five Days and a half, at 1 *s.* 8 *d. per Diem*	1	18	0
Their Sum Total is	30	12	5 ½

But if Bricklayers do not work by the Day, then they write their Bills after another Manner: For then they either undertake the Work by the Great, *viz.* to do all, and to find all the Materials belonging to *Bricklayer's-Work*; or else they are to do it by Measure, and to do all the Work, and to find all the Materials, at such a Price, by the Rod, for Walling; by the Square, for Tiling; and by the Yard, for Paving, *&c.*

But if the *Bricklayer* does not find any Materials, he may then work by Measure; and in this Case, his Bill may be made after the following Manner, *viz.* for so many Rods of Walling, at so much *per* Rod, *&c.* according as he has made his Agreement.

Sometimes Chimney-Work is agreed for with the *Bricklayer*, by the Hearth; either only to find the Workmanship, or that and Materials too; and in this Case, the Bill is made according to the Agreement.

There are likewise other Things which come into a *Bricklayer's* Bill *viz.* all kind of Ornamental Work in Brick; which is usually set down and rated at so much *per* Foot, or at so much *per* Piece. Or there may be a Sum of Money allow'd over and above the Price or Value of the Rod-Work; and then the Ornamental Work will be included in it.

You are to understand by ornamental Work, Arches, either Straight, or Circular, over Windows

dows or Doors; Fascia's, either with, or without Moldings; Architraves, Round Windows, or Rubb'd Returns, Friezes, Cornices of all Sorts, Water-Tables wrought, and Water-Courses: All which are valued by the Foot, running Measure.

To these may be added Base-Mouldings, and Plinths, and the the Splaying of the Jaumbs of Windows and Doors on the Inside of Buildings.

Also Pilasters, Peers, Pediments, Grotto's, and Rustick Quoins.

These five last mention'd are valu'd at so much *per* Piece, according to the Largeness and Goodness of the Work and Materials. And thus all ornamental Work ought to be valu'd.

By ornamental Work, in *Bricklayer's Work*, is to be understood, all kind of Brick-Work hew'd with an Ax, or rubb'd on a Rubbing-Stone, or of Stone wrought with Chissels, or rubbed with Stones or Cards: All such is ornamental Work, and ought to be paid for besides the Rod-Work.

BRICK-WORK, as to the measuring, &c.] 1. Sometimes Brick-Walls are wrought Part of the Way two Inches thinner than the rest of the Work, which two Inches serve for a Water-Table to the Wall, which is commonly set off about two Foot above the Ground; and therefore the *Brick-Work* may be measured at the same Thickness, which is above the Water-Table; and then the Two-Inch Work may be measured, as follows:

After the Dimensions of the Wall have been taken, (from the Bottom to the Height, that it is to be taken at two Bricks,) then add twenty Feet in Length, by the Height of the Two-Inch Work, *viz.* from the Bottom or Setting-off of the Water-Table, the Half which is so much Four-Inch Work; and afterwards reduce that to a Brick and half Work.

2. As to the measuring of Gable-Ends, in *Brick-Work*, that is to be done after the same Manner that Carpenters measure their Gables, (saving that this is reduced to Rod-Work.) See GABLE-END.

3. Be sure to take Notice in taking Dimensions of Walls which join to an Angle, that the Length of one Wall be taken on the Outside of the Angle, and the other's Length to the Inside of the Angle.

4. If there is a Gable-End to measure, and the Breadth of the House be given, (or known which is the Base of the Gable-End, and the Length of the Perpendicular is required,) Measurers have a short Way of finding it; which that I may render the plainer, I shall give the following Example:

Suppose the Base of the Gable be twenty-four Foot, and the Length of the Perpendicular be requir'd; take the Length of the Rafter, (which will be eighteen Feet,) and add to it half of itself, which is nine Feet, and it will make twenty-seven Foot, the Half of which is thirteen Feet six Inches, the Length of the Perpendicular.

 But

But though this Way is com-
nly practifed, yet it is not
ct; for it makes the Perpen-
ular a little too much.

This, indeed, is practifed in
ofs that are three quarters
h; but will not be exactly
e in any other.

But there are two other Ways
finding the Perpendicular,
ich are exact: The First is by
portion, *viz.* as 30 to 22. 35.
is the Length of the Rafter to
Perpendicular required; or
tract the Square of one half
the Base, or one half of the
eadth of the House, from the
uare of the Length of the Raf-
, and there will remain a Num-
r, the square Root of which is
Length of the Perpendicu-
.

5. If any Deductions for Doors, Windows, *&c.* are to be taken out in *Brick-Work* of two Bricks, or two Bricks and a half thick, then add one Third to the Lengths of Doors or Windows in two Bricks, or two Thirds to the Length; for those of the two and a half *Brick-Work*, (or it may be two Thirds, or one Third to the Breadth, and not the Length, according as which will be soonest or easiest divided;) and the Lengths and Breadths being afterwards multiply'd one into the other, the Product will be the proper Directions in a Brick and half Work, without any farther Trouble; and neither Master nor Workman, will be wrong'd.

TABLE, *Shewing by Inspection the Price of any Number of odd Feet of* BRICK-WORK, (*or other, perform'd by the Rod,*) *calculated from one Foot to thirty-four Feet,* (*or half a quarter of a Rod,*) *and at any Price from* 1 s. *to* 10 l.

The

Price of the												
Rod	1 s.			2 s. 6 d.			5 s.			10 s.		
$\frac{3}{4}$ Rod	9 d.			1 s. 10 d. 2 q.			3 s. 9 d.			7 s. 6 d.		
$\frac{1}{2}$ Rod	6 d.			1 s. 3 d.			2 s. 6 d.			5 s.		
$\frac{1}{4}$ Rod	3 d.			7 d. 2 q.			1 s. 3 d.			2 s. 6 d.		
$\frac{1}{8}$ Rod	1 d. 2 q.			3 d 3 q.			7 d. 2 q.			1 s. 3 d.		
The Price of any Number of Feet under 34, or half quarter of a Rod.	s.	d.	q.	s.	d.	q	s.	d.	q.	s.	d	q.
1				0	0	0	0	0	0	0	0	1
2				0	0	0	0	0	1	0	0	3
3				0	0	1	0	0	2	0	1	1
4				0	0	1	0	0	3	0	1	1
5				0	0	1	0	0	3	0	1	3
6				0	0	2	0	1	0	0	2	1
7				0	0	2	0	1	1	0	2	3
8				0	0	3	0	1	2	0	3	0
9				0	0	3	0	1	3	0	3	2
10				0	1	0	0	2	0	0	4	0
11				0	1	0	0	2	1	0	4	2
12				0	1	1	0	2	2	0	5	0
13				0	1	1	0	2	2	0	5	1
14				0	1	1	0	2	3	0	5	3
15				0	1	2	0	3	0	0	6	1
16				0	1	2	0	3	1	0	6	3
17				0	1	2	0	3	2	0	7	0
18				0	1	3	0	3	2	0	7	2
19				0	1	3	0	3	3	0	7	0
20				0	2	0	0	4	1	0	8	2
21				0	2	0	0	4	1	0	9	0
22				0	2	1	0	4	2	0	9	1
23				0	2	1	0	4	3	0	9	3
24				0	2	2	0	5	0	0	10	1
25				0	2	2	0	5	1	0	10	3
26				0	2	3	0	5	2	0	11	0
27				0	2	3	0	5	3	0	11	3
28				0	3	0	0	6	0	1	0	0
29				0	3	0	0	6	1	1	0	2
30				0	3	1	0	6	2	1	1	0
31				0	3	1	0	6	3	1	1	2
32				0	3	2	0	7	0	1	1	3
33				0	3	2	0	7	1	1	2	1

Price of the			
Rod	20 *s.*	30 *s.*	40 *s.*	50 *s.*
$\frac{3}{4}$ Rod	15 *s.*	22 *s.* 6 *d.*	30 *s.*	37 *s.* 6 *d.*
$\frac{1}{2}$ Rod	10 *s.*	15 *s.*	20 *s.*	25 *s.*
$\frac{1}{4}$ Rod	5 *s.*	7 *s.* 6 *d.*	10 *s.*	12 *s.* 6 *d.*
$\frac{1}{8}$ Rod	2 *s.*	3 *s.* 9 *d.*	5 *s.*	6 *s.* 3 *d.*

The Price of any Number of Feet under 34, or half quarter of a Rod.

Feet	*s.*	*d.*	*q.*	*s.*	*d.*	*q.*	*s.*	*d.*	*q.*	*s.*	*d.*	*q.*
1	0	0	3	0	1	0	0	1	2	0	1	3
2	0	1	2	0	2	1	0	3	0	0	3	3
3	0	2	1	0	3	2	0	4	2	0	5	3
4	0	2	3	0	4	0	0	5	2	0	6	3
5	0	3	3	0	5	2	0	7	2	0	9	1
6	0	4	2	0	6	3	0	9	0	0	11	1
7	0	5	2	0	8	1	0	11	0	1	1	0
8	0	6	1	0	9	1	1	0	2	1	3	2
9	0	7	1	0	10	3	1	2	2	1	6	0
10	0	8	0	1	0	0	1	4	0	1	8	0
11	0	9	0	1	1	2	1	6	0	1	10	2
12	0	10	0	1	3	0	1	8	0	2	1	0
13	0	10	3	1	4	0	1	9	2	2	2	3
14	0	11	2	1	5	1	1	11	0	2	4	3
15	1	0	2	1	6	3	2	1	0	2	7	1
16	1	1	2	1	8	1	2	3	0	3	9	3
17	1	2	1	1	9	1	2	4	2	2	11	0
18	1	3	1	1	10	3	2	6	2	3	2	
19	1	4	0	1	11	0	2	8	0	3	3	0
20	1	5	0	2	1	2	2	10	0	3	6	2
21	1	6	0	2	3	0	3	0	0	3	9	0
22	1	6	3	2	4	0	3	1	2	3	10	3
23	1	7	2	2	5	1	3	3	0	4	0	3
24	1	8	2	2	6	3	3	5	0	4	3	1
25	1	9	1	2	8	0	3	6	2	4	5	1
26	1	10	1	2	9	1	3	8	2	4	7	2
27	1	11	1	2	11	0	3	10	2	4	10	1
28	2	0	0	3	0	0	4	0	0	5	0	0
29	2	1	0	3	1	2	4	2	0	5	2	2
30	2	1	3	3	2	3	4	3	2	5	4	2
31	2	2	3	3	4	1	4	5	2	5	7	0
32	2	3	2	3	5	1	4	7	0	5	8	3
33	2	4	2	3	6	3	4	9	0	5	11	1

Price of the

Rod	3*l.*	*s.*	*d*	4*l.*	*s.*	*d.*	5*l.*	*s.*	*d.*	10*l.*	*s.*	*d.*	
$\frac{3}{4}$ Rod	2	5	0	3	0	0	3	15	0	7	10	0	
$\frac{1}{2}$ Rod	1	10	0	2	0	0	2	10	0	5	0	0	
$\frac{1}{4}$ Rod	0	15	0	1	0	0	1	5	0	2	10	0	
$\frac{1}{8}$ Rod	0	7	6	0	10	0	0	12	6	1	5	0	

The Price of any Number of Feet under 34, or half a quarter of a Rod.

Feet	*l.*	*s.*	*d.*	*l.*	*s.*	*d.*	*l.*	*s.*	*d.*	*l.*	*s.*	*d.*	*q.*
1	0	2	1	0	2	0	0	3	3	0	0	7	2
2	0	4	2	0	6	0	0	7	2	0	1	3	0
3	0	6	3	0	9	0	0	11	1	0	1	10	2
4	0	8	1	0	11	0	1	1	3	0	2	3	2
5	0	11	1	1	3	0	1	6	3	0	3	1	2
6	1	1	2	1	6	0	1	10	2	0	3	8	0
7	1	4	2	1	10	0	2	3	2	0	4	7	0
8	1	6	3	2	1	0	2	7	1	0	5	2	2
9	1	9	3	2	5	0	3	0	1	0	6	0	2
10	2	0	0	2	8	0	3	4	0	0	6	8	0
11	2	3	0	3	0	0	3	9	0	0	7	6	0
12	2	6	0	3	4	0	4	2	0	0	8	4	0
13	2	8	1	3	7	0	4	5	3	0	8	11	0
14	2	10	2	3	10	0	4	9	2	0	9	6	0
15	3	1	2	4	2	0	5	2	2	0	10	5	0
16	3	4	2	4	6	0	5	7	2	0	11	2	0
17	3	6	3	4	9	0	5	11	1	0	11	10	2
18	3	9	3	5	1	0	6	4	1	0	12	8	2
19	4	0	0	5	4	0	6	8	0	0	13	4	0
20	4	3	0	5	8	0	7	1	0	0	14	2	0
21	4	6	0	6	0	0	7	6	0	0	15	0	0
22	4	8	1	6	3	0	7	9	3	0	15	7	2
23	4	10	2	6	6	0	8	1	2	0	16	3	0
24	5	1	2	6	10	0	8	6	2	0	17	0	0
25	5	3	3	7	1	0	8	10	1	0	17	8	2
26	5	6	3	7	5	0	9	3	1	0	18	6	2
27	5	9	3	7	9	0	9	8	1	0	19	4	2
28	6	0	0	8	0	0	10	0	0	1	0	0	0
29	6	3	0	8	4	0	10	5	0	1	0	10	0
30	6	5	1	8	7	0	10	8	3	1	1	5	2
31	6	8	1	8	11	0	11	1	3	1	2	3	2
32	6	10	2	9	2	0	11	5	2	1	2	11	0
33	7	1	2	9	6	0	11	10	2	1	3	9	0

The

he Explanation of the preceding TABLE.

First, At the Head of the Table, over each Column, is the Price of one Rod, three quarters of a Rod, half a Rod, a quarter, and half a quarter of a Rod.

Secondly, In the first Column is the Number of odd Feet, from one descending to thirty-three; and against these odd Feet, in the other Columns, stands the Price or Value which the odd Feet come to, at the several Rates over each Column.

The Method of using it is as follows:

Find the Price *per* Rod agreed on in the Head of the Tables, and just under it you have the Price of three quarters, half, a quarter, and half quarter; and under them, in the same Columns, you will find the Price of any Number of odd Feet from one to thirty-four, or one Eighth of a Rod.

EXAMPLE I.

At 5 l. the Rod, what is the Value of three quarters, an half, a quarter, and half-quarter of a Rod, and thirty-three Feet?

	l.	*s.*	*d.*
The Price of the whole Rod is ———	5	0	0
The three quarters is ———	3	15	0
The half is ——— ———	2	10	0
The quarter is ——— ———	1	5	0
The half quarter is ——— —	0	12	6
The thirty-three Feet is ———	0	11	10½
The Sum is ———	13	14	04½

EXAMPLE II.

At 2 l. 10s. per Rod, what comes thirty Feet to?

Look at the Head of the Table for 50*s.* and you will find 5*s.* 4*d.* ½; which is the Price of 30 Feet. at 50*s.* *per* Rod.

EXAMPLE III.

What comes 25 Feet to, at 4l. 15s. per Rod?

Because you cannot find 4*l.* 15*s.* at the Top of the Table, therefore first find out what 25 Feet comes to at 4*l.* *per* Rod, which will be 7*s.* and 1*d.* next find what 25 Feet comes to at 10*s.* *per* Rod, which is 10*d.* ¾; and in the next Place, what 25 Feet comes to at 5*s.* *per* Rod, which is 5*d.* ¼.

Set them down as follows:

	l.	s.	d.
25 Feet at 4*l.* *per* Rod is ————	0	7	1
25 Feet at 10*s.* *per* Rod is ————	0	0	10¾
25 Feet at 5*s.* *per* Rod is ————	0	0	5¼
The Sum is ————	0	8	5

EXAMPLE IV.

What comes 29 *Feet to at* 7l. 10s. per *Rod?*

	l.	s.	d.
29 Feet at 5*l.* *per* Rod is ————	0	10	5
29 Feet at 2*l.* 10*s.* *per* Rod is ————	0	5	2½
The Sum is ————	0	15	7½

EXAMPLE V.

What comes 32 *Feet to at* 6l. 17s. 6d. per *Rod?*

	l.	s.	d.
32 Feet at 5*l.* *per* Rod is ————	0	11	10½
32 Feet at 1*l.* *per* Rod is ————	0	2	3½
32 Feet at 10*s.* *per* Rod is ————	0	1	1¾
32 Feet at 5*s.* *per* Rod is ————	0	0	7
32 Feet at 2*s.* 6*d.* *per* Rod is ————	0	0	3½
The Sum is ————	0	16	2¼

A Table for reducing *Brick-Work* of any Thickness, to the Standard Thickness of a *Brick* and half.

	½ Brick.			1 Brick.			1½ Brick.			2 Bricks.		
	R.	*Q.*	*F.*	*R.*	*Q.*	*F.*	*R.*	*Q.*	*F.*	*R.*	*Q.*	*F.*
1 Quarter	0		22	0	0	45	0	1	0	0	1	22
2 Quarters	0	0	45	0	1	22	0	2	0	0	2	45
3 Quarters	0	1	0	0	2	0	0	3	0	1	0	0
1	0	1	22	0	2	45	1	0	0	1	0	22
2	0	2	45	1	1	22	2	0	0	2	2	45
3	1	0	0	2	0	0	3	0	0	4	0	0
4	1	1	22	2	2	45	4	0	0	5	1	22
5	1	2	45	3	1	22	5	0	0	6	2	45
6	2	0	0	4	0	0	6	0	0	8	0	0
7	2	1	22	4	2	45	7	0	0	9	1	22
8	2	2	45	5	1	22	8	0	0	10	2	45
9	3	0	0	6	0	0	9	0	0	12	0	0
10	3	1	22	6	2	45	10	0	0	13	1	22
11	3	2	45	7	1	22	11	0	0	14	2	45
12	4	0	0	8	0	0	12	0	0	16	0	0
13	4	1	22	8	2	45	13	0	0	17	1	22
14	4	2	45	9	1	22	14	0	0	18	2	45
15	5	0	0	10	0	0	15	0	0	20	0	0
16	5	1	22	10	2	45	16	0	0	21	1	22
17	5	2	45	11	1	22	17	0	0	22	2	45
18	6	0	0	12	0	0	18	0	0	24	0	0
19	6	1	22	12	2	45	19	0	0	25	1	22
20	6	2	45	13	1	22	20	0	0	27	2	45
21	7	0	0	14	0	0	21	0	0	28	0	0

Rods contained upon the superficial Measure of the Wall.

A Table for reducing *Brick-Work* of any Thickness, to the Standard Thickness of a *Brick* and half.

		2 Bricks½.			3 Bricks.			3 Bricks½.			4 Bricks.		
		R.	*Q.*	*F.*	*R.*	*Q.*	*F.*	*R.*	*Q.*	*F.*	*R.*	*Q.*	*F.*
1 Quarter		0	1	45	0	2	0	0	2	22	0	2	45
2 Quarters		0	3	22	1	0	0	1	0	45	1	1	22
3 Quarters		1	1	0	1	2	0	1	3	0	2	0	0
Rods contained upon the superficial Measure of the Wall.	1	1	2	45	2	0	0	2	1	22	2	2	45
	2	3	1	22	4	0	0	4	2	45	5	1	22
	3	5	0	0	6	0	0	7	0	0	8	0	0
	4	6	2	45	8	0	0	9	1	22	10	2	45
	5	8	1	22	10	0	0	11	2	45	12	1	22
	6	10	0	0	11	0	0	14	0	0	16	0	0
	7	11	2	45	14	0	0	16	1	22	18	2	45
	8	13	1	22	16	0	0	18	2	45	21	1	22
	9	15	0	0	18	0	0	21	0	0	24	0	0
	10	16	2	45	20	0	0	23	1	22	26	2	45
	11	18	1	22	22	0	0	25	2	45	29	1	22
	12	20	0	0	24	0	0	28	0	0	32	0	0
	13	21	2	45	26	0	0	30	1	21	34	2	45
	14	23	1	22	28	0	0	32	2	45	37	1	22
	15	25	0	0	30	0	0	35	0	0	40	0	0
	16	26	2	45	32	0	0	37	1	22	42	2	45
	17	28	1	22	34	0	0	39	2	45	45	1	22
	18	30	0	0	36	0	0	42	0	0	48	0	0
	19	31	2	45	38	0	0	44	1	22	50	2	45
	20	33	1	22	40	0	0	46	2	45	53	1	22
	21	35	0	0	42	0	0	49	0	0	56	0	0

A

A Table for reducing *Brick-Work* of any Thickness, to the Standard Thickness of a *Brick* and half.

		4 Bricks ½.			5 Bricks.			5 Bricks ½.	
	R.	*Q.*	*F.*	*R.*	*Q.*	*F.*	*R.*	*Q.*	*F.*
1 Quarter	0	3	0	0	3	22	0	3	45
2 Quarters	1	2	0	1	2	45	1	3	22
3 Quarters	2	1	0	2	2	0	2	3	0
Rods contained upon the superficial Measure of the Wall. 1	3	0	0	3	1	22	3	2	45
2	6	0	0	6	2	45	7	1	22
3	9	0	0	10	0	0	11	0	0
4	12	0	0	13	1	22	14	2	45
5	15	0	0	16	2	45	18	1	22
6	18	0	0	20	0	0	22	0	0
7	21	0	0	23	1	22	25	2	45
8	24	0	0	26	2	45	29	1	22
9	27	0	0	30	0	0	33	0	0
10	30	0	0	33	1	22	36	2	45
11	33	0	0	36	2	45	40	1	22
12	36	0	0	40	0	0	44	0	0
13	39	0	0	43	1	22	47	2	45
14	42	0	0	46	2	45	51	1	22
15	45	0	0	50	0	0	55	0	0
16	48	0	0	53	1	22	58	2	45
17	51	0	0	56	2	43	62	1	22
18	54	0	0	60	0	0	66	0	0
19	57	0	0	63	1	22	69	2	45
20	60	0	0	66	2	45	73	1	22
21	63	0	0	70	0	0	77	0	0

The Explanation of the TABLE.

At the Head of the Table, is ſet the Thickneſs of the Wall, in Bricks and half Bricks, from half a Brick in Thickneſs, to ſix Bricks thick.

In the firſt Column is placed the Number of the Rods contain'd in any Wall, from one quarter of a Rod to twenty-one Rods; and againſt thoſe Numbers, in the other Column, ſtands the Quantity of *Brick-Work*, in Rods, Quarters, and Feet, which a Wall contains of any of theſe Thickneſſes, at the Head of the Table.

EXAMPLE I.

If a Wall upon the Superficies contain twelve Rods, and it be three Bricks and a half in Thickneſs, how many Rods does that Wall contain of the ſtandard Thickneſs of a Brick and a half?

Find twelve Rods in the firſt Column, and under 3 Bricks ½, at the Head of the Table, in the Angle of Meeting, you will find 28 Rods; which is the true Quantity required at the ſtandard Thickneſs of a Brick and half thick.

EXAMPLE II.

If a Wall be four Bricks and a half thick, and contain upon the Superficies nineteen Rod, how many Rod of Brick-Work *are contain'd in that Wall, at the ſtandard Thickneſs?*

Look for nineteen Rod in the firſt Column, and under 4 Bricks ½, you will find fifty-ſeven Rod, the Content required.

And the ſame may be done as to any other Wall.

BRIDGE, a Work of Maſonry or Timber, built over a River, Canal, *&c.* for the Conveniency of croſſing the ſame.

To build a *Bridge* of Timber over any Brook, Rill, or ſmall River, which does not exceed forty or fifty Feet in Length, and that without ſetting any of the Timber down in the Water, which is both a cheap and ſafe Way, proceed according to the following Directions:

Let the Timber be ſo jointed, as in ſome meaſure, to reſemble an Arch of Stone or Brick: Let the Joints be well made, and ſtrongly ſhut together with Cramps and Dogs of Iron.

This *Bridge* muſt be made to reſt upon two ſtrong firm Pillars of Wood at each End of the *Bridge*, and both well propped with Spurs or Braces; and there ought to be two good Buttreſſes of Brick, for theſe Wooden Pillars and Spurs to ſtand in, that they may not give way or ſlip. When this has been done, the *Bridge* may be planked over and gravelled, and it will laſt a long Time. Sir *Hugh Plat* ſays this Method has been practiſed.

Of all the Contrivances that Men have uſed for facilitating Commerce, ſays M. *Gautier*, this of building Bridges over Rivers, both ſmall and large, has been none of the leaſt Conſequence.

Of all the Bridges that have ever been, if we will give Credit

dit to Hiſtory, it is agreed, that the *Bridge* built by *Trajan* over the *Danube*, was the moſt grand, and the fineſt.

As that River, over which it was built, was very wide, ſo of Neceſſity the *Bridge* muſt be very long: For it was compos'd of twenty Arches of a hundred and fifty Foot in Height, and their Opening from one Pile, or Peer, to another, was an hundred and ſixty Feet, or about twenty-five Fathom; which makes the Length of the *Bridge* about ſix hundred Fathom, or about five hundred forty-ſix Fathom of *Paris* Meaſure; for the ancient *Roman* Foot is but eleven Inches *Paris* Meaſure.

The Dimenſions of a like Work are almoſt beyond all the Ideas of the Architects of our Times, if what is ſaid be true. This *Bridge* being finiſh'd, the *Romans* invaded the *Barbarians* on this Side the *Danube*. *Adrian*, *Trajan*'s Succeſſor, afterwards caus'd it to be demoliſh'd, to hinder the ſame People, being vanquiſh'd, from making uſe of it in paſſing the River *Danube*, to carry their Arms into the *Roman* Empire.

The Peers of this fine *Bridge* are ſtill to be ſeen in the Middle of the *Danube*. It was erected between *Servia* and *Moldavia*, a little above *Nicopolis*. Alſo among the famous *Bridges* renown'd in Hiſtory, is reckon'd that of *Darius*, over the *Boſphorus* of *Thrace*; and alſo that of *Xerxes*, over the *Helleſpont*, and of *Pyrrhus*.

The *Romans* have ſtill at *Rome*, three very fine *Bridges* over the River *Tyber*. The Emperor *Adrian* cauſed the firſt to be built, which is the *Pont Ælius*, at preſent called *Pont St. Angelo*, or *Angel's-Bridge*, the fineſt of all thoſe that are at this Day at *Rome*. It was called the *Bridge Angelo*, on Account of an Angel pretended to have been ſeen at the Entrance of it. It was garniſhed on the upper Part with a Covering of Braſs, ſupported by forty Pillars of the ſame.

The ſecond was the *Triumphal Bridge*, of which the Ruins are ſtill to be ſeen in the River *Tyber*. Over this *Bridge* the Emperors and Conſuls uſed to paſs, when they had a Triumph decreed them; which was at that Time adorn'd with all imaginable Art.

The third was the *Pons Janiculenſis*, at preſent called *Sixtus's-Bridge*; becauſe Pope *Sixtus* IV. cauſed it to be rebuilt in the Year 1475. This *Bridge* was antiently built of Marble.

The fourth, was that called *Pons Cæſtius*; at preſent, *St. Bartholomew's-Bridge*; which was re-edify'd in the Time of the Emperor *Veſpaſian*.

The fifth was that nam'd *Fabricius*, or *Tarpeius*; at preſent called *Ponte Caſpi*, or *Quatro Capi*.

The ſixth was the *Senatorian*, or *Palatine Bridge*, now called *Sancta Maria*.

The ſeventh was the *Pons Horatius*, or *Sublicius*; one of the fineſt *Bridges* of *Rome*, and of which the Ruins are ſtill to be ſeen in the *Tyber*, and which has not yet been re-edify'd.

I ſhall relate the Elevation of it, according to the Relation of an *Italian* Author in his Works concerning the Antiquities of the City of *Rome*.

The Figure of it ſeems extraordinary and fantaſtical: To ſee a ſecond *Bridge* raiſed over a firſt, and in a like Work, Columns, and other Ornaments of Architecture over that; ſo that it appeared more like a Triumphal Arch, or Portico, than a *Bridge*. This *Bridge* was demoliſh'd in the Reign of the Emperor *Otho*; but rebuilt by *Antoninus Pius*.

The eighth and laſt, was that without the City of *Rome*, and about two Miles diſtant; which was called *Milvius*, in the *Flaminian-Way*.

Beſides theſe *Roman Bridges*, we have modern ones, who are not without their Merit.

Among thoſe of *France*, may be reckon'd thoſe of *Avignion*, of *Saint Eſprit*, and of *Lyons* upon the *Rhone*. The firſt of theſe is demoliſh'd; there remaining only ſome Arches on the Side of *Avignion*.

The ſecond is ſtill ſubſiſting; and may be ſaid to be one of the fineſt *Bridges* in the Univerſe. One Thing particular in theſe three *Bridges*, is, That their Plan is not in a Straight Line; eſpecially thoſe of *Avignion* and *Saint Eſprit*.

The Angle is little ſenſible in that of *Lyons*; but nevertheleſs perceivable; and is on the upper Side of the Stream.

But as for the two preceding, it is certain, that they have an Angle, or a Sort of Bending, the Convexity of which oppoſes the Waters of the *Rhone*, as though being by this Diſpoſition centred, and are bouted, the better to reſiſt the Weight and Current of the Waters.

The *Bridge* of *Avignion* was compos'd of eighteen Arches, in Length one thouſand three hundred and forty Paces, making about five hundred Fathom. It was begun in the Year 1176, and finiſh'd in 1188.

The Schiſm between *Benedict* and *Boniface*, was fatal to it in the Year 1385; but a greater Calamity happen'd to it in the Year 1602; by the Negligence of repairing, one Arch fallen down, which cauſed the Fall of three others. In fine, in the Year 1670, the Cold was ſo violent, that the *Rhone* was frozen ſo as to bear Waggons with the heavieſt Loads for ſeveral Weeks; and upon the Thaw happening, ſuch Mountains of Ice daſhed againſt the Piers, as ſhook them, and cauſed ſeveral Arches to fall down.

But nevertheleſs, the third Pile on *Avignion* Side, with the Chapel of *St. Nicholas* built upon it, has always born up againſt all Accidents.

Pont Saint Eſprit is finer and more bold than that of *Lyons*, or *Avignion*: It conſiſts of nineteen great Arches, beſides ſeven ſmaller ones. It has Arches from fifteen to twenty Fathoms, opening, rather more than leſs, which make the Length of the *Bridge* upwards of four hundred Fathoms.

The *Bridge* of *Lyons*, upon the *Rhone*, has twenty Arches. It is obſerv'd farther of theſe *Bridges*, that they are defended by Towers to ſecure the Paſſages.

Among

Among the fine *Bridges* of *France*, are reckon'd the *Pont Royal* of the *Thuilleries*, that of *Tolouse* upon the *Garonne*, and one Arch of *Pont-Neuf* at *Paris*.

I ſhall now quit *France*, and paſs over to *England*, and take a View of *London-Bridge*

London-Bridge was begun in the Reign of *Henry* II. in the Year 1176, and finiſh'd in the Reign of King *John*, in the Year 1209. Since that Time, it has been divers Times burnt, and ruin'd by Ice, and as often repair'd, or re-built. The King and the City contributed to the Charge. This *Bridge* is made of hewn Stone: It has nineteen Arches of a hundred and twenty-five Fathoms, *Paris* Meaſure, or eight hundred Feet long, and thirty or twenty-eight Feet and one Eighth broad. Some ſay ſixty or fifty-ſeven Foot one quarter; for the *London* Foot is fourteen Sixteenths of that of *Paris*. In Height, on the two Sides of the *Bridge*, are built two Rows of ſtately Houſes; and a conſiderable Fund is ſettled for the maintaining it. This *Bridge* is continually beaten and overflow'd by the Flux and Reflux of the Tide. Large Veſſels, which come up the *Thames*, don't go above the *Bridge*; but ſmall ones paſs through it. Its Piers are perfectly well guarded by Starlings.

If we paſs into *Italy*, there we ſhall ſee very fine *Bridges*. That of *Alexander Farneſe*, Duke of *Parma*, is accounted a very fine one.

Palladio gives us many Deſigns of ſeveral fine *Bridges*; and mentions the greateſt Part of thoſe which the *Romans* built; as that of *Rimini*, in the *Flaminian-Way*; thoſe of *Vicenza*, upon the *Bachiglione* and the *Kerone*. He gives us two *Bridges* of Stone, of his own Invention; which are very fine: But upon which, he does not pretend, that Carriages may paſs; it being compos'd of Lodges, many Streets, Porticoes, Pediments, ſupporting Statues of Marble, or Braſs, for the finiſhing of the Work.

There is alſo a very fine *Bridge* at *Madrid*, hard by one of the City Gates, called the *Bridge* of *Segovia*, on the River *Manzaneres*.

In the Relations of the *Levant*, by *Poulet*, there is Mention made of one ſingle Arch in the little City of *Munſter*, upon the *Narante* in *Botnia*, a Building infinitely more bold than that of the *Rialto* at *Venice*; which is alſo one ſingle Arch; and paſſes for a Maſter-Piece of Art, built in 1591, the Deſign of *Michael Angelo*; and one Part of an Arch, which is upwards of thirty-two Fathom at the Baſe.

There is no City in the World that has ſo many *Bridges* as that of *Venice*; which are to the Number of three hundred and thirty-nine.

One of the Things which impoſes moſt upon a Man, is a ſtately *Bridge* over a large River: The Boldneſs of huge Arches compos'd of an infinite Number of ſmall Materials, either Stones, or Bricks, ſo firmly united together, that they at laſt ſeem to form but one Piece, and by their Weight afford a ſafe Paſſage to Men, and all heavy Carriages, in the paſſing Brooks, and the largeſt and moſt rapid Rivers.

Men have invented many different Sorts of *Bridges* and they are made after many different Manners, according to the Situation of the Places, the Neceſſity, and the Materials employ'd. They are for the moſt part of Stone in ſome Places, and of Carpentry in others, according as they have the Conveniency of Stones for the firſt, and of Wood for the latter.

Palladio is the only Perſon who has treated any Thing largely of *Bridges*; and what he ſays in general, is reduced to the giving us to underſtand, that *Bridges* are the principal Parts of a Way.

That it is ſurprizing to ſee that they form properly a Way upon the Water; and that the Properties of a *Bridge* are,

I. That they be well deſign'd.
II. Commodious.
III. Durable.
IV. Well-adorn'd.

Bridges are well-deſign'd, when they are placed over a River, upon the Square, and not ſlanting; and are well laid out by a Line.

Bridges are commodious, when they are upon a Level upon the grand Highway that abuts upon them; when the Ramps or Aſcents be eaſy and Imperceptible, and the Way broad.

They are durable, when they have good Foundations, and are built according to Art with good Materials:

And they are well-adorn'd, when they are decorated according to the Rules of good Taſte in Architecture; which are agreeable to thoſe Ruſtick Works, and clouter and heavy Maſſes of Maſonry, with which *Bridges* are built.

Palladio goes farther, and gives Precepts, but ſuch as have no Place in all Sorts of *Bridges*, ſo as to be a general Rule.

A Man muſt often conform to the Situation of Places and Circumſtances in erecting *Bridges*. Some Difficulties are to be met with. Whereas *Palladio* directs, That in order to build a *Bridge*,

Firſt, To make choice of that Place in a River that has the leaſt Depth of Water, to the End that it may be laſting; and that the Foundation may be even and firm, as Rock and Gravel-Stone.

Secondly, That thoſe Places be avoided, where the Water turning, makes Vortexes or Whirle-Pools, and where the Bottom is ſoft Sand or Gravel; becauſe theſe Matters are eaſily carry'd away by the Violence of great Waters, which commonly alter the Bed of a River, and ſap the Foundation of the Piers, and are often the Occaſion of the Ruin of *Bridges*.

Thirdly, and *Laſtly*, The Stream of the Water muſt be ſtraight, and without Elbows or Sinuoſities in the Banks; becauſe theſe Turnings and Windings will come in Time to be deſtroy'd by the Force of the Current, and the *Bridges* will become inſulated, and without Epaulments; and beſides, that there will be amaſs'd in theſe Places a thouſand filthy Things which the River er will carry thither; which being ſtopp'd by the Piers, will at laſt choak up the Opening of the *Bridge*.

All theſe Difficulties which *Palladio* mentions often, happen to be in a Place where one would project a *Bridge*; and it is the Addreſs of an Artiſt to ſurmount them by Art.

The *Bridge* called *Pont-Neuf*, and that of the *Thuilleries*, had never been erected in the Places where they are, if all theſe could have been had: But when one can have his Choice, it will be very well to follow *Palladio*'s Advice.

This Author, beſides, ſays that there are two Sorts of *Bridges*; the one made of Wood, and the other of Stone.

That which is erected over the Torrent called *Ciſmoné* at the Foot of the *Alpes*, between the Cities of *Trent* and *Baſſane* in *Italy*, is form'd by ſix equal Bays of Joiſts, and carried up entirely to the Length of near ſeventeen Fathom, between the Abutments built upon the Brinks of the River. *Fig*. I. *Plate* I.

The Joiſts lying along upon the Beams, and being cover'd, make the Planking or Flooring and Way of the *Bridge*. See the *Figures*.

Palladio ſays, that there is no *Bridge* made after the ſecond Manner. See *Plate* I. *Fig*. II. Although we are aſſur'd there is one in *Germany*. And in Effect, M. *Blondel*, who relates all that *Palladio* has ſaid, aſſures us, that he has ſeen the like at *Nerva*, a City belonging to the King of *Sweden*, in the Gulf of *Finland*, in the *Baltick-Sea*.

The Aſſemblage of the Third, is that of a Scheam-Arch. The Diviſions are unequal in Number; and it has at each End a long Brace at the End below, in the Wall of the Abutment.

The Fourth Figure is in the Form of a Vault, or Mold for an Arch; and the Aſſemblages between the two Poincons are diſpoſed after the Manner of Vouſſoirs. The Diviſions are in Number unequal, that there may be one Vouſſoir in the Middle, ſerving as a Key.

Says M. *Blondel*, if there were another Aſſemblage equal to that of the *Bridge* underneath it, the Work would be infinitely ſtronger.

Upon this Notion, it was projected, to make a *Bridge* croſs the River *Seine*, overagainſt the *Seine*, above *St. Cloud*, to abridge the Way of *Verſailles*.

As to *Stone-Bridges*, *Palladio* obſerves four Things:

I. As to the Heads of *Bridges*, or their Abutments

II. As to the Piers.

III. As to the Arcades.

IV. As to the Pavement upon the Arcades.

The Abutments ought to be very ſolid, and to be made on ſuch Places of the Rivers that are rocky, hard Gravel, or good Ground; otherwiſe, they muſt be ſecured by Art, with other Piers, or other Arches.

The Piers ought to be equal in Number, to the End that there may be one Arch in the Middle, where commonly the Water has the greateſt Current; which renders the Work ſtronger, more equal,

equal, and more agreeable to the Sight; the Foundations ought to be laid at that Time of the Year, during which the Waters are at the lowest, as in Autumn: And if the Foundation is rocky, hard Gravel, or Stony Ground, the first Stones of the Foundation may be laid without hollowing or digging any deeper.

But if it be soft Sand, or Gravel, then it will be proper to carry it off, till you come to find a solid or firm Bottom; or, if this be too difficult, you must at least carry off one Part of it, and pile the rest. But in the first Place, that Side of the River where you are to work, must be inclos'd in Dams; and the Current must have its Liberty on the other Side.

As to the Piers, they ought not to be less in Bigness than one sixth Part of the Breadth of the Arcade; nor commonly more than a Fourth. Their Structure ought to be large good Stones, well joined together with Cramps of Iron, or other Metal; to the End that by this Enchainment, they may be as one entire Stone.

It has been a Custom to make Advances or Projectures at the End of Piers, at Right Angles; and sometimes in Semicircles, the better to throw off the Water, and to resist the Shocks of Tices, or other Things, which the River carries when it is large.

Thirdly, The Arches ought to be made of very long Stones, and well jointed. Those are the strongest, that are semicircular; because they bear entirely upon the Piers, without pushing the one against the other.

When one is constrain'd by the too great Height, the Arches may be diminish'd, or made Scheam-Arches; so that the perpendicular Height upon the Line of their Chord may be one Third of the same Chord; in which Case, the Abutments must be extreamly well fortify'd.

After this, *Palladio* gives the Designs of some antique *Bridges*, or of his own Invention.

The first is that of *Rimini*, built by *Augustus*, over a River twenty-nine Fathoms broad, made with five Arches; the three middlemost are equal, and each twenty-five Feet; and the two others, but twenty Feet; the Abutments are each seven Foot and a half; the Piers are eleven Feet; and the Arches semicircular.

The Projecture of the Piles rises no higher than the Imposts, above which are Tabernacles and Niches for placing of Statues. The whole Length of the Work is crown'd with a Cornice; and above that, a Parapet, adorn'd from its Zocle, its Base and Cornice, with *Tuscan* and massive Work.

He afterwards gives the Description of the *Bridge* over the *Bachiglione*, of sixteen Fathoms wide, compos'd of three Arches, each twenty-two Foot and a half; the Abutments, two Foot and a half in Breadth, and the Piers five Feet.

The Arches are Scheam ones; and their Height is one Third of the same Chord, as well the middle Arch, as the other two.

He also gives an Account of the *Bridge* of *Rerone*, whose River is sixteen Fathoms wide. The

Bridge

Bridge is composed of three Arches; that in the Middle, being twenty-nine Feet; and the other two, twenty-five each; the Abutments of which, are but three Foot and a half, and the Piers, five Feet; their Projecture, at Right Angles; the Arcades are Scheam ones.

Palladio also gives the Design of a *Bridge* after his own Manner, over a River thirty Fathoms wide, between the Brinks of the River, and the Abutments, which consists of but three Arches; that in the Middle, being ten Fathoms, and the two others, eight apiece; the Piers two Fathoms, or one Fifth of the Breadth of the great Arch.

The Arches are Scheam ones, and their perpendicular Height above the Imposts, is one Third of their Breadth.

Leon Baptista Albert tells us, that the Parts of a *Bridge*, are the Piers and Arches; and the Pavement above the upper Part of the *Bridge* has a large Way for the Passage of Cattle and Waggons; and little Banks on each Side, for the Conveniency of Foot Passengers, inclos'd on the Outside by their Breast-Works, or Parapets.

In some Places, he says, *Bridges* are cover'd, as antiently *Adrian's Bridge* at *Rome* was, now call'd *Pont St. Angelo*, which was the finest and most magnificent of them all; the Ruins of which cannot be beheld without Veneration.

As for the Structure of a *Bridge*, he says, it must be allow'd the same Breadth as the grand Highway that abuts upon it: That the Piers ought to be equal in Number and Size. Their Breadth should be one Third of that of the Opening of the Arch. That there must be before the Piers Juttings-out in the Form of the Prows of Gallies, against the Current of the Water; which Juttings-out, in their Projecture, must have one half of the Breadth of the Pier itself; and which must be raised above the greatest Heights that the Water rises; and that there must be made on the other Side, others in Form of Poups; which will not be disagreeable, if their Points are cut off, and made more blunt than the other.

He says, it will not be amiss, if on the Right of the Juttings-out, there be Countersorts on each Side, or Pilasters reaching up the Height of the *Bridge*, the better to sustain the Flanks; and their Breadth below, not to be less than two Thirds of that of the Pier; the Impost of the Arch ought to be entirely out of the Water: The Ornaments of the *Ionick* or rather *Doric* Architecture.

Serlio tells us, that at *Pont Sixtus*, the Piers have one Third of the Breadth of the great Arches; the greatest Arch but half a Circle of Height of one Sixteenth of the Diameter.

At *Pont St. Angelo*, the Piers are one half of the great Arch, and is semicircular; the Bandeau, or Head-Band, the Height of one Ninth of the Diameter of the Arch; the Piers bear upon a grand Base, or Patten of the Pillar, in Form of a Zocle quartered, raised some Feet above the

ordinary

ordinary Level of the Water, by a Projecture on the Outside, round the whole Pier.

Its Spur, or Counterfort, is a Semicircle, which rises to the Middle of the Arch; a square Pilaster above its Parapet, with Pedestals, at equal Distances; which serve to sustain, according to the Opinion of *Albertus*, the forty-two Columns which support the Covering of the *Bridge*, the Arches being semicircular.

The *Bridge de Quatro Capi*, antiently *Fabricius*; of which the Author relates, that there are but two Arches remaining, which are equal and semicircular, have a Pier the Breadth of the Arch, with a Spur, or Counterfort, round it, and a Niche above. The Bandeau of the Arch is rustick, and its greatest Height is one Tenth of the Breadth of the Arch.

Pont Milvius has semicircular Arches, born upon Imposts of the Height of one Third of their Diameter; the Piers are half the Breadth; and upon them, there are Niches without Ornaments.

M. *Blondel*, of the *Royal Academy of Sciences*, an accomplish-ed Man, caused to be built at *Xaintes*, upon the *Charente*, near the Place where the Ebb of the Tide commences, a *Bridge* of Stone, in the Year 1665; the Piers of this *Bridge* are in Proportion, as 3 to 8, as to the Breadth compar'd with the Opening of the Arches; the Pier at the End, towards the *Pont Levis*, and which serves for an Abutment, has one Sixth of the greatest Width; because it is to sustain on that Side the Push of all the Arches, (which are Scheam ones,) in order to carry the Height of the Imposts above the common Height of the Waters of the River, without making any Alteration of the Level of the old *Bridge*.

This is in a Manner the Substance of what the most celebrated Architects have given us in writing, as to the Proportion of *Bridges*; but no body has given us as yet the demonstrative Reasons: They have not acquainted us with the Fusts of their Columns; what Measures we shall give either to the one or the other; which may be helpful to us in imitating them: They have given us no Reason why they do after that Manner, rather than any other.

The ablest Architects are not agreed among themselves as to the Proportions they give to Buildings, not only as to their Solidity, but even not in respect to their Ornaments.

So true it is, That Arts and Sciences are still imperfect. All these depend upon a certain Taste and certain Ideas that Men have had different from one another, and in different Ages. So many Architects, so many different Manners.

It may be seen, as to all that has been before related, they give us no Reason why they make their Piers, their Abutments, their Arches, &c. of such a Largeness, or such a Thickness: and those who now work according to these ancient Examples, know no more than the Authors themselves, for what Reasons they do so.

They

They conduct themselves only by Ideas that they cannot demonstrate; but which appear to be imitable, by the Example of so many others who have succeeded otherwise; for which Reasons they say the Work is beautiful and solid, because the Proportions between the Parts which compose it, are there observ'd.

Although I have made a diligent Search into this Affair, says M. *Gautier*, I have not found what has satisfy'd me.

It were to be wish'd that some accomplish'd Person would set himself upon the solving of these Difficulties, in order to render them easy to the Publick.

M. *De la Hire*, of the *Royal Academy of Sciences*, has labour'd for this Purpose; but those who are not so learned as himself, cannot comprehend him, for want of being acquainted with *Algebra*, he having expressed himself in Terms drawn from this Science; which, Workmen, and Persons of but a moderate Share of Learning, know little or nothing of, and consequently understand not how to be profited by them.

Of the Projection of Bridges.

The Sieur *Gautier*, Architect, Engineer, and Inspector of the *Bridges* to the *French* King, says, in his *Traité des Ponts*, there are so many Things to be known in relation to the building of *Bridges*, either of Carpentry, or Masonry, that it is hard to find one Man that is equally qualified with the Knowledge of them all. And it is sufficient in a Work of Consequence, if many Persons can be found, who, all of them together, understand well what is best to be done.

A Carpenter or Mason of Experience, cannot be to highly prov'd.

These two Persons are ordinarily the Head, the Workmen, the Arms; and a well-accomplish'd Engineer, or Inspector, the Soul of the Work, for either the carrying on, and the ready Execution or good Manner of it. And it is impossible, this Conductor, who shall be an Engineer, Architect, or Inspector, be so fitly qualify'd for that Office, as that he may be depended upon, unless he knows also the working Part.

Nor is it possible, he should know the working Part, if he does not know the Parts and Materials to be used in the Work; and also the Utensils, Scaffolding, Plummets, Engines for raising great Weights, Pit-Wheels, Pumps, Buckets, &c. for emptying and clearing the Foundations, Dams, &c. of so many different Forms; the Manner of piling the Foundations, great Borers for boring the Rocks, according to their Consistence; Centres, or Molds for Arches, Assemblages, the Cut of Stones, and an infinite Number of other Things which cannot be foreseen: So that for the erecting of a considerable *Bridge*, he ought to be a Person of universal Knowledge, and not ignorant of any Thing that relates to the Mystery of Architecture, which supposes the Knowledge of all those

thoſe Things, if he would ſucceed.

When any one projects a *Bridge*, he ought to begin

1. With making an exact local Plan; which Plan ſhall preciſely lay down the Extent of the Water, the Sands, (if it has any,) the Banks, or Brinks of the River, and the Ways or Streets that abut upon this *Bridge*.

2. He muſt project upon this Plan the *Bridge* deſign'd, whether of Maſonry, or Carpentry; with the Number of Arches, and Quantity of Piles, Bays, or Joiſts. He muſt always lay down the *Bridge* over the River upon the Square, and never ſlanting.

3. He muſt, upon this Plan, trace a Line which ſhall cut the *Bridge* in the Middle, and there ſound the Depth of the Water from Fathom to Fathom, or from two to two, or from three to three, according as there ſhall be Occaſion.

This Sounding is to be made either by a Pole, divided into Feet, at the End of which is a Leaden Weight, according as the Current of the Water ſhall require.

If this ſhall not be ſufficient, he muſt make uſe of a Cannon-Ball, put into a little Bag, tyed to the End of a Cord, which has been before divided into Feet and Fathoms.

He muſt make uſe of theſe, or other Methods, which ſhall be found to be moſt proper, according to the Rapidity of the Water that is to be ſurmounted.

All this is to be done by means of a Boat, which may be conducted in different Manners; either by a Cable, which is carried a-croſs the River, or by other Cords made faſt to Trees, or Stakes on the Bank, or to Stakes drove down for that Purpoſe; round which the cable that is to hold it is to be many Times turned, and ſlacken'd, according as Occaſion requires, to guide the Boat more to one Side than the other.

4. The Soundings of the Water being made, and ſet down on the Plan, they ſerve for making a Profile of the River, which marks or ſets out the Depth of the Water that has been found; and the Line under the Water, whether it be ſandy or rocky, to which Attention muſt be given, marking the Difference on the Profile.

Upon this Profile is marked by a Line the Depth of the Water, at the loweſt it is at any Time of the Year, which the Bridge-Maſters of great and navigable Rivers will acquaint you with; and the Peaſants or Inhabitants of the neighbouring Places to ſmall Rivers, will inform as to the Height of thoſe Inundations, which have happen'd in their Memory.

There may alſo be drawn in the Profile, which ſhall ſhew the a Mean of the Height of the Waters.

All theſe Lines being drawn by a perfect Level, parallel the one to the other, may be waſhed with a Water Colour.

5. The Profile being thus raiſed, a Sounding-Iron may be made, of a convenient Length, for ſounding below the Depth of the Water, the Ballaſt, or Sand; and no Certainty can be attained till this is done, and the Depth of the Water be known: And, in

in order to this, there are two Methods uſed, either by a Sounding Inſtrument of Iron, made on purpoſe, having a large Ring at its Head for a Crowning, croſs which there goes the Arm of a Borer, larger or ſmaller, in order to turn it with; and having at the Top a Head to be driven down, to make it enter till it comes to a firm Bottom below the Sands.

This Sounding-Iron is made pointed and barbed at the End with four Angles; ſo that being bored or forced through the Sand, or Part of the Rock, or ſolid Ground, that it meets with below the Sand, by being turned ſeveral Times, in order to bring up in the Hollows of the Barbes ſome ſmall Quantities of the ſolid Ground it meets with, and thus being drawn up, the Quality of it is to be entered down in the Memorial that is provided for this Purpoſe, in order to know what kind of Ground the Bottom is.

There are Inſtruments for ſounding of another ſort, which have a little Pocket in the Form of a Snail-ſhell at the End in the Shape of a Borer, which receives nothing but Sand in turning one way, and the Earth under the Sand by turning it the other.

Theſe Sounding Inſtruments muſt be all of one Piece, that they may be as ſtrong as may be; ſometimes they are adjuſted according to the Hardneſs or Eaſineſs of the Ground to be penetrated; ſometimes they are of no Uſe, eſpecially when the Sand is too groſs, and meets with Flints that the Sounder cannot remove.

In this Caſe, they make uſe of a Stake of Oak made round, of the ſtraiteſt Piece of a Tree of three, four, five, or ſix Inches Diameter; which having determin'd to what Depth of the Earth they would ſound, they arm with a Lardoir, or pointed Iron at the End, for removing the Flints; with a Ferrel at the Head, to be able to reſiſt the Strokes of a Beetle with two or three Helves, with which the Sounder is driven down.

All this requires a great deal of Pains, Care, and Expence too; but the Satisfaction of doing the Work well, and making a faithful Relation of it on the Profile, of the Depth of the Sand or Gravel that is to be piled, or which ought to be removed for the Foundation of the Piles, in order to ſettle the Dams neceſſary, will make amends for them; and ſo long as a Perſon is ignorant of this Depth, he can neither project a *Bridge*, nor know how to compute the Expence, ſince he cannot tell what Timber it will take up, nor what Precautions ought to be taken for ſecuring the Work.

6. When a Knowledge of the Conſiſtence of the Ground has been obtained, as whether Sand, Earth, Rock, *&c*, then a perſon may proceed with Safety upon the Profile he has made, to lay down the Projection of the *Bridge*, whether of Maſonry or Carpentry; then the Length and Thickneſs of the Piers and Piles may be known, according as the Foundation muſt be more or leſs in Depth.

7. This being done, and the Height of the highest Inundations being known by the Information of the antientest neighbouring Inhabitants, Marks are to be made at this Height; and supposing three Feet upwards to be the Intradosse, or inward Face of the Arches of the *Bridge* that one would lay down a Projection of, and also the Bays, Joists, Beams, *&c.* of a *Wooden Bridge*, which is the same: The Work may be so regulated, that it may be known to what Height the greatest Inundation will rise, and to what Depth the Foundation of the Piles, *&c.* must be carried.

8. The next Thing to be done, is to provide the Materials which are to be employ'd in the Work.

For a Stone Bridge.

It ought to be considered from whence the Free Stone may be had, its Distance, the Easiness or Difficulty of cutting it, its Carriage, its Nature, as to its being strong or weak, in regard to the Effort it is to sustain, being pressed by the Reins of the Arches, if it will be able to sustain the Weight; for there is some Stones so tender, especially having been but lately taken out of the Quarry, as one may say bleeding, that they will crack, split, or shiver to Pieces.

As to the Lime, it is to be considered from whence that is to be had, its Nature, and when it takes hold, whether as soon as it is used, or a long Time after; the Wages of Workmen, the Cheapness and Conveniency of having Provisions, the Conveniency of the Place, and Number of the Workmen required to finish the Work in a certain Time, before the Rains of Autumu, which make Rivers overflow; and a many Precautions are necessary to be taken, that cannot well be enumerated.

For a Bridge of Carpentry.

The Builder must inform himself from whence the Timber must be had: if it be sound and good; the Time for procuring it; the Charge of it, as to what it will cost, laid down at the Place; what Quantity must be fashion'd for Piles, and what for Centres, Scaffolding, *&c.* that all the Materials may be ready in Time, to begin the Work without Hinderance, and to be able to finish it before the contrary Seasons for the compleating of the *Bridge* commence.

The Breadth of *Bridges* are to be regulated according to the Multitude or Throng of People who pass over them, and the large Streets or Roads that abut upon them: The Height and Breadth of the Arches, according to the Commerce and Navigation of the River.

All these are necessary for the laying down an exact Projection. To these may be added, other Things necessary, according as the Circumstances of the Place shall require, and according to the Discretion of the Master-Builder.

Of the Largeness of Bridges, *in Proportion to the Quantity of Water they must receive when Inundations happen.*

It has been already said, when any one lays down the Projection of a *Bridge*, he must inform himself as to the Quantity of Waters

Waters which paſs in the River over which the *Bridge* is to be built, in order to make the Arches, the croſs Beams, Joiſts, &c. of a ſufficient Largeneſs, to be able to give them a Currency.

The common Rule is to make the Intradoſſes of the Arches at the Places of the Keys; and the croſs Beams or Joiſts of the *Bridges* of Carpentry, three Feet higher than the higheſt Inundations.

This Rule is not obſerv'd in all the Arches of a *Bridge*, which has many; but it is look'd upon ſufficient, if the middlemoſt be ſo, and moſt of the others diminiſhing, ſo as to make an eaſy Ramp above, to gain the Height of the Arch.

There are many *Bridges* which are after this Manner; but the ſureſt Method is to have all the Intradoſſes of the Arches of the ſame Height, three Feet or upwards higher than the greateſt Inundations, although they be not ſo wide; made, if you pleaſe, by elevating the Beginning of the Centres higher, to hinder the Waters from being forced to paſs under them, which cauſes Hollows, or waſhes away the Foot of the Piles, and at length throws them up; and too often all the Work, by reaſon of this Error.

The Piers of *Bridges* do very much diminiſh the Largeneſs of the ordinary Bed of Rivers; and this occaſions the Waters to be very much preſſed in the Arches at the Times of Inundations.

The Rivers then make Hollows among the Piers, and alſo under the Arches, after ſuch a Manner, that they throw into the Depth what they have diminiſhed or taken from their Breadth; and this is one of the principal Cauſes of the Ruin of *Bridges*.

Bridges ſhould never be projected in narrow Places, at leaſt when they cannot be founded on a Rock, and when extraordinary Precautions are not taken.

If one Third of the Breadth of a River be diminiſhed in the making of a *Bridge*, by the Spaces or Compaſs of the Piers, and this River is but two Fathom deep in this Place when it has its ordinary Courſe, one may reckon that it acquires one Fathom more in Depth at the Times of Inundations, it having been contracted by reaſon of the Maſonry of the Piers uſed in it one third Part.

Rivers do neither augment or diminiſh, but according as it rains more or leſs. There are ſome Countries in which it rains more than in another, that is its Neighbour.

The Quantity of Rain that falls at *Paris*, according to *M. de la Hire*, is about nineteen Inches or thereabouts, one Year with another.

M. the Count *de Pont Briand*, who has made the like Obſervations in his Caſtle near *St. Malo*, has found it to be twenty-four Inches ſix Lines: And Father *Fulchiron*, at *Lyon*, has found it thirty-ſix Inches nine Lines.

If theſe three Quantities (very different one from the other) be joined together, there will be a Reduction of twenty-ſix Inches nine Lines, which falls upon the Surface of the Earth from *Lyons* to *St. Malo*, in the Space of a Year.

The Wind, the Sun, the Earth, the Plants, &c. consume a good Part of this Water, and the rest runs down the Declivities of Hills into the Valley, into the Streams, Rivers, and the Sea, and pass under the *Bridges* built over the Rivers.

If one measures upon a good Map the Extent of Land which amasses all the Waters which flow in a River, over which a *Bridge* is built, we shall find that those which pass in the River *Rhone*, under the *Bridge* of *St. Esprit*, they come from an Extent of Country containing 2800 Leagues square.

Those of the Royal *Bridge* of *Paris*	1700
Those of the River *Tiber* at *Rome*	1100
Those of the *Rône* at *Lyons*	800
Those of the *Garonne* at *Toulouse*	440
And those of the *Thames* at *London*	430

By this Means, upon a Map, one may see the Difference of all Rivers, more or less, and the greater or lesser Quantity of Water that they can furnish, and that passes under the *Bridges*; the Figures of which are given by M. *Gautier*.

If the Extent of square Leagues be cubed, at the Rate of twenty-six Inches nine Lines in Height, we have the Quantity of cubick Fathoms of Water which passes every Year under these *Bridges*, a Deduction being made for what the Sun, Winds, Plants, &c. have not dissipated.

Thele Remarks may be of Use to a Person who lays down the Projection of a *Bridge*, in order to regulate the Opening of the Arches, larger in one Place than another, in Proportion to the Quantity of Rain that falls more or less in one Place than in another, in regard to the Extent of Land the Rivers run through; but this Calculation ought not to be made but by the Testimony of the most antient Inhabitants of the Country.

Many have been of Opinion, that the Inundations that have happened from Time to Time, were caused by certain Revolutions, and regular Periods, which return, after many Ages, as they did before.

The Example of the Observations which have been made at *Rome* upon the *Tiber*, since the building of this City, which was formerly the Capital of the Universe, and since the Time of its first Kings, confirms the contrary; since it is proved by an Account, and by exact Remarks which have been made of it, that never any Inundation, compared with another which went before, has had any Relation to the Precedent. See what follows.

In the Year of the building of *Rome* 240, the River *Tiber* overflow'd after an extraordinary Manner:

In the Year
- 391
- 563
- 546
- 557 it overflow'd twice.
- 591 twice.

In

In the Year
- 600 — it was like a Deluge during twice overflowing.
- 765
- 875 — this last *Tiber* overflow'd twice.

In the Reign of
- Vespasian.
- Nerva.
- Adrian.
- Antoninus Pius.
- Marcus Aurelius.
- Mauritius.

In the Time of the Pontificate of Pope
- G. III.
- Gregory II.
- Adrian I.
- Nicolas I.
- Gregory IX.
- Nicolas III.
- Urban VI. in 1379.
- Martin V.
- Sixtus IV.
- Alexander VI.
- Leo X.
- Clement VII.
- Paul IV. in 1557.
- Pius V. and Sixtus V. 1589.
- Clement V. in 1598.

The Water under the Arches of a *Bridge* ſhould never be forced, ſo as to cauſe any Diſorders or Turbulencies under them, more than they ordinarily have at the Banks of the River within which they ordinarily flow; if there be given to theſe Arches between the Piers a Paſſage equal to that the River had in its natural Bed, ſo that if its Breadth, for Example, were one hundred Fathoms, the Void of the Arches between the Piers and the Abutments ſhould have the like Breadth; to the End, that the Friction or Beating of the Water againſt the Piers of the *Bridge*, when built, be equal to that which they made on the Banks of the River before the building of the *Bridge*.

A *Bridge* may be made equal to the Breadth of the River, by making their Abutments to enter the Land beyond the Banks of the River, as much Space as the Piers occupy in its bed, and by that Means rendering the Preſſure of the Water under the Arches equal to what they were before the building of the *Bridge*.

Of the Rapidity of Waters under Bridges, *and the Means to avoid it.*

It is certain, that the Piers o *Bridges* don't become deſtitute of Gravel, nor oftener fall to Ruin, by any Thing more, than by the Rapidity of thoſe Waters which undermine their Foundations.

If the Current of a River can be diminiſhed, it is certain tha the Piles of a *Bridge* will not be in danger of being ſo often overthrown: And there are but two Ways of leſſening the Current of Rivets; the firſt is by lengthening their Courſe, in making them circulate in a Plain, if it be poſſible, and the great Turnings and Windings that they are made to have, diminiſhing their Declivity, cauſes them to loſe their Force, by Reaſon of their great Compaſs.

This is the Means that the Anticnts made uſe of, in rendering their Rivers navigable, where the Diſpoſition of the Country would permit it, they being unacquainted with the Art of making Sluices.

But this Method is not practicable in ſtopping the Courſe of a River, at a Place where there is Occaſion for a *Bridge*.

The ſecond and laſt Method of diminiſhing the Courſe or Current of a River at the Place of a *Bridge*, and which is what the Antients knew nothing of, is by ſtopping ſhort the Funds of the moſt rapid Rivers, by Rows of Banks, Stakes, or Piles, which cut the Current of the Water in the Bottom of the Bed, and raiſe it to what Height one pleaſes.

Waters augment and diminiſh in Rivers, in Proportion to their greater or leſſer Declivity; which they find in gliding in their Bed, which they hollow or wear by little and little: From the firſt Ages that they have begun to flow, they hollow them more and more every Day, according to the Force they have of carrying along the Sand and Stones in their Inundations.

All theſe Bodies deſcending, rub or wear the Banks of the Rocks, which contain thoſe Rivers, and aggrandize them, or bring them to that Largeneſs in which we ſee them at this Day.

It is in common at theſe rocky Places, where the Rivers are moſt kept in above, and the moſt quiet or ſmooth; and from whence they paſs with greater Rapidity, becauſe of their Fall. Theſe Rocks have given Men the Notion, and put them upon imitating them, and rendering Rivers calm and navigable, artificially, by Retentions, ſo as to make the Waters loſe their Rapidity on their Surface, by giving it to them underneath by their Fall, which they have to leap from above the Sluice made by Art. And this is the Method that ought to be uſed to hinder the waſhing away of the Foundation of a *Bridge*, when it is not founded very low.

The *Bridge* of *Courſan*, in *Languedoc*, one of the fineſt *Bridges* in this Province, upon the River *Aude*, in the Dioceſe of *Narbonne*, has ſhaken, or fallen by this Default.

Rivers may be made to flow with more or leſs Rapidity, as they are more or leſs pent up. I ſhall explain myſelf.

When a Projection of a *Bridge* over a River is made, it is certain that the Piers of Maſonry, or the Rows of Piles or Stakes that are projected there, diminiſh the Bed of a River over which a *Bridge* is to be made.

Now ſuppoſing this Diminution to be but one Fifth; we may conclude for certain, that when Inundations happen, they will hollow the Bed one fifth Part more than they did before the building of the *Bridge*; becauſe the Waters gain in Depth what they have loſt in Breadth.

Again, it is certain that the Bed of a River having been rendered narrower by one Fifth, the Waters which are always the ſame in Quantity in their Current, from their Sources to the Sea, being divided into three Streams or Rivulets, or re-united into Rivers, paſs with greater Quickneſs by one Fifth, in the Place where they are contracted within a ſmaller Compaſs, in order to erect a *Bridge* there, and by Conſequence, waſh away the Foundation; from whence they have more Hold by one Fifth; and they hurry away with this firſt Fifth with the greater Quickneſs, the Flints, *&c.* and ſuch other Bodies, which they

they had not Force enough to carry along with one Fifth leſs of their Weight, or the Quickneſs of their Motion, which I eſteem as equal the one to the other. If the Current of an entire River be retrench'd the one half of its whole Length, there is no Doubt but that the Waters which this River contain'd before, would flow with a double Rapidity; and, on the contrary, that they would diminiſh their Quickneſs by one half, if they ſhould be enlarged one half more than they were.

For Example: The *Pont Royale*, or the Royal *Bridge* of the *Thuilleries* in *Paris*, over the *Seine*, ſeventy Fathoms long, its Piles bar the Breadth of the River, diminiſhing it about one Twelfth; there is no Doubt to be made, but that the Waters paſs one Twelfth quicker through the Arches, than they paſſed before the *Bridge* was made.

For the ſame Reaſon I conclude, that *Pont Neuf*, above that of the *Thuilleries*, being, for Example, twice as large as that of the *Thuilleries* in the Opening of its Arches; for they have about ninety-ſix Fathoms void Space in their Length; whereas thoſe of the *Bridge* of the *Thuilleries* have not above fifty-ſix; the Water of the *Seine*, which paſſes through both theſe, at the Time of great Inundations, muſt paſs ſlower by one half at *Pont Neuf*, than it paſſes at that of the *Thuilleries*.

And, in fine, if the Waters of the *Seine* carried a Flint but of one Pound Weight, under *Pont Neuf*, to the moving of which, they contributed all their Rapidity or Swiftneſs, and Weight; the ſame Waters paſſing under *Pont Royal* of the *Thuilleries*, would carry a Flint of two Pound Weight, according to the Proportion of the Greatneſs of the Opening of the Arches of the one *Bridge* and the other, and the Retrenchment of the Waters at the Time of their Overflowings.

All theſe Notions are eſſential to a Man who projects a *Bridge*.

If we would examine further the Force Waters have upon Bodies of the ſame Matter, but of different Bigneſſes, the Reaſon will eaſily be perceived, why they carry away Sand ſooner than Ballaſt; and the latter, than Flints, than great Blocks of Stone, although all compoſed of one and the ſame Matter.

And when it ſhall be known, that the Movement join'd to all theſe Bodies different in Size, takes ſuch as have the moſt Surface, in reſpect to their Weight, that the Ballaſt, which is the largeſt, the Water raiſes rather the firſt than the laſt, a Perſon will not be ſurprized at all theſe Effects, and will preſently ſee the Reaſon of it.

Thus Sand having more Surface, in reſpect to its Weight, than the Ballaſt which is larger; the Water raiſes the firſt ſooner than the laſt, becauſe it has more Hold of it.

It may be ſeen by this, that the more Bodies are diminiſh'd, the more their Surfaces are augmented; which, in reſpect to the Movement, having ſcarce any Thing but Surface, and very little of Body, they become ſo right, that the least Motion carries them away; as is seen when they are reduced into Dust: And Gold, for Example, being reduced into Leaves, is carried away by the least Puff of Wind.

This, which has been already said, being known, we shall next pass to the other Means used for the building of *Bridges*; which is the lowering of the Waters of the Rivers.

Of Lowering the Waters of Rivers, and the Manner of diverting them, for laying the Foundations of a Bridge.

When it is intended to work upon the Foundations of a *Bridge*, the most proper Season of the Year must be taken for this Work. as the Summer-Time, after the Snow has been all melted.

If the River, in which Piers are to be erected, is situated between two Mountains, and the Course of it cannot be possibly diverted, and carry'd through a Plain, the Architect must content himself with laying the Foundation by fixing first one Pier by Dams, or Water-Stops of Piles, which may direct the Current of the Waters of the River to the opposite Bank, or so that it may run round the Work.

He must render the Waters calm in that Place where he would erect the Piers.

After the Waters of a River have been so diverted, as to have fixed one half of the Piers of the Foundation, then the Current of the Waters are to be turn'd to that Side which has been erected by another Dam, contrary to the first, that has been demolish'd, in order to finish the other Side of the *Bridge*.

When this has been contriv'd, the Architect must examine if there be any Mill-Dams below, which may raise the Course of the Waters, which he ought to cause to be broken down in the Place of the Bank, or Dam; which will be least prejudicial, or will do the least Damage; and to give the River a Currency that Way, to lower the Waters as much as may be.

These Ruptures are made by stripping the Dam or Dike of all its Straightenings, or narrow Passages; and of all those Things that retain the Water in one Place: Where any Opening has been made, there should be nothing left but Piles and Stakes, to be able to be us'd for the shutting in these Openings after the Piers of the *Bridge* have been fixed, and raised higher than the Waters of a Mill-Dam or Pond.

But when the Water of a River, over which you would erect a *Bridge*, may be easily diverted or turn'd off; as when an Island or Islet is to be met with, and that the River may be diverted with one of these Currents; this Facility does infinitely forward the Work: And it is the same, when there happens to be a Plain, where the River has a large Extent; when there is any Inundation, and afterwards retreats again into the former Channel, where it is reduced again to its common Quantity of Waters; then the Foundation of the *Bridge* may be piled the whole Extent of the Plain where the River does not run, when its Waters are at the lowest; and when

when the Foundations of theſe Spaces are laid, a Canal is made through all that Part which is finiſh'd, through which, the Current of the Waters may by little and little be deriv'd, by very ſimple Works, according to the Diſpoſition of the Places, by cutting the Current of the River as deep as may be, and in that Place of its Courſe where it has leaſt Depth.

As if I have but three Foot Depth of Water to engage with, to divert the Courſe of a River, and to conduct it into another Channel, made by the Hands of Men, I make uſe of other very eaſy Works, without Piles, according to the Difficulty to be encountred in that Place.

Theſe eaſy Works are nothing but Racks, in Form of Ladders, which carry the Surface of the Water of the River which is to be turn'd; which are placed on the Side perpendicularly and vertically croſs the Courſe of the Waters, overagainſt and a little below the Channel of Derivation, which has been made by Art, into which the River is to enter into a new Bed. There are made many Ranges of theſe Racks, which do thus croſs the River in Form of a Mole; and a-croſs all the Vacuities of the Bars, the Waters paſs without Interruption.

The Sides of theſe Racks being faſt bound with Cords, at their Croſſings at the croſs Quarters of the Timber, and at the Diſcharges which ſecure them on all Sides, the Channel of Derivation being hollowed, and ready to receive the Waters of the River, many Faſcines are thrown between two of theſe Racks, good Store of which have been provided beforehand, at the Foot of the Work; or of Flints or Stones, for to make them ſink to the Bottom of the Racks; which will cauſe the River to ſwell, and conſtrain it, by little and little, to enter into the little Channel prepared for it.

And one may have the Satisfaction to ſee, according as the ordinary Courſe of the River is obſtructed, and the Waters being abridg'd of their uſual Current, will increaſe in the new Channel of Derivation; ſo that this laſt being commonly but one Tenth, or one Twentieth of that of the River that has been ſtopp'd, it may be perceiv'd viſibly to grow larger, and the Water to carry along with it all that it meets in its Way; as Rocks which they could not remove out of the Way; Stumps and Roots of Trees, that the Workmen were not able to get out of the Way; ſo that by that Time the Waters have paſſed that Way for twenty-four Hours, it becomes ſpacious, and proper to receive all the Waters of the River, were they twice as many as they really are.

This Method, ſays M. *Gautier*, I made uſe of twenty-five Years ſince, upon the River *de Neſte*, in the Plain of *Aventignan*, which empties itſelf into the *Garonne*, at the Foot of the *Upper Pyrenees*, at *Montrejaux*; where the Maſts could not float, but againſt the Rocks; which render'd every Day the Ragers in Danger of periſhing, and where a great many had before periſh'd, running no Risk in croſſing

croſſing the Plain; with Works as eaſy, and even ſlighter, than I had turned the Courſe of the *Aude*, in the *Lower Pyrenees*, within the ſame Bounds, and in as many Places as its Courſe changed, eſpecially where theſe ſimple Works were conſtructed.

Laſtly, I alſo made uſe of the ſame Expedient in turning the River *Orb* above *Beſiers*, for the Uſe of a Mill; where I had five Foot of Water to turn off.

The Sides of theſe Racks are nothing but Trees cleft with Wedges, and bored in the Manner of Ladders.

Elm, or Poplar, &c. are Trees proper for this Work. The Holes are made with a large Boring-Iron, or with ſmall Chiſſels, about the Space of ten or twelve Inches the one from the other.

And the Bats of theſe Racks are only Pieces of Wood, and the Ends of Branches of all Sorts of the like Trees, made into Stakes, from two, to three Inches Diameter, more or leſs; for you muſt have of all Sorts.

Laſtly, The Waters of a River may be lower'd one or two Feet, in reſpect to its Fall, by hollowing the Bed.

For this Purpoſe, ſo much of the Sand is cleared away at the Sides of the River as is judg'd neceſſary, in order to enlarge it by lowering the Waters; for this is certain, the more it is enlarged, the more are the Waters lower'd.

Alſo a Number of Stakes may be ſtuck in thoſe Places, where the Current of the Water is not very rapid, to clear the Bottom of Ballaſt; againſt which Planks may be nailed, which may force the Water to paſs under them with the greater Weight, and conſequently with the greater Rapidity; and by that Means, remove and hollow the Heap of Sand, which it would be very laborious to take away any other Way.

And alſo loaded Boats may be uſed for the ſame Uſe; which faſten'd with Cords, and placed on the Places that are to be cleared of the Ballaſt, by letting them lie there ſome Time, the Water which theſe Boats preſs over them, by ſtopping the Courſe of the River, cauſes it to paſs with a Rapidity according as they are more heavily loaded, ſo that at Length the Waters hollow a Bed for themſelves.

All theſe Methods are made uſe of, more or leſs eaſy, according as Occaſion requires, which the Prudence of him who directs the Work makes uſe of, as he finds moſt for his Purpoſe, that he may have the leaſt Trouble to pile the Foundations of a *Bridge*, having the leſs Height of Water to lower.

When a Perſon has laid down the Projection of a *Bridge*, either of Carpentry, or Maſonry, over any conſiderable River, and there is a Difficulty in bringing to the Place the Materials for the Work, the Ways neceſſary for bringing them are to be prepar'd.

For a *Bridge* of Carpentry, a large Boat is uſed; upon which, Scaffolding is erected, or an Engine is placed, and [a Sonnetti] Inſtrument proper for ramming down

down the Piles; to which Boat are brought in other leſſer Boats, the Piles, Flood-Gares, Stones, Planks, and other Materials.

1. Of what Thickneſs or Subſtance the Butments of all Sorts of *Stone-Bridges* ought to be, or what Proportion they ought to bear to the Arches, and the Weights they are to ſupport.

2. What Proportion the Piers or Pilaſters ought to bear to the Apertures or Heights of the Arches, and the Weights they are laden by.

3. What ought to be the Length of the Vouſſoirs, from their *Intradoſſe* to their *Extradoſſe*, or from their inner Face to their outer Face, for Arches of all Sizes, quite up to their Key-Stone.

4. What Sort of Arches fixed upon one and the ſame Diameter, would be capable of ſuſtaining the greateſt Burthens, and the ſeveral and exact Degrees of Strength, whether they be of an Elliptical, Circular, or Third Point; or *Gothick*, and carried up to what Height you pleaſe.

Theſe Principles, which are here to be laid down, muſt be well known and indubitable.

Which Principles muſt be explain'd by evident Terms, and in a known Language, that they may be intelligible to every body, and lie open to their Judgment.

Now it will require no extraordinary Genius to comprehend what I have to offer towards the clearing up of the four Difficulties here propoſed; and I am not without Hope, that the meaneſt Workman, who has but common Senſe on his Side, will unravel and demonſtrate what I have here to advance.

Firſt, By the Aſſiſtance of common Senſe, and by the Help of ſome ſmall Smattering in Phyſicks, and ſome Skill in Argument together, you may readily comprehend the Combination or Conjunction of the ſeveral Materials employed in the building of Arches and Vaults: For they will give to underſtand, that were it not for the ſeveral Methods of ſquaring or cutting of Stone, or for the Mortar which binds them together, it would be altogether impoſſible to conſtruct an Arch of any Sort.

The Antients only practiſed the firſt Method above, in their fineſt Works, without making uſe of any Mortar; as may be ſeen in the Archivolts of their Arches, whoſe Vouſſoirs have no Mortar at all between their Joints.

An Example of this is ſtill viſible in the antique Aqueduct *Dugard*, in *Languedoc*, and elſewhere; and alſo in the Amphitheatre at *Niſmes*; where the Joints of the Stone are perfectly free from Mortar; as alſo in the Vaulting of the Temple of *Diana* at the ſame Place.

But theſe Examples are not always to be regarded or followed, and eſpecially where the Smalneſs of the Materials (ſuch as Bricks) do abſolutely require to be firmly bound with Mortar, to ſtrengthen the Work, and to enable it to ſtand.

Secondly, It will be neceſſary, that you ſhould be well enough acquainted with Staticks, to know that whatever turns upon an Axis, after the Manner of the Scales of a Balance, whose Brachia are either equal or unequal, will never be *in æquilibrio* with another Body, if that Body be not of equal Weight with the Body at first proposed; or if by a reciprocal Distance from the Centre or Axis, it be not brought to have a Pressure downwards, to counter-balance the Pressure of its opposite Weight.

By the Knowledge of this, we provide against the Push or Impulsion of Arches, by keeping them within due Bounds, by the Assistance of Forces of a Power equal to their Push or Effort to spread or part asunder.

Thirdly, You are to have Recourse to Mechanicks, to judge of the Strength of all these Bodies, and the Power of moving Forces; and in what Respect, and after what Manner, these Bodies, which are supported or born up in the Air, act upon those that are fixed upon Earth, and which are looked upon as immoveable; such as the Arch of a *Bridge*, or a Vault, which have very various Powers upon the Butments and Piers which support them, and which are supposed to be immoveable.

And upon this Account, a Knowledge of Stone-Cutting is necessary; by which you may determine the several Powers of the Voussoirs, one upon another: For there is no Voussoir, which being differently inclin'd, (though it may be cut by the very same Moulds,) but what acts in a Manner peculiar to itself, upon those upon which it is placed, or upon which are placed over it; so that they all act differently, according to the different Inclinations of their Planes.

Fourthly, Geometry will be necessary for the right Understanding of these four Propositions, that you may be able to determine or calculate the Surfaces and solid Content of these Bodies in Question, thereby to come at their Powers, and to compare them with one another.

A small Portion of these above specified Sciences, acting in Concert with common Sense, will be sufficient for the clear conceiving of what I shall here advance.

I shall omit nothing that can contribute to render what I have to say plain and intelligible to Persons of the meanest Capacity, for whom alone I have undertaken this Task.

The Learned M. *De la Hire* pretends to have demonstrated the Push of Vaults, and determin'd the Thickness of Piedroits which support them.

The Vaults in his Work are Arches of *Bridges*; and the Piedroits are the Culees or Abutments.

It is, says he, a Problem, that is one of the most difficult in Architecture, to know the Force of Piedroits of Vaults for sustaining their Push; and Architects have not yet found out any certain Rule for determining it.

This Problem appertains to Me-

Mechanicks; and by the Help of that Science, we may be able to solve it, by making some Suppositions, which are agreed upon as to the Construction of these Sorts of Works.

The Push of Vaults, is the Effort that all the Stones which are form'd, cut into Quoins, and called Voussoirs, make to remove or disperse the Jaumbages or Piedroits which sustain these Vaults.

And as those Persons who have been less bold in their Enterprizes, have allow'd an extraordinary Force to these Piedroits, to render their Works the more durable, as has been the Practice of most of the Antients; and others, on the contrary, have been too dating in making their Piedroits too weak, and so delicate, as not to seem to be able to support the incumbent Weight; I have thought it necessary, by the Help of Geometry, to search after a Rule by which we may arrive at a Certainty, in determining the Force or Power that they ought to have.

It has been generally observ'd, that when the Piedroits of a Vault are too weak to sustain the Push of it, the Vault cracks or splits about the Middle, between its Impost and the Middle of the Key-Stone; and therefore it may be suppos'd, that in the upper Half of the Demi-Arch, the Voussoirs are so firmly united the one with the other, that they form, as it were, but one entire Stone: And upon this Supposition, and the Solidity of the Foundation on which these Piedroits are placed, the Demonstration of the following Rule is establish'd. See *Page* 70. *of the Memoirs of the Academy in the Year* 1712.

M. *De la Hire* enters upon the Business, gives the Figure of a Vault of which he undertakes to shew the Push, and determine the Largeness the Piedroit ought to have that is to support it.

Upon which, our Author M. *Gautier* remarks as follows:

I own ingenuously, says he, that I have not Genius enough to comprehend it. I have never been able to follow his Operation, as he has composed it; and I look upon all that he has told us, as what those who are but half-learn'd, and especially Workmen, are not able to comprehend: For if Algebra, from which he borrows his Assistances, be absolutely necessary to be known, in order to conceive what he says, I believe, scarce any Stone-Cutter, Mason, or Architect, for whom Treatises of this kind ought to be rendred as easy as possible, will be any Thing profited by it; because such Persons in common, don't apply themselves to that Science, as unnecessary to their Profession, and their Time being wholly taken up in the working Part.

If M. *De la Hire* had well solved these Difficulties, in such a Manner as had rendred them intelligible and easy to those who are employed in Building, as he was able to have done, being better qualify'd for it than any Man, he would have done a Work very much desired, and would have prevented me.

We find in *the Memoirs of the Royal Academy of Sciences, in the Year* 1704. upon the Figure of the Extradoſſe or outward Face of a circular Vault, whoſe Vouſſoirs are *in æquilibrio* among themſelves, that a Vault, or a Semicircular Arch, being placed upon the two Piedroits, all the Stones or Vouſſoirs which compoſe this Arch, being made and placed, one with the other, ſo that their Joints being prolonged, all meet together at the Centre of the Arch.

It is evident that all theſe Vouſſoirs have the Form of a Coin larger above than below; by Vertue of which they reſt, and are ſuſtain'd by one another, and reciprocally reſiſt the Effort of their Weight, which would incline them to fall.

The Vouſſoir in the Middle of the Arch, perpendicular to the Horizon, and which is called the Key, or Key-Stone of the Vault is ſuſtain'd on both Sides by the Vouſſoirs next to it, exactly as by inclin'd Planes; and conſequently the Effort it makes towards falling, is not equal to its Weight, but is a certain Part of it, by ſo much greater, than the inclin'd Planes which ſuſtain it are the leſs inclin'd.

So that if they were but never ſo little inclin'd, that is to ſay, perpendicular to the Horizon, as well as the Key of the Vault, it would tend to falling by its own Weight, and would not be any more ſuſtain'd, but would actually fall, if the Cement, which is not here conſider'd, did not hinder it.

The ſecond is ſuſtain'd by the third, which is on the Right or Left of the Key of the Vault, is ſuſtained by a third Vouſſoir; which, by Reaſon of the Form of the Vault, is neceſſarily more inclin'd, in reſpect to the ſecond, than the ſecond is in reſpect to the firſt; and conſequently the ſecond Vouſſoir, in the Effort it makes towards falling, exerts a leſſer Part of its Weight than the firſt.

For the ſame Reaſon all the Vouſſoirs, beginning from the Key of the Vault, proceed always exerting a leſs Part of their whole Weight: And, in fine, this laſt, which is placed upon a Horizontal Surface of a Piedroit, does not exert any Part of its Weight; or, which is the ſame Thing, makes no Effort towards falling, becauſe it is entitely ſuſtained by the Piedroit.

If you will have it, that all theſe Vouſſoirs make an equal Effort towards falling, where they be *in æquilibrio*, it is viſible, that every one, from the Key of the Vault, to the Piedroit, do continually exert a leſſer Part of their whole Weight. The firſt, for Example, exerting but one Half, the ſecond one Third, the third one Fourth, &c. there is no other Way of making theſe different Parts equal, but by augmenting in Proportion the Wholes of which they are Parts; that is to ſay, that the ſecond Vouſſoir be heavier than the firſt, and the third than the ſecond, and ſo on to the laſt; which ought to be of an infinite Weight, that it may make no Effort towards falling; and that one Part being void of Weight, may not be equal to the finite Efforts of the other Vouſſoirs, at leaſt, that this Weight be not infinitely great.

To

To have a clearer Idea of this Matter, there needs only a Reflecting upon this, that all the Voussoirs, except the last, can't let any other Voussoir fall without its being raised; and that they resist this Elevation to a certain Point determin'd by the Greatness of their Weights, and by the Part of it which they exert; that there is none but the the last Voussoir, that can suffer another to fall, without being in some sort elevated or raised itself; and that only gliding Horizontally, that the Weights, as far as they are bounded, bring not a Resistance to the Horizontal Movement; and that they do not begin to have there any finite one, but as when they are conceived as infinite.

M. *De la Hire*, in his *Treatise of Mechanicks*, printed in 1695, has demonstrated, that this was the Proportion upon which the Weight of the Voussoirs of a Semicircular Arch are to be augmented, to the End that they may be all *in æquilibrio*.

Which is the surest Disposition that can be given to a Vault, to make it durable.

Till this Time, Architects had no certain Rule, and wrought in the Dark, or by Guess.

If we number the Degrees of a Quadrant, or quarter of a Circle, from the Middle of the Key of a Vault, to a Piedroit, the Extremity of each Voussoir will appertain to, or be Part of an Arch, by so much greater as it is farther distant from the Key.

According to M. *De la Hire*'s Rule, the Weight of one Voussoir must be augmented above that of the Key, as much as the Tangent of the Arch of this Voussoir exceeds the Tangent of the Arch from the Half of the Key; the Tangent of the last Voussoir necessarily becomes infinite, and so consequently does its Weight.

But as Infinity has no Place in Practice, so the Matter comes to this, To load the last Voussoirs as much as possible, to the End that they may be able to resist the Effort the Vault makes towards Splitting; which is what is called the Push.

M. *Parent* has made an Enquiry what should be the outward Bending, or the Extradosse of a Vault, whose Intradosse, or interior Face, is a Semicircle; and all the Voussoirs *in æquilibrio* by their Weight, according to M. *De la Hire*'s Rule.

For it is manifest, that all these Voussoirs, unequal in a certain Proportion, would bear or push outwards a certain regular Curvature. He has not found it but by Points, but in a very plain Manner.

So that by his Method a Vault may easily be constructed, of which one may be sure, that all the Voussoirs will be *in æquilibrio*.

One considerable Advantage may be gain'd by the Inquisition of M. *Parent*, is, that he has at the same Time discovered the Measure of the Push of a Vault, or what Relation this Push has to the Weight of the Vault.

He only knew, that this Effort was very great, and therefore opposed to it great Masses of of Stones or Abutments; rather too ftrong, than too weak; but we knew not precifely what Proportion to keep to: But this, we may come to a Knowledge of at prefent, that Arts always are improving by the Help of Geometry, *&c.*

See, by what Rules, all thofe who have been Builders, have conducted themfelves, or have been guided by others until this Time, in making Piedroits for the Support of Vaults, as well as the Abutments of *Bridges.*

The Learned Father *Deran*, in his Treatife *De la Coup des Pierres*, and M. *Blondel*, Architect to the King of *France*, as skilful Men as any the preceding Age has afforded, in his Treatife of Architecture proceeded on the fame Foot as thofe who fucceeded them.

In any kind of Vault or Arch of a *Bridge* whatfoever, fay they, either Elliptick, Semicircular, or of the third Point, and a Portion of a Circle divided in the Circumference in the Intradoffe, or interior Face, into three equal Parts, *Plate* I. *Fig.* I. as may be feen in the femicircular Figure at A O, O P, and L M, prolonged in on, P M in S, fo that M S may be equal to P M. Let fall a Diameter S R prolonged from the Point S upon A R, which will determine the Thicknefs of the Abutment by M R, prolong this Perpendicular to Q, until it meets or touches E Q, being continu'd; and this will give the Thicknefs of the Mafonry below, or to the Bottom of the Intradoffe.

This Operation is not proved to make it evident, that it is juft and true: So there is nothing to be faid of it; and fo what follows muft be put to the Venture.

That which is the moft remarkable in this Conftruction of all thefe different Arches, is, fays M. *Blondel*, the Differences of their Pufhes, that is, the Force which they have each in particular to charge the Piers or Piedroits, which bear them, more or lefs: For it is certain, the higher an Arch is carry'd, the lefs does it pufh. When on the contrary, Flat or Straight Arches, are fuch whofe Pufh is the ftrongeft; which is either augmented or diminifh'd, according to the Difference of their Straightnefs, more or lefs.

Therefore it is to the Purpofe, to give the different Thickneffes of Piers or Piedroits, according to the Difference of their Pufhes, conformable to the Operation which I fhall produce anon.

See what he has follow'd upon this Difficulty in the Execution of his *Bridge* of *Xaintes.*

He makes his Piers as three to eight, in Proportion to the Opening of the Arches, and to the Pier at the End of the *Pont-Levis*, which ferves for an Abutment, and which is cut off, he has allowed one Sixth of Breadth more; becaufe it fhould fuftain on this Side the Pufh of all the ches which are at the Centre; by which it will be eafy to judge whether he has follow'd the Method which he has prefcrib'd to us, and which he makes a general Rule.

Arches,

Palladio tells us, that the Heads of *Bridges*, by which Name, he calls the Abutments, ought to be very ſolid, and adviſes to make them at thoſe Places where the Brinks of Rivers are rocky or ſoft Gravel, or good Ground; or that elſe they muſt be made firm by Art, by other Piers, or by other Arches.

He afterwards ſpeaking of Arches, ſays, that thoſe that are ſemicircular, are the ſtrongeſt; becauſe they bear entirely upon the Piers, without puſhing one another.

When one is conſtrain'd by a too great Height, one may make them diminiſh'd Straight Arches, ſo that the perpendicular Height upon the Line of their Chord may be the Third of the ſame Chord; in which Caſe, the Abutments muſt be extremely well fortify'd.

In the Deſcription that he has given of the *Bridge* of *Rimini*, built by *Auguſtus*, he gives ſeven Feet and a half to the Abutments, which ſupport Arches of an Opening of twenty Feet. He afterwards gives the Proportions of that built upon *Bachiglione*, an antique *Bridge*, where the Abutments are three Foot and a half wide, ſupporting Arches of an Opening of twenty Feet.

The Abutments of the antient *Bridge* of *Rerone*, have no more Breadth than three Foot and a half, and ſupport Arches of an Opening of twenty-five Feet, as he tells us.

And this all the Account that Architects have given us concerning the Abutments of *Bridges*: From which, certainly, we cannot take any Meaſures for making a general Rule, and coming to any Certainty, nor to be able to give any Reaſon for what we do.

But I ſhall, ſays M. *Gautier*, undertake one without any Preamble.

If we examine the firſt Figure, we ſhall ſee that AM is the Diameter of a Semicircular Arch AEM; or it may be flat, *Gothick*, or any other that you pleaſe, whoſe Puſh you are deſirous to know, in order to proportion a Butment to it, or a Power equal to it.

Continue the Diameter MA indefinitely, and upon a Level towards C.

Raiſe the Perpendicular AD indefinitely, upon the Foot of the Arch A, whoſe Point A muſt be look'd upon as immoveable.

Then from the Point of Support, or the Foot of the Arch A, to the Middle of the Bottom of the Key, produce the Line AE, and from the aforeſaid Point of Support A, with the Opening AE, deſcribe the Quadrant DEB, which will interſect the Line AM in the Point B, and AD indefinitely in D. Now it is certain, that AB, AE, and AD, are equal in this Caſe, as they are all Radii of one and the ſame Circle. Produce AC alſo equal to AB.

Then produce the Hypotheneuſe BD, which ſhall interſect AE in the Point I.

From the Point I, let fall the Perpendicular IL upon AD, which will be the half of AB.

From the Summit or Top of E, produce indefinitely the Line EG parallel to BC, which will intersect AD in H, and set off the Line IL from H to G, which will serve for a Butment to the Arch AEM, by letting fall the Perpendicular GU.

DEMONSTRATION.

1. If you examine the Disposition of this Figure, you will find that CB being upon a Level, and AB being considered as the Half of the Platband or Face of a Beam, &c. it could not remain in that Posture, if it was not counterbalanced by AG which is equal to it, and opposite with it on the other Side of the Point of Suspension, support A, which is immoveable.

Now AC is equal to AB, whether in Length, in Power, or in Weight, &c. then AC must equiponderate, or be *in æquibrio* with AB. Let us here suppose AB to have a Force of ninety Degrees; thereby the better to illustrate our Demonstration, and make it familiar to every Body.

2. But this Half of the Platband or Beam AB is raised perpendicularly upon itself in the Point of Support A, as may be observ'd in AD.

So that neither leaning to one Side nor the other, it can have no Power nor Pressure towards B or C, as it had before, so that there is no Need of any Thing to keep it *in æquilibrio*; for the Point of Support A is sufficient for that, inasmuch as we here suppose it to be immoveable: And thus the Platband or Beam perpendicularly placed in AD will have no Force; which we will express by a Cypher, instead of having a Force which we have here called ninety Degrees, when its Direction was Horizontal, as AB.

3. In short, this same Platband or Beam AB, whose Force is ninety Degrees, being placed in AE, with an Inclination of forty-five Degrees, between AD oo, and AB 90. It is certain, that its Power will be mix'd, or to speak plainer, it will be a mean Proportion between AB and AD; so that if the first has a Push of ninety Degrees, and the Force of the other is nothing, AE will observe a Medium between them, and consequently will have a Force of forty-five: And thus the Half of AB, which is AN, will be sufficient to counterbalance it, which is the same with IL, or HG, which will be able to counter-balance AE, which was to be prov'd: So that AB, the Half of the Platband or Beam, whose Force is ninety Degrees, being to AD oo, as IL, the Half of AB, whose Force is forty-five, is to LD oo, and in a reciprocal Proportion to each other, the Push AE of the Arch AEM will be HG, as that of the Platband AB will be AC, which was to be demonstrated.

Thus you are to proceed in investigating the Pushes of Scheam-Arches, which are less than a Semicircle, as well as those of *Gothick* Arches, which are more, by comparing an Arch of any Sort with the Demonstration here laid down in relation to a Semicircular Arch, whose Push is always determin'd by the Line BD; and in this is equal to IL.

From what I have said above, it will be evident, that MR, which

which is the Butment, according to the Method of Father *Deran*, and M. *Blondel*, is widely different from AV, which I have juſt now demonſtrated to be the proper Butment.

That if the Arch AEM was converted into a Platband, or Straight Arch AM, the Half of it would be counterbalanced by AT, which is equal to it.

I apprehend, that the meereſt Smatterers in Geometry, ſuch as the greateſt Part of Maſter-Maſons and Stone-Cutters are, will underſtand what I have been ſaying, and be able to trace it out, and demonſtrate it to their own Satisfaction.

Of the Straight Arch, Fig. II.

Let GA and HC be the Sides of the Walls of the Butments, or rather of the Piedroits, which are to ſuſtain a Straight Arch.

Let AC be the Space between the Walls; which alſo will be the Length of the Straight Arch; and which we will ſuppoſe to be ten Feet, or ten Fathoms, or any Meaſure elſe you pleaſe.

Draw the Line CA, and with the Opening of the Compaſſes AG, conſtitute the equilateral Triangle AFC, and from the Point F with the ſame Opening of the Compaſſes FA, deſcribe the Segment of a Circle AEG, which ſhall be biſected equally in E, upon the aforeſaid Segment AEC, the Height BE, which is about one Eighth of AC, will give the Height of the Straight Arch, for ſquaring or cutting of the Stones which are to compoſe it.

Then continue CA to D, ſo that AE may be equal to BE, the Half AC: Now it muſt be evident, that the Puſh, Preſſure, or Impulſion of AB being equal to the Reſiſtance or Force of DA, the Straight Arch cannot ſhove or puſh the Wall GD beyond the Axis CA; and thus AD will be the Thickneſs of the Butment and Piedroit, or the Walls which are to ſuſtain the Impulſion or Puſh of the Platband or Straight Arch; which, indeed, is more frequently uſed in Civil Structures, to ſupport Platforms, Ceilings, or Roofs of Galleries, or other Paſſages a-croſs a Court; and in Churches, to ſerve for Tribunes, than for *Bridges*. There is a very fine one in the Church of the Jeſuits at *Niſmes*, which was conſtructed by the Direction of the Father *Mourgues*, and after the Deſign of the late Sieur *Cubiſol*, who was a very skilful Architect, to which he gave leſs Height at BE than what we allow here; whether it was, that he was aſſured of the Soundneſs or Solidity of the Stones, or the Truth of the Work, &c. As his Deſign is perfectly bold, I ſhall here give ſome Account of it.

The Platband or Straight Arch we are here ſpeaking of, is four Fathoms, two Feet, and ſix Inches in Length; the Stones of it are one Foot in Thickneſs; their Height BE is two Foot towards the Key, and at each End AGCH.

They begin with two Foot four Inches. This Platband or Straight Arch had a Riſe about ſix or ſeven Inches at B, when its Stones were ſet together upon the Centre; but it afterwards sunk down three Inches, when its Joints came to settle, upon the taking away of the Centering or Stay; so that at this Day, it rises about four Inches above B.

The Practice of this Sort of Work, and a Knowledge of the Soundness or Hardness of the Stone, are what must guide you in regulating the Height BE; but for want of sufficient Experiments, we are still in the Dark as to this Matter; there not having yet been found out any Rule in Mechanicks to determine it.

The skilful Architect must do in this according to the best of his Knowledge: If he succeeds, he will be esteem'd and applauded; but if he fails, he is despised and laughed at.

However, by repeated Experiments, and Proofs of the Consistence of Stones, it would not be impossible to ascertain some sure Rules, with regard to this Matter.

For it is of the greatest Importance to know the different Solidity and Firmness of Stones, and other Bodies, which vary considerably, according to their several Climates, and Grains, that we may be able to proportion their Substance to the Efforts, Pressures, or Pushes of Straight Arches, or others, which being once calculated (*viz.*) the Weights of the Efforts of the Bodies, which the Haunses of the Platbands of the Voussoirs of Arches are to sustain.

We may imagine the Strength of Stones we are to use by taking a Piece of them of a Cubical Inch in Dimension, that is, in Form of a Dye, and by loading it with Weights, till it disperses, and yields to the Pressure of its super-incumbent Burthen; from whence some certain Rules may be established with regard to this Matter. Thus,

If such a Cube of Stone, as we have mention'd, supports a Weight a Thousand or a Million of Times heavier than its own Weight, and bigger than itself, we will allow it but one quarter of that Resistance or Force, when we come to apply it to the Building of our Works, whether they be Piers, or *Bridges*, which are to sustain the greatest Burthens; or whether they be Voussoirs of Vaults, or Archivolts of *Bridges*; which are the Parts that give or exert the greatest Efforts or Push, as well as Stones of Straight Arches, whether they are to support Towers, or Steeples, *&c.*

Thus I allow three Fourths of the Strength of those Bodies, to make Amends for the Imperfections of the Work; for there is no Man whatever that can set or join them together in any Building, with that exact Arrangement in which they were placed by Nature in the Beds or Strata of the Quarries from whence they were raised.

Clumsey Joints, stuff'd up with Mortar, and Shells, which do not bear throughout equally, and consequently yield to the Pressure of the superincumbent Weight, is the Reason why Buildings every Day split and settle from the Difference of the Binding,

Binding, which is not equally firm in all its Parts; from whence arise very disagreeable Deformities, and Eye-sores, as well as very prejudicial Accidents to the Work itself.

The Joints between B and C of the Platband or Straight Arch, are indented, (*Plate* II. which are sometimes ordered after a very different Manner; for there are those who like a plain uniform Joint best, such as you see between B and A; but this depends upon the Skill, or at least the Fancy of the Architect: For the more complex Joints are (say some,) the more they are confus'd, and the more subject to Defects; but the more simple and plain they are, the truer and stronger is the Work; as those Joints between A and B, which are plain.

What I have here offered by Way of discovering the several Degrees of the Strength and Solidity of Materials, is something like the Experiment made by the *Gentlemen of the Royal Academy of Sciences*, upon a twisted Cord, composed of twenty Yarns or Threads; each of these Yarns, when single, were capable of sustaining one Pound Weight without breaking; but being twisted together, and converted into a Cord, they could not bear up a Weight of twenty Pounds; for it broke with a Weight of between sixteen and eighteen Pounds. For it is impossible for human Art to twist these twenty Yarns so nicely together, as that all of them shall sustain an equal Share of the Burthen suspended by them, when in a Cord; so that some of them loaded with a Weight superior to their Strength, they are all in general unequally or disproportionally laden; from whence it must appear, that they must be unable to bear the Weight of the twenty Pounds before mention'd.

Thus supposing a Cubical Inch of Stone to be able to bear a hundred Pound, it would not from thence follow, that ten such Cubes would sustain a Burthen of a thousand Pounds; because a Mass or Body of that Weight would not bear alike upon all the Ten; wherefore some of them being more heavily laden than others, they would all be crush'd one after another; from whence happens the flying or breaking of the Stones of an Archivolt, and of all Bodies; which are not upon an Equality throughout the whole Plane of their Superficies, or Beds.

The following Table gives the Proportions of the Length of the Voussoirs, or the Heights of the Archivolts, as they can be gathered and ascertain'd from the best Authors Works of Antiquity; which cannot be reduced to an exact Geometrical Demonstration, and has been merely calculated from the Experience of the more solid or coarse Consistence of Stones; upon which Point, this whole Article turns.

So that Physicks bear a greater Share of this, than either Mechanicks or Geometry.

What ought to be the Largeneſs of Piers in reſpect to the Openings of the Arches, and the Weights they ſuſtain.

The Size of Piers ought to be preciſely determin'd according to the Spring of the Arches.

No Perſon yet has laid down any certain Rule as to this. I ſhall relate what the moſt accompliſh'd Architects have told us in relation to this Matter.

Leon Baptiſte Albert would have the Piers of a *Bridge* be equal in Number and Size, and their Largeneſs, one third Part of the Opening of the Arch.

Palladio ſays, the Piers ought to be equal in Number, to the End that there may be one Arch in the Middle, where there is commonly the greateſt Current of the Water.

That the Piers ought not to be leſs than a ſixth Part, nor commonly more than a fourth Part of the Width of the Arch.

He next gives ſome Examples of antient *Bridges*, and ſays, that thoſe of the ancient *Bridge* of *Rimini* are eleven Feet, and the Opening of the Arches twenty-five Feet: That the *Bridge* of *Bachiglione*, which is alſo antient, the Piers are five Feet, and the Arches thirty in their Opening: That the *Bridge* over the *Rerone* has its Piers alſo five Feet, which ſupport an Arch whoſe Opening is twenty-nine.

He, after this, gives the Projection of a *Bridge*; where he makes the Piers two Fathoms wide, to ſupport an Arch whoſe Opening is ten Fathoms.

Serlio ſays, that the Piers at

of Opening from fifteen to sixteen Fathoms, or thereabouts.

Too great a Variety in all these Works, gives us Ground to think, that Authors have not yet observed any certain or general Rule, founded upon demonstrable Principles for establishing the Piers of *Bridges*.

But nevertheless, one may draw from all these Models something that may be serviceable to us on Occasion; it is not to be doubted, but that these able Architects, who have conducted all these Works, had a Reason for all the Projections of these Piers, before they erected them.

I do not all doubt, but that a Pier of two Fathoms wide, wholly built of large Blocks of hewn Stone, as is that of the Antique *Pont Degard*, which bears more, perhaps, than any other in the World besides, will not be sufficient to support the Effort of an Arch of an Opening of twenty Fathoms; when another Pier of four Fathoms will not support an Arch of ten, which shall be constructed only of hewn Stone, and the Inside of the Body of the Work of simple rough Walling, or Shards or Pieces of unhewn Stone.

This would break sooner than the first: this would heap up all the Weight of its Charge or Load, and the last would be immoveable.

It is upon these Principles, and the using Materials more or less solid, and differently arranged, which we should have Regard to, and determine as to the Width of Piers in all Sorts of *Bridges*.

The same Materials used in different Countries, some of which have a firmer Consistence than others used for building Gates of Cities, Fortifications, Towers, Steeples, *Bridges*, &c. might be examin'd, in order to gain from thence the Advantage one desires for projecting a *Bridge*, greater or smaller, in these Places.

And although by all these Examples I have form'd a Rule for determining the Piers of all Sorts of *Bridges*, according to the foregoing Tables, which is one Fifth in an Arch of ten Fathoms Opening; a larger or lesser Breadth may be allow'd to Piers, in regard to the greater or lesser Solidity of the Materials that are made use of.

With this Remark, that if the Bed of a River is very wide, then a greater Breadth may be allow'd; because there is no need of fearing of penning up the Waters, in Case of Inundations, but on the contrary, in the Beds of Rivers that are too close pent up and straighten'd, it will be of great Importance, not to give the Piers of *Bridges* which one builds, but as little Width as can be, that we may be able to support without Fear the Load of the Arches; especially when at the same Time we are ty'd up by the unfavourable Disposition of the Places.

In constructing the following Table, I have observed the Proportion of one Fifth of the Width of the Piers, in respect to the Openings of the Arches, from those of twenty Feet Opening, and upwards; and those which are under ten, a small Arch of three Feet Opening, and even to that of one Foot.

It is found by this Table, that a Pier ought to be one Foot ten Inches wide for an Opening of three Feet; and one Foot six Inches for an Opening of one Foot; which may be practised upon any Common-Sewer, or on any Stream of Water, how small soever, whose Piers are made all of hewn Stone, when by Reason of the bad Situation of the Place, we shall be obliged to it; and the Whole is proportion'd not only to the Mass of Masonry, which the Piers ought to support, but also to the Stream of Water that is to pass under it.

As we cannot come at the Knowledge of the Solidity of the Materials, but by making Trials, in order to know how far their Effort will be able to sustain the Weight that they are to be charged with, Experiments may be made, as in the preceding Chapter, upon which this Table is fram'd; since there is no Rule, which exactly determines the Breadth of the Piers more after one Manner than another.

The Table which I have given of the Breadth of the Piers of all Sorts of Arches, from that of twenty Fathoms Opening, is proportion'd as well as possibly I could do it, from all that has been done; on which I have thought proper to make Observations as to this Matter.

What Bearing Voussoirs should have from their Intradosse or interior Curvity or Face of the Arch, to their Extradosse or exterior Curvity in Arches of all Largenesses, to the Place of the Key-Stone.

See what the most able Architects have said, who have written on this Matter.

M. *Blondel* says, that they have not so solid Stone at *Paris*, for the building of *Bridges*, as the *Romans* have in *Italy*.

And that to supply this Defect, they have made at *Pont Neuf*, and the *Thuilleries*, Voussoirs of a great Length, and at the same Time very well secured by the Returns and Courses of the Crossettes, to make an infinitely greater Binding, and to take a much better Hold.

The *Bridge* of *Toulouse* may, without Difficulty, be put upon a Foot parallel with the finest in *Europe*.

Nevertheless it is built with nothing else but Bricks, except that at the Angles or Heads of the Arches, there are some Rows in the Intradosse, or interior Face of the Arch, where hewn Stones are us'd; which are certainly no more than the principal Parts of its Ornaments; and it may be said, that although the Arches which are about sixteen, and so many Fathoms the Opening, are, for all that, made only of Brick, situated in the Cut, according to the Bearing which Voussoirs and Pendants should have.

After that Manner, that this Disposition, so well establish'd together with good Mortar, which they used, and that makes the Binding, forms a Work which seems to be all of one Piece, although composed of very small Materials.

And this is the Reason that the Arrangement join'd to the Solidity of these Materials, is the Cause of the Whole of its Goodness.

Leon-

Leon-Baptista Alberti says, that the Height of the Headband or Fillet of Arches in considerable *Bridges*, which is what the *French* call Voussoirs, or their Bearing from the Intradosse to their Extradosse, when they are thus determin'd on their Extradosse Arches, ought never to be less than one Fifteenth of the Width of the Opening of the Arches which they form.

It is upon this, I have establish'd the Column of Voussoirs in the following Table, supposing the Masonry to consist entirely of large Blocks of very hard hewn Stone.

And it is upon the same Principles, that the antient *Bridge Dugard* was made.

But neverthelefs I have not omitted to determine the same Voussoirs, when there are none but soft or tender Stones to be used.

Palladio says, that the Voussoirs of the Arches of *Bridges* should be made of very long and well jointed Stones, but does not determine their Length. When he speaks of the *Bridge* of *Rimini*, whose Arches are semicircular, the Voussoirs, where the Headband has one Tenth of the Opening of the Arches, which are twenty Feet in Diameter.

In that of *Bachiglione*, whose Arches are Scheam ones, that in the Middle of the Opening of thirty Feet, the Height of its Headband is one Twelfth of the Diameter; and the Space above the Key-Stone of the great Arch, which is between the Headband and the Cornice, is equal to half the Headband.

In the ancient *Bridge* of *Verone*, as the two before mentioned upon an Arch of an Opening of twenty-nine Feet, the Headband has the same Proportion as the preceding.

In a particular Design which *Palladio* gives of a very fine *Bridge* which he has projected, the grand Arch of which has ten Fathoms for the Diameter [flatted] makes not the Headband, or the Length of the Voussoirs, but one Seventeenth of the Width of the grand Arch, and one Fourteenth of the smaller, which have an Opening of eight Fathoms.

See what *Serlio* says as to the *Palatine Bridge* at *Rome*, antiently called *Senatorius*, he remarks, that the Headband of the Arch at its greatest Height, is one Twelfth of the Breadth of the Arch.

As to the *Bridge* of *Quatro Capi*, antiently called *Fabricius*, of which there are but two antique Arches remaining, the Headband of the Arches which are of the Rustick Form, and whose Voussoirs are the one longer than the other alternately, that which has the most Bearing, is one Tenth of the Breadth of the Arch.

The *Bridge Milvius* has its Headband in Projecture, in Form of a Plinth quite plain, and whose Height is one Tenth of the Diameter of the Arch.

This is the Substance of all that the most accomplished Architects have left us as to the Proportions of Voussoirs.

A Table

A Table of the Proportion of all the principal Parts of Semicircular *Bridges*, from an Arch of one Foot opening to an Arch of twenty Fathoms, or one hundred and twenty Feet; the Differences of their Culees, or Abutments, Piers, and Vouſſoirs.

The Openings of Arches.	Culees or Abutments.			Piers.			Vouſſoirs of hard Stones.			Vouſſoirs of ſoft Stones.		
Feet.	Ft.	Inc.	Lin.	Ft.	Inc.	Lin.	Ft.	Inc.	Lin.	Ft.	Inc.	Lin.
1	2	6	6	1	6	0	1	0	6	1	6	0
2	2	9	0	1	8	0	1	1	0	1	7	2
3	2	11	6	1	10	0	1	1	6	1	8	4
4	3	2	0	2	0	0	1	2	0	1	9	6
5	3	4	6	2	2	0	1	2	6	1	10	8
6	3	7	0	2	4	0	1	3	0	2	0	0
7	3	9	6	2	6	0	1	3	6	2	0	8
8	4	0	0	2	8	0	1	4	0	2	1	6
9	4	2	6	2	10	0	1	4	6	2	2	3
10	4	5	0	3	0	0	1	5	0	2	3	0
11	4	6	9	3	1	3	1	5	6	2	4	0
12	4	8	6	3	2	6	1	6	0	2	4	6
13	4	9	9	3	3	9	1	6	6	2	5	0
14	5	0	0	3	5	0	1	7	0	2	6	0
15	5	1	9	3	6	3	1	7	6	2	6	9
16	5	3	6	3	7	6	1	8	0	2	7	0
17	5	5	3	3	8	9	1	8	6	2	8	0
18	5	6	0	3	9	0	1	9	0	2	9	0
19	5	7	9	3	10	3	1	9	6	2	9	3
20	5	10	0	4	0	0	1	10	0	2	9	6
21	6	0	11	4	2	5	1	10	6	2	9	9
22	6	4	0	4	5	0	1	11	0	2	10	0
23	6	6	6	4	7	0	1	11	6	2	10	3
24	6	9	7	4	9	7	2	0	0	2	10	6
25	7	0	6	5	0	0	2	0	6	2	10	9
26	7	3	5	5	2	5	2	1	0	2	11	0
27	7	6	6	5	5	0	2	1	6	2	11	3
28	7	9	0	5	7	0	2	2	0	2	11	6
29	7	11	7	5	4	7	2	2	6	2	11	9
30	8	3	0	6	0	0	2	3	0	3	0	0

The Openings of Arches.	Culees or Abutments.			Piers.			Vouffoirs of hard Stones.			Vouffoirs of foft Stones.		
Feet.	Ft.	Inc.	Lin.	Ft.	Inc.	Lin.	Ft.	Inc.	Lin.	Ft.	Inc.	Lin.
31	8	5	11	6	2	5	2	3	6	3	0	10
32	8	9	0	6	5	0	2	4	0	3	1	8
33	8	11	0	6	7	0	2	4	6	3	2	6
34	9	2	7	6	9	7	2	5	0	3	3	0
35	9	5	6	7	0	0	2	5	6	3	3	10
36	9	6	6	7	2	6	2	6	0	3	4	0
37	9	9	6	7	5	0	2	6	6	3	4	6
38	10	0	0	7	7	0	2	7	0	3	5	0
39	20	3	1	7	9	7	2	7	6	3	5	6
40	10	8	0	8	0	0	2	8	0	3	8	0
41	10	11	3	8	2	5	2	8	10	3	8	10
42	11	2	8	8	5	0	2	9	8	3	9	8
43	11	5	6	8	7	0	2	10	6	3	10	6
44	11	8	11	8	9	7	2	11	4	3	11	4
45	12	0	0	9	0	0	3	0	0	4	0	0
46	12	3	3	9	2	5	3	0	10	4	0	10
47	12	6	8	9	5	0	3	1	8	4	1	8
48	12	10	0	9	7	0	3	2	6	4	2	6
49	13	0	11	9	9	7	3	3	4	4	3	0
50	13	4	0	10	0	0	3	4	0	4	3	10
51	13	7	3	10	2	5	3	4	10	4	4	8
52	13	10	8	10	5	0	3	5	8	4	5	6
53	14	1	6	10	7	0	3	6	6	4	6	4
54	14	4	11	10	9	7	3	7	4	4	7	2
55	14	8	0	11	0	0	3	8	0	4	8	0
56	14	11	3	11	2	5	3	8	10	4	8	10
57	15	2	8	11	5	0	3	9	8	4	9	7
58	15	5	6	11	7	0	3	10	6	4	10	3
59	15	8	11	11	9	7	3	11	4	4	11	2
60	16	0	0	12	0	0	4	0	0	5	0	0
61	16	3	3	12	2	5	4	0	10	5	0	10
62	16	6	8	12	5	0	4	1	8	5	1	8
63	16	9	6	12	7	0	4	2	6	5	2	6
64	17	0	11	12	9	7	4	3	4	5	3	0
65	17	4	0	13	0	0	4	4	0	5	3	10
66	17	7	3	13	2	5	4	4	10	5	4	8
67	17	10	8	13	5	0	4	5	8	5	5	6
68	18	1	6	13	7	0	4	6	6	5	6	4
69	18	3	11	13	9	7	4	5	4	5	7	2
70	18	6	0	14	0	0	4	6	0	5	8	0
71	18	11	3	14	2	5	4	8	10	5	8	10
72	19	3	8	14	5	0	4	9	8	5	9	7
73	19	5	6	14	7	0	4	10	6	5	10	3
74	19	8	11	14	9	7	4	11	4	5	11	2
75	20	0	0	15	0	0	5	0	0	6	0	0
76	20	3	3	15	2	5	5	0	10	6	0	10

The Openings of Arches.	Culees or Abutments.			Piers.			Vouſſoirs of hard Stones.			Vouſſoirs of ſoft Stones.		
Feet.	Ft.	Inc.	Lin.	Ft.	Inc.	Lin.	Ft.	Inc.	Lin.	Ft.	Inc.	Lin.
77	20	6	8	15	5	0	5	1	8	6	1	8
78	20	9	6	15	7	0	5	2	6	6	2	6
79	21	0	11	15	9	7	5	3	4	6	3	0
80	21	4	0	16	0	0	5	4	0	6	3	10
81	21	7	6	16	2	5	5	4	10	6	4	8
82	21	10	8	16	5	0	5	5	8	6	5	6
83	22	1	6	16	7	0	5	6	6	6	6	4
84	22	4	11	16	9	7	5	7	4	6	7	2
85	22	8	0	17	0	0	5	8	0	6	8	0
86	22	11	3	17	2	5	5	8	10	6	8	10
87	23	2	8	17	5	0	5	9	8	6	9	7
88	23	5	0	17	7	0	5	10	6	6	10	3
89	23	8	11	17	9	7	5	11	4	6	11	2
90	24	0	0	18	0	0	6	0	0	7	0	0
91	24	3	3	18	2	5	6	0	10	7	0	10
92	24	6	8	18	5	0	6	1	8	7	1	8
93	24	9	6	18	7	0	6	2	6	7	2	6
94	25	0	11	18	9	7	6	3	4	7	3	0
95	25	4	0	19	0	0	6	4	0	7	3	10
96	25	7	3	19	2	5	6	4	10	7	4	8
97	25	10	8	19	5	0	6	5	8	7	5	6
98	26	1	6	19	7	0	6	6	7	7	6	4
99	26	4	11	19	9	7	6	7	4	7	7	2
100	26	8	0	20	0	0	6	8	0	7	8	0
101	26	11	3	20	2	5	6	8	10	7	8	10
102	27	2	8	20	5	0	6	9	8	7	9	7
103	27	5	6	20	7	0	6	10	6	7	10	3
104	27	8	11	20	9	7	6	11	4	7	11	2
105	28	0	0	21	0	0	7	0	0	8	0	0
106	28	3	3	21	2	5	7	0	10	8	0	10
107	28	6	8	21	5	0	7	1	8	8	1	8
108	28	9	6	21	7	0	7	2	6	8	2	6
109	26	0	11	21	9	7	7	3	4	8	3	0
110	29	4	0	22	0	0	7	4	0	8	3	10
111	29	7	3	22	2	5	7	4	10	8	4	8
112	29	10	8	22	5	0	7	5	8	8	5	6
113	30	1	6	22	7	0	7	6	6	8	6	4
114	30	4	11	22	9	7	7	7	4	8	7	2
115	30	8	0	23	0	0	7	8	0	8	8	0
116	30	11	8	23	2	5	7	8	10	8	8	10
117	31	2	8	23	5	0	7	9	8	8	9	7
118	31	5	6	23	7	0	7	10	6	8	10	3
119	31	8	11	23	9	7	7	11	4	8	11	2
120	32	0	0	24	0	0	8	0	0	9	0	0

N. B. *That which I have frequently rendered here,* Headband *or* Fillet, *M.* Gautier *expresses by* Bandeau, *which is the common* English, *according to* A. Boyer: *But M.* Gautier *makes it the same as* Intradosse; *which he explains to be the interior Curvity of a Vault, Arch, or Voussoirs of a* Bridge.

Experiment.

In order to be fully confirm'd in what I have here advanced concerning the Push or Pressure of Arches, Vaults, the Length of Voussoirs, or the Height of Archivolts, says M. *Gautier*, I had Recourse to the following Experiments.

I got a small Semi-Arch of ten certain Measures in Diameter, and semicircular, as in the last Table.

This Semi-Arch I composed of nine Voussoirs (See *Fig.* II. AB CD, *Plate* I.) which were made of Wood, and turn'd it up against a Wall in ABE, as against an immoveable Key upon the Half-Centre BCE.

Having thus placed the nine Voussoirs CB upon the Half-Centre CB, I loaded them behind with other Pieces of Wood, equal to them in Bulk, and alike in Weight: I then placed nine more one upon another, after the same Manner, as you may see in the Figures FG, and behind them, I ranged the four others at HI.

This done, I uncentred the Semi-Arch EBC, and it remained unmoveable in the Position you see it in the *Figure* I.

I afterwards took away the Pieces of Wood which form the Butment, and keep in the Reins or Sides of the Arch one after another, beginning at the Top, according to the Numbers 9, 8, 7, 6, 5, 1, 13, 4, 12, 3; so that there remained but the four undermost, *viz.* 11, 2, 10, 11; which four supported the Semi-Arch without falling, ABCD; But as soon as I began to remove the 11th, the Voussoirs fell asunder.

This Experiment gave me to understand, in the first Place, that the Mass of Stone-Work, which the Sides or Declivity of Vaults are laden by, serve them for a Support to keep the Voussoirs *in æquilibrio*, that they may not deviate or depart from the Curve form'd by their Centre, and likewise that their strongest Push is in that Place.

2. Though this Experiment was made with Pieces of Wood (which consequently were very light,) that had nothing between their Joints to keep them together, they nevertheless keep each other up by means of their Mould or Cut; the Pattern of which I had given to a Joiner to make them by.

Thirdly, This Experiment confirms me in the Opinion, that if we made our Archivolts without any Mortar, Cement, or Cramps, in Imitation of the Antients, and afterwards ran a kind of fine Mortar, or Cement made of pulveriz'd Stones into the several Abreuvoirs, the Work would be much more durable than it can be with thick beds of Mortar, which yield to the immenſe Weight of the ſuperincumbent Stones.

Fourthly, From this Foot of the Semi-Arch I raiſed the Perpendicular CK, and found that the Vouſſoirs BC, which are interſected by the Line CK, never attempted to fall aſunder till the Part of the Butment CH was made lighter than the Semi-Arch CB, which confirms what I ſaid before.

Thus the Vouſſoirs, together with the Materials they are laden with, ought to be *in æquilibrio* with the Butment, which you would have to reſiſt their Puſh, or the Work muſt infallibly fall to pieces.

I have calculated the above Table, in order to make theſe Things familiar to Perſons who are ignorant of Geometry.

Which of the ſeveral Sorts of Arches, fixed or erected upon one and the ſame Diameter, would be capable of ſuſtaining the greateſt Weight; and in Proportion their Efforts or Puſhes are to one another, whether they be Scheam, Elliptical, Semicircular; or, in ſhort, of the third Point.

If due Attention be given to what I have already laid down, it may be allow'd, that the Puſhes of all Arches of different Degrees of Flatneſs, are to the Weight by which they are laden, as their ſeveral Degrees of Inclinations are to the Breadth of the Butment which is to ſecure them; and it will alſo be found that thoſe, whoſe Puſh is the leaſt inclin'd, (or oblique.) will be able to ſuſtain greater Burthens than thoſe which border upon the ſtraight Arch, which is the weakeſt of all.

Let us ſuppoſe the ſtraight Arch (A, Fig. I. Plate II.) to be a Beam tranſported or remov'd to AC, Fig. IX. Now it is certain, that if in this Situation it was laden with a Weight of 100 *lb.* and it ſhould happen to break it, it would nevertheleſs be able to bear not only a Weight of 100 *lb.* but alſo a Weight infinitely greater, if it was raiſed perpendicularly, as AB.

But I ſhall put the Caſe here, that it would bear 200; and therefore 150 being the main Proportional between 100 and 200, or between the firſt Puſh (as we will call it) and the ſecond, it will be found that the Beam, being elevated to 45 Degrees in AD, it will bear that Weight, (*viz.*) 150 *lb.* form'd as a ſemicircular Arch.

If you lower it down to AF, to form the Elliptical Arch, it will be found, that if you take the main Proportional between the ſemicircular Arch AD and the Beam (or ſtraight Arch) AC, which is AF, the Ellipſis AF will bear but 125.

In ſhort, the main Proportional between the ſemicircular Arch, which is 150, and the Power or Force of AB, which 200, will be 175, and that will be expreſſly the Strength of the *Gothic* Arch AE.

Thus the higher an Arch is carried up, the ſtronger will it be; and the flatter or lower it is, the weaker.

This Argument may be determined to the niceſt Exactneſs, when

when you ſhall have diſcovered or agreed upon the Difference of the Strength of a Beam, when laid down like a ſtraight Arch, as A C, and when erected perpendicularly A B.

From this Demonſtration I infer, that the *Gothic* Arch is ſtronger than the ſemicircular, the ſemicircular Arch ſtronger than the elliptical, and this laſt than the ſtraight Arch.

This Figure IX. in Plate II. helped me to make an Experiment, whereby to find the Weight of all ſorts of Bodies at different Inclinations, as may be learned from the following Paragraph.

Of the Weight of Bodies at different Inclinations, and the Manner of calculating it.

A Body AB, (Plate II. Fig. II.) uniform throughout its whole Length, Breadth, and Thickneſs, no matter whether round or ſquare, weigh'd 100 upon the Point of Support A, and erected perpendicularly, as A B.

But when this ſame Body A B was laid down horizontally, or on a level upon the Points A and C E, which were equidiſtant from the Centre or Middle, and Extremities of it, it preſſed upon the firſt Point of Support A with a Weight of 50 only; becauſe as the whole Weight is equally ſupported by the Points A and C, the whole Weight aforeſaid is equally divided between them: Thus, as it bears 50 upon A, and 50 upon C, the Sum of thoſe two Numbers is 100 the whole Weight.

When I alter'd the Poſition of this Body A C, by giving it an Elevation of 45 Degrees in A D, I found that weighing in the whole 100 Parts that preſſed upon D with a Weight of 25 only; wherefore it muſt preſs upon A with a Weight of 75, in as much as the Sum of thoſe two Numbers gives 100, the whole Weight which was to be.

I then made two other Experiments, by elevating the aforeſaid Body from A D to $67\frac{1}{2}$ in E, by depreſſing or declining to 22 Degrees $\frac{1}{2}$ in F; upon the Point E it preſſed with a Weight of $12\frac{1}{2}$ only, the Point F with a Weight of $37\frac{1}{2}$; and conſequently I found that E A, at an Elevation of $67^\circ\ 30'$, preſſing after the Rate of $12\frac{1}{2}$ only upon the Point E, muſt preſs with a Weight of $87\frac{1}{2}$ upon its Point A; and alſo that A F, at an Elevation of $22^\circ\ 30'$ muſt preſs upon its Point A with a Weight of $62\frac{1}{2}$. And thus you will find it in continued Proportion by ſubdividing the Parts B E, E D, D F, and F C.

From hence I calculated a Table; which ſhews that the Body A B preſſes the greater upon its Point of Support A, the higher it is elevated above the Level A C in F, or towards D, &c. and inverſely, that the aforeſaid Point of Support A is the leaſt preſſed the more the Body A B quits its Perpendicularity, and is inclined towards E D F and C, and ſo on, beneath the Level A C, by a reverſe Poſition, indeed, but, however, in the very ſame Proportion.

In

In order to this, I have ſup‐pos'd the Body AB to be uniform throughout its whole Dimenſions, and to weigh 100 equal Parts, (no matter whether Pounds, or any other Weight,) and to be one hundred equal Parts in Length.

Having laid down this for a Foundation, I found no Difficulty in inveſtigating the Effort and Power of all Bodies whatſoever, more or leſs inclin'd, whether in Lines or Curves; for by reducing the Curves to ſtraight Lines, or at leaſt by ſuppoſing the Curves to be equally ſupported by the Extremities of their Chords; and by comparing the Chords with one another, I reſolve the Strength and Weight of all Arches and Vaults, of whatſoever Figure they be, whether regular or irregular.

But you muſt previouſly conſult the following Table, where it will be ſeen, at the firſt Column, the various Degrees of the Inclination of Bodies, whoſe Weight we would calculate.

You muſt again ſuppoſe the Body to weigh an hundred Pounds, or any Thing elſe, or any other Number, which anſwers to the equal Parts into which it is divided, and the Inclination of the Body ſhall be determined by a certain Number of Degrees.

This being premiſed, let there be an Arch of any other inclin' Body, to whoſe Weight or Preſ ſure at the Key you woul know.

Let us ſuppoſe it to be the Hal of the ſemicircular Arch, (Fig. I Plate I.) whoſe Preſſure o Weight you would know at the Key E; calculate the Stone or Brick Work EHA of the Semi-Arch, which you may eaſily do by a previous Knowledge of the Weight of a cubic Foor of either of thoſe materials, and you may come to the Knowledge of the reſt. For ſuppoſe that EHA weighs 9750 *lb.* the Half of which will be 4875 *lb.* but the Chord of the Semi-Arch is at an Inclination of 45°, therefore (from the Table below,) ſay, If 50, the laſt Number or Term, gives the Weight of 25, at an Inclination of 45 Degrees, how much will 4875, the Half of the Weight of the whole Triangle AHE, give? And it will come come out 2437 ½; ſo that the Body EHA, weighing in all 9750, will weigh at the Key at an Inclination of 45° 2437 ½; which laſt Number being ſubſtracted from 9750, it will be found to weigh or puſh at the Foot or Couſſinet 7312 ½.

It is after this Manner that you are to determine, with regard to the Strength of Buttreſſes, for the Support of Vaults, Walls, &c.

A Table

be an Arch of any other inclin'd Body, to whose Weight or Pressure at the Key you would know.

Let us suppose it to be the Half of the semicircular Arch, (Fig. I. Plate I.) whose Pressure or Weight you would know at the Key E: calculate the Stone or Brick Work EHA of the Semi-Arch, which you may easily do by a previous Knowledge of the Weight of a cubic Foot of either or those materials, and you may come to the Knowledge of the rest. For suppose that EHA weighs 9750 lb the Half of which will be 4875 lb but the Chord of the Semi-Arch is at an Inclination of 45°, therefore (from the Table below.) say, If 50, the last Number or Term, gives the Weight of 25, at an Inclination of 45 Degrees, how much will 4875, the Half of the Weight of the whole Triangle AHE, give? And it will come out 2437 ½; so that the Body EHA, weighing in all 9750, will weigh at the Key at an Inclination of 45° 2437 ½; which last Number being substracted from 9750, it will be found to weigh or push at the Foot or Coussinet 7312 ½.

It is after this Manner that you are to determine, with regard to the Strength of Buttresses, for the Support of Vaults, Walls, &c.

A Table

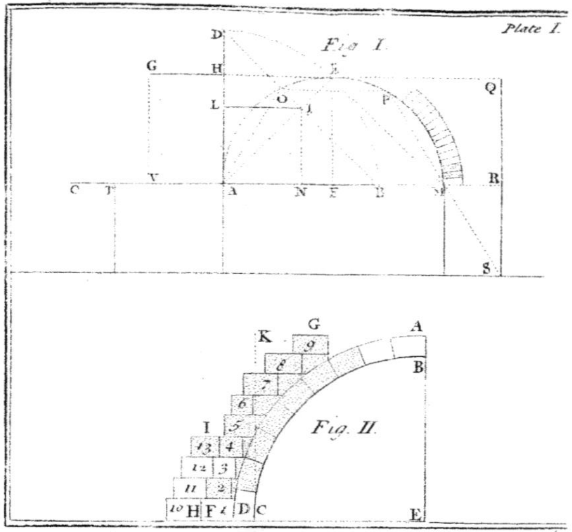

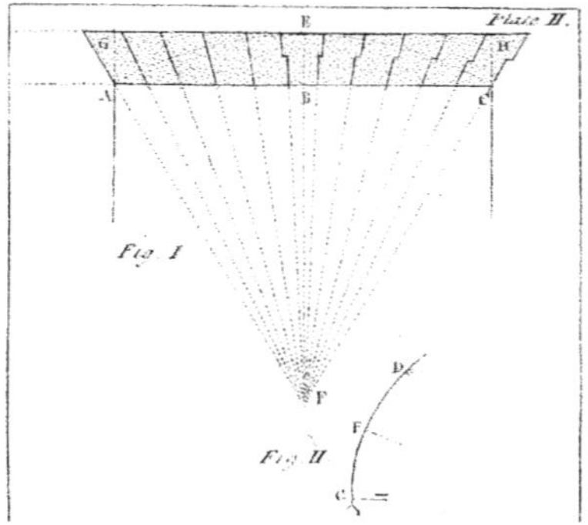

Plate I.

Fig. I

Fig. II

Plate II.

Fig. I

Fig. II

A Table of the Proportions of the Weights and Pushes of Regular Bodies, at all Degrees of Inclinations.

The Number of Quantities of an inclin'd Body to be determined.

Deg.	lb.	Oz.	Deg.	lb.	Oz.	Deg.	lb.	Oz.	Deg.	lb.	Oz.
1	0	8 8/9	24	13	3/9	47	26	1/9	70	38	8/9
2	1	1/9	25	13	8/9	48	26	6/9	71	39	4/9
3	1	6/9	26	14	4/9	49	27	2/9	72	40	0
4	2	2/9	27	15	0	50	27	7/9	73	40	5/9
5	2	7/9	28	15	5/9	51	28	3/9	74	41	1/9
6	3	3/9	29	16	1/9	52	28	8/9	75	41	6/9
7	3	8/9	30	16	6/9	53	29	4/9	76	42	2/9
8	4	4/9	31	17	2/9	54	30	0	77	42	7/9
9	5	0	32	17	7/9	55	30	5/9	78	43	3/9
10	5	5/9	33	18	3/9	56	31	1/9	79	43	8/9
11	6	1/9	34	18	8/9	57	31	6/9	80	44	4/9
12	6	6/9	35	19	4/9	58	32	2/9	81	45	0
13	7	2/9	36	20	0	59	32	7/9	82	45	5/9
14	7	7/9	37	20	5/9	60	33	3/9	83	46	1/9
15	8	3/9	38	21	1/9	61	33	8/9	84	46	6/9
16	8	8/9	39	21	6/9	62	34	4/9	85	47	2/9
17	9	4/9	40	22	2/9	63	35	0	86	47	7/9
18	10	0	41	22	7/9	64	35	5/9	87	48	3/9
19	10	5/9	42	23	3/9	65	36	1/9	88	48	8/9
20	11	1/9	43	23	8/9	66	36	6/9	89	49	4/9
21	11	6/9	44	24	4/9	67	37	0	90	50	0
22	12	2/9	45	25	0	68	37	7/9			
23	12	7/9	46	25	5/9	69	38	3/9			

Draw-Bridge, is one that may be drawn or taken up by Means of a Sweep, or Counterpoiſe, and which ſhuts up againſt a Gate. There are others with Pitfals and Beams, ſuſtained by two large Stakes fifteen Foot high; one Part of which lowers as the other riſes.

TO BRING UP, a Term uſed among Workmen, eſpecially among Carpenters, when they are talking with Bricklayers: Thus they ſay, *Bring up* the Foundation ſo high; *Bring up* ſuch a Wall; *Bring up* the Chimneys, *&c.* which is as much as to ſay, Build the Foundation ſo high, Build the Wall; Build the Chimneys, *&c.*

BROAD STONE, is the ſame with Freeſtone; only this is ſo called, becauſe raiſed broad and thin out of the Quarries, *viz.* not more than two or three Inches thick.

As to its Uſe: The Uſe of this Sort of Freeſtones, which are called *Broad-Stones*, is for paving Court-Yards and Paſſages, and before Shop-Doors, as in Walks or Paths in the City of *London*, to ſeparate them from the Highway.

As to their Price: If the Breadths and Lengths are promiſcuous, then the common Price for fitting and laying the Stone in Mortar, from 6*d.* to 8*d. per* Foot ſquare, or from 4*s.* to 6*s. per* ſuperficial Yard.

But ſome of theſe Stones are cut into perfect Squares, like Paving-Tiles, but much larger, as eighteen, twenty, or twenty-four Inches ſquare or more; but as theſe are nearer, ſo they are dearer; ſome Pavements of theſe being worth 18*d. per* Foot; and if the Stones are good, and well poliſh'd, as they ought to be for Kitchens, Dairy-Houſes, Brew-Houſes, *&c.* they will be worth 15, or 16*d. per* Foot.

SPANISH BROWN is a dark dull Red, of a Horſe-Fleſh Colour. It is an Earth that is dug out of the Ground: But there is ſome of it pleaſant enough to the Eye, conſidering the Deepneſs of it.

It is of great Uſe among Painters; being generally uſed as the firſt and priming-Colour, which they lay on upon any kind of Timber-Work, being cheap and plentiful, and a Colour that works well, if it be ground fine; which may be done with leſs Labour, than ſome better Colours do require. That which is of the deepeſt Colour, and the freeſt from Stones, is the beſt.

The other Sorts are not ſo good to give a Colour to the Eye, but yet they ſerve as well as any other for the Priming Colours, to ſeaſon the Wood to lay other Colours upon.

BUFFET, } a little Apart-
BUFET, } ment, ſeparated from the reſt of the Room, by ſlender Wooden Columns for placing China, Glaſs-Ware, *&c.* Called alſo a Cabinet.

The *Buffet*, among the *Italians*, called *Credenza*, is incloſ'd within a Baluſtrade, Elbow-high.

BUILDING, is uſed to ſignify both the Conſtructing and Raiſing of an Edifice; in which Senſe, it comprehends as well the Expences, as the Invention and Execution of the Deſign.

In

In *Building* there are three Things to be considered; *viz.* First, Commodity or Conveniency. Secondly, Firmness. Thirdly, Delight.

To accomplish which Ends, Sir *Henry Wotton* considers the whole Subject under two Heads, *viz.* the Seat or Situation, and the Work.

1. As for the Seat: Either that of the Whole is to be consider'ed, or that of its Parts.

2. As to the Situation, Regard is to be had to the Quality, Temperature, and Salubrity or Healthfulness of the Air; that it be a good healthy Air, not subject to foggy Noisomeness from adjacent Fens or Marshes; also free from noxious Mineral Exhalations: Nor should the Place want the sweet Influence of the Sun-Beams; nor be wholly destitute of the Breezes of Wind, which will fan and purge the Air; the Want of which would render it like a stagnated Pool, or standing Lake of Air, and would be very unhealthy.

Pliny advises not to build a Country-House too near a Fen or Standing-Water; nor yet over-against the Stream and Course of a River; because the Fogs and Mists which arise from a large River, early in a Morning, before Day-Light, cannot chuse but be very unwholesome.

Dr. *Fuller* advises chiefly to chuse a wholesome Air: because, says he, the Air is a Dish one feeds on every Minute; and therefore it had need be salubrious.

Cato advises, that a Country-House have a good Air, and not lie open to Tempests, seated in a good Soil, and let it exceed therein, if you can; and let it stand under a Hill, and behold the South, in a healthy Place.

As to Commodiousness, or Conveniency, Sir *Henry Wotton* advises, that the House or Seat have the Conveniency of Water, Fuel, Carriage, &c. that the Way to it be not too steep, and of an incommodious Access, which will be troublesome both to the Family, and Visitants. And as for the Conveniency of being supply'd with Necessaries, it should not be seated too far from some Navigable River, or Arm of the Sea.

Wood and Water, says Dr. *Fuller*, are two Staple Commodities.

As for Water; the Want of it is a very great Inconveniency, the Detriment of many Houses to which Servants must bring the Well upon their Shoulders.

And as to Wood; where a Place is bald of Wood, no Art can make it a Perriwig in Haste.

Optical Precepts, or Maxims Such I mean, says Sir *Henry Wotton*, as concern the Properties of a well-chosen Prospect; which may be stiled the Royalty of Sight. For as there is a Lordship (as it were) of the Feet whereon a Man walk'd with much Pleasure about the Limits of his own Possessions, so there is a Lordship likewise of the Eye which being a ranging and imperious (I had almost said) usurping Sense, cannot endure to be circumscrib'd within a small Space, but must be satisfy'd both with Extent and Variety: Yet on the other Side, I find vast and indefinite Prospects, which drown all Apprehenſions of very remote Objects condemn'd by good Authors, as if ſome Part of the Pleaſure (whereof we are ſpeaking) did thereby periſh.

Agreeableneſs and Pleaſantneſs of Proſpect is to be valu'd.

Dr. *Fuller* ſays, a Medley View (ſuch as of Land and Water, at *Greenwich*) beſt entertain the Sight, refreſhing the weary Beholder with Change of Objects: Yet, ſays he, I know a more profitable Proſpect, where the owner can only ſee his own Land round about him. And to this Head of Situation, he adds, as follows:

A fair Entrance with an eaſy Aſcent, gives a great Grace to a Building, where the Hall is a Preferment out of the Court, the Parlour out of the Hall; not (as in ſome old Buildings,) where the Doors are ſo low, that Pigmies may ſtoop; and the Rooms ſo high, that Giants may ſtand a-tiptoe.

A political Precept: That great Architect Sir *Henry Wotton* ſays, One private Caution, which I know not well how to rank among the reſt of the Precepts, unleſs I call it political, is this, *viz.* By no Means to build too near a Great Neighbour; which were to be as unfortunately ſeated on the Earth, as *Mercury* is in the Heavens; for the moſt part ever in Combuſtion or Obſcurity, under brighter Beams than his own.

Contrivance.: The Situation being fix'd on, the next Thing in Order, is Contrivance; which being a Thing of great Moment in this Affair of *Building*, before it is entred upon, it will be ne-

Breadth, and Thickneſs of Ground-Plates or Cells, Breſt-Summers, (and in all, whether Timber, Brick, or Stone *Buildings*,) the Dimenſions of Summers, Girders, Trimmers, or Joiſts.

Alſo in the upper Floor, the Scantling of Dragon-Beams, Raiſons, or Railing-Pieces, or Wall-Plats, &c.

And alſo the Thickneſs of Partitions, Walls, &c. in Brick or Stone Fabricks.

All which, and all other Parts (whether in the Ichnography, or Orthography) of *Buildings*, ought to be repreſented; as alſo Ovens, Stoves, Broilers, Furnaces, Coolers, Fats for Brewing, &c. with their juſt Meaſures to the beſt Advantage, as to Conveniency, Health, Strength, and Ornament.

All which Dimenſions ought to be ſet in the proper Places to which they belong in the Diagrams in Characters; becauſe if the Schemes be not very large, it will be very difficult to take the Dimenſions of the ſmaller Parts nicely, if not of the great ones themſelves; for it will ſcarce be practicable, to take either of them to an Inch, nor perhaps, to two, three, or four Inches, according as the Diagram may be in Amplitude.

In the Orthographical Schemes, there muſt be true Delineations and Dimenſions of each Face, and all its Concomitants, as Doors, Windows, Balconies, Turrets, or Cupola's, Chimney-Shafts, Faſcias, Ruſtick Quoins, Architraves, Friezes, Cornices, Pediments, Pilaſters, Columns, Shells over Doors, Lanthorns, and all other Ornaments.

If it be a Timber *Building*, all the Members in that Face ought to have their ſeveral Sizes in Characters, and true Poſitions by the Scale.

As for Example: The Ground-Plates or Cells, Introduces, Breſt-Summers, Beams, Principal Poſts or Braces, Quartets, Prick-Poſts, or Window-Poſts, Jaumbs, or Door-Poſts, or Puncheons, King-Pieces, or Joggle-Pieces, Struts, Collar-Beams, Door-Heads, Principal Rafters, Shreadings, &c.

The Ichnography, Orthography, and Stenography of the Stair-Caſe may alſo be delineated, and all its Parts, as Hand-Rails, Riſers, Noſeing of the Cover or Top, String-Board and Mouldings on it, as Cartouzes, Balluſters, Pendents, &c. with their true Poſitions, Forms, and Dimenſions; all which, if they be carefully done by an ingenious Surveyor, a Workman will hardly be like to commit any Blunder.

Sir *Henry Wotton* adviſes as to this Matter, as follows:

Firſt, Let no Man who intends to build, ſettle his Fancy on a Draught on Paper or Vellum of the Work or Deſign, how exactly ſoever delineated or ſet off in Perſpective, without a Model or Type of the whole Structure, and of every Parcel and Partition, either of Paſteboard, or Wainſcot.

Secondly, Let the Model be as plain as may be, without Colours, or other Beautifying, leaſt the

Pleaſure of the Eyes preoccupate the Judgment.

Thirdly, and *Laſtly*, The bigger this Type is, ſo much the better it is: Not that I would perſwade any Man to ſuch an Enormity as that Model made by *Antonio Labaco*, of St. *Peter's* Church in *Rome*, containing twenty-two Feet in Length, ſixteen in Breadth, and thirteen in Height, which coſt four thouſand one hundred and eighty-four Crowns, the Price of a reaſonable Chapel; yet in a Fabrick of forty or fifty thouſand Pounds, there may very well be thirty Pounds expended in procuring an exact Model; for a little Penury in the Premiſes, may eaſily create ſome Abſurdity or Error of a far greater Charge in the Proſecution, or at the Concluſion of the Work.

What Sir *Henry Wotton* here adviſes, is very requiſite, eſpecially for large and ſumptuous *Buildings*, either publick, or private; but it is not worth the while to be at the Charge of a Model for every little Dwelling-Houſe which Men build for their own Conveniency.

I ſhall here add as to the Conveniency, what is recommended by Sir *Henry Wotton*, that the Chief Rooms, Studies, Libraries, &c. ſhould lie towards the Eaſt; that thoſe Offices which require Heat, as Kitchens, Brew-Houſes, Bake-Houſes, and Diſtillatories, to the South; thoſe which require a cool freſh Air, as Cellars, Pantries, Granaries, to the North; as alſo Galleries for Paintings, Muſæums, &c. which require a ſteady Light.

He tells us, the antient *Greeks* and *Romans* generally ſituated the Fronts of their Houſes towards the South; but the modern *Italians* vary much from this Rule.

And, indeed, as to this Matter, Regard muſt ſtill be had to the Country, each being forced to provide againſt its Inconveniencies: So that a good Parlour in *Ægypt*, might make a good Cellar in *England*.

The Situation being fixed on, and the Deſign and Contrivance deſign'd, the next Thing to be conſider'd, is the Work itſelf; under which the principal Parts are firſt to be conſider'd; next, the Acceſſories and Ornaments.

Under the Principals are, firſt, the Materials; next, the Form or Diſpoſition.

As for the Materials, they are either Stone, as Marble, Freeſtone, Brick for the Walls, Mortar, &c. or of Wood, as Fir, Cyprus, Cedar for Poſts and Pillars of upright Uſe; Oak for Summers, Beams, and Crop-Work, or for joining and Connection.

As to the Form or Diſpoſition of a *Building*, it is either ſimple, or mix'd.

The ſimple Forms are either circular or angular; and the circular ones either compleat, as juſt Spheres; or deficient, as oval ones.

The Circular Form is very commodious, and the moſt capacious of any, ſtrong and durable beyond the reſt, and very beautiful; but is the moſt chargeable of all others; and much Room is loſt by the Bending

ing of the Walls, when it comes to be divided into Apartments, besides an ill Distribution of Light, unless it be from the Centre of the Roof.

For these Reasons, it was, that the Antients used this Form only in their Temples and Amphitheatres, which had no need of Compartitions. There are the same Inconveniencies attending Oval Forms, without the same Conveniencies, being less capacious.

As for Angular Forms, Sir *Henry Wotton* observes, that *Building* neither loves many nor few Angles. The Triangle is condemn'd above all others, as wanting both Capaciousness and Firmness; as also not being capable to be resolved into any other regular Figure in the inward Partitions besides its own.

As for Forms of *Building* of five, six, seven, or more Angles, they are much fitter for Fortifications, than Civil *Buildings*.

There is indeed, a celebrated *Building* of *Vignola* at *Caparole*, in the Figure of a Pentagon; but the Architect had very great Difficulties to grapple with as to the Disposition of the Lights, and the saving the Vacuities.

So that such *Buildings* seem rather for Curiosity than Conveniency. And for this Reason, Rectangles are generally chosen, as being a Medium between the two Extreams.

But then Authors are in Dispute, whether the Rectangle should be an exact Square, or an Oblong. Sir *Henry Wotton* prefers the Oblong, provided the Length exceeds not the Breadth by more than one Third.

As to Mixed Forms or Figures, a Judgment may be made of them from what has been already said of simple ones; only that they have this particular Defect, that they offend against Uniformity.

Indeed, Uniformity and Variety may seem to be Opposites: But Sir *Henry Wotton* observes, that they may be reconciled; by Instances in the Structure of the Human Body, where they both meet together.

Some observe, that in building Houses long, the Use of some Rooms will be lost; and they will take up more for Entries and Passages, and will require more Doors: And if a *Building* be a Geometrical Square, if the House be any Thing large, the middle Rooms will want Light more, than if they were built in the Form of an H, or some other such like Figure, unless there be a Court in the Middle of it: Which was the Method of building great Houses in former Times.

Some much commend this Way of building an House in the Form of a *Roman* H: For they say, this Form makes it stand better, and firmer against the Winds; and that the Light and Air comes to it every Way, and every Room is nearer one to the other.

Some approve of this Form very much; because in it, the Offices may be remote from the Parlour and Rooms of Entertainment, and yet in the same House, which may serve very well for a Country Gentleman's House.

In *Buildings* of this Form, ſome propoſe the Diſpoſition of the Apartments thus:

In the Front of one of the long Part of the H, is to be the Kitchen, the Bake-Houſe, Brew-Houſe, and Dairy-Houſe.

In the ſame Part behind it, the Hall, in the Middle of the H; which ſeparates the Parlours which are in the other long Part, and Rooms of Entertainment from the Offices.

Thus much for the firſt grand Diviſion, *viz.* the Whole of a *Building*.

As for the Second Diviſion, or the Parts of a *Building*, they are comprized by *Baptiſta Alberti*, under five Heads, *viz.* the Foundation, the Walls, the Apertures, the Compartitions, and the Covering.

1. As for the Foundation, *Vitruvius* orders the Ground to be dug up, to examine its Firmneſs, that an appearing Solidity is not to be truſted, unleſs the whole Mold cut through be ſound and ſolid: 'Tis true, he does not ſay to what Depth it ought to be dug. But *Palladio* determines it to a ſixth Part of the Height of the *Building*.

And this is called by Sir *Henry Wotton*, the Natural Foundation, whereon the Subſtruction or Ground-Work is to ſtand to ſupport the Walls; which he calls, the Artificial Foundation.

This then is to be level; the loweſt Ledge or Row of Stone, only cloſe laid with Mortar; and by how much the broader it is, by ſo much will it be the better; but at leaſt, it ſhould be twice the Breadth of the Wall.

Some adviſe, that the Mate-

both for the Eafe of Weight and Expence. *Fourthly*, That certain Courfes or Lodges, of more Strength than the reft, be interlaid like Bones, to fuftain the Fabrick from total Ruin, if fome of the under Parts chance to decay. *Fifthly*, and *Laftly*, That the Angles be firmly bound, they being the Nerves of the whole Fabrick: Which are ufually fortify'd by the *Italians* on each Side the Corners, even in Brick *Buildings*, with fquared Stones; which add both Beauty and Strength.

The Intermiffions of Walls, as has been before mention'd, are either Colums, or Pilafters; of which there are five Orders, *viz.* the *Tufcan*, *Doric*, *Ionic*, *Corinthian*, and *Compofite.* All which are diftinctly treated under their refpective Articles.

Columns and Pilafters are frequently form'd Archwife, both for Beauty and Grandeur.

As for Apertures, they are either Doors, Windows, Stair-Cafes, Chimneys, or Conduits for the Suillage, &c. All which you may fee confidered under their proper Heads.

And as to the laft, Art fhould imitate Nature in thefe ignoble Conveyances, and conceal them from the Sight, where a running Water is wanting, into the moft remote, loweft, and thickeft Part of the Foundation, with fecret Vents paffing up through the Walls like a Tunnel to the open Air. Which is recommended by all the *Italians*, for the Difcharge of noifome Vapours.

As to Compartitions or Diftribution of the Ground-Plot into Apartments, &c. Sir *Henry Wotton* lays down thefe Preliminaties, That the Architect do never fix his Fancy on a Paper-Draught, be it fet off never fo exactly in Perfpective, much lefs on a meere Plan, without a Model or Type of the whole Structure, and every Part of it, either in Pafteboard or Wood.

In the Compartition itfelf there are two general Views; *viz.* the Gracefulnefs and Ufefulnefs of the Diftribution for Rooms of Office and Entertainment, as far as the Capacity of it, and Nature of the Country will allow.

The Gracefulnefs will confift in a double Analogy or Correfpondency. *Firft*, Between the Parts and the Whole; by which a large Edifice fhould have large Partitions. Entrances, Doors, Columns, and, in fhort, all its Members large, proportional to the *Building*.

The fecond Analogy is between the Parts themfelves, not only confidering their Lengths and Breadths, as we fpeak of Doors and Windows. But here, fays Sir *Henry*, enters a third Refpect of Height. a Point; I confefs, faith he, fcarce reducible to any general Precept.

The Antients determined their Rooms which were oblong, by double their Breadth, and their Height by half their Breadth and Length added together.

When they wou'd have the Room a perfect Square, they made their Height half as much more as their Breadth: But the Moderns difpenfe with thefe Rules, fometimes fquaring the Breadth, and making the Diagonal of it the Meafure of the Height

Height, and ſometimes more. This Deviation from the Rules of the Antients is aſcrib'd to *M. Angelo.*

Sometimes they ſquar'd the Breadth, and doubled that ſquare Number; and taking the Root of that ſquare Number, for the Height, and ſometimes more; but ſeldom lower for the Breadth.

But what is here mention'd is ſcarcely now practiſed, unleſs it be in a Nobleman's Houſe; who will have a Hall, &c. higher pitch'd than the reſt of the Rooms in the *Building*; and ſometimes a Dining-Room; or elſe, for the moſt Part, all the Rooms of a Floor are of an equal Height: And this ſeems to be the moſt commodious Method; becauſe in this Caſe, there is no Loſs of Room, as there muſt neceſſarily be, where one Room is open almoſt to the Top of the Houſe, as may be obſerved in ſome old *Buildings*; and then the Floor of the ſecond Story will lie level and even, and not in the old Method of Steps out of one Room into another.

As to the Height of Rooms, that is various amongſt us, according to the Perſons for whom they are built, and the Cuſtom of the Place. In the Country, common Timber *Buildings* are uſually about ſeven Feet one Third, or eight Foot at the moſt, betwixt the Floors.

The ſecond Sort of Houſes in the Country is about nine Feet between the Floors; which for the moſt part is the Pitch of the Rooms at *Tunbridge-Wells*.

The third Sort in the Country, (*viz.* in *Kent* and *Suſſex*,) are Gentlemens Seats; which for the moſt part are ten or twelve Feet high, ſuch as are new *Buildings*. But in old Stone *Buildings*, 'tis common to be much higher, *viz.* fourteen or ſixteen Feet.

By Act of Parliament, for *building* of *London*, there were reckon'd four Rates of Houſes, *viz.*

The	1	Rate	2	Foot
	2		3	
	3		4	
	4		5	at Diſcretion, &c.

The	1	Rates, Cellars in Height, betwixt Floor and Ceiling.	6	and a half Foot
	2		6	
	3		6	
	4			at Diſcretion, &c.

The	1	Rate, Firſt Story.	9	Foot
	2		10	
	3		10	
	4			at Diſcretion, &c.

The

	Rate		
The 1	Rate, Second Story,	9	
2		10	
3		10	Foot and a half,
4			at Diſcretion, &c.

	Rate		
The 1	Rate, Third Story,	9	
2		9	
3		9	Foot and a half
4			at Diſcretion, &c.

	Rate		
The 3	Rate, Fourth Story,	$8\frac{1}{2}$	Foot and a half high
4			at Diſcretion, &c.

The ſecond Conſideration, as to the Compartiments, is the Uſefulneſs; which conſiſts in having a ſufficient Number of Rooms of all Kinds, with their proper Communications, and without Diſtraction.

Here the chief Difficulty will be in the Lights and Stair-Caſes.

The Antients were pretty eaſy as to both theſe, having generally two cloiſtered open Courts; one for the Women's Side, and the other for the Men. Thus then the Reception of Light was eaſy into the Body of the *Building*, which muſt be ſupply'd among us either by the open Form of the *Building*, or by graceful Refuges or Breaks, by terraſſing a Story in danger of Darkneſs, and by Abajours or Sky-Lights.

As for caſting the Stair-Caſes, it may be obſerved, that the *Italians* frequently diſtribute the Kitchen, Bake-houſe, Buttery, &c. under-grouud, next above the Foundation; and ſometimes level with the Foot of the Cellar, raiſing the firſt Aſcent into the Houſe fifteen Feet, or more; which, beſides the removing of Annoyances out of Sight, and gaining ſo much Room above, adds a Majeſty to the whole Fabrick, by elevating the Front.

Sir *H. Wotton* obſerves, that in *England* the natural Hoſpitality will not allow the Buttery to be ſo far out of Sight; beſides, that a more luminous Kitchen, and a ſhorter Diſtance, are required between that and the Dining-Room than that Compartition will admit.

In the Diſtribution of Lodging-Rooms, 'tis a popular and antient Fault, eſpecially among the *Italians*, to caſt the Partitions ſo, that when the Doors are all open, a Man may ſee through the whole Houſe; which is founded on an Ambition of ſhewing a Stranger all the Furnirure at once; which is an intollerable Hardſhip upon all the Chambers, except the innermoſt, into which a Perſon cannot come, but through all the reſt, unleſs the Walls be extreme thick for ſecret Paſſages: Nor will this ſuffice, unleſs there be three Doors to each Chamber, a Thing inexcuſable, except in hot Countries.

Beſides, it is a Weakening to the *Building*; and the Neceſſity which it occaſions, of making as many

many common great Rooms as there are Stories, which devour a great deal of Room, which might be employ'd in Places of Retreat; and also must likewise be dark, as running through the Middle of the House.

In the Compartition, the Architect will have Occasion for frequent Shifts, through which his own Sagacity, more than any Rules, must conduct him.

Thus he will be frequently put to struggle with Scarcity of Ground; sometimes to damn one Room for the Benefit of the rest, as to hide a Buttery under a Stair-Case, &c. At other Times, to make those the most beautiful which are most in sight, and to leave the rest, like a Painter, in the Shadow, &c.

As for the Covering of a *Building*, this is the last in Execution, but the first in Intention: For who would build but to shelter?

In the Covering or Roof, there are two Extremes to be avoided; which are the making it too heavy, or too light. The first will press too much on the under Work; and the latter has a more secret Inconvenience; for the Covering is not only a bare Defence, but a Bond or Ligature to the whole *Building*.

Indeed of the two Extremes, a House top-heavy is the worst.

Care ought to be taken, that the Pressure be equal on each Side: And *Palladio* wishes that the whole Burthen may not be laid upon the outward Walls; but that the inner may likewise bear their Share.

The *Italians* are very curious in the Proportion and Gracefulness of the Pent or Slopeness; and divide the whole Breadth into nine Parts, whereof two serve for the Height, or highest Top or Ridge from the lowest; but in this Point, Regard must be had to the Quality of the Region: For, as *Palladio* intimates, those Climates which fear the falling of much Snow, ought to have more inclining Pentices than others.

As to the Accessories or Ornaments of a *Building*, they are fetch'd from Sculpture and Painting.

As for Sculpture, Care ought to be taken that there be not too much of it, especially at the first Approach of a *Building*; or at the Enterance, where a *Doric* Order is much preferable to a *Corinthian* one; that the Niches, if they contain Figures of white Stone, be not colour'd in their Concavity too black, but rather dusky, too sudden Departments from one to another being displeasing to the Sight.

Fine Sculptures ought also to have the Advantage of Nearness, and coarser of Distance; and likewise in placing the Figures aloft, they should be inclin'd or lean a little forward, because the visual Ray extended to the Head of the Figure is longer than that reaching to its Feet, which will necessarily make that Part to appear farther off; so that in order to reduce it to an erect Posture, it must be made to stoop a little forwards.

But M. *Le Clerc* does not allow of this Inclination, but will have every Part in its just Perpendicular.

As

As to Painting; the chief Things that are to be regarded are, that no Room have too much, which will surfeit, except in Galleries, &c. that the best Pieces be placed where there are the fewest Lights. Rooms with several Windows are Enemies to Painters; nor can any Pictures be seen in Perfection, unless illuminated, like Nature, with a single Light.

That in the disposing of them Regard be had to the Posture of the Painter in working, which is the most natural for the Postute of the Spectator; and that they be suited to the Intention of the Rooms they are used in.

To make a Judgment of a *Building*, Sir *Henry Wooton* lays down the following Rules.

1. That before a Person fixes upon any Judgment, he be informed of its Age; that if the apparent Decays be found to exceed the Proportion of Time, he may thence conclude, without farther Inquisition, that either the Situation is naught, or that the Materials or Workmanship are too slight.

2. If the *Building* be found to bear its Years well, then let the Viewer run back from the Ornaments and Things that strike the Eye, to the more essential Members, till he is able to form a Conclusion, that the Work is commodious, firm, and delightful; which are the three Qualities of a good *Building*, which have been laid down at first, and agreed on by all Authors.

This he accounts the most scientifical Way of judging.

Vassari proposes a third, *viz.* by passing a running Examination over the whole *Building*, according to the Structure of a well-made Man; as whether the Wall stand upright upon a clean Footing and Foundation; whether the *Building* be of a beautiful Stature; whether it appear well burnish'd as to the Breadth; whether the principal Enterance be in the middle Line of the Front or Face, like our Mouths; whether the Windows be set in equal Number and Distance on both Sides like our Eyes; whether the Offices are usefully distributed, &c. like our Veins.

Vitruvius recommends a third Method of judging, summing up the whole Art under these six Heads.

1. *Ordination*, or the settling the Model or Scale of the Work.

2. *Disposition*, *i. e.* the just Expression of the first Design of the *Building*, (which two Sir *Henry Wotton* is of opinion he might have spared,) as belonging rather to the Artificer than the Censurer.

3. *Eurithmy*, *i. e.* the agreeable Harmony between the Length, Breadth, and Height of the several Rooms, &c.

4. *Symmetry*, or the Agreement between the Parts and the Whole.

5. *Decor*, which is the true Relation between the *Building* and the Inhabitants: From whence *Palladio* concludes, the principal Enterance ought never to be limited by any Rule; but the Dignity and Generosity of the Master.

6. *Distribution*, *i.e.* the useful casting of the several Rooms for Offices,

Offices, Entertainment, or Pleasure.

The last four of these are always to be run over, before a Man passes any determinate Censure: And Sir *Henry Wotton* says, are sufficient to acquit or condemn any *Building* whatever.

Dr. *Fuller* presents us with two or three good Aphorisms, or general Maxims, for Contrivance in *Building*, which are as follow:

First, Let not the common Rooms be several, nor the several Rooms common; that the common Rooms should not be private or retired, as the Hall, (which is a Pandochæum,) Galleries, &c. which are to be open; and the Chambers, Closets, &c. retired and private, provided the whole House be not spent in Paths.

Light (God's eldest Daughter) is a principal Beauty in a *Building*; yet it shines not alike from all Parts of the Heavens.

An East Window gives the Infant Beams of the Sun, before they are of Strength to do any Harm, and is offensive to none but a Sluggard.

A South Window, in Summer, is a Chimney with a Fire in it, and stands in need to be skreen'd by a Curtain.

In a West-Window, the Sun grows low, and ever familiar towards Night in Summer-Time, and with more Light than Delight.

A North Window is best for Butteries and Cellars, where the Beer will be sour, because the Sun smiles upon it.

Thorough Lights are best for Rooms of Entertainment, and Windows on the Side for Dormitories.

Secondly, As to Capaciousness: A House had better be too little for a Day, than to big for a Year; therefore Houses ought to be proportion'd to ordinary Occasions, and not to extraordinary. It will be easier borrowing a Brace of Chambers of a Neighbour for a Night, than a Bag of Money for a Year: Therefore 'tis a Vanity to proportion the Receipt to an extraordinary Occasion; as those do, who by overbuilding their Houses, dilapidate their Lands, so that their Estates are pressed to Death under the Weight of their House.

Thirdly, As for Strength: Country-Houses must be Substantives, able to stand of themselves, not like City *Buildings*, supported and flanked by those of their Neighbour, on each Side.

By Strength, is meant such as may resist Weather and Time, but not Attacks; Castles being out of Date in *England*, except on the Sea-Coasts, &c.

As for Motes round Houses: 'Tis questionable whether the Fogs that arise from the Water, are not more unhealthful than the Defence that the Water gives countervails, or the Fish brings Profit.

In working up the Walls of a *Building*, let not any Wall be work'd up above three Foot high, before the next adjoining Wall be brought up to it, that so they may be joined together, and make good Bond in the Work. For some Bricklayers have an ill Custom of carrying or working up a whole Story of the

the Party-Wall, before they work up the Fronts, or other adjoining Work, that ought to be bonded or work'd up together with them, which is often the Cause of Cracks, and Setlings in the Walls of the *Building*, which very much weakens it.

The Strength of a *Building* is sometimes much impair'd by the erecting it, by Reason of the Masters not having prepared either sufficient Materials, or Money, before he set about *building*. For when *Buildings* are erected by Fits and Pauses, by doing first one Piece, and then another, the Work dries, and sinks unequally; by which Means the Wall becomes full of Chinks and Crevices: And therefore this Way of *building* by Fits, is condemned by all Authors.

Fourthly, As for Beauty: Let not the Front look asquint upon a Stranger, but accost him right at his Entrance. Uniformity and Proportion are very pleasing to the Eye. And 'tis observable, that Freestone, like a fair Complexion, grows old, whilst Bricks keep their Beauty longest.

Fifthly, Let the Offices keep their due Distance from the Mansion-House; those are too familiar which presume to be of the same Pile with it.

The same may be said of Stables and Barns; without which, a House is like a City without Works: It can never hold out long.

It is not only very inconvenient, but rather a Blemish than a Beauty to a *Building*, to see the Barns and Stables too near the House; because Cattle, Poultry, and such like, must be kept near them; which will be an Annoyance to an House.

Gardens ought also to be disposed in their proper Place. When God planted a Garden Eastward, he made to grow out of the Ground every Tree pleasant to the Sight, and good for Food. Sure, (says Dr. *Fuller*) he knew better what was proper to a Garden, than those who now-a-days only feed their Eyes, and starve their Taste and Smell.

Mr. *Worlidge* advises, that the Garden join to one, if not more Sides of the House: For what can be more pleasant and beautiful for the most Part of the Year, than to look out of the Parlour and Chamber-Windows into Gardens.

For Beauty, says he, let there be also Courts or Yards kept from Cattle, Poultry, &c. and planted with Trees, to shade, defend, and refresh your House; and the Walls also planted with Vines, and other Wall-Fruit: All which will add Pleasure and Beauty to your Habitation.

In Architecture, says a certain Author, there seems to be two opposite Affectations, *viz.* Uniformity and Variety: Yet these seeming Opposites may be very well reconciled; as may be observ'd in our own Bodies, the great Pattern of Nature; which is very uniform in the whole Configuration, each Side agreeing with the other in Number, Quality, and the Measure of the Parts: And yet some are round, as the Arms; others flat, as the Hands; some prominent, and others indented or retir'd. In like Manner, the Limbs or Members of a noble Fabrick may be correspondent enough, altho' they be various, provided Persons do not run out into extravagant Fancies, when they are contriving how to divide and cast the Work.

Enormous Heights of six or seven Stories are to be avoided, as well as irregular Forms; and so on the contrary, should low distended Fronts, they being unseemly. And again, when the Face of a *Building* is narrow, and the Flanks deep.

As to the modern Way of *building* in *England* compar'd to the antient:

In comparing the modern *English* Way of *building* with the old, one cannot but wonder at the Genius of those Times. Nothing is, or can be more pleasant than Height; nor any Thing more conducive to Health, than a free Air: And yet in old Times, they were wont to dwell in Houses, most of them with a blind Stair-Case, low Ceilings, and dark Windows; the Rooms built at Random, without any Thing of Contrivance, and often with Steps from one to another; as if the People of former Ages were averse to Light and good Air; or were pleased with playing at Hide-and-seek.

Whereas on the contrary, the the Genius of our Times is altogether for light Stair-Cases, fine Sash-Windows, and lofty Ceilings.

And such has been of late the Industry of our Builders, in relation to Compactness and Uniformity, that a House, after the new Way, will afford as many more Conveniencies, upon the same Quantity of Ground.

The Contrivance of Closets, in most Rooms and painted Wainscot, now in so common Use, are likewise two great Improvements, the one for Conveniency, and the other for Cleanliness and Health: And, indeed, for so damp a Country as *England* is, nothing could be better contriv'd than Wainscot, to ward off the moist Effluvia of damp Walls.

In a Word, for handsome Accommodations, and Neatness of Lodgings, *London* has undoubtedly gain'd the Pre-eminence of all Places in *Europe*.

The greatest Objection against the Houses in the City of *London* (being for the most part of Brick) is their Slightness, occasion'd by the Fines exacted by the Landlords.

So that few Houses, at the common Rate of *Buildings*, last longer than the Ground-Lease; and that is commonly about fifty or sixty Years. And if there happens to be a long Continuance of excessive Heat in Summer, or of Cold in Winter, (though, indeed, those Extreams happen but seldom with us,) the Walls being thin, become at last so penetrated with the Air, that it must needs make the Inhabitants uneasy.

But then this Manner of *building* is very much to the Advantage of Builders, and such Trades as have Relation to them; for they scarce ever want Work in so large a City, where Houses are here and there always either repairing or rebuilding.

Again, the Plaister'd Ceilings, which are so much more used in *England*, than in other Countries,

tries, do, by their Whiteneſs, make the Rooms much lighter, and are alſo excellent againſt the Ravages of Fires; they alſo ſtop the Paſſage of the Duſt, and leſſen Noiſe over Head, and render the Air ſomething cooler in Summer-Time, and warmer in Winter, becauſe they keep out cold Air better than Boarded Floors alone can do.

Dr. *Fuller* ſays, he who alters an old Houſe, is ty'd as a Tranſlator to the Original, and is confin'd to the Fancy of the firſt Builder. Such a Man would be unwiſe to pull down a good old Building, perhaps, to erect a worſe new one. But thoſe who erect a new Houſe from the Ground, are worthy of Blame, if they make it not handſome and uſeful, when Method and Confuſion are both of a Price to them.

I ſhall here propoſe a cheap Contrivance in *building*, which ſome approve of, *viz.*

Raiſe the Walls with Bricks, where Bricks may be had, making firm and ſtrong Quoins, at the Corners of the Houſe, of ſufficient Strength to ſupport the Floors and Roof, or the main Beams of it; the Walls may be built ſquare, and the Walls between them, built of the ſame Materials, and work'd up together with the Quoins, leaving one Half of the extraordinary Breadth of the Quoins without, and the other, within the Wall, whereby there will be much Charge ſav'd, both in Materials and Workmanſhip, and yet the *Building* be firm and ſtrong.

Some General Rules to be obſerved in building.

Theſe which follow, were eſtabliſh'd by Act of Parliament, before the *Rebuilding* of the City of *London* after the Fire.

Firſt, In every Foundation within the Ground, you muſt add one Brick in Thickneſs to the Thickneſs of the Wall next the Foundation to be ſet off, in three Courſes equally on both Sides.

Secondly, No Timber muſt be laid within twelve Inches of the Foreſide of the Chimney-Jaumbs.

Thirdly, That all Joiſts on the Back of any Chimney, be laid with a Trimmer at ſix Inches Diſtance from the Back.

Fourthly, That no Timber be laid within the Funnel of any Chimney, upon Penalty of ten Shillings to the Workman, and ten Shillings every Week it continues unreform'd.

Fifthly, That no Joiſts or Rafters be laid at greater Diſtances from one to the other, than twelve Inches; and no Quarters at greater Diſtance than fourteen Inches.

Sixthly, That no Joiſts bear at longer Length than nine Foot.

Seventhly, That all Roofs, Window-Frames, and Cellar-Floors be made of Oak.

Eighthly, That the Tile-Pins be made of Oak.

Ninthly, That no Summers or Girders in Brick *Buildings*, do lie over the Heads of Doors or Windows.

 Tenthly,

Tenthly, That no Summers or Girders do lie leſs than ten Inches into the Brickwork; nor no Joiſts leſs than eight Inches, and that they be laid in Lome.

Alſo ſome adviſe that all Tarſels for Mantle-Trees to lie on, or Lintels over Windows, or Templers under Girders, or any other Timber, which muſt lie in the Wall, be laid in Lome, which is a great Preſerver of Timber; whereas Mortar eats and corrodes it.

Some Workmen pitch the Ends of Timber which lie in the Walls, to preſerve them from the Mortar.

As to the ſurveying of *Building*, I ſhall touch briefly on it. As to the Method by which the Manner and Form of taking Dimenſions, will appear, that is, as follows:

The SURVEY *of a* BUILDING *erected by* HENRY GAINES, *for Mr.* WILLIAM BLAKEWAY. *The Thickneſs of the Walls (as by Agreement,) Brick and half, at* 3 l. per *Rod. For Mortar and Workmanſhip, the Dimenſions were taken as follows:*

	Feet.	Parts.		
1. The Length of one Side	40	50	648	0
From the Foundation to the Raiſing	16	00		
2. The Breadth at one End	17	16	283	14
The Height to the Croſs-Beam	16	50		
3. A Partion-Wall within	17	16	180	18
Height to the Firſt Story	10	50		
4. The Length of the other Side	39	33	275	31
From an old Wall, to the Raiſing	7	0		
5. The Breadth at the other End	17	0	82	11
From the Floor, to the Croſs-Beam	4	83		
6. A Water-Table, 30 Foot reduced to	7	50	23	70
From the Foundation, to the Table	3	16		
7. A Setting off on the other Side of the Houſe	16	83	16	83
8. A Gable End	66	7	66	0
The total Area or Content of theſe Dimenſions			1575	27

Particulars

Particulars to be deducted.

		Feet.	Parts.		
1. One Door-Case	Broad	8	66	81	58
	High	9	42		
2. Another Door-Case	Broad	4	33	32	13
	High	7	42		
3. A third Door-Case	Broad	4	33	22	34
	High	5	16		
4. A Window-Case	Broad	4	50	20	25
	High	4	50		
5. Another Window-Case	Broad	4	50	20	25
	Deep	4	50		
The Total of these Deductions		176	55		
Taken from the whole Content		1575	27		
There remains due		1398	72		

Which reduced into square Rods, is five Rods, thirty-eight Feet; and so, according to the Contract, there will be due to the Bricklayer 15*l.* 8*s.* 3*d.*

Mr. *Ven. Maudey* gives us the following Method of surveying *Buildings*, and taking Dimensions, and setting them down in a Pocket-Book.

2. Before you begin to set down the Dimensions, it will be proper to divide the Breadth of the Page into so many several Columns as you shall think you shall have Occasion for, either with Lines drawn with Ink, or a Pencil. Your Pocket-Book should be of the broadest Size, *viz.* four Inches broad; which may be divided into four Columns.

2. Before you set down any Dimensions, you must first set down the Names of the Masters of the Work, and the Workmen; also the Place where, and the Day of the Month, and Date when you measure.

As for Example: Suppose you are to measure Glaziers Work: Then you must observe, if it were glazed with square Glass, you must write Squares over the Dimensions; and if there is any Part or all in Quarry-Glass, you must write *Quarries*; that when you come to make the Bill of Measurement, you may express them severally; because they are of several Prizes.

For the clearer Understanding of this, I shall give you this Bill of Measurement of Glaziers Work, as follows:

Squares. F. I. P.	Products. F. I. P.	Squares. F. I. P.	Products. F. I. P.
5 8 6 / 5 7 6	31 11 30	4 3 0 / 1 2 0	4 11 6
5 3 6 / 2 4 6	12 6 0	2 0 0 / 1 6 0	3 5 0
2 6 0 / 1 2 0	(3) 8 9 0	6 0 9 / 5 0 3	3 5 3
2 1 0 / 1 8 6	(2) 7 2 4	1 2 0 / 3 0 0	(2) 7 0 0
	60 5 11		45 4 9

An Explanation of the Column.

In the first Column, towards the Left Hand, are the Dimensions of Glazings done with Quarrels; which are to be cast up by Cross Multiplication. See Multiplication.

In the second Column, are the Products of each Dimension just against it:

In the third Column, you have the four Dimensions of Glazing done with Squares; and in the last, you have the Product of each Dimension just against it.

At the Bortom of the second Column, you have the Sum total of the Products of the Dimensions done with Quarrels, which is 60 Feet, 5 Inches, and 11 Parts.

At the Bottom of the last Columu, there is the total Sum of the Products of those Dimensions of the Glazing that was done with Squares, which is 45 Feet, 4 Inches, and 9 Parts: As for tho odd Parts, they signify but little; if they be left out in the Sum total of Measurement, they amount to but very little in Value, unless there be many and large Articles of them to be added together.

N. B. When you are taking Dimensions, and setting them down in your Pocket-Book, whether of the Work of a Glazier, Carpenter, Bricklayer, &c. you must remember to leave every other Column vacant, that when you have set down all the Dimensions in the Book, (which is to be done before you cast any up, and which is to be done in another Book, or Sheet of Paper,) the Product of each Pair of Dimensions may be entered down just against them, as is to be seen in the foregoing Examples.

If there be another Person to measure against you, and there should happen a Mistake in either of your Castings up, this should be corrected by one Reading over the Dimensions to the other looking on his Book, that the Error may be found out and rectify'd, that both the Accounts may agree exactly together.

When

When you make your Bill of Measurement, you must set your Name to it, at the lower End of the Bill.

An Example of a Bill of Measurement.

Glaziers Work done for A. B. *of* Stepney, *by* C. D. *of* Ratcliff. *Measured* June 22, 1733.

For sixty Foot five Inches of Glazing done with Quarrels, at 5 *d. per* Foot	1	5	1	$\frac{1}{4}$
For forty-five Feet and four Inches of Glazing done with Squares, at 7 *d. per* Foot.	1	6	2	$\frac{1}{4}$
Measured the Day and Year abovewritten, by E. F. Sum total is	2	11	3	$\frac{1}{2}$

The Method of taking the Dimension of Bricklayers Work, which is the most troublesome of the Work of any Artificer concern'd in *Building*; I shall give an Instance of it

Although I before advised to divide the Page of your Measuring Book into four Parts, or Columns, yet in the Mensuration of Bricklayers Work, it will not be necessary to divide the Page into any more than three Columns; one large, for the Appellations, and two smaller, for the Dimensions; the other for the Products:

As in this Example following:

Appellations	*Dimen.*		*Products.*		
	Br. 3		*Br.* 3		
	Ft.	*In.*		*Ft.*	*In.*
Basis of the Front and Rear	25 0	0 6	(2)	25	0
	Br. 2 $\frac{1}{2}$		*Br.* 2 $\frac{1}{2}$		
Front and Rear	25 11	0 0	(2)	550	0
	2 *Br.* $\frac{1}{2}$		2 *Br.* $\frac{1}{2}$		
Basis of both the Flank-Walls	36 0	2 6	(2)	36	2

Appellations.	*Dimen.*			*Products.*	
	2 *Br.*			2 *Br.*	
				Ft.	*In.*
	2 *Br.*			2 *Br.*	
Both the Flanks	36	2			
	11	0	(2)	795	8
	1 *Br.* ½			1 *Br.* ½	
The Wall between the Chimneys	11	6			
	9	10	(2)	113	1
	1 *Br.*			1 *Br.*	
The Falling-back of both Chimneys	5	0			
	4	0	(2)	40	0
	2 *Br.*			2 *Br.*	
The four Jaumbs	14	0			
	11	0	(2)	161	0
	2 *Br.*			2 *Br.*	
The Fore-Part or Breasts of both Chimneys	11	6			
	5	0	(2)	115	0

Having set down the Dimensions with their Products, you must in the next Place set down the Deductions for the Windows and Doors, with their Products.

The Deductions.

Appellations.	*Deduct.*			*Products.*	
	2 *Br.* ½			2 *Br.* ½	
	Ft.	*In.*		*Ft.*	*In.*
The four Windows	6	6			
	4	0	(4)	104	0
	2 *Br.* ½			2 *Br.* ½	
The two Doors	6	0			
	4	0	(2)	72	0

The

The next Thing you have to do, is, to add the Products of each several Thickness of the Sum.

The Products of the several Thicknesses.

3 *Br.*	2 ½ *Br.*	2 *Br.*	1 ½ *Br.*	1 *Br.*
25 0	555 0	755 8	113 1	40 0
	36 2	161 0		
		115 0		
	586 2			
		1071 8		

The several Products of each Thickness being added, in the first column on the Left Hand, is 25 Feet of three Bricks. In the second, 586—2 of 2 ½ Bricks, &c.

Now to find these Products, see *Cross Multiplication*, N. 2.

Having found the total Sum of the Products of the Deductions, each total Sum must be subtracted from the total Sum of the Products of the Dimensions that are of the same Thickness.

E. gr. The Deductions in —— 2 ½ *Br.*

104 00
73 00

The Total Product in 2 ½ *Br.* is —— 176 00

Which 176 Feet of 2 ½ Brick-work, being contained in the Windows and Doors, must be subtracted from the 586 Feet 2 Inches, being the toral Product of all the Dimensions of that Thickness, *viz.* 2½ Bricks, *viz.* 2 ½ Brickwork.

This is manifest to Reason, because when the Dimensions of the Front and Rear were taken, the Whole Length was taken over the Doors and Windows, not allowing an Abatement for them.

N. B. Whatsoever Doors or Windows, or any other Vacancies, are measured over when the Dimensions were taken, you must remember to deduct them out of the whole Product of the Dimensions of the same Thickness, wherein they were situated.

In order to render this plainer, take the following Example:

The Doors and Windows being in 2 ½ Brick-Work, you must set down the total Product of all the Dimensions of that Thickness, which is —— 586 02

The Total Product of all the Deductions of that Thickness, which are to be subtracted, is —— 179 00

The Remainder is 410 02

The like must have been done, if there had been any other Deductions

ductions in any other Thicknesses: All such Deductions must have been subtracted from the Products of the Dimensions, before you went about to reduce your Thicknesses to the Standard Thickness of a Brick and half.

More of this Nature, *viz.* of surveying *Buildings*, or taking Dimensions, *&c.* may be seen under the Heads Carpenters Work, Joiners, Bricklayers, Plasterers, Masons, Painters, Thatchers, *&c.*

Of measuring Buildings.

I shall in this Place only mention the Artificers relating to *Building*, who usually work by Measure; which are, *First*, Bricklayers; *secondly*, Carpenters; *thirdly*, Plasterers; *fourthly*, Painters; *fithly*, Glaziers; *sixthly*, Joiners; and *seventhly*, Masons.

Some of which work by the superficial Yard, some by the Rod, some the Square, and some by the Foor: Of all which Works the Dimensions are taken either with a ten-foot Rod, or a five-foot one; or else with a two-foot Rule, and sometimes with a Line.

But however the Dimensions are taken, they are generally set down in Feet, Inches, and Parts of Inches; or else in Feet, and centesimal Parts of Feet; which last Way is the easiest cast up: And as to the Centesimal, *i.e.* hundredth Parts, the following Table will shew them.

A Table

A Table of Centesimal Parts for every Inch, and quarter of an Inch, in a Foot.

		1 Quarter of an Inch.	2 Quarters of an Inch.	3 Quarters of an Inch.
Inches.	100 Part of a Foot.	100 Part of a Foor.	100 Part of a Foot	100 Part of a Foot.
0	. 00	. 02	. 4	. 0[illegible]
1	. 08	. 10	. 12	. 12
2	. 16	. 18	. 20	. 22
3	. 25	. 27	. 29	. 31
4	. 33	. 35	37	39
5	. 42	. 44	45	47
6	. 50	. 52	53	55
7	. 56	. 60	62	64
8	. 66	. 68	70	72
9	. 75	. 77	79	81
10	. 83	. 85	87	40
11	. 92	. 94	96	98
1 *Foot.*	. 100			

To ſet down any Number of Feet, Inches, and Parts; as ſuppoſe 40 Feet, 6 Inches, and 3 quarters; you muſt firſt ſet down 40 Feet with a Period or Comma after it thus, 40, and then look in the firſt Column for 6 Inches, and at the Head of the Table for 3 Quarters, you will find 55; which ſet down beyond the 40 to the Right Hand, and it will ſtand thus, 40,55.

Of the valuing of Buildings.

In order to the eſtimating the Charge of erecting any Houſe, as near as can be, or to value one that is already built, to come pretty near the Truth, provided it be built of Brick and Timber.

1ſt, Find the Dimenſions in Length, Breadth, and Height, in reſpect to the Number of Stories.

2dly, By the Length and Breadth, the Quantity of Squares upon each Floor may be found; and alſo the Squares of Roofing in Carpenters Work, and alſo of Tiling in Bricklayers Work.

3dly, By the Height you may give a near Eſtimate of the Rows of Brickwork contain'd in the Walls round about, and in the Partition Walls, if there be any; and alſo in the Chimneys. Then,

4thly, Conſider how many Pairs of Stairs, and of what Sort.

5thly,

5thly, What Number of Partitions of Timber, with Doors.

6thly, What Timber Front,

7thly, What Number of Window Frames, and Lights.

8thly, What Iron Work.

9thly, What Lead, &c.

Of which see the particular Heads

Mr. *Leybourn* puts the Question, What will be the Charge of erecting a *Building* of Brick Walls and Timber, which shall be 20 Feet in Front, and 44 deep, and for the Front to be shorter than the Flanks, and to consist of Cellars, three Stories, and Garrets, which is one of the second Rate Houses. Now supposing the Price of Materials (in *London*) to be as follows, *viz.*

		l.	*s.*	*d.*
For	Bricks *per* Thousand	00	16	00
	Tiles *per* Thousand	01	05	00
	Lime *per* Hundred	00	10	00
	Sand *per* Load	00	03	00
	Oak or Fir Timber *per* Load	02	15	00
	Deal Boards *per* Hundred	07	10	00
	Laths *per* Bundle	00	01	06

As for Plasterers Work.

		l.	*s.*	*d.*
For	Lathing, Plastering, Rendring, and Washing with White and Size, *per* Yard	00	01	02
	Lathing and Plastering *per* Yard	00	00	10
	Plastering and Sizing *per* Yard	00	00	06

Smiths Work.

		l.	*s.*	*d.*
For	Iron for Balconies *per* lb.	00	00	05
	Folding Casements *per* Pair	00	16	00
	Ordinary Casements *per* Piece	00	04	06

For Painting.

	l.	*s.*	*d.*
Window Lights	00	00	06
Shop Windows, Doors, Pales, *per* Yard	00	01	00

Now, says he, from these Rates of Materials for *Building*, and for Workmanship, such a House will amount to about 360 *l.* which is about 41 *l. per* Square.

Mr. *Phillips* proposes the following Method to find the Value of a *Building*, viz. Suppose a House to be one Rod, or sixteen Feet and a half in Front, and two Rods deep, back in the Flanks, the Compass of this House will be six Rods: And if this House stands in a high Street, having a Cellar, four Stories, and a Garret, (which is one of the

third

third Rate Houfes,) the Height thereof will be fifty Feet, or three Rods; fo that there will be eighteen Rods of Brickwork in the Walls, which may all be reduced to a Brick and a half thick, and (fuppofing each Rod of Brickwork to contain 4500 Bricks,) will coft about 7*l.* the building, *viz.* Bricks, Mortar, and Workmanfhip; then the whole eighteen Rods of Brickwork will coft about 126*l.*

The Timber Work for Floors, Windows, Roofs, &c. about as much more.

The Tiling, Plaftering, Lead, Glazing, and Painting will be about as much more; fo that the Whole will amount to 378*l.*

The Allowance for the Party-Walls will very well pay for the Chimneys; fo that this Houfe cannot amount to above 400*l.* the building, which is not full 13*l.* *per* Square: But this is a very great Price, in comparifon of Mr. *Leybourn*'s; but he fays it will be worth more or lefs, according to the Market Price of the Materials.

The *Friendly Society* of *London* for infuring Houfes, have two Rules by which they value them, *viz.* Either by the Rent, or Number of Squares contained on the Ground-Plot.

The laft is the general Rule by which they value all *Buildings*; which is grounded on an Act of Parliament for rebuilding the City of *London*, made about *Anno* 18. *Car.* II.

The *Buildings* of the City of *London* are valued according to their Rates, of which Rates there are four, *viz.*

Firft Rate	2	Stories, Cellars, and Garrets.
Second Rate	3	
Third Rate	4	
Fourth Rate	5	

And the naked *Building* or Shell of a Brick Houfe (the Floors being finifhed) is thus valued, by the Square or 100 Feet in the High-Streets, *viz.*

Firft Rate	25*l.*	*per* Square.
Second Rate	35*l.*	
Third Rate	45*l.*	
Fourth Rate	50*l.*	

But thefe may be augmented at the Difcretion of the Surveyor, or according to the finifhing of the Houfe.

Vitruvius, *Lib.* 1. *Cap.* 2. determines fix Confiderations, in order to the judging or cenfuring of a *Building*, viz.

1. *Ordination.* 2. *Difpofition.* 3. *Eurythmy.* 4. *Symmetry.* 5. *Gracefulnefs.* 6. *Diftribution.*

The two firft of thefe might have been very well fpared, fince he feems to mean no more by Ordination, than but a well-fettling the Model or Scheme of the whole Work; nor any Thing by *Difpofition*, but a neat and full Expreffion of the firft Idea and Defign of the *Building*, which feems more properly to belong to the Artificer than the Cenfurer.

The other four are fufficient to approve or condemn any *Building* whatfoever.

Eurithmy is that agreeable Harmony which is between the Length, Breadth, and Height of all the Rooms of the *Building*, which will be very pleafing to Beholders;

Beholders; which is always ſo to all, by a ſecret Power, that is in Proportion.

Here it may be proper to obſerve, that though Exceſs of Height is the leaſt Error or Offence that can be committed againſt the Sight; yet even that Error is no where of ſmall Importance, becauſe it is the greateſt Treſpaſs upon the Purſe.

Symmetry, which is a due Proportion of each Part, in reſpect to the Whole; whereby a great *Building* ſhould have large Apartments or Rooms, great Lights or Windows, great Stair-Caſes, great Pillars or Pilaſters, *&c.* In ſhort, all the Members and Parts large, in Proportion to the *Building*.

For as it would be but an odd Sight to ſee a large Man with little Legs, Feet, Arms, Hands, *&c.* ſo alſo would it be uncomely to ſee a large *Building* conſiſting of little Apartments, Lights, Stair-Caſes, Entrances, *&c.* and ſo, on the contrary, it will be as odd to ſee a little Man with large Limbs; the ſame will it be in a ſmall *Building* ſo contrived, as to have large Rooms, large Stair-Caſes, large Lights, large Entrances, *&c.*

But again, as it is an unbeſeeming Sight to ſee either a great or little Man with ſome of his Limbs or Parts proportionable to his Body, and others ſome ſo little, as if they did belong to a Pigmy; and others ſo large, as if they did belong to a Giant: So would it be equally as ugly and offenſive to the Sight, to ſee a ſmall Houſe have ſome of its Parts monſtrous, *viz.* large in ſome Parts of the Apartments, and by conſequence others muſt be as ſmall, or elſe ſome muſt be annihilated, and ſo conſequently will be wanting; or large Stair-Caſes, large Windows, and large Doors, or any other Parts larger than they ought to be, in reſpect to the Symmetry of the Parts with the Whole.

It is likewiſe unſeemly to ſee ſome of the Parts too little, and not proportionable to the whole Structure, as to ſee a Man with one leg proportional to his Body, and the other very ſmall, or to have one Eye of a Man, and the other of a Bird.

Many Errors are committed by Workmen in Symmetry, for want either of due Conſideration, or Skill.

Sometimes it may be obſerved in the Courſe of cenſuring, that a Door or Chimney has been ſo miſplaced, either to the Right or Left, as to ſpoil the intended Uſe of a Room: And though ſometimes it is not totally ſpoiled, yet it ſhews the want of Contrivance in the Artificer.

Sometimes you may obſerve a Chimney ſo ſituated in the Angle of a Room, (which though it was deſigned for Conveniency, becauſe it could not well be carried up otherwiſe from the Chimney below it,) yet this Chimney ſhall in ſome meaſure ſpoil the intended Uſe of two Rooms, *viz.* that in which the Chimney is, and the next adjoining to the Chimney Jaumb. Thus two Chambers have been in great meaſure ſpoiled by a Chimney being ſet in the Angle of the inner one, the Door coming into it from the Chamber without, juſt by one Jaumb, and of conſequence

ſequence that whole Chimney was carried a Foot too far out in the Room, which might as well have been carried farther the other Way, by which Means the Door was placed too far towards the other Wall; ſo that the Partition Wall, by this Means, was made ſo narrow between the other Wall and the Door, (at the Chimney Jaumb,) that it was thereby rendered unfit to ſet a Bed in againſt it, though it was the fitteſt Place in the Room for that Purpoſe.

Sometimes little diminutive Stair-Caſes are made in a handſome ſpacious Structure; and, on the contrary, in a ſmall or middling Houſe, Stair-Caſes ſo large, that if you ſee them before any other Apartment, you might well conjecture that the Rooms of that *Building* were proportionable to the Stair-Caſe, twice or three Times larger than you find them.

Nay, perhaps this ſhall not be all the Error; for theſe Gueſs-Workmen do ſo manage the Matter, as to ſpoil the Conveniencies of Cloſets under them, (for any other Conveniency;) though it be now the Faſhion to to make ſome little Conveniencies under a Stair-Caſe; for Cloſets are accounted an Improvement in our Way of *Building*.

Sometimes you may obſerve an ill Poſition of Lights, (or Windows) to a Stair-Caſe, not out of Neceſſity, but Want of Skill and Contrivance.

And again, as to Lights, (or Windows,) you will ſometimes ſee an ill Poſition, as well as an irregular Diſpoſition in them, *viz.* either in regard to Uniformity, or as to ſecuring them from the Weather; that is, when they are placed too near the Surface of the *Building*, that the Walls do not project over them, the better to carry the Wet from them, which runs down the Walls in ſtormy Weather.

And then again, as to Uniformity in placing them, it ſometimes ſo happens, that they cannot place them on the Garrets exactly over thoſe in the Stories below; and therefore when they will not be brought into Uniformity with thoſe below them, they ought to be placed as uniformly as poſſibly can be within themſelves.

This has been obſerved in a Fabrick which ſtood in the Form of a Roman Capital L, having two Fronts on the Outſide of the L, confronting two Streets which croſs one another at Right Angles: The Foot or ſhort Part of the L in the *Building*, was not ſo wide, but that it might be ſpanned by one Roof; but the long Part was too wide to be ſpanned by one Roof, unleſs it were carried up a great deal higher than the other Part, which would have been very unſightly: And therefore, three Roofs were ſet on the long Part of the L, parallel with that on the ſhort Part; ſo that there were three Gutters, and four Gable-Heads, on that Part which was the long Part of the L; and in each of theſe Gable-Heads, there was a Window.

Now according to the Diviſions of the Apartments in the Stories below, the Windows in them would not fit, to be placed (any of them) perpendicular under the Middle of theſe Gables,

the

the Workman thinking to render it something nearer to Uniformity, places three Fourths of these Windows all towards, nay, very near one Side of the Gables, pretending, they were without Doubt, nearer directly over the others; and therefore it was nearer to Uniformity, whereas at the same Time, they are farther from it: For by this Means they are not in an uniform Position, neither in respect to the Stories below them, nor yet within themselves; which last they would have been, if they had been placed in the Middle of each Gable, and would have been more decent and handsome, both with Regard to the Front without, and the Room within.

These, and many more are the Blunders committed for want of Contrivance or a good Judgment as to Symmetry. I shall only add, that it ought to be observed, whether Doors have their due Symmetry as to their Dimensions, as well as their Positions, *viz.* That they be not too high, as if they were for a Barn, nor too low, as if they were made for Houses in *Sophia* in *Bulgaria*; where both Jews and Christians have their Doors of their Houses but a little above three Foot high; which are therefore so contriv'd, because the *Turks* should not bring in their Horses; which they would do, and make use of them for Stables in their Travels, if it were not for this Contrivance. This, as well as all other Parts of a *Building*, ought to be analogous to the rest of the Fabrick. I shall next speak of

Decor or Becomingness, or rather Suitableness; which is the keeping a due Respect between the Inhabitant and the Habitation: Whence *Palladio* concludes, that the principal Entrance was not to be regulated by any certain Dimensions, but by the Dignity of the Master; yet to exceed rather in the more, than in the less, is a Mark of Generosity, and may always be excused with some noble Emblem or Inscription.

Distribution, is that useful Casting, or Contriving of all Rooms for Office, Entertainment or Pleasure, which has been already sufficiently treated of under this Head of *Building*.

These are the four general Heads which every Man ought to run over, before he pretends to pass his Judgment upon a *Building*, or undertakes to censure a Work that he views.

Dr. *Fuller* advises rather to believe any Man than an Artificer in his own Art, as to the Charges of a *Building*, especially, if either himself, or any Friend of his, is to be concern'd in the *Building* that is designed to be erected: Not but that Builders can tell nearly the Charge, when they know the Design, but it is very rare that they will give a just Estimate of it according to their Judgment; because they think if they should acquaint a Gentleman with the full Expence at first, it would discourage him from prosecuting it; and for that Reason, they sooth him up, 'till it will cost him something considerable; after which, he must go through with it, or lose what has been expended.

The

The Spirit of *Building* first possess'd People after the Flood; which then caused the Confusion of Languages, and since, the Confusion of many a Man's Estate: And hence, when some Persons would wish a Curse upon one with whom they are angry, they wish them to be possessed with the *Spirit of Building*, or, as others term it, the *Italian Plague*.

BUST, } in Sculpture, *&c.*
BUSTO, } is a Term used for the Figure or Portrait of a Person in Relievo, shewing only the Head, Shoulders, and Stomach, the Arms being lopped off; which is ordinarily placed on a Pedestal or Consiole.

Felibien observes, that though in Painting, one may say, a Figure appears *in Busto*, yet it is not proper to say, in a *Bust*. The *Bust* is the same with that the *Latins* call *Herma*, from the *Greek Hermes*, *Mercury*, the Image of that God being frequently represented in this Manner by the *Athenians*.

Bust is also used, especially by the *Italians*, for the Trunk of a human Body from the Neck to Hips.

BUST, or } a Pyramid, or
BUSTUM, } Pile of Wood, whereon the Bodies of the Deceased were antiently placed in order to be burnt.

BUTMENTS, [of *Bouter*, *French*, to abut or terminate any Thing] are those Supporters or Props, on or against which, the Feet of Arches rest. Also little Places taken out of the Yard or Ground-Plot of a House, for a Buttery, Scullery, *&c.*

BUTTERY, in the Houses of Noblemen and Gentlemen, is the Room belonging to the Butler; where he deposites the Utensils belonging to his Office; as Table-Linnen, Napkins, Pots, Tankards, Glasses, Cruets, Salvers, Spoons, Knives, Forks, Pepper, Mustard, *&c.*

As to its Position, Sir *Henry Wotton*, says, it ought to be placed on the North Side of the Building, which is designed for the Offices.

We, in *England*, generally place it near the Cellar, *viz.* the Room commonly just on the Top of the Cellar-Stairs.

BULLEN-NAILS, are a Sort of Nails with round Heads, and short Shanks, lin'd and lacquered. There are several Sizes of them.

They are used in hanging Rooms, fitting up Beds, covering of Stools, Chairs, Couches, Desks, Coffins, *&c.*

BUTTRESS, a kind of Butment built Archwise, or a Mass of Stone, or Brick, serving to prop or support the Sides of a Building, Wall, *&c.* on the Outside, where it is either very high, or has any considerable Load to sustain on the other Side, as a Bank of Earth, *&c.*

They are also used against the Angles of Steeples, and other Buildings of Stone, *&c.* on their Outside and along the Walls of such Buildings, as have great and heavy Roofs, which would be subject to thrust the Walls out, if they were not thick, if no *Buttresses* were placed against them.

Buttresses

Buttresses are alſo placed for a Support and Butment, againſt the Feet of ſome Arches, that are turn'd acroſs great Halls in old Palaces, Abbeys, *&c.* and generally at the Head of Stone Buildings, when there are large Crocket Windows; and they are alſo placed for Butments to the Arches of theſe Windows.

The Theory and Rules of *Buttreſſes* are one of the *Deſiderata* in Architecture; but it is not improbable, but that a ſagacious Architect and Mathematician, who would apply himſelf diligently to examine into the Matter, might bring it within the Bounds of Reaſon and Rules, whereby it might be known very near, of what Size, and conſequently of what Weight, a *Buttreſs* or Butment ought to be; which muſt be various, according to the Dimenſions and Form of the Arch, and the Weight which is ſuper-incumbent on it.

As to the Weight of the Materials, both on the Arch, and in the *Buttreſs* or Butment, it is not difficult to calculate. But it may probably be objected, there may be a ſenſible Difference as to the Strength and Goodneſs of the Mortar; which may in a Meaſure compenſate for the Weight of the *Buttreſs* or Butments; for where there is a ſtrong firm Mortar uſed, of leſs Weight (or Magnitude,) of Brick or Stone, ſhall be capable to reſiſt the Preſſure of an Arch with its ſuperincumbent Materials, than where the Mortar is bad and weak. To which it may be anſwered, it would not be difficult to make Experiments of the Strength of Mortar, both as to the direct and oblique Force, by ſhoving of it out of its Poſition, or pulling it the ſhorteſt Way from its Adherents, I mean, by lifting it directly up.

It ſeems to be very feaſible, and it would certainly be very uſeful, to try what Butment would be requiſite for Arches of all Dimenſions or Forms, whether Straight, Semicircular, Skeen or Scheme, or of the Third or Fourth Point, or Elliptical, *&c.* See the Article BRIDGES.

Dr. *Hook*, Geometry-Reader in *Greſham College*, in his *Treatiſe of Helioſcopes*, did promiſe to publiſh ſomething to the foregoing Pupoſe, whether he ever did do it, I know not; but what he promiſed in that Treatiſe, was as follows: *viz.* A true Mathematical and Mechanical Form of all manner of Arches, with the true Butment neceſſary to each of them, a Problem, ſaith he, which no Architectonick Writer hath ever yet attempted, much leſs perform'd. A Treatiſe of this Nature would be extreamly uſeful for the want of a certain Rule in Arching, with its neceſſary Butment, hath often proved the Ruin of ſome Structures which have been of no ſmall Charge, as to Bridges, *&c.*

Of the Price of Building Buttreſſes.

If this Work be not put out to be done by the Day, it is uſually done by the Cubical Foot. Some reckon the Workmanſhip at two Pence Halfpenny *per* Cubick Foot, which, reckoning the Materials,

terials and Workmanſhip may be done for about 6*d.* and 7*d.* a Foot.

C A

CABINET, the moſt retir'd Place in the fineſt Apartment of a Building, ſet apart for writing, ſtudying, or preſerving any Thing that is precious. A compleat Apartment conſiſts of a Hall, Antichamber, Chamber, and *Cabinet*, with a Gallery on one Side.

Hence we ſay, a *Cabinet* of Paintings, Curioſities, &c.

CABLED-FLUTE, ſuch Flutes in Architecture, as are filled up with Pieces in the Form of a Cable.

CALCULATION, the Act of computing ſeveral Sums, by adding, ſubtracting, multiplying, and dividing, &c.

CALIDUCTS, a kind of Pipes or Canals, diſpoſed along the Walls of Houſes and Apartments, uſed by the Antients for conveying Heat to ſeveral remote Parts of the Houſe, from one common Furnace.

CALOTTE, in Architecture, a round Cavity, or Depreſſure, in Form of a Cap or Cup, lathed and plaiſtered, uſed to diminiſh the Riſe or Elevation of a moderate Chapel, Cabinet, Alcove, &c. which, without ſuch an Expedient, would be too high for other Pieces of the Apartment.

CAMBER-BEAM, a Piece of Timber in an Edifice, cut Arching, or Archwiſe, or with an Obtuſe Angle in the Middle, commonly uſed in Platforms, as Church Leads, and on other Occaſions, where long and ſtrong Beams are required. A *Camber-Beam* being much ſtronger than another of the ſame Size; and being laid with the hollow Side downwards, (as they generally are) they repreſent a kind of Arch.

CAMBRING. The Seamen ſay a Deck lies *Cambring*, when it does not lie level, but higher in the Middle, than at either End.

CAMES, the ſmall ſlender Rods of Caſt Lead, of which the Glaziers make their Turn'd Lead. For their Lead being caſt into ſlender Rods, of twelve or fourteen Inches in Length, are called *Cames*; and ſometimes they call each of thoſe Rods a *Came*, which, when it has been afterwards drawn through their Vice, makes their turn'd Lead.

CAMERATED, vaulted, or arched.

CANT, a Term uſed by ſome Carpenters of a Piece of Timber, when it comes the wrong Way in their Work, they ſay, *Cant it*, *i. e.* Turn it about.

CANTALIVERS, Pieces of Wood fram'd into the Front or other Sides of a Houſe to ſuſtain the Moulding and Eves over it. Theſe ſeem, in Effect, to be the ſame with Modilions, except that the former are plain, and the latter carv'd; they are both a kind of Cartouzes, ſet at equal Diſtances, under the Corona of the Cornice of a Building.

As to the Price of making, Mr. *Leybourn* ſays, they are commonly

monly made by the Piece, at different Rates, according to the Curiosity of the Work: And some Workmen say, they have 2*s*, 6*d.* for making and carving each. But they will carve them in *London* for twenty Pence *per* Piece.

As for their Painting, Mr. *Leybourn* likewise tells us, they are usually painted by the Foot, running Measure, *i.e.* by the Number of Feet in Length only, at different Rates, according to the Curiosity of the Workmanship. And some Workmen say, they have commonly 1*s. per* Foot for the Cornice, if plain without Carving, and 3*s.* 6*d. per* Foot with the *Cantalivers*.

CANTING-STAIRS. See STAIRS.

CANTONED, in Architecture, is, when the Corner of a Building is adorn'd with a Pilaster, an angular Column, Rustick Quoins, or any Thing that projects beyond the Naked of a Wall.

CAPITAL, [of *Caput*, *L.* the Head] is the uppermost Part of a Column or Pilaster, serving as the Head or Crowning thereof, placed immediately over the Shaft, and under the Entablature.

Capital of a Column, is properly that whose Plan is round.

Capital of a Pilaster, is that whose Plan is square, or at least rectilineal.

The *Capital* is the principal and essential Part of an Order of Column, or Pilaster: It is of a different Form in the different Orders; and is that which chiefly distinguishes and characterizes the Orders. Such of these as have no Ornaments, as the *Tuscan* and *Doric*, are called Capitals with Mouldings, and the rest which have Leaves and other Ornaments, Capitals with Sculptures.

The *Tuscan Capital* is the most simple and unadorn'd: Its Members or Parts are but three, *viz.* an Abacus, under this an Ovolo, or Quarter-Round; and under that, a Neck or Colarino; the Neck terminates in an Astragal, or Fillet belonging to the Fust or Shaft.

M. Le Clerc says, this *Capital* only consists of three Parts, an Abacus, a Quarter-Round or Boultin, and a Gorge or Neck, which terminates under the Quarter-Round in a Fillet; the Astragal underneath belonging to the Shaft.

The Character of the *Capital* whereby it is distinguished from the *Doric*, &c. is, that the Abacus is square, and quite plain, having no Ogee, or other Moulding; and that there are no Annulets under the Ovolo. It is true, Authors do vary a little as to the Character of the *Tuscan Capital*.

Vignola gives the Abacus a Fillet, instead of an Ovolo. *Vitruvius* and *Scamozzi* add an Astragal and Fillet between the Ovolo and Neck: *Serlio*, only a Fillet; and *Philander* rounds the Corners of the Abacus.

In the *Trajan* Column there is no Neck, but the Astragal of the Shaft is confounded with that of the Capital.

The Height of this *Capital* is the same with that of the Base, *viz.*

viz. one Module or Semidiameter.

Its projecture is equal to that of the Cincture at the Bottom of the Column, *viz.* five Eighths of the Module.

According to *Vitruvius*, the Height of the *Tuscan Capital* (by the Astragal at Bottom) must be half the Diameter of the Body of the Column below.

And this Height being divided into three Parts; the first and uppermost Part goes to the Abacus (which is a square or flat Moulding;) the second Part goes to the Boultin and Fillet under it; the Boultin is a quarter of a Circle; the Fillet, a narrow flat Moulding; and this Part is subdivided into four Parts; three of which go to the Boultin, and one to the Fillet; and the third and last Part go to the Neck, which is flat and straight.

Again, the Neck is divided into two Parts, one of which is the Breadth of the Astragal under it (which consists of a Semicircle and a Fillet under it.)

The Astragal is again divided into three Parts, of which, two go to the Semicircle, and one to the Fillet.

The Projecture of the *Capital* is to be the half Part of the Diameter of the Body of the Column below.

The Astragal projects in a Square.

Scamozzi makes the Height of the *Capital* by the Astragal at the Bottom, also half the Diameter of the Column below; and this Height being divided into sixty Parts, twenty of them go to the Abacus or Plinth (as he calls it,) fifteen to the Echinus or Half-Round (which is called the Boultin, by *Vitruvius*,) and five to the Rondel or Bead-Moulding (which is a Semicircle,) three to the List, (which by *Vitruvius* is called a Fillet,) and seventeen to the Neck or Frieze. Again, seven such Parts go to the Rondel of the Astragal, and three to its List.

Palladio also makes the Height of the *Capital* half the Diameter of the Body of the Column below (*viz.* by the Astragal, which is by none of them reckon'd a Part of the *Capital*, though properly speaking, it ought to be so accounted;) and this Height he divides into three equal Parts, the uppermost of which goes to the Abacus (which by him is also called the *Dado* or Dye,) the next Part goes to the Ovolo or *Echinus* (which by *Vitruvius* is called the Boultin.) the other Part he divides into seven, one of which he makes the Listella (which *Vitruvius* calls the Fillet) under the Ovolo, and the other six Parts go to the Collarino or Neck (which is also called by him the Hypotrachelium, or Frieze of the *Capital*.

The *Doric Capital* has three Annulets, or little square Members underneath the Ovolo, instead of the Astragal in the *Tuscan*, besides an Abacus, an Ovolo, and a Neck, in common with the *Tuscan*; and a Talon, Cyma, or Ogee, with a Fillet over the Abacus. Authors also vary as to the Characters of this *Capital*.

Palladio, *Vignola*, &c. put Roſes under the Corners of the Abacus, and in the Neck of the *Capital*.

Vitruvius makes the Height of the Capital (by the Aſtragal at the Bottom) equal to half of the Diameter of the Body of the Column below.

And this Height being divided into three Parts, the firſt and lowermoſt goes to the Neck, the next to the Boultin (under which Term, ſeveral Members are comprehended;) and this Part is by him deſcribed in two Forms; the firſt of which is a Boultin (as 'tis deſcribed before,) and three Fillets under ir, and the other is a Boultin, and an Aſtragal under it; and this Part is divided into three Parts, two of which go to the Boultin, properly ſo called, and the other to three Filles, or to the Aſtragal.

The Fillets are all of an equal Size: In the Aſtragal, the Fillet is one Third of the whole. The third and uppermoſt Part of the *Capital* is again divided into three, the two lowermoſt of which go to the Square, and the other to the Cymatium (which is an Ogee with the Hollow downwards,) and a Fillet over it.

An Ogee is a Moulding which bears ſome Reſemblance to an S; which *Vitruvius* makes of two quarter Circles joined together; and this Cymatium being alſo divided into three Parts, two of them go the Ogee, and one to the Fillet.

The Aſtragal under the *Capital* is equal to one half of the Neck.

Scamozzi makes the *Capital* of the ſame Height, and divides it into ſixty Parts, three of which go to the Fillet of the Cymatium, five to the Ogee of the Cymatium, twelve to the Square, fourteen to the Boultin, five to the Rondel, and two to the Fillet of the Aſtragal under the Boultin, and nineteen to the Neck.

The Aſtragal under the Neck contains ten ſuch Parts, ſix and a half of which go to the Rondel, and three and a half to the Fillet. Theſe are deſcribed according to *Vitruvius*'s Terms, *Scamozzi* not mentioning any of them.

Palladio likewiſe makes this *Capital* of the ſame Height, as *Vitruvius*, and divides it into three Parts; the uppermoſt of which, he ſubdivides into five Parts, two of which he allows to the Cymatium, and is ſubdivided again into three Parts, one of which he gives to the Liſtella or Annulet, and which by *Vitruvius* is called a Fillet; and the other two to the Cima recta, (which is an Ogee, as here deſcribed;) and the other three of the firſt Subdiviſions of this Part go to the Abacus, which by *Vitruvius*, in this Number, is called the Square.

The ſecond of the three grand Diviſions of the *Capital* is ſubdivided in three Parts, two of which go to the Ovolo or Echinus (which is by *Vitruvius* called the Boultin;) and the other to the Annulets under it, which are three, and are all equal.

The third principal Hypotrachelium or Frize L, (which is by *Vitruvius* called the Neck.) The

The Aſtragal under the Neck is as high as all the three Annulets.

The *Ionic Capital* is compoſed of three Parts: An Abacus, conſiſting of an Ogee; under this, a Rind, which produces the Volutes or Scrolls, the moſt eſſential Parts of this *Capital*; and at the Bottom, an Ovolo or Quarter-Round: The Aſtragal under that Ovolo belongs to the Shaft: The middle Part is called a Rind or Bark, from its ſuppoſed Reſemblance to a Bark of a Tree, laid on a Vaſe, whoſe Brim is repreſented by the Ovolo, and ſeeming to have been ſhrunk up in drying, and to have been twiſted into the Volutes.

The Ovolo is adorn'd with Eggs, as they are ſometimes called from their oval Form, the the *Greeks* call it Εχινος.

The Height of this *Capital*, M. *Perrault* makes eighteen Minutes; its Projecture, one Module, ſeven Tenths.

The Differences in the Character of this *Capital* flow moſtly from the different Management of the Volutes, and conſiſt in this: That in the Antique, and ſome of the Modern, the Eye of the Volute does not anſwer the Aſtragal of the Top of the Shaft, as *Vitruvius*, and ſome of the Moderns make it: That the Face of the Volutes, which uſually makes a Flat, is ſometimes curved and convexed ſo that the Circumvolutions go advancing outwards, as is frequent in the Antique.

2. That the Border or Rim of the Scroll in the Volute, is ſometimes not only a plain Sweep, as it ordinarily is, but the Sweep is accompany'd with a Fillet.

3. That the Leaves which inveſt the Baluſter, are ſometimes long and narrow, ſometimes larger and broader.

4. That the two Faces of the Volutes are ſometimes joined at the outward Corner, the Baluſters meeting in the inner, to make a Regularity between the Faces on the Front and Back of the Building with thoſe of the Sides.

5. That among the Moderns, ſince *Scamozzi*, the *Ionic Capital* has been altered, and the four Faces made alike, by taking away the Baluſter, and hollowing all the Faces of the Volute inwards, as in the *Compoſite*.

6. That *Scamozzi*, and ſome others, make the Volutes to ſpring out of the Ovolo, as from a Vaſe, after the Manner of the modern *Compoſite:* Whereas in the Antique, the Bark paſſes between the Ovolo and Abacus, quite ſtraight, only twiſting at its Extremities, to form the Volute.

7. *Laſtly*, That of late Years, the Sculptors have added a little kind of Feſtoons, ſpringing from the Flower whoſe Stalk lies on the Circumvolution of the Volute; and ſuppoſed to repreſent the Locks of Hair hanging down both Sides of the Face.

The *Ionic*, according to *Vitruvius*, is made thus: Divide the Semidiameter of the Body of the Column below into eighteen Parts, take nine ſuch Parts, of which, three muſt go to the Cymatium, one ro the Fillet, and two to the Cima or Ogee under it it; then take four Parts for the Trochilus of the Volute or Scroll [the Trochilus is that Member from whence the Scroll begins] thence take four Parts from the Boultin, which is the fourth Part of a Circle, and which must be carved with Eggs and Anchors Then take two Parts for the Astragal is carv'd with Eggs and Anchors; then take two Parts for the Astragal under the Boultin: The Astragal is carved with Beads, and has a Fillet on each Side of it, each one fourth of the Whole: Then the six Parts remaining must go to the half of the Volute below. Then take eight more such Parts, which must go to make the Remainder of the Frize or Neck of the *Capital*; and three more such Parts for the Astragal, under the Neck of which one Part goes to the Fillet.

Palladio's Description of this *Capital* agrees with that of *Vitruvius*; and *Scamozzi*'s is so blind, that I believe scarce any are the wiser for it.

M. *Le Clerc* says, the most essential Part of this *Capital* is the Volute; which several Architects imagine to have been intended to represent the Rind or Bark of a Tree inclosed between the Abacus and Quarter-Round, having its two Extremes twisted into Scrolls, and those two Scrolls bound with a large Rope in the Middle; which comes pretty near the Figure that the Antients gave on the two Sides of the *Capital*.

Other Architects considering that this *Capital* bears some Resemblance to the Head-Dress of a *Greek* Lady, believe it to have taken its Origin from thence: But this being a Matter of no great Use, we leave every one to judge of it as he pleases.

The *Capital* of the Antients being found improper in angular Columns, by reason of the Diversity of its Faces, *Scamozzi* composed a new one with four similar Faces.

Some Architects, however, won't allow the Volutes to spring out of the Vase of the *Capital*, but will have them consist of one and the same Rind continued under the Abacus; which, by this Means, will appear the better supported; an Instance of which we have in the Five Orders of M. *Perrault*. And they would have Reason on their Sides, were there the same good Taste in this, as in the other Design; but as that can't be, we must be contented with the other, which is easily design'd, and has a beautiful Appearance.

'Tis true, the new Abacus, which it has in his 49th Figure, being better proportion'd to the Largeness of the Volute than that of *Scamozzi*, renders it more graceful; besides, that it is further inrich'd with little festoons falling from the Volutes, which some modern Sculptors have been pleased to add.

He observes, that when there are Eggs cut in its Quarter-Round, their Number should be 24; and the Shaft or Fust should be channelled with an equal Number of Flutings.

He adds, we sometimes also cut Pearls and Olives in the Astragal, over the Ovolo, though it it belong to the Shaft: But were the *Capital* made of a Matter different from that of the Shaft, then muſt the Aſtragal be conſidered as a Bauguette, making Part of the *Capital*, and not of the Shaft: To which laſt the Fillet underneath would be left; were it otherwiſe, the *Capital* would be but poorly terminated by its Ovolo or Quarter-Round; beſides that, it would be too flat and ſquab.

He adds, that the Antients having made the Baluſter of this *Capital* very ſhort, there is ſome Difficulty in adjuſting the Volutes to the Quarter-Round in the *Capitals* of the Pilaſters.

This has occaſioned ſeveral Architects to flatten or diminiſh the Convexity of the Quarter-Round; which is a very conſiderable Irregularity, which they might have avoided, by lengthening out the Baluſter, ſo as to go beyond the Quarter-Round; at the ſame Time making the Circumvolution of the Volute advance a little further.

However, ſays he, if a Perſon has a mind to follow the Cuſtom, he means, if he chuſes to diminiſh the Convexity of the Quarter-Round, as is here done, he commits a Fault, that has good Authority on its Side: Which, however, he would do well to avoid, eſpecially as it may be done without much Trouble.

A Difficulty of the ſame Kind may be met with in the Quarter-Round of the modern *Capital*, which our Architects have likewiſe diminiſhed, in order to receive the Volutes more favourably, which ſhould further have a Curvity, like that of the Abacus; but from which a man is under a Neceſſity of receding and of opening the Volutes, ſo as to be above the Quarter-Round, after it has run perpendicularly acroſs the Face of the Pilaſter: And the ſame, he ſays, may be underſtood of the *Roman Capital.*

The *Corinthian Capital* is much the richeſt; it has no Ovolo, and its Abacus is very different from thoſe of the *Tuſcan*, *Doric*, or *Ionic*; as having its Faces circular, hollowed inwards, with a Roſe in the Middle of each Sweep.

Inſtead of Ovolo and Annulet, here is only a Brim of a Vaſe; and the Neck is much lengthened and inrich'd with a double Row of each Leaves, bending their Heads downwards, and between them ſmall Stalks ariſing, whence ſpring the Volutes, which do not reſemble thoſe of the *Ionic Capital*; and which inſtead of the four in the *Ionic*, are here ſixteen; four on each Side under the four Horns of the Abacus, where the Volutes meet in a ſmall Leaf, which turns back towards the Corner of the Abacus.

The Leaves are divided, each making three Ranges of leſſer Leaves, of which they are compoſed; each leſſer Leaf is again moſt commonly parted into five, and are called Olive Leaves; but ſometimes into three, and are called Laurel Leaves: The middle Leaf, which bends down, is parted into eleven. In the Middle, over the Leaves, is a Flower ſhooting out between the Stems and Volutes, like the Roſe in the Abacus.

The Height of this *Capital* is two Modules $\frac{1}{3}$, and its Projecture 1 $\frac{1}{3}$.

The Differences in the Character of this *Capital* are, that in *Vitruvius*, &c. the Leaves are in the Form of the *Acanthus*; whereas in the Antique, they are more uſually Olive Leaves.

2. That their Leaves are uſually unequal, and the undermoſt commonly being made the talleſt; but ſometimes the ſhorteſt; tho' they are ſometimes all equal.

3. Sometimes the Leaves are ruffled, ſometimes quite plain: The firſt Row generally bellies out towards the Bottom; but at other Times they are ſtraight.

4. Sometimes the Horns of the Abacus are ſharp at the Corners, which ſeems to agree to the Rules of *Vitruvius*; but they are more uſually cut off.

5. There is alſo ſome Difference in the Form and Size of the Roſe.

6. The Volutes are alſo ſometimes joined to each other, and at other times wholly ſeparate.

7. Sometimes the Spires of the Volutes continue twiſting, even to the End, in the ſame Courſe; and ſometimes they are turned back again, near to the Centre, in the Form of an S.

According to *Vitruvius*, the Height of this *Capital* (by the Aſtragal at the Bottom) is equal to the Diameter of the Body of the Column below, one ſeventh Part of which goes to the Abacus, which conſiſts of a Boultin, a Fillet, and a Plinth, [which is no other than a larger Fillet.]

The Abacus being ſubdivided into three parts, one of them goes to the Boultin, and a third Part of the next goes to the Fillet, and the reſt to the Plinth.

The Height of the Aſtragal below the *Capital*, is one twelfth Part of the Diameter of the Body below, and is divided into three Parts; of which the Fillet contains one, and the Boultin two.

Scamozzi makes this *Capital* in Height 1 $\frac{1}{6}$ of the Diameter of the Column; which being divided into 75 Parts, four of them go to the Boultin, one to the Fillet, nine to the Plinth, and the reſt to the Neck.

Palladio makes this *Capital* equal in Height to the whole Diameter of the Body of the Column below; and one ſixth Part more, which is allow'd to the Abacus, by which he is ſuppoſed to mean all the Mouldings above the Acanthus Leaves.

The Leaves of this *Capital*, M. *Le Clerc* ſays, are in Number 16, 8 in each Row.

Each Leaf, ſays M. *Le Clerc*, is divided into ſeven or nine Plumes; two of which, or to ſpeak more properly, one whole and an half on each Side go to form the Return or Deſcent.

Sometimes the Return conſiſts of three Plumes almoſt intire, each Plume being divided according to the Nature of the Leaf.

Upon this he remarks, that the Leaves of the *Capital* ate ordinarily thoſe of Olive, thoſe of Acanthus, or thoſe of Smallage; but he gives the Preference to the two laſt, and particularly when the *Corinthian* is raiſed over any other Order.

For

For its Leaves being flat and plain, they reflect more Light than the others, which are more wrought and uneven; for which Reason the first have a better Effect, when seen at a Distance, than the last; which are only fit to be viewed near at hand.

He likewise observes, that in making the Leaves either of this or the *Roman Capital*, great Care must be taken that they be well designed; particularly, that dividing them into Plumes, those Plumes don't run too far off from one another, but shall all together appear to form one single Leaf; which must not be too narrow towards the Top, that each Plume direct to its Origin, &c. Without such Precaution, the Leaves will lose all their Grace and Beauty.

If a *Corinthian* Order were to be placed very high, as in the Lanthorn of a Dome, he would rather chuse not to divide the Leaves of its *Capital* at all, but to preserve the Mass intire.

In some *Capitals* we find Leaves that are finely wrought, which nevertheless are of an extreme ill Taste, as those of Olives, for instance, in some Pilasters. This, he says, he mentions for the sake of those, who, having no great Share of Judgment, think they can't fail of doing well, if they do but imitate what they find in Buildings of Reputation.

The *Composite Capital* is so called, as partaking of the *Doric* in its Quarter-Round, of the *Ionic* in its Volutes, and of the *Corinthian* in its double Row of Leaves underneath, which are in Number 16.

M. *Le Clerc* says, the Leaves he gives it are of Laurel, which not being much edged or indented, are less delicate; and for that Reason more suitable to the Volutes of this Capital, which are tolerably massive, but agreeable to the Modilions of the Entablatures.

In the Middle of the Abacus, is a Flower, and under the Horns, Leaves which return upwards, as in the *Corinthian*: But instead of Stalks in the *Corinthian*, the *Composite* has small Flowers, which lie close to the Vase or Ball, twisting round towards the Middle of the Face of the *Capital*, and terminating in the Rose.

The Height of the *Composite Capital* is two Modules one Third, and its Projecture one Module two Thirds, as in the *Corinthian*.

1. The Differences of the Character of this *Capital* consist in this; that the Volutes which ordinarily descend and touch the Leaves, are in some Works of the Antique separated from them; that the Leaves which are generally unequal in Height, the lower Rank being the tallest, are sometimes equal.

2. That the Volutes of the Moderns generally spring out of the Vase; whereas they do in the Antique ordinarily run straight the Length of the Abacus, over the Ovolo, without striking into the Vase.

3. That the Volutes, whose Thickness is contracted in the Middle, and enlarg'd above and below in the Antique, in the

Works

Works of the Moderns have their Sides parallel.

4. That the Volutes which have been hitherto made, as tho' solid, both by the Antients and Moderns, are now made much lighter, and more airy; the Folds standing hollow, and at a Distance the one from the other.

This *Capital*, called the *Composite* or *Roman*, is made and divided by *Vitruvius*, *Scamozzi*, and *Palladio*; except that the Carving of this is different from that.

Some Architects distinguish the *Tuscan* and *Doric Capitals*, which have no Ornaments, by the Title of *Capitals of Mouldings*; and the three other, which have Leaves and Ornaments, by the Title of *Capitals of Sculpture*.

An *Angular Capital*, is a *Capital* which bears the return of an Entablature, at the Corner of the Projecture of a Frontispiece.

Capital of a Balluster, is that Part which crowns the Ballustter, which sometimes does somewhat resemble the *Capitals* of some Columns, and particularly the *Ionic*.

M. *Le Clerc* gives the *Capital* of the *Spanish* Order eight large Leaves, simple, but a little wav'd with grenate Stalks, or Flowers rising among them, which may be managed in various Manners, according to the various Places where this Order is used.

The Horns of the Abacus are supported by little Volutes; the Middle of the Abacus being adorned with a Lion's Snout instead of a Rose, which noble Animal is the Symbol of *Spain*, and expresses the Strength and

CARINA, a Term used in antient Architecture, a Name given by the *Romans* to all Buildings in the Form of a Ship, [from *Carina*, the Keel of a Ship,] as we still use the Word *Nave* for *Navis*, a Ship, the middle or principal Vault of our Churches, because it has that Figure.

CARPENTERS *Work*, in a Building, includes the Framing, Flooring, Roofing; the Foundation, Carcase, Doors, Windows, *&c.*

Of *Carpenters Work:* The several kinds of it, (in relation to Building) with their Prices and Methods of measuring them, *&c.* are too many to be comprehended under this so general a Term as *Carpenters Work*; for which Reason they shall be referred to their Particulars, (as Ftaming, Flooring, Roofing, *&c.*)

The Measuring of Carpenters Work.

Carpenters Works, which are measurable, are Flooring Partitioning, or Roofing; all which are measured by the Square of ten Feet long, and ten broad; so that one Square contains 100 square Feet.

1. *Of Flooring.*

If a Floor be 57 Feet 3 Inches long, and 28 Feet 3 Inches broad, how many Squares of Flooring is there in that Room?

Multiply 57 Feet 3 Inches by 28 Feet 6 Inches, and the Product is 1631 Feet, *&c.* which divide by 100, (this is done by cutting off two Figures towards the Right Hand with a Dash of the Pen,) and the remaining Figures are the Quotient, and the Figures cut off are Feet: Thus 1631 divided by 100, by cutting off 31 from the Right Hand thereof, the Quotient is 16 Squares, and the 31 cut off is 31 Feet.

See the Work both by Decimals, and also by Feet and Inches.

```
      57  25
      28   5
   ---------
    2862   5
    5800
   11450
   ---------
 1631.62   5
```

```
      F.  I.
      57  3
      28  6
   ----------
     456
    114
      28  7  6
       7  0  0
   ----------
   16|31  7  6
```

Facit 16 Squares, and 31 Feet.

Note, That 5 is the Decimal for Half of any Thing, 25 is the Decimal for a Quarter, or 3 Inches; and 125 is the Decimal of 1 Inch and half, or $\frac{1}{8}$; because 3 Inches is a quarter of a Foot, and 5 is the Decimal of 6 Inches; because 6 Inches is half a Foot.

Example

Example 2. Let a Floor be 53 Feet 6 Inches long, and 47 Feet 9 Inches broad; how many Squares are contained in tha Floor?

```
  47.75
  53. 5
 ------
  23875
 14325
13875
 ------
25|54.625
```

```
 F.  I.
 53  6
 47  9
 ------
 371
 212
 ------
 26  9
 13  4  6
 23  6
 ---------
25|54  7  6
```

Facit 25 Squares, and 54 Feet.

By Scale and Compasses.

In the first Example, extend your Compasses from 1 to 28.5; and that Extent will reach from 57.25 to 16 Squares, and near a third Part.

In the second Example, extend the Compasses from 1 to 47.75, and that Extent will reach from 53.5 to 25 Squares, and above a half.

Of Partitioning.

Example 1. If a Partition between Rooms be in Length 82 Feet 6 Inches, and in Height 12 Feet 3 Inches; how many Squares are contained therein?

The Length and Breadth being multiplied together, the Product will be 10|10.625; which divide by 100, (as before has been shewn,) and the Answer is 10 Squares 10 Feet; the Inches or Parts in these Cases.

```
  12.25
  82. 5
 ------
   6125
  2450
 9800
 ------
10|10.625
```

```
 F.  I.
 82  6
 12  3
 ------
990  0
 20  7  6
 ---------
10|10  7  6
```

Facit 10 Squares 10 Feet.

If

If a Partition between Rooms be in Length 19 Feet 9 Inches, and in Breadth 11 Feet 3 Inches; how many Squares are contain'd theteiu?

```
   61.75
   11.25
  ------
   45875
  18350
 2175
9175
-------
```

The Length and Breadth being multiplied together, the Product is 1032 Feet; which divided by 100, the Answer will be 10 Squares and 32 Feet.

```
 F.    I.
  91   9
  11   3
----------
1009   3
  22  11   3
-------------
10|32   2   3
```

Of Roofing.

It is a Rule among Workmen, that the Flat of any House, and half the Flat thereof, taken within the Walls, is equal to the Measure of the Roof of the same House; but this is when the Roof is true pitch'd; for if the Roof be more flat or steep than the true Pitch, it will Measure to more or less accordingly

Example 1. If a House within the Walls be 44 Feet 6 Inches long, and 18 Feet 3 Inches broad; how many square Feet of Roofing will cover that House?

Multiply the Length and Breadth together, and the Product will be 812 Feet the Flat; the Half of which is 406, which being added to the Flat, the Sum is 1218; which being divided by 100, the Answer is 12 Squares, and 18 Feet.

```
       18.25
       44. 5
      ------
        9125
       7300
      7300
     --------
Flat  812.125
Half  406
     --------
       12|18
     --------
```

```
           F.  I.
           44  6
           18  3
          --------
          352
           11  1  6
            9  0  0
          ----------
The Flat  812  1  6
          406
          -----
Sum       12|18
```

Facit 12 Squares, 18 Feet.

By Scale and Compasses.

In the first Example of Partitioning, extend the Compasses from 1, to 12. 25, and that Extent will reach from 82. 5. to ten Squares and one Tenth.

In

In the ſecond Example, extend the Compaſſes from 1 to 11.25, and that Extent will reach from 91.75 to ten Squares, and a little leſs than a third Part.

In the Example of Roofing, extend the Compaſſes from 18.25, that Extent will reach from 44.5 to 812, the Flat; to which add the half thereof, and the Sum is 12.18, which is 12 Squares 18 Feet, as above.

There are other Works about a Building done by the *Carpenter*, that are meaſured by the Foot, running Meaſure; that is, by the Number of Feet in Length only, as Cornices, Doors and Caſes, Window-Frames, Gutterings, Lintels, Summers, Skirt-Boards, &c.

Note 1. In the meaſuring of Flooring, after the whole Floor has been meaſured, there are to be deducted out of it the Well-Holes for the Stairs and Chimneys; and in Partitioning, for the Doors, Windows, &c. except (by Agreement) they are to be included.

Note 2. In meaſuring of Roofing, ſeldom any Deductions are made for the Holes for the Chimney-Shafts, the Vacancies for Lanthorn-lights and Sky-lights; for they are more Trouble to the Workman, than the Stuff that would cover them is worth.

CARPENTRY [of *Carpentum*, *L.* a Car or Cart] is the Art of cutting, framing, and

Mr. *Thomas Johnson*, of *London*, his Bill of Materials had of, and Work done by *John Robinson*, this 23d of *June* 1733.

	l.	*s.*	*d.*
For 15 Loads of Oaken Timber, at 21*s.* the Load	15	15	00
For 24 Loads of Fir Timber, at 33*s.* the Load	42	12	00
For 150 Feet of Oaken Planks, two Inches thick, at 3*d. per* Foot	01	17	06
For 15 M. 10*d.* Nails, at 6*s. per* M.	04	10	00
For 6 *C.* of Deals, at 6*l.* the *C.*	36	00	00
For 30 *lb.* of large Spikes, at 4*d. per lb.*	00	10	00
For 7 Weeks Work for myself at 3*s. per* Day	06	06	00
For 7 Weeks Work for my Man, at 2*s.* 6*d. per* Day	05	10	00
The Total is	113	00	06

But here it is to be noted, if the *Carpenter* does not work by the Day, then he writes or makes his Bill for so many Square of Roofing (at the Price agreed upon *per* Square) so much Money.

Likewise, for so many Square of Flooring, at so much *per* Square, so much Money.

Also, for so many Squares of Partitioning, at so much *per* Square, so much Money.

Also, for so many Square of Cieling Joists, &c.

Windows are either set down at so much *per* Light, or so much *per* Window.

The Door Cases at so much *per* Piece, either inward or outward.

Mantletrees, Tassels, &c. at so much *per* Piece.

The Lintelling, Guttering, Cornice, Winder-Boards, &c. at so much *per* Foot.

Stairs, either at so much *per* Pair, or so much *per* Step, &c.

CARTON, } a Design in
CARTOON, } Painting, made on strong Paper, to be afterwards calked throug , and transferred on the fresh Plaister of a Wall, to be painted in Fresco.

CARTOUCHES, } [of *Car-*
CARTOUSES, } *toccio, I-*
CARTOUZES, } *talian*] an Ornament in Architecture, Sculpture, &c. representing a Scroll of Paper. It is usually a Table, or flat Member with Wavings, on which is some Inscription, or Device, Ornament of Armoury. Cypher, or the like. They are sometimes made of Stone, Brick, Plaister, Wood, &c. for Buildings.

The

They are, in Architecture, often much the same as Modilions; only these are set under the Cornice, in Wainscotting, and those under the Cornice, at the Eaves of a House.

Perrault says, *Cartouch* is an Ornament of carved Work, of no determinate Form, whose Use is to receive a Motto or Inscription.

CARTRIDGES, in Architecture, as some Workmen call them, are the same as Cartouches.

CARYATIDES, } [so called
CARIATES, } from the *Caryatides* a People of *Caria*] are in Architecture, a kind of Order of Columns or Pilasters, under the Figures of Women dress'd in long Robes, after the Manner of the *Carian* People, and serving instead of Columns to support the Entablement.

Vitruvius relates the Origin of the *Caryatides*. He observes, that the *Greeks* having taken the City of *Caria*, led away their Women Captives; and to perpetuate their Servitude, represented them in their Buildings as charged with Burdens, such as those supported with Columns.

M. *Le Clerc*, aptly enough, calls these, Symbolical Columns and tells us, that the antient *Greeks* had a Custom in the Columns of their publick Buildings, to add Figures and Representations of the Enemies they had subdued, to preserve the Memory of their Victories.

That they having reduced the rebellious *Carians* to Obedience, and led away their Wives Captives; and that the *Lacedemonians* having vanquish'd the *Persians* at *Platæa*, they were the first Subjects of these Columns; which have preserved to late Posterity both the Glory of the Conquerors, and the Dishonour of the Conquered.

And hence originally came the Names *Caryatides*, and *Persian* Columns; which Names have been since apply'd to all Columns made in human Figures, though with Characters very different from one another.

M. *Le Clerc* likewise observes, that the *Caryatides* are not now represented among us, as they were among the Antients, *viz.* as Subjects of Servitude and Slavery, *viz.* with Hands tied before and behind, such Characters seeming injurious to the fair Sex; and for that Reason we give them others entirely opposite, never using them in building but as singular Beauties, and such as make the greatest Ornament thereof.

Among us, they are represented under the noble Symbols, or Images of *Justice*, *Prudence*, *Temperance*, *&c.*

The *Caryates* should always have their Legs pretty close to each other, and even across, or the one athwart the other, their Arms laid flat to their Bodies, or to the Head, or as little spread as possible; that as they do the Office of Columns, they may as near as possible bear the Figures of them.

When the *Caryatides* are insulated, they should never have any great Weight to support; nor greater than those of Balconies, ules, little Galleries, or slight Crownings, and their Entablature and Pedestal are not to be thought so proper to bear great Loads.

If the *Caryatides* have a Projecture beyond the Wall, in the Manner of Pilasters, they may be used in the Architecture of a Gallery or Salon, provided they may be not made to sustain any Thing but an Entablature, the Weight of the Vault being born by the Wall behind, which serves them as a Ground or Bottom.

The *Caryatides* ought always to appear in Characters proper to the Places they are used in. As for Instance, those which support the Crowning of a Throne, should be Symbols or Representations of Heroick Vertues.

Those which are set in a Place of Devotion, should bear the Characters of Religion, and those in Halls, and Banquetting-Rooms, should carry the Marks of Mirth and Rejoicing.

Caryatides, and common Columns, should never be used together under the same Entablature; for besides that, there can never be a just Symmetry between them. The Figures of Women, as tall as common Columns, would appear monstrous, and make all the rest of the Architecture appear low and mean.

Again, the *Caryatides* should never be made of an immoderate Stature, lest being too large, they might become frightful to Ladies; and for this Reason, one would chuse to confine them sometimes under the Impost of a Portico, such Imposts serving them for an Entablature.

They may also upon Occasion be raised upon Pedestals, which ought not to be lower than one Third of their Height: And besides this, if there be Consoles placed over their Heads, the Figures may be made of a reasonable Height.

Sometimes the Arms of the *Caryatides* are cut off for the greater Delicacy, as those for Instance, in the Halls of the *Swiss* Guards in the *Louvre*. But M. *Le Clerc* does not approve of such Mutilations.

These Kinds of Mutilations, which are only used to make the Figures more light and delicate, or rather to make them more conformable to the Columns, are only proper for *Termini*, or Forms, which are a kind of Half-human Figures, seeming to proceed out of a Vagina or Sheath.

The *Caryatides* are sometimes represented in the Form of Angels; which, *M. Le Clerc* is of Opinion, should not be, except at Baldaquins, and Altars. And such as do appear under that holy Form, ought, in his Opinion, to support the Entablature with their Hands; or as others say, with their Heads, as hearing it easily and without Trouble.

The Entablature supported by Angels, M. *Le Clerc* would have to be of the *Corinthian* Order; and that by the Vertues, of the *Ionic*; and both the one and the other somewhat less massive than the ordinary.

The Antients made the *Caryatides* frequently to support Corbels or Baskets of Flowers; and these

these they called Caniferæ, and Cistiferæ.

CASCADE, is a Cataract or Water-Fall; either natural as that at *Tivoli*, &c. or artificial, as those of *Versailles*, and that either falling with a gentle Descent, as those of *Sceaux*; in Form of a Buffette, as at *Trianon*; or by Degrees, in Form of a Perron, as at *St. Cloud*; or from Bason to Bason, &c.

CASING *of Timber-Work*, is the Plaistering a House all over on the Outside with Mortar, and then striking it wet by a Ruler, with the Corner of a Trowel, or the like Instrument, to make it resemble the Joints of Freestone; by which Means, the whole House appears as if built thereof.

As to the Method of doing it: Some direct it to be done upon Heart Laths; because that the Mortar would in a little Time decay Sap-Laths: And although it will require more Labour to lath it with Heart, than with Sap Laths, yet will be better for the Mortar to hang to, because Heart-Laths are the narrowest; and Laths ought to be closer together for Mortar, than for Loam. They also say, that they commonly lay it on in two Thicknesses, *viz.* a second before the first is dry.

As to the Price: It has been done for 3*d.* or 4*d.* *per* Yard, including Doors and Windows, i. e. measuring it as if there were none, and for 6*d.* *per* Yard excluding Doors and Windows, i. e. deducting them from the Whole.

CASE *of Glass*, as of Crown Glass.

A *Case* of Crown Glass contains usually twenty-four Tables; each Table being circular, or nearly so, and about three Feet six Inches, or three Foor eight Inches Diameter.

Newcastle Glass: A *Case* contains thirty-five Tables, and each Table does, or ought to contain six Feet of Glass. Some that have been measured, contained three Feet and a half on the upper or circular End, and about eighteen or twenty Inches at the lower End. and the Perpendicular Height about three Feet.

Mr. *Leybourn*, and from him, Mr. *Wing*, says, that a Table of *Newcastle* Glass contains about five Feet; and that forty-five of those Tables go to a *Case*.

Mr. *Wing* also says, that twenty-five Tables make a *Case* of *Normandy* Glass.

CASEMATE, } in Building,
CASEMENT, } is a hollow Moulding, which some Architects make one Sixth of a Circle; and others, one Fourth.

CASEMENTS, in Architecture, are Windows to open.

As to the Price: Mr. *Leybourn* says, they are valued according to their Largeness and the Goodness of their Workmanship in their Locks and Hinges, from 3*s.* to 20*s.* a *Casement*.

Casements about two Feet and a half long, are worth about 4*s.* or 4*s.* 6*d.* apiece.

Folding Casements, of the like Size, with Bolts, Hinges, &c. about 12, or 13*s.* the Pair.

Mr. *Wing* says, they are worth 7*d.* or 8*d.* the Pound, and some of them 9*d.*

Some Smiths in *London* have asked 6*d* a Pound for *Casements*,

ments, and ſaid they were worth more if they had Turnbouts (or Turn-buckles, as ſome called them,) or Cock-ſpurs, and Pull-backs at the hind Side, to pull to with.

Some Smiths in the Country make them by the Foot, meaſuring the whole Circumference round by the outer Edge of the *Caſement*; ſo if a *Caſement* be two Feet long, and ½ broad, they reckon it to make ſeven Feet.

In *Suſſex*, they have been offer'd to be made for 6*d. per* Foot, if ordinary; but if ſomething extraordinary (as *Folding Caſements*, &c.) 8*d.* a Foot.

Of Painting them: They are uſually painted at three Halfpence, two Pence, or three Pence apiece, according as they are in Largeneſs.

Of Hanging *Caſements*: Country Glaziers ſay, Hanging of *Caſements* is Smiths Work, and that if they don't do it, they pay the Glaziers for doing it, at the Rate of two Pence for hanging a ſmall *Caſement*, and three Pence apiece for large ones.

CASTING, with Founders, is the Running of a melted Metal into a Mould prepared for that Purpoſe.

Caſting of Lead on Cloth, is the Uſing a Frame or Mould cover'd with Woollen Cloth, and Linnen over it, to caſt the Lead into very fine Sheets.

Caſting in Sand or Earth, is the Running of a Metal between two Frames or Moulds filled with Sand or Earth, wherein the Figure the Metal is to take, has been impreſs'd *in creux* by Means of a Pattern.

Caſting in Stone or Plaiſter, is the filling a Mould with fine liquid Plaiſter, that had been taken in Pieces from off a Statue, or other Piece of Sculpture, and run together again.

There are two Things to be minded in reſpect to the Mould.

Firſt, That it be well ſoak'd in Oil, before the Plaiſter be run, to prevent its Sticking.

Secondly, That each Piece of which it conſiſts, have a Packthread to draw it off the more eaſily when the Work is dry.

Caſting in Joinery, &c. Wood is ſaid to caſt or warp, when either by its own Drought, or Moiſture of the Air, or other Accidents, it ſhoots or ſhrinks, and alters it Flatneſs and Straitneſs and becomes crooked.

CATACOMBS, Grotto's, or ſubterraneous Places for the Burial of the Dead.

CATADROME, a kind of Engine like a Crane, uſed by Builders in lifting up and letting down any great Weights.

CATAFALCO [in *Italian*, a Scaffold] in Architecture, it is uſed for a Decoration of Architecture, Sculpture, or Painting, raiſed on a Timber-Scaffold, to ſhew a Coffin or Tomb in a Funeral Solemnity.

CATENARIA [in the Higher Geometry] a Curve Line, which a Chain or Rope forms itſelf into, when hung freely between two Points in Suſpenſion.

CATHETA, a Perpendicular or Plumb Line falling from the Extremity of the Underſide of the Cymatium (of the *Ionic* Capital) through the Centre of the Volute.

CATHETUS, in Geometry, a Perpendicular, or a Line, or Radius, falling perpendicularly on another Line or Surface.

Cathetus of Incidence, in Catoptricks, is a right Line drawn from a radiant Point perpendicular to the Plane of the Speculum or Mirrour.

Cathetus of Reflection, or of the Eye, is a Right Line drawn from any Point of a reflected Ray, perpendicular to the Plane of Reflection, or of the Speculum.

Cathetus of Obliquation, is a Right Line drawn perpendicular to the Speculum in the Point of Incidence or Reflection.

Cathetus, in Architecture, is a perpendicular passing along through the Middle of a Column. See CATHETA.

CATOPTRICKS [of κάτοπτρον, *Gr.* a Speculum] the Science of Reflex Vision, or that Branch of Opticks, which treats of, or gives the Laws of Light reflected from Mirrours or Specula.

CAVETTO [the Word is *Italian*, and signifies the same as hollow] a hollow Member, or round concave Moulding containing a Quadrant of a Circle, and having a quite contrary Effect to that of a Quarter-Round, it is used as an Ornament in Cornices.

Mr. *Felibien* takes Notice, that Workmen confound the *Cavetto* with a Scotia, but to ill Purpose, the *Cavetto* being indeed only half a Scotia.

When it is in its natural Situation, it is by Workmen frequently called Gula, or Guele, or Mouth, in *English*; and when inverted, Gorge or Throat.

CAVASION, in Architecture, a Term used to signify the Under-digging or Hollowing of the Earth for the Foundation of a Building. This, *Palladio* says, ought to be the sixth Part of the Height of the whole Building.

CAULICOLES, } are 8 lesser
CAULICOLI, } Branches or Stalks in the *Corinthian* Capital, springing out from the four greater or principal Caules or Stalks.

The eight Volutes of this Order are sustained by four Caules or primary Branches of Leaves, and from which these Caulicoles or lesser Foliages do arise.

Some Authors confound these with the Volutes themselves; some with the Helices in the Middle; and some with the principal Stalks whence they arise.

Some define them to be carved Scrolls (under the Abacus) in the *Corinthian* Capital.

CEILING, the upper Part or Roof of a lower Room, or a Lay or Covering of Plaister over Laths nail'd on the Bottom of the Joists, which bear the Floor of the upper Room, or on Joists put up for that Purpose, and called Ceiling-Joists, if it be in a Garret.

These Plaister'd *Ceilings* are much used in *England*, more than in any other Country, nor are they without their Advantages, they making the Rooms lightsome, are excellent in Case of Fire, stop the Passage of the Dust, and lessen the Noise over Head, and in the Summer-Time, make the Air of the Rooms cooler.

Of

Of Meaſuring *Ceilings*: This Sort of Work is uſually done by the Yard (containing nine ſuperficial Feet,) and in taking the Dimenſions, if the Room be wainſcotted, they conſider how far the Cornice bears into the Room, by putting a Stick perpendicular to the *Ceiling*, cloſe to the Edge of the uppermoſt Part of the Cornice, and meaſuring the Diſtance from the perpendicular Stock to the Wainſcot; twice which Diſtance is deducted from the Length and Breadth of the Room taken upon the Floor, and the Remainder gives the true Length and Breadth of the *Ceiling*; which if it be taken in Feet, as it moſt uſually is, the one being multiply'd into the other, and the Product divided by 9, gives the Content in Yards ſquare.

As to the Price: The Workmanſhip, *viz.* Lathing, Plaiſtering, and Finiſhing, is commonly reckon'd at *London*, at about two Pence three Farthings *per* Yard.

In ſome Places of the Country, as *Kent*, &c. they have three Pence a Yard. And in ſome Parts of *Suſſex*, ſome Workmen ſay they have four Pence for Workmanſhip only.

But if the Plaiſterer finds all Materials, and lath it with Laths of Heart of Oak, they uſually reckon 1 *s.* a Yard, and if Fir-Laths, about 8 *d*

CEILING-*Joiſts* or Beams.

Of meaſuring them: The Workmanſhip of putting up *Ceiling-Joiſts*, is meaſured by the Square; ſo that the Length in Feet being multiplied by the Breadth in Feet, and two Places of Figures being cut off on the Right Hand, what remains on the Left Hand, is Squares; and what is cut off is odd Feet, 25 of which make a Quarter; 50, a half; 75, three quarters of a Square.

As for the Price: The putting up *Ceiling-Joiſts* is valued at 4, 5, or, ſome ſay, 6 *s.* a Square.

CELERITY [in Mechanicks] is the Velocity or Swiftneſs of a moving Body, or that Affection of a Body in Motion, by which it is enabled to paſs over a certain Space in a certain Time.

CELLS, are little Houſes, Apartments, or Chambers; particularly thoſe wherein the Antient Monks and Hermits, &c. liv'd in Retirement.

CELLARS, are the loweſt Rooms in a Houſe; the Ceilings of which lie level with the Surface of the Ground on which the Houſe ſtands, or at moſt, but very little higher.

As to their Situation: Sir *Henry Wotton* ſays, they ought, unleſs the whole Houſe be cellared, to be ſituated on the North Side of the Houſe, as ſtanding in Need of a cool and freſh Air.

Of Digging them: They are uſually dug by the Solid Yard, which contains twenty-ſeven ſolid Feet; and therefore the Length, Breadth, and Depth being all multiply'd together, and the Product divided by 27, the Quotient will give the Content in ſolid Yards.

CEMENT, CÆMENT, CIMENT, } in the general of the Word, ſignifies any Compoſition of glutinous, or tenacious

tenacious Nature, proper for binding, uniting, and keeping Things in Cohesion.

Cement, in Architecture, is a strong Sort of Mortar, used to bind or fix Bricks or Stones together for some kind of Mouldings; or in cementing a Block of Bricks (as they call it;) for the carving of Capitals, Scrolls, or the like.

It is of two Sorts; one called *Hot Cement*, and the other *Cold Cement*; because the *Hot Cement* is made and used with Fire; and the *Cold Cement* is made and used without Fire.

To make the *Hot Cement*, take half a Pound of Bees-Wax, an Ounce of fine Brick-Dust, an Ounce of Chalk-Dust or powdered Chalk; sift both the Brick-Dust and Chalk through a fine Hair-Sieve, (the Brick and Chalk may be beat in a Mortar, before it is sifted.) Let all these be boiled together in a Pipkin, or other Earthen Vessel, for about a quarter of an Hour, keeping it continually stirring with a Piece of Iron or Lath, then take it off, and let it stand for four or five Minutes, and it is fit for Use.

The Bricks which are to be cemented with this Kind of *Cement*, must be made hot by the Fire, before the *Cement* is spread on them; and after that, be rubbed to and fro one upon another, after the same Manner that Joiners do, when they glue two Boards together.

The *Cold Cement* is less used; and is accounted a Secret known but to few Bricklayers.

It is made after the following Manner;

Take a Pound of old *Cheshire* Cheese, pare of the Rind, and throw it by, then cut or grate the Cheese very small, put it into a Pot with a Quart of Cows Milk; let it stand all Night, and in the Morning, take the Whites of twenty-four or thirty Eggs, and a Pound of the best unslack'd or quick Lime, and beat it in a Mortar to a very fine Powder, sift it in a fine Hair-Sieve, put the Cheese and Milk to it in a Pan, or Bowl, and stir them well together with a Trowel or such like Thing, breaking the Knobs of the Cheese, if there be any, then add the Whites of Eggs, and temper all well together, and it will be fit for Use.

This *Cement* will be of a white Colour; but if you will have it of the Colour of Brick, put into it, either some very fine Brick-dust, or some Almegram, but not too much, but just enough to give it a Colour.

CENTRE, [in Geometry] of a Circle, is a Point in the Middle of a Circle, or circular Figure, from which all Lines drawn to the Circumference, are equal.

Centre of a Parallelogram or *Polygon*, is the Point wherein its Diagonals intersect.

Centre of Magnitude is a Point equally remote from the extream Parts of a Line, Figure, or Body; or the Middle of a Line or Plane, by which a Figure or Body is divided into two equal Parts.

Centre of a Sphere is the Point from which all the Lines drawn to the Surface, are equal.

Centre

Centre of an Ellipsis is that Point where the two Diameters, the Transverse, and the Conjugate, intersect each other.

Centre of Gravity, in Mechanicks, is a Point within a Body, through which, if a Plane pass, the Segments on each Side will equiponderate, i. e. neither of them can move the other.

CENTRAL, something relating to a Centre.

CHALK, a white Substance usually accounted as a Stone: Though Dr. *Slare* thinks there is not sufficient Reason for it; since it having been examin'd by the Hydrostatical Balance, it is found to want much of the Weight and Consistence of a real Stone.

Chalk is of two Sorts, the hard dry strong Stone, used in making Lime; the other is a soft unctuous Chalk, used for manuring Lands; easily dissolving with Rain and Frost.

CHAMBER, in a House, or Building, is any Room situate between the lowermost (excepting Cellars,) and the uppermost Rooms. So that there are in some Houses two, in others three or more Stories of Chambers.

Sir *Henry Wotton* directs, that the principal Chambers for Delight be situated towards the East.

As to the Proportions: The Length of a well-proportion'd Lodging should be the Breadth and half of the same, or some small Matter less; but should never exceed that Length. As for the Height, three Fourths of the Breadth will be a fit Height.

Palladio directs that *Chambers*, Antichambers, and Halls, either flat, or arched, be made of the following Heights.

If they be flat, he advises to divide the Breadth into three Parts, and to take two of them for the Height of the Story from the Floor to the Joist, as in the Figure.

Let the Figure represent the Chamber, and whose Height you would find, which suppose to have in Breadth 24 Feet within the Work, which shall be divided upon the Line AB into three equal Parts within Points, where

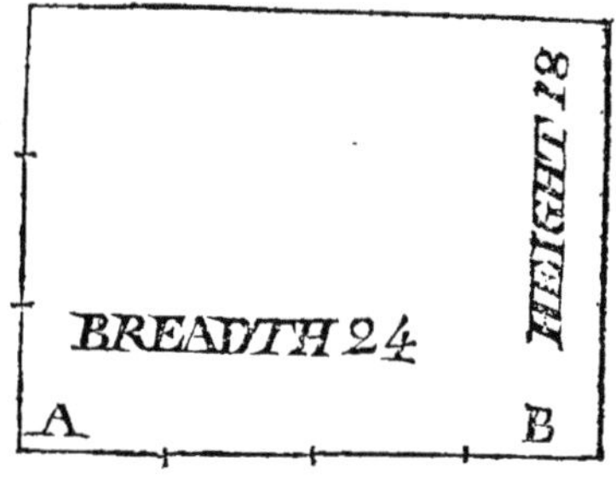

is mark'd the Numbers 1, 2, 3, each Part being eight Feet; two of each Parts shall be the Height of the Chamber, &c. to wit, sixteen Feet from the Floor to the Joist.

If you would have it higher, the Breadth must be divided into seven Parts, of which take five for the Height.

Let the Figure be of the ſame Breadth with the foregoing, to wit, twenty-four Feet within the Work, which ſhall be divided

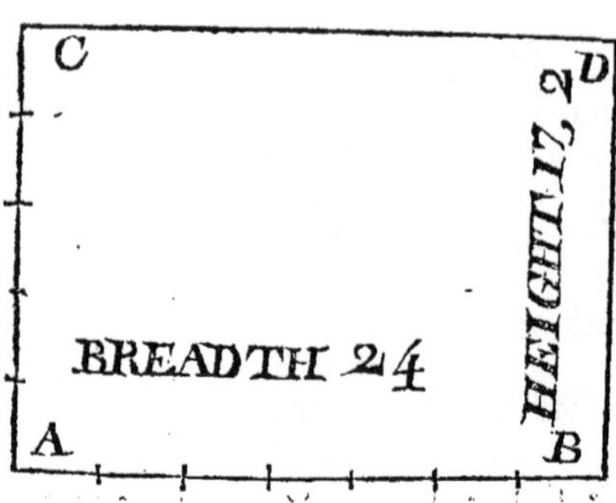

upon the Line AB into ſeven equal Parts; take five of them to make the Height of the Story AC and BD, and the ſaid Height will be ſeventeen Feet two Inches from the Floor unto the Joiſts.

Or divide the ſaid Height into four Parts; and three of thoſe Parts will likewiſe give a greater Height.

Let the Figure be of the ſame Breadth, *viz.* twenty-four Feet within the Work; which divide upon the Line AB into four e-

HEIGHT 16
BREADTH 24
A 8 2 8 3 B
8 8

qual Parts; three of which you muſt take for the Height of the Story, which will be eighteen Feet from the Floor to the Joiſt.

The Height of the *Chambers* of the ſecond Story ſhall be a twelfth Part leſs than the Chambers below; as ſuppoſe its firſt Story be ſixteen Feet from the Floor to the Joiſt, divide the ſixteen Feet into twelve equal Parts; and take eleven of them, which will make fourteen Feet eight Inches for the Height of the ſecond Floor to the Joiſt.

Again, ſuppoſe the firſt Story be ſeventeen Feet two Inches, and in twelve equal Parts; take eleven of them which will make fifteen Feet, ſeven Inches for the Height of the ſecond Story from the Floor to the Joiſt.

If you would make above the Second Story and *Attic*, or third Story, the ſecond muſt always be divided into twelve equal Parts; nine of which will give the Height from the Floor to the Bottom of the Joiſts.

In

In the building of *Chambers*, Regard ought to be had as well to the Place of the Bed, which is uſually ſix or ſeven Feet ſquare; and the Paſſage, as well to the Situation of the Chimney; which, for this Conſideration, ought not to be placed juſt in the Middle, but diſtant from it about two Feet, or two and a half, to the End it may make Room for the Bed; and by this Means the Inequality is little diſcern'd, if it be not in Buildings of the Breadth at leaſt of 24 Feet within the Work; and in this Caſe, it may be placed juſt in the Middle.

CHANCEL [ſo called, of *Cancellæ*, Lettices or croſs Bars, wherewith the *Chancels* were anciently encompaſſed, as they are now with Rails] a Part of the Choir of a Church, between the Altar, and the Communion-Table, and the Baluſtrade or Rails that incloſe it; where the Miniſter ſtands at the Celebration of the Holy Communion.

CHANNEL in Architecture, is particularly uſed for a Part of the *Ionic* Capital, a little hollow'd, in Form of a Canal, lying under the Abacus, and running the whole Length of the Volute, incloſed by a Liſtel; or it is otherwiſe deſcribed to be that Part of the *Ionic* Capital, which is under the Abacus, and lies open upon the Echinus or Eggs, which has its Centres or Turnings on every Side, to make the Volutes.

Channel of the Larmier is the Soffit of a Cornice which makes the Pendant Mouchette.

Channel of the Volute, in the *Ionic* Capital, is the Face of its Circumvolution, incloſed by a Liſtel.

CHANTLATE, in Building, a Piece of Wood faſtened near the Ends of the Rafters, and projecting beyond the Wall to ſupport two or three Rows of Tiles, ſo Placed, to hinder the Rain Water from trickling down the Sides of the Walls.

CHAPEL } A ſort of little
CHAPPEL } Church, where an Incumbent, under the Denomination of a Chaplain, officiates.

CHAPLET, a String of Beads, called *Pater-noſters*, uſed in the *Romiſh* Church.

CHAPLET, in *Architecture*, is a ſmall Ornament cut or carved into round Beads, Pearls, Olives, and Pater-noſters, as is frequently done in *Baguettes*.

A *Chaplet* is, in truth, little elſe but a *Baguette*, inrich'd with Carving.

CHARGE, in Painting. *Over-Charge* is an exagerated Repreſentation of any Perſon, wherein the Likeneſs is preſerved, but ridiculed.

The Method is to puſh out and heighten ſomething already amiſs in the Face, wheher by way of Defect or Redundancy: Thus, *v. gr.* if Nature has given a Man a Noſe a little larger than ordinary, the Painter fails in with her, and makes the Noſe extravagantly long; or if the Noſe be naturally too ſhort, in the Painting, it ſhall be made a mere Stump; and ſo of the reſt.

CHARNEL, a Building, a kind of a Portico or Gallery, uſually in or near the Church-Yard, over which were antiently ly laid the Bones of the Dead, after the Flesh was totally consumed.

CHESNUT. The *Horse-Chesnut*, says a certain Author, ought to be universally propagated, being easily increased from Layers, and grows in goodly, Standards, and bears a most glorious Flower: It is much used for Avenues in *France*, and was brought into these Parts of *Europe* from *Turkey*, and has been raised from Nuts brought from thence; which grow well with us, and in Time to fair large Trees, full of Boughs and Branches, green-leaved, and streaked on the Edges; with Threads in the Middle, that in their native Country turn to *Chesnuts*; but rarely with us.

It is valued for the fair green Leaves and Flowers; and for want of Nuts is propagated by Suckers: Its Name comes from the Property of the Nuts, which in *Turkey* are given to Horses for their Provender, to cure such as have Coughs, or are broken-winded.

Mr *Chomel* says, that nothing seems to him more agreeable, or that would bring more Profit to a Country, than *Chesnuts* planted in Rows, well managed, and kept in good order; which would not only be pleasing to the Eye, but the Flower would be agreeable to the Smell, and the Taste in Time will also be gratified.

These Trees are of quick Growth, they shoot up in a little Time, and their Leaves, which are very fair and beautiful, will form a Shade, which will invite People to retire under them.

In some Places he tells us, that *Chesnut-Trees* grow like Oaks, and make Forest Trees; they likewise plant them at a Foot Distance one from the other, like young Oaks for Coppice and Underwood; but this is rarely done, because they are not good for burning, by reason of their crackling in the Fire, and Aptness to burn Peoples Clothes that sit at it.

As to the particular Uses of *Chesnut* Timber, they are next to the Oak, most coveted by Carpenters and Joiners; and formerly most of our antient Houses in *London* were built of it, there being a large Forest of them not far from this City in the Reign of our King *Henry* II.

It makes the best Stakes and Poles for Palisadoes, Pediments for Vine Props and Hops; it is also proper for Mill-Timber and Water-Works, or where it may lie buried.

It is so prevalent against Cold, that *Chesnut-Trees* defend other Plantations from the Injuries of the severest Frosts.

The *Chesnut-Tree* is also proper for Columns, Tables, Chests, Chairs, Stools, Bedsteads, and Wine-Casks, and those for other Liquors, giving the Liquor the least Tincture of the Wood of any whatsoever; and having been dipped in scalding Oil, or well pitched, is extremely durable. It will look fair without, indeed, when rotten within; however, the Beams give warning of a fall of a House by their cracking.

The Coals of this Wood are excellent for the Smith, soon kindled, and as soon quenched.

CHIMNEY, that Part of a Room, Chamber, or Appartment, wherein the Fire is made.

Chimneys consist of these following Parts: The Jaumbs or Sides coming out perpendicularly, sometimes circularly, *&c.* from the Back, the Mantletree, which rests on the Jaumbs; the Tube or Funnel which conveys away the Smoke; the *Chimney-*Piece or Moulding, which is on the Foreside of the Jaumbs over the Mantletree, and the Hearth or Fire-Place.

Palladio lays down the following Proportions for the Breadths and Depths of *Chimneys* on the Inside, and for that Height to the Mantletree.

Chimneys in	Breadth.	Height.	Depth.
Halls,	6, 7, or 8 Feet.	4½, or 5 Feet.	2½ or 3 Feet.
Chambers,	5½, 6, or 7 Feet.	4, or 4½ Feet.	2, or 2½ Feet.
Studies and Wardrobes.	4, 4½, or 5 Feet.	4, or 4½ Feet.	2, or 2½ Feet.

Wolfius orders the Breadth of the Aperture at Bottom to be to the Height as three to two, to the Depth as four to two.

In small Apartments the Breadth is three Feet, in Bed-Chambers four, in larger Apartments five, in small Banquetting Rooms five and a half, in large six.

But the Breadth must never exceed two and a half, lest there being too much Room for Air and Wind, the Smoke be driven into the Room: Nor must the Height be too little, lest the Smoke miss its Way, and be check'd at first setting out.

The same Author advises to have an Aperture, through which the internal Air may, on Occasion, be let into the Flame to drive up the Smoke, which the internal Air would otherwise be unable to do.

Some make the Funnel twisted, to prevent the Smoke's descending too easily; but the better Expedient is to make the Funnel narrower at Bottom than at Top, the Fire impelling it up more easy when contracted at the Bottom; and in mounting it finds more Space to disengage itself, and therefore will have less Occasion to return into the Room.

Mr. *Felibien* orders the Mouth of the Tube, or that Part joined to the *Chimney-Back*, to be a little narrower than the rest; that the Smoak coming to be repelled downwards, meeting with this Obstacle, may be prevented from getting into the Room.

To prevent smoaking *Chimneys*, Mr. *Lucar* advises to leave two Holes, or make two Pipes in the *Chimneys* one over the other on each Side, the one slo-

ping

ping upwards, and the other downwards; thro' these Holes or Pipes, says he, the Smoke will easily pass out of any Funnel which way soever the Wind blows.

Philip D'Orme advises to provide a hollow Brass Ball, of a reasonable Capacity, with a small Hole in one Side for the putting in Water, to be hung up in the *Chimney*, at a Height a little above the greatest Flame, (with the Hole upwards,) by an Iron Wire that shall traverse the *Chimney* a little above the Mantletree; where, as the Water grows hot, it will rarefy and drive thro' the Aperture or Hole in a vapoury Steam, which will drive up the Smoke that would otherwise linger in the Funnel.

Some think it would be better if this Brass Ball were made with a short Nose to screw off when it is to be fill'd with Water; and then the Hole at the End of this Nose need not be bigger, than that at the small End of a Tobacco-Pipe.

It also may be proper to have two of these Balls, one of which may supply the Place of the other when it is exhausted; or, upon Occasion, to blow the Fire in the mean Time.

Others place a kind of moveable Vane or Weather-Cock on the Top of the *Chimney*; so that what Way soever the Wind comes, the Aperture of the *Chimney* will be skreen'd, and the Smoke have free Egress.

Indeed the best Prevention of a smoking *Chimney* seems to lie in the proper placing of the Doors of a Room, and the apt falling back of the Back, and due gathering of the Wings and Breast of the *Chimney*.

Rules about Timbers near *Chimneys*: It is a Rule in Building, that no Timber be laid within twelve Inches of the Foreside of the *Chimney* Jaumbs; that all the Joists on the Back of any *Chimney* be laid with a Trimmer, at six Inches distance from the Back; that no Timber be laid within the Funnel of any *Chimney*.

Chimney Hooks, are Hooks of Steel and Brass, put into the Jaumbs of *Chimneys*, into each Jaumb one, for the Handle of the Fire Tongs and Fire Pan to rest in.

Their Price: The Steel Hooks are about 1 *s.* the Pair, and the Brass about 2 *s.* the Pair in *London*.

Chimney Jaumbs, are the Sides of a *Chimney*, commonly standing out perpendicularly (but sometimes circularly) from the Back, on the Extremities of which the Mantletree rests. See CORNER-STONE.

Chimney-Piece, is a Composition of certain Mouldings of Wood or Stone, standing on the Foreside of the Jaumbs, and coming over the Mantletree.

The Price: *Chimney-Pieces* of Free-Stone, wrought plain, are worth 10 *s.* but there may be such Mouldings wrought in them, as with their Coves, and other Members, may be worth 20, 30, or 40 *s.* *per* Piece.

Chimney-Pieces of *Egyptian* Marble, or black-fleak'd Marble, or of *Rome*, or liver-coloured Marble, (of an ordinary Size,) are worth twelve or fourteen Pounds *per* Piece.

Chimney-

Chimney-Pieces of Wood, are also of different Prizes, as from ten to twenty Shillings a Piece, more or less, according to their Size, Goodness of the Stuff, and Curiosity of Workmanship.

The Price of painting *Chimney-Pieces*: They are usually painted by the Piece, at about two Shillings each, more or less, according to their Magnitude, and Goodness of the Work.

In the Year 1713 was publish'd a *French* Book, intitled, *La Mechanique du Feu*; or, The Art of augmenting the Effects, and diminishing the Expence of Fire, by *M. Gauger*, which was since published in *English* by Dr. *Desaguliers*; in which the Author examines what Dispositions of *Chimneys* are most proper to augment the Heat; and also proves geometrically, that the Dispositions of parallel Jaumbs, with the Back inclined as in the common *Chimneys*, is less fitted for reflecting Heat into a Room, than parabolical Jaumbs, with the Bottom of the *Tablette* horizontal.

He also gives several Constructions of his new *Chimneys*, and the manner of executing or making them.

Of measuring *Chimneys*: Bricklayers commonly agree for building of *Chimneys* by the Hearth; yet they sometimes work them by the Rod, as in other Brick-Work, and then their Method of taking the Dimensions is as follows:

If you are to measure a *Chimney* standing alone by itself, without any Patty Wall adjoin'd, then girt it about for the Length, and the Height of the Story is the Breadth; the Thickness must be the same the Jaumbs are of, provided that the *Chimney* be wrought upright from the Mantletree to the Cieling, not deducting any thing for the Vacancy between the Floor (or Hearth) and the Mantletree, because of the Gatherings of the Breast and Wings, to make Room for the Hearth in the next Story.

If the *Chimney-Back* be a Party Wall, and the Wall be measured by itself, then you must measure the two Jaumbs and the Breast for a Length, and the Height of the Story for the Breadth, at the same Thickness the Jaumbs were of.

When you measure *Chimney-Shafts*, girt them with a Line round about the least Part of them for the Length, and the Height will be your Breadth.

And if they be four Inch Work, then you must set down your Thickness at one Brick-Work; but if they be wrought nine Inches thick, (as sometimes they are, when they stand high and alone above the Roof,) then you must account your Thickness one Brick and half, in consideration of Wyths and Pargetting, and Trouble in Scaffolding.

It is customary in most Places, to allow double Measure for *Chimneys*.

For Example: Suppose the following Figure A B C D E F G H I K L to be a *Chimney* that has a double Funnel towards the Top, and a double Shaft, and is to be measured according to double Measure.

I first begin with the Breast-Wall I L, and the two Angles

LK

LK and HI, which together are eighteen Feet 9 Inches; then I take the Height of the Square HF twelve Feet six Inches, which multiplied together, produce 234 Feet 4 Inches, six Parts for the Content of the Figure F G H K. As for the Square D *a* E *b*, the Length of the Breast-Wall, and two Angles, is fourteen Feet six Inches, and Height D *a* nine Feet; which multiplied together, make 130 Feet six Inches for the Content of the Part D *a* E *b*. Then the Height of the next Square seven Feet, and the Length of the Breast-Wall and two Angles is ten Feet three Inches; which multiplied together, produce seventy Feet nine Inches for the Content of the Square B *c c d*.

The Compass of the *Chimney* Shafts is thirteen Feet nine Inches, and the Height six Feet six Inches; which multiplied together, produce eighty-nine Feet four Inches, six Parts, the Content of the Shafts.

The Depth of the middle Fetter that parts the Funnels is twelve Feet, and its Width one Foot three Inches; which mul-

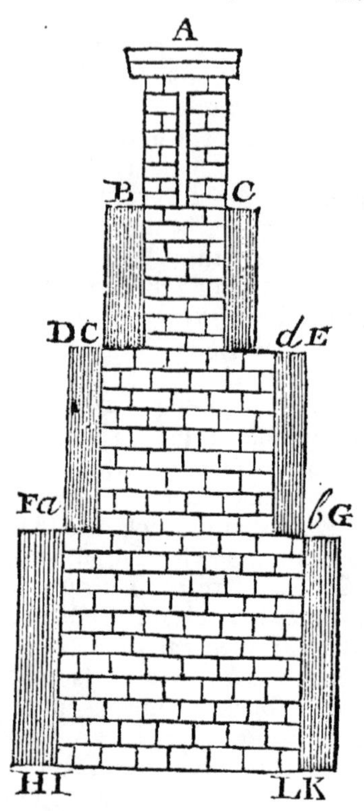

tiplied together make fifteen Feet the Content thereof.

The Work.

	F.	I.	
	18	9	
	12	6	
	225	0	
FGHK	9	4	6
	234	4	6
	F.	I.	
	14	6	
	9	0	
D a E *b*	130	6	

	18.75
	12.5
	9375
	3750
FGHK	1875
Product	234.375
	14 5
D *a* E *b*	130 5

F. I.

	F.	I.	
	10	3	
	7	0	
Bc Cd	71	9	
	F.	I.	
	13	9	
	6	6	
	82	6	
	6	10	6
The Shaft	89	4	6
	F.	I.	
	1	3	
	12	0	
The Fetter	15	0	

	1025
	7
Bc Cd	71.75
	13.75
	6.5
	6875
	8250
The Shaft	89 375
	1.25
	12
	15.00

	F.	I.	P.
FGHK	234	4	6
Da Eb	130	6	0
Bc Cd	71	9	0
The Shaft	89	4	6
The Fetter	15	0	0
The Sum	541	0	0
The Double	1082	0	0

272) 1082 (3 Rods.
68) 266 (3 quarters.

Rem. 62 Feet.

When the five Products have been produced together, the Sum is to be doubled, and that double Sum is the Content of the *Chimney* in Feet, according to the double or customary Measure; which Feet must be reduced to Rods, as directed in reducing Feet to Rods.

So the Feet in the foregoing Example being reduced to Rods, (the Thickness being supposed one Brick and a half,) it makes three Rods three quarters, and sixty-two Feet; that is, four Rods wanting two Feet. This is all the Measure that can be allowed, when a *Chimney* stands in a Gable or Side-Wall; in which Case the Back of the *Chimney* (here not mentioned) is accounted as Part of the Gable; but if the *Chimneys* stand by themselves, as all Stacks of *Chimneys* in great Buildings do, in such Case it is all *Chimney-Work*, and therefore ought to be measured double on all Sides.

The Price: *Chimneys* are sometimes measured and paid for by the Rod, like other Brick-Work; and sometimes, by the Fire-Hearth, at at ſo much *per* Hearth, and the Price is various, from 20 to 50*s*. *per* Hearth.

Mr. *Wing* ſays, that building of *Chimneys* for ordinary Buildings, with Architrave, Frize, and Corniſh, is worth from 15 to 20*s*. *per* Hearth, according to their Height and Subſtance; and without Architrave and Frize, from 10 to 20.

He adds, that in great Buildings they are uſually done by the Feet, *viz*. at about 6 *d. per* Foot.

They are commonly built in *London*, and ſome other Places, for about fifteen Shillings a Hearth; and ſome ſay, they have twenty and twenty-five Shillings *per* Hearth for building in *Suſſex*.

M. *Gauger* has given us a new Treatiſe of *Chimneys*, and has ſhewn a Way how to build them for the moſt Conveniency. He has ſhewn you how you may readily light a Fire, if you have it always blaze, what Wood you ſhould burn, how to warm you you on all Sides, though never ſo cold, and and yet without ſcorching; how always to breath freſh Air, and of what Degrees of Height you pleaſe; how to keep the Room ever free from ſmoaking, aud without any Damp; and how to put out a Fire that has catched the Funnel of a *Chimney* in a Moment.

All theſe Conveniencies depend upon the Diſpoſition of the Hearth, Jaumbs, and the Funnel upon an Iron or Copper-Plate, apply'd in ſuch a Manner, that it leaves a void Space behind, through which the external Air that ſhould go into the Room, paſſes, and warms, upon a Trap which ſerves inſtead of a Pair of Bellows, upon a Baſcule or Swipe, which is fitted to the Funnel of the *Chimney*; and the particular Way of forming the upper End of the Funnel of ſome *Chimneys*.

A Model of a Hearth and Jaumbs for the Increaſe of Height.

Suppoſe the Space between the Extremities of the Jaumbs, taken on the Side of a Room be 4 Feet, and the Depth of the *Chimney* twenty Inches, which is the common Size of *Chimneys*; and if there are thoſe which are larger or ſmaller, they increaſe or diminiſh the Lines by which they would determine.

Take a Board, ſuppoſe AB, *b a*, four Foot long and twenty Inches broad, whoſe Sides muſt be drawn by a Rule one upon another, or a ſquare Draught made in the Middle of M of the Side B*b*, mark the Length MC eleven Inches, and from C, mark upon the ſame Side, the Leugth CG, which muſt be four or five Inches long.

From the Point H, draw H*p*, by a Rule upon the Line GHA: From the Point G, draw alſo C*p*, by your Rule upon the Line BM, upon the Point P, where theſe two Lines drawn by the Rule, meet as in a Centre; and from the Diſtance PH, or PC, deſcribe the Arch HC: Do the ſame Thing on the other Side M*b*, in order to deſcribe the Line *c h a*.

Then

Then, within three Inches of this rectangular Figure, trace another, as at Z, 3 Inches long, and two and a half broad. These two rectangular Figures ought to answer to the Middle M of C*c*, cut off the Draught upon the Board marked AH, C*m*, *cha*; and so you will have your Model for the *Chimney*.

The Great Rectangle X will serve as a Model for the Ash-Pan, which must be dug in the Hearth, of a convenient Depth, if you have a Mind to have one.

The small rectangular Figure Z serves to be a Model for a Pair of Bellows of a new Invention.

The Hearth is to be opened here; and this Opening is to yield a Passage to the Wind that comes from the Street, or some other convenient Place, by the Means of a Funnel or Pipe concealed in the Floor of the Room.

This Hole or Opening is to be furnished with an Iron, or Copper Frame; to which is fastened a small Trap-Door, that shuts close, and lies open towards the Fire; the Sides of the Frame and Trap are made slope and Bevel-wise; on the Side opposite to the Turning-Joint or Hinge, with which the Trap-Door is fastened to the Frame, is placed a small Button, that you may lift this Trap-Door with the Tongs; and you may put on a Button: There will be below on both Sides the Trap, a small Part of a Circle, whose Centre must touch the Hinge, that the Wind may not get out another Way than before, and towards the Fire, when the Trap-Door is lifted up; and to the End it may be kept open to such a Height as you think proper, and yield more or less Wind: Two small Springs must be fastened under the Frame, each of which must rest upon some Parts of the Circle, and press them so, that the Trap-Door may be kept up.

Let the Bottom of the Tablet or little Board be placed parallel to the Horizon, according to its Breadth, or Level that Way; for it may be arched; and it must not be above ten or twelve Inches distant from the Bottom of the *Chimney*, to the End that the Funnel of the *Chimney* may have no more Breadth in that Place.

If the Funnel is loose, you must have Languets or Tenons on the Sides, in some Parts of the Circles, from the Top of the Jaumb to the Floor.

In building or forming the Bottom of the *Chimney*, so that the Air may come into the Room hot, you must make use of a single Copper or Iron Plate or Back made of several Sheets about four Feet long, and three and a half high; furnish'd with several Iron Bands, which must be five Feet broad, and not so high by ten Inches, as the great; apply them to it in such a Manner, that the first Band may reach from the Top to within ten Inches of the Bottom, that the second may have the same Distance from the Top as the first has from the Bottom; that the third be placed in the same Manner as they first, are represented in the First Figure.

It would be convenient, if you can, to hollow the Wall as much as is neceſſary, that the Back may not be too forward; but be that as it will, there muſt be made, as it were, two Gutters an Inch deep in the Wall, which may anſwer the Tenons that may enter in, which are to be filled with very freſh Mortar, and a Space muſt be left between the Wall and the Back, four Inches deep.

It would, perhaps, be more convenient to make a Caiſſe or Box of Iron, furniſhed with Tenous of the Dimenſions aforeſaid, and to faſten it in the Bottom of the *Chimney*; you may alſo order as many little Cells as you pleaſe, but there muſt not be fewer than ten or twelve Inches Diſtance between the Tenons; and Matters muſt be ſo contriv'd, that the ſecond little Cell be bigger than the firſt, and the third than the ſecond and ſo of the reſt.

This Box ſhould have but two Openings; one at the Bottom, at D; and the other on the oppoſite Side above, at R.

In framing the *Chimney*, you muſt make a Conduit-Pipe, which muſt be open to a Street or Court, and be about a Foot ſquare; this Pipe will convey the cold Air as far as D, yet not without the Uſe of the particular Inſtrument at R before deſcribed. From D, it enters into the Box, where it runs windingly through all the Cells formed by the Tenons or Languets: It grows warm there, and comes out at the Hole R, at the Corners of the Tablet; inſomuch, that the Heat of the Room may be augmented or diminiſh'd, according as you partly ſtop, or open the Hole, which need be but two Inches Diameter.

If you have a Mind to heat ſome particular Part of a Room, ſuppoſe a Perſon ſick in Bed, you may appiy a Tin Pipe to this Hole, by which you may alſo convey the warm Air into another Room; perhaps, a Leather, or Paſteboard one may do.

Laſtly, if the Heat is not ſufficient, you may cauſe the little Cells of this Box to paſs under the Hearth, and under the Tablet.

When once the Work deſcrib'd is underſtood, there will be no Difficulty to make it ſerve in all Parts of the Hearth, where you think it may contribute to increaſe the Heat.

But if you cannot poſſibly be able to adjuſt the little Cells in the Bottom of the *Chimney*, you muſt content your ſelf to do it in the Jaumbs, under the Hearth and little Board.

As to the forming of the upper Part of the *Chimney*, to prevent its ſmoking, you muſt firſt obſerve, that your *Chimney* be not commanded by any Thing; that is, that there are no Buildings about it higher than the Funnel: You muſt alſo place your Funnels oue hy the Sides of another, as the common Practice is. Suppoſe that the Funnel within be thirty Inches long, and the Breadth ten, make a Ledge of two Inches, ſloping underneath, quite round and within; the Opening will be no more than twenty-ſix Inches long, and ſix broad; divide this Length into to three Parts by two Partitions, each of four Inches; the lowermost Part of which will descend Anglewise into the Pipe; each of the three Openings will be six Inches Square.

Make three curtail'd square and hollow Pyramids, the Basis of each of which within will be eleven or twelve Inches square, and the Height from twelve to fifteen Inches; divide this upper Opening by a small Languet of two or three Inches in Height, which you are to place different Ways: You are also to apply and fix these three Pyramids near one another, over the three Openings you have contriv'd on the Top of the Funnel of the *Chimney*. If the Opening of the *Chimney* is too small, which is scarcely to be supposed, you must lessen the Apertures of the Pyramids; and if it be too big, you must enlarge them, or instead of three, use four.

These Pyramids may be made of Tin, Clay, or Potters Earth, bak'd, as you do other Earthern Ware.

You may fit a Cap to these Pyramids, made in such a Manner, that being higher, it may serve to suspend a Body above the Opening of the Pyramid, made in the Form of a triangular Prism, one of the Angles of which must be turned towards the upper Openings of the Pyramids, and the Smoak gets out thro' the Sides. It will be best to make all those Pieces of Tin.

The Swipe is an Iron-Plate, placed in some Part of the Funnel of the *Chimney*: It should be exactly of the Length and Breadth of that Place, where you put it, that it may stop it exactly.

To the Middle of this Swipe two Trunnions or Knobs are to be fitted, which are put into the Wall; by the Help of which, you may fit it where you please to have it; and draw it with two two Wires that are fastened to both Ends.

This Swipe being shut, keeps the Heat in the Room, when the Fire is covered, and there is no Smoke: It likewise hinders the Smoke of the neighbouring *Chimneys*, to enter in, as it very often happens when there is no Fire in the Hearth: You may likewise use it to extinguish it, when a *Chimney* is set on Fire, having no more to do, than to take out the Coals or hot Embers, and shut the Swipe.

The Wood most proper to burn, is that which is called Float-Wood; which has less Heat, and burns quicker than new Wood.

Float, Beech, or Bakers Billets, burn faster than the other.

Green Wood will not burn so well as dry; it grows black in the Fire, causes much Smoke, and is hard to be lighted. Whitewood, and the Poplar, Birch, Aspen, &c. are the worst of all Woods to burn.

If there is a Distinction to be made between Oak, young Oaks burn and heat much, the old grows black in the Fire, makes a sort of scaled Coal, that yields no Heat, and is soon put out.

P 2 Thus

Thus in uſing Oak for Firing, you muſt chuſe Billets of three or four Inches Diameter. The Oak, whoſe Bark is taken off for the Tanners Uſe, burns well enough, but yields very little Heat.

Yoke-Elm burns well, makes a good Fire, and a good many red-hot Brands, which laſt long.

The beſt of all Woods, is new Beech, which makes a good clear Fire, and but little Smoke, when well ordered: It yields a good deal of Heat, and many good Embers.

CHISSEL, an Inſtrument much uſed in Carpentry, Joinery, Maſonry, Sculpture, *&c.*

Chiſſels are of different Kinds; and have different Names, according to the different Uſes they are apply'd to. As,

1. The *Former*, which is uſed in Carpentry and Joinery, firſt of all before the *Paring Chiſſel*, and juſt after the Work is ſcrib'd.

2. The *Paring Chiſſel* which has a fine ſmooth Edge and is uſed in paring off, or ſmoothing the Irregularities, which are made by the Former. This is not ſtruck with a Mallet, as the Former is, but is preſſed with the Workman's Shoulder.

3. The *Skew-Former*, which is uſed in cleanſing acute Angles with the Point or Corner of its narrow Edge.

4. The *Mortice-Chiſſel*, which is narrow, but very thick and ſtrong, to endure hard Blows; and 'tis cut to a very broad Baſil; its Uſe is cutting deep ſquare Holes in Wood for Mortices.

5. The *Gouge*; which is a *Chiſſel* with a round Edge, one Side of which ſerves to prepare the Way for an Augre; and the other to cut ſuch Wood as is to be rounded, hollow'd, *&c.*

5. *Socket-Chiſſels*; which are chiefly uſed by Carpenters, *&c.* having their Shank made with a hollow Socket at Top, to receive a ſtrong Wooden Sprig fitted into it with a Shoulder.

Theſe *Chiſſels* are alſo diſtinguiſh'd according to the Breadth of the Blade, into Half-Inch *Chiſſels*, quarter of an Inch *Chiſſels*, *&c.*

7. *Ripping Chiſſel*; which is a *Socket Chiſſel*, an Inch broad, having a blunt Edge with no Baſil to it, for ripping and tearing two Pieces of Wood aſunder, by forcing in the blunt Edge between them.

CHOIR, is that Part of a Church, Cathedral, *&c.* where the Prieſts, Choriſters, and Singers ſit.

The *Choir* is diſtinguiſh'd from the Chancel, or Sanctuary, where the Communion is celebrated; as alſo from the Nave or Body of the Church, where the People aſſiſt.

CHORD [in Geometry] is a Right Line connecting the two Extreams of an Arch; or it is a Right Line terminated at each Extream in the Circumference of a Circle, without paſſing through the Centre, and dividing the Circle into two unequal Parts called Segments, as the Line A B in the Figure.

The

... *Gauge*; which is a ... a round Edge, one ... serves to prepare ... for an Augre; and the ... such Wood as is ... hollow'd, &c.

... *Chissels*; which are ... by Carpenters, &c. ... Shank made with a ... ocket at Top, to re... Wooden Sprig fit... with a Shoulder.

... *Chissels* are also distin... to the Breadth ... into Half-Inch ... quarter of an Inch ...

... *Chissel*; which is a ... an Inch broad, ha... Edge with no Basil ... and tearing two ... Wood asunder, by for... blunt Edge between

...R, is that Part of a ... Cathedral, &c. where ... Choristers, and Singers

... is distinguish'd from ..., or Sanctuary, where ... is celebrated; ... the Nave or Body ..., where the People

...) [in Geometry] is a ... connecting the two ... an Arch; or it is a ... terminated at each ... the Circumference of ... passing through ... dividing the Cir... unequal Parts called ... the Line A B in the

The

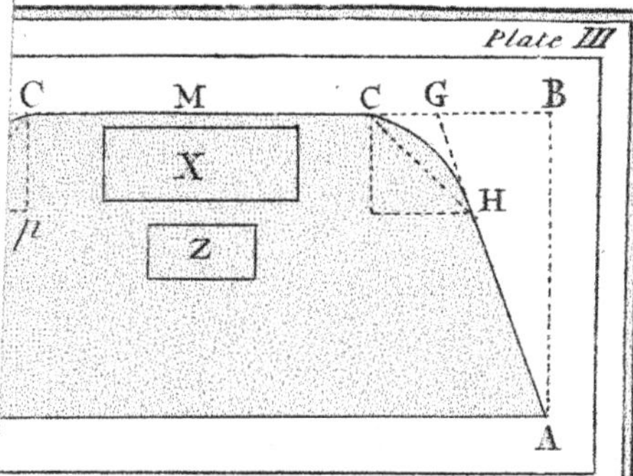

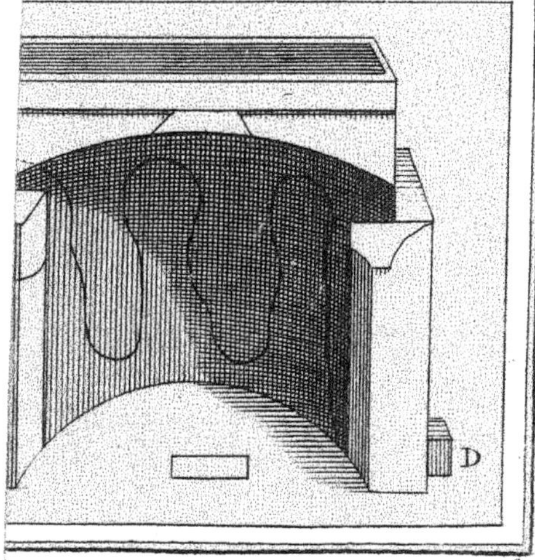

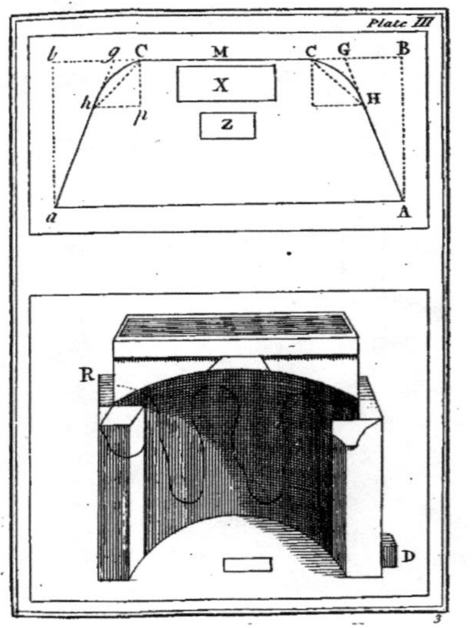
Plate III
l
g
C
M
C
G
B
X
h
p
H
Z
a
A
R
D
3

The *Chord of an Arch* is a Right Line drawn from one Extremity of an Arch to the other; called also the Subtense.

To make a Line of Chords for the Mensuration of the Quantity of Angles.

First, Draw a Right Line at Pleasure, as AB, and from any Point, as D, raise the Perpendicular DN, and compleat the Quadrant DNB of any given Magnitude.

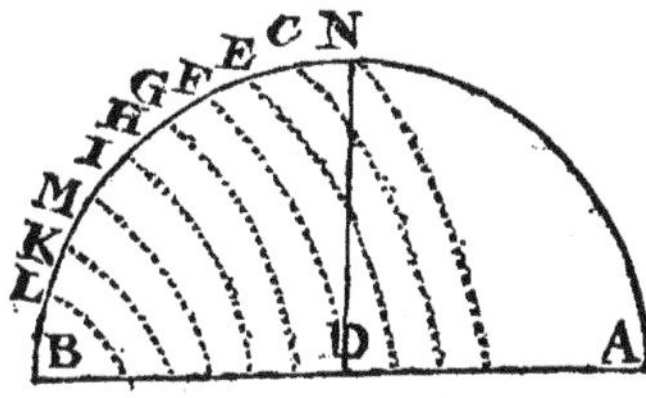

Secondly, Divide the Arch NB into ninety equal Parts, and then setting one Foot of the Compasses in B, and extending the other to the several Divisions of the Arch, transfer them to the Line AB, as the several Divisions of the Arch, aud transfer them to the Line AB, as the several pricked Arches L 10, 20, 30, 40, &c. exhibits; which will compleat the Line of Chords requir'd.

N. B. The larger these Scales are made, the better they are for Practice.

CHURCH, is defined by *Daviler*, to be a large Vessel extended in Length, with Nave, Choir, Isles, Chapel, and Belfry.

A Simple Church, a *Church* so called, having only a Nave and a Choir.

A *Church with Isles* is one which has a Row of Porticoes in Form of Vaulted Galleries, with Chaplets in the Pourtour.

Church in a Greek Cross is one the Length of whose Cross is equal to that of the Nave; so called, because most of the *Greek Churches* were built in this Form.

Church in a Latin Cross, is one whose Nave is longer than the Cross, as most of the *Gothick* Churches.

Church in Rotondo, one whose Plan is a perfect Circle, in Imitation of the Pantheon.

CIMA, *or* SIMA, a Member or Moulding; called also *Cymatium*, and *Gula*: Which See.

CIMBLY. See PEDESTAL.

CILERY, a Term in Architecture, signifying the Drapery or Levage that is wrought upon the Heads of Pillars.

CIMELLARC, in Architecture, is a Vestry, or Room, where the Plate, Vestments, and other rich Things belonging to the Church, are kept.

CINCTURE, } i.e. a Girdle
CEINTURE, } in Architecture, a Ring, List, or Orlo, at the Top and Bottom of the Shaft, at one End from the Base, and at the other from the Capital.

That at the Bottom is particularly called *Apophyges*, as if the Pillar took its Height from it; and that at Top, *Colarin*, or *Collar*, or *Colier*, and sometimes Annulus.

The *Cincture* is supposed to be in Imitation of the Girts or Ferrils,

Fertils, which where used by the Antients to strengthen and preserve the primitive Wooden Columns.

CIPHER, } one of the Numeral Characters or Figures, in this Form o.
CYPHER, }

The *Cypher*, in itself, implies a Privation of Value; but when placed with other Characters on the Left Hand of it, in the common Arithmetick, it serves for augmenting each of their Values by ten; and in Decimal Arithmetick, lessens the Value of each Figure at the Right thereof in the same Proportion.

CIRCLE, is a plain Figure contain'd under one Line, which is called a Circumference, unto which all Lines drawn from a Point in the Middle of the Figure, call'd the Centre, and falling upon the Circumference of it, are all equal the one to the other. The Circle contains more Space than any plain Figure of equal Compass.

Problem I. both the Diameter and Circumference to find the Area.

The RULE.

Every *Circle* is equal to a Parallelogram, whose Length is = to half the Circumference, and the Breadth equal to half the Diameter; therefore multiply half the Diameter, and the Product is the Area of the *Circle*.

Thus if the Diameter of a Circle (that is the Line drawn cross the *Circle* through the Centre,) be 22. 6 and if the Circumference be 71, the Half of 71 is 35.5. and the Half of 22.6 is 11.3, which multiplied together, the Product is 401.15, which is the Area of the *Circle*.

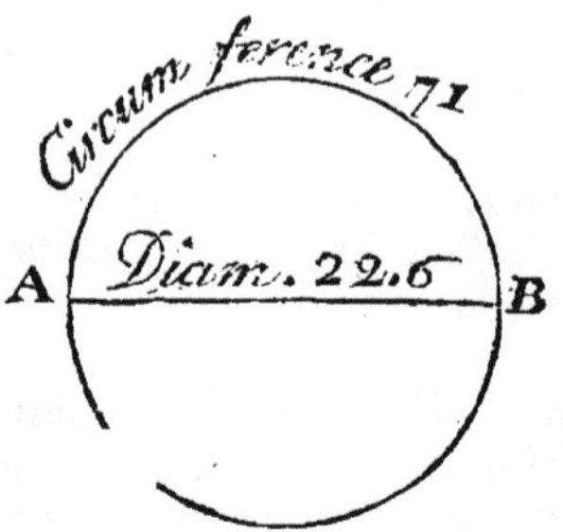


```
 35.5 Half Circumf.
 11.3 Half Diamet.
 ----
 1065
 255
355
------
401.15
```

DEMONSTRATION.

Every *Circle* may be conceiv'd to be a Polygon of an infinite Number of Sides, and the Semidiameter must be equal to the Perpendicular of such a Polygon, and the Circumference of the *Circle* equal to the Periphery of the Polygon; therefore half the Circumference multiply'd by half the Diameter, gives the Area as aforesaid.

Or (with *F. Ignatius Gaston Pardies,*) every *Circle* is equal to a Rectangle Triangle, one of whose Legs is the Radius, and the other a Right Line equal to the Circumference of the *Circle:* For such a Triangle will be greater than any Polygon inscribed, and less than any Polygon circumscribed (by the 24th, 25th, 26th, and 27th Articles of the *Fourth Book of his Elements of Geometry*)

Geometry) and therefore muſt be equal to the *Circle*.

For, ſays he, ſhould it be greater than the *Circle*, be the Exceſs as little at it will, a Polygon may be circumſcribed, whoſe Difference from the *Circle* ſhall be yet leſs than the Difference between that *Circle*, and the Rectangle-Triangle; and that that Polygon will be leſs than the Triangle, is abſurd: And if it be ſaid, that this Rectangled-Triangle is leſs than the *Circle*, an inſcribed Polygon may be made, which ſhall be greater than that Triangle; which is impoſſible

This cannot but be admitted as a Principle, That if two determinate Quantities A and B, are ſuch, that if every imaginable Quantity, which is greater or leſs than A, is alſo greater or leſs than B, theſe two Quantities, A and B, muſt be equal.

And this Principle being granted, which is in a manner, ſelf-evident, it may be directly prov'd, that the Triangle before mentioned is equal to the *Circle*; becauſe every imaginable inſcrib'd Figure, which is leſs than the *Circle*, is alſo leſs than the Triangle; and every circumſcrib'd Figure greater than the *Circle*, is alſo greater than the Triangle.

Problem II. having the Diameter of a *Circle* to find the Circumference.

As 7 to 22, ſo is the Diameter to the Circumference; or as 113 to 355, ſo is the Diameter to the Circumference.

To deſcribe a Circle, *whoſe Circumference ſhall paſs through any three given Points, provided they are not in a Right Line, as the Points* A B C.

Firſt, Draw two Right Lines from A to B, and from B to C, it matters not which; then divide thoſe two Right Lines con-

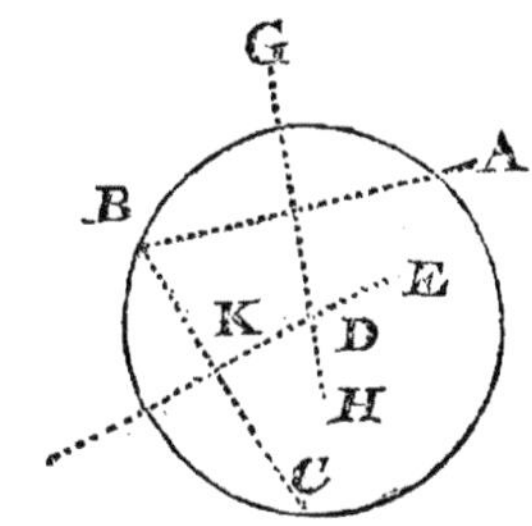

tain'd between the three Points, each into two equal Parts, by the Perpendicular G H, and F E, which will interſect each other in D, the Centre of the *Circle* that will paſs through the Points.

Secondly, Set the Compaſſes in D, and extend the Foot to A, and then deſcribe the *Circle* requir'd.

A *Circle* is generated by the Fluxion of a Line, whoſe Centre being exactly in the Middle of it, all Right Lines drawn from thence to the Circumference, are equal.

2. All Right Lines which paſs through the Centre of a *Circle*, as D C, or E G, are called Diameters, and divide its Superficies into two equal Parts, which are called Semicircles; as the Diameters D B C, which dividing

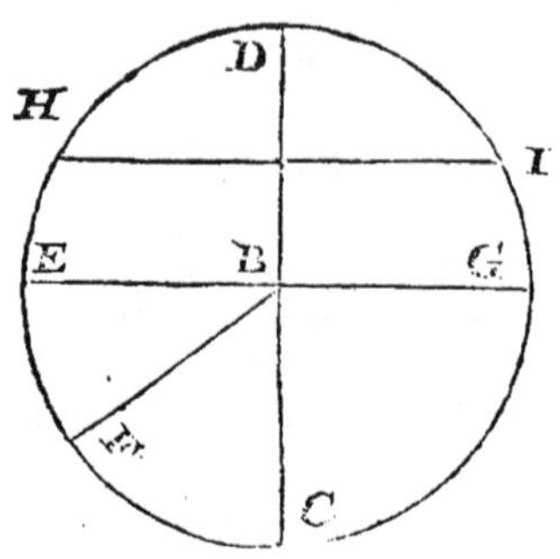

the *Circle* EDGC into two equal Parts, do thereby constitute the two Semicircles EDG, and GCE, therefore a Semicirle is a Figure contained under the Diameter, and that Part of the Circumference which is cut off by the Diameter.

3. Half the whole Diameter of a *Circle* is called the Semidiameter, or Radius, of the *Circle*, as EB, DB, GB, CB, *&c.*

4. A Segment, Section, or Portion of a *Circle*, is a plain Figure or Superficies contained under a Right Line (which is less than the Diameter,) and Part of the Circumference, either greater or less than a Semicircle.

As for Example: The Right Line HI cutting off the upper Part of the *Circle* HDI, does thereby divide the whole *Circle* into two unequal Parts, which are called Segments, Sections, or Portions; as HDI, the lesser contained under the Line HI, and Part of the Circumference HDI, which is less than the Semicircle ADC and HCI under the Line HI, and the remaining Part of the Circumference HECI, which is greater than the Semicircle ECG.

5. The fourth Part of a *Circle*, as EBD, or DBG, *&c.* is called a Quadrant, being contain'd under two Semidiameters (which are called the Sides) and the fourth Part of the Circumference, which is called the Limb of the Quadrant.

6. A Sector of a *Circle* is a Figure contained under two Semidiameters, (as BF and BE,) and Part of the Circumference FE contain'd between the Circumference of every *Circle* is supposed to be divided into 360 equal Parts, which are called Degrees; therefore as a Quadrant is one fourth Part of a *Circle*, the Degrees contained in it, are 90; and a Semicircle therefore contains 180 Degrees.

8. Every Degree of a *Circle* is supposed to be subdivided into 60 equal Parts, which are called Minutes; therefore one Fourth of a Degree contains 15 Minutes, half a Degree 30 Minutes, three Fourths of a Degree 45 Minutes, *&c.*

9. All Right Lines (less than the Diameter,) which divide Circles into Portions as HI are called Chord-Lines, or Subtense Lines, of those Arches, which they so subtend; because they subtend both Segments, that is, the Line HI is common as well to the Segment HECGI, as to the Segment HDI.

10. Those Parts of the Circumference of *Circles*, which are contain'd between the Extreams of Chord-Lines, as HDI, or EDG are called Arches, or Arch-Lines.

11. The

11. The Complement of an Arch to a Quadrant or 90 Degrees is ſo much as an Arch wants of 90 Degrees: As for Example, the Complement of the Arch G I is the Arch I D, and the Complement of the Arch H E is H D; alſo in Number of Degrees, the Complement of 40 Degrees is 50 Degrees; becauſe that 40 Degrees is leſs by 50 Degrees, than 90 Degrees: So likewiſe is 40 Degrees the Complement of 50 Degrees, and 10 Degrees the Complement of 80 Degrees; and the like of any other Quantity of Degrees whatſoever.

12. The Complement of any Number of Degrees and Minutes to 90 Degrees is the Quantity of Degrees and Minutes which are wanting to make their Sum 90 Degrees compleat.

As for Example: The Complement 27 Degrees 16 Minutes, being ſubtracted from 90 Degrees, the Remainder will be 62 Degrees, 44 Minutes; and the like of any other Quantity.

13. The Exceſs of an Arch, greater than a Quadrant, is ſo much as the ſaid Arch exceeds 90 Degrees.

As for Example: The Exceſs of the Arch E D I is the Arch D I; becauſe when the Arch E D (which is a Quadrant) is taken from E D I, the Remainder is D I, which is the Exceſs more than the Quadrant E H D.

14. The Complement of an Arch leſs than a Semicircle, is ſo much, as that Arch wants of a Semicircle or 180 Degrees.

As for Example: The Arch E H is the Complement of the Arch H D G: So likewiſe is C E H the Complement of the Arch H D, and H D G is the Complement of the Arch H E.

To divide the Circumference of the *Circle* C A E D into 360 equal Parts or Degrees, by which the Quantities of all Angles are meaſured: *Firſt*, Deſcribe a *Circle* of the given Magnitude, as A C E D, and draw the two Diameters C E and A D at Right Angles, to each other through

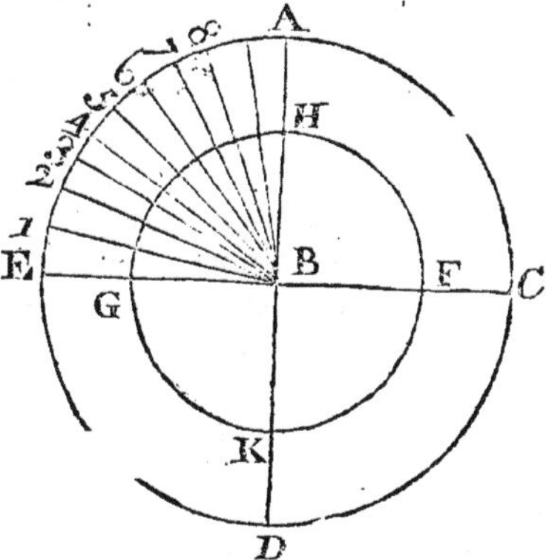

its Centre; then will the Circumference of the *Circle* be divided into four equal Parts, at the Points C A E D, and conſequently the *Circle* into four Quarters, each of which is called a Quadrant, as A B C, or A B E, &c.

Secondly, Make A 3 and E 6 equal to the Radius B E, or A B, then will the Arch A E be divided into three equal Parts, each = 30 Degrees.

And here obſerve, That as this Diviſion of the Arch A E was made by the Radius A B being ſet from A to 3, and from E to 6; therefore it is plain, that as thereby the Arch is divided into three equal Parts, each containing a Third of 90 Degrees, the Radius A B muſt = 60 Degrees of the Arch A E.

There-

Therefore hereafter, when the Radius of a *Circle* is mention'd, the Arch of 60 Degrees is to be underſtood by it at the ſame Time.

Thirdly, With the Compaſſes divide E 3, 36, and 6 A, each into three equal Parts; and each of thoſe Parts into ten equal Parts (there not being yet any Geometrical Method of dividing an Arch otherwiſe,) then will the Quadrant A E be divided into 90 equal Parts; and conſequently if the other three Quadrants A C B, C B D, and B E D, be divided in like manner, the Whole will be divided into 360 equal Parts, as required.

It is by the Number of Degrees contained in every Arch of an Angle, that its Quantity is meaſured.

Thus the Quantity of the Right Angle A B C is 90 Degrees, the Acute Angle 6 B E 60 Degrees, and the Obtuſe Angle C B 6 120 Degrees; that is the the Arch or Quadrant C A 90 Degrees, and the Arch A 6 30 Degrees; which taken together, are = 120 Degrees.

Hence 'tis plain, that all Acute Angles contain leſs than 90 Degrees, or a Right Angle, and all Obtuſe Angles more than ninety Degrees, and leſs than a hundred and eighty Degrees, or two Quadrants, which taken together, are = to a Semicircle; therefore a Semicircle contains a hundred and eighty Degrees.

If from the Points, 8, 7, 5, 4, 3, 2, 1, of the Quadrant there be right Lines drawn to the Centre B; and if on the Centre B. another *Circle* be deſcrib'd, as the *Circle* F H G K, the Quadrant H G will be divided by the Lines 8 B, 7 B, 6 B, *&c.* unto the like Number of equal Parts or Degrees, as the Quadrant A C; wherefore 'tis plain, that all *Circles*, either great or ſmall, have their Circumferences alike, divided into three hundred and ſixty equal Parts or Degrees, and each of thoſe Degrees are ſuppoſed to be again divided into ſixty equal Parts called Minutes.

Every *Circle* is nearly equal to a Parallelogram, whoſe Length is equal to half its Circumference, and Breadth to the Semidiameter: Or every *Circle* is equal to the Parallelogram, whoſe Length is equal to the Diameter, and Breadth to eleven Fourteenths thereof.

Suppoſe the Circle A B I to be fourteen Feet Diameter, then I ſay, that the Oblong B H I M, whoſe Length B H is equal to

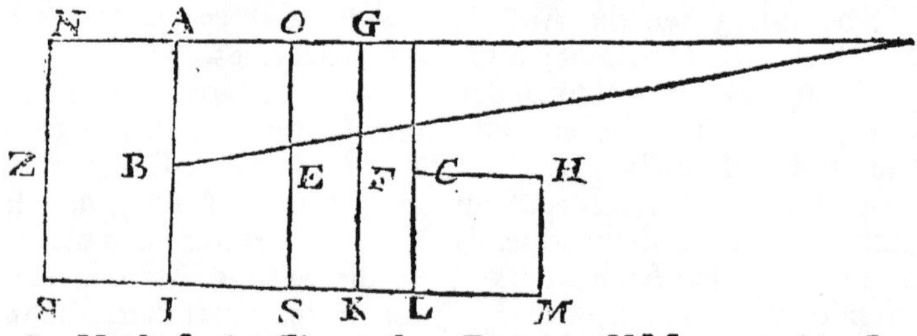

22, the Half of the Circumference of the *Circle* A E I, and Breadth H M, to the Semidiameter B L, is equal to the Parallelogram

rallelogram ACIK, whoſe Length CK is equal to the Diameter AI, and the Breadth to eleven Fourteenths thereof.

For HM, which is equal to ſeven Feet, being multiply'd into BH, twenty-two Feet, the Product or Area thereof is one hundred and fifty-four Feet. And again, AI the Diameter fourteen Feet, being multiplyed into AC $\frac{11}{14}$ thereof, *viz.* eleven Feet, the Product or Area thereof is one hundred and fifty-four Feet, as before.

Therefore thoſe two Parallelograms are equal, and either of them nearly equal to the Area of the *Circle* ABI.

N.B. *In the preceding Figure, there ſhould have been a* Circle *deſcrib'd on* B, *as a Centre, whoſe Diameter is* BA.

Corollary 1. Hence appears the general Rule for the Menſuration of *Circles*; to multiply half the Circumference by half the Diameter.

Corollary 2. Hence every Semicircle is nearly equal to the Oblong or Parallelogram, whoſe Length is equal to half the Curve or Arch, and Breadth to the Radius or Semicircle;

That is, the Semicircle AEI is nearly equal to the Parallelogram BHIM, whoſe Length IM is nearly equal to half the Arch Line AEI, and Breadth HM, to the Semidiameter BI; for the Parallelogram ACBF is equal to half the Paralellogram ACIK.

Corollary 3. Hence it appears, that every *Circle* is nearly equal to that Right-angled Triangle, whoſe Baſe is equal to its Circumference, and perpendicular to the Semidiameter thereof.

For if the Side AQ of the Parallelogram AQBH be continued to X, and made equal to the whole Circumference AEZI, and the Hypothenuſe BX being drawn, I ſay, that the Triangle BAX is equal to the Parallelogram AQBH.

For as AX is equal to twice AQ, and AB is perpendicular to AX, therefore QX the Side continu'd will be equal to BH, which is parallel to AQ, and conſequently QH will be biſected in P by the Hypothenuſe BX.

Now ſeeing that QX is equal to AQ, and PX to PB, and the Angles XQP, and BHP, both right-angled; therefore the Triangle AXB is equal to the Parallelogram AQBH; becauſe the Triangle QXP is equal to the Triangle BPH.

Theorem. Every Sector of a *Circle* is nearly equal to a Parallelogram, whoſe Length is equal to the Semidiameter of the *Circle*, and Breadth to half the Curve or Arch Line thereof; or whoſe Length is equal to half the Semidiameter, and Breadth to the whole Curve or Arch Line thereof.

Let NMX be the Sector of a *Circle*, whoſe Arch Line NX is equal to five Feet, and Semidiameter NM to ſeven Feet, I ſay, that the Oblong AN NM, whoſe Length is equal to the Semidiameter NM, and Breadth, to half the Arch NX, is equal to the Oblong HBVM, whoſe Length VM is equal to the whole Curve NX, and Breadth, to half the Semidiameter BM.

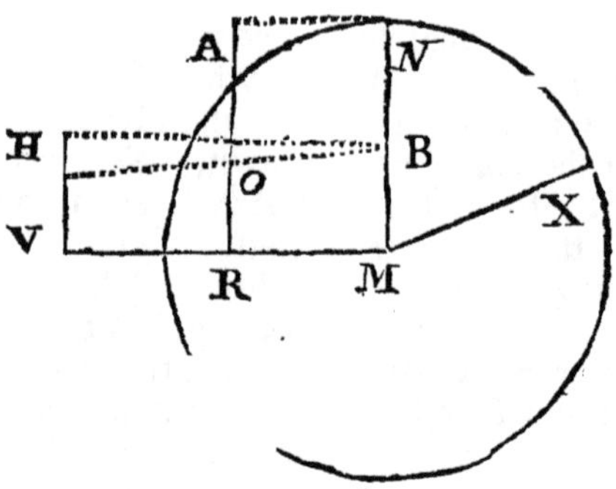

For AN, which is equal to half the curved Line NX, *viz.* two Feet; one half being multiply'd into ſeven Feet, the Semidiameter, the Area or Product will be ſeventeen one half: So likewiſe HB, which is equal to the whole Curve NX, *viz.* five Feet, being multiply'd into half the Semidiameter BM, three Feet and a half, the Area or Product will be alſo ſeventeen and a half, as before..

For as NX is equal to HB, and HV equal to BM ſo alſo is AR equal to NM, and AN to half NX, and as all the Angles of the three Parallelograms ANRM, OBRM, HOVR, are all Right Angles, and their reſpective oppoſite Sides equal; therefore the two Parallelograms ANOB, and OBRM, which compoſe the Parallelogram ANRM, made by the Half Arch multiply'd into the whole Semidiameter, is equal to the two Parallelograms OBRM, and HOVR, which compoſe the Parallelogram HBVM, made by the whole Arch multiply'd into half the Semidiameter QED.

Corollary. Hence 'tis evident that the Axis of any *Circle*, Semi-Circle, Quadrant, or Sector, may be found by this one general Rule, *viz.*

Multiply half the Curve, or Arch-Line, by the Semi-Diameter, or multiply half the Semi-Diameter by the whole Curve, or Arch-Line; and the Product will be the Area required.

Theorem. If the Proportion that the Diameter of a *Circle* hath to its Circumference, be allowed to be as 7 are to 22, then the Square made of the Diameter of any *Circle* is in Proportion to that *Circle*, (as 14 is to 11;) and therefore every inſcribed *Circle* within a Square is $\frac{11}{14}$ thereof.

Let the *Circle* be inſcribed within the Square, I ſay that the *Area* of that *Circle* is equal to $\frac{14}{11}$ of the Square.

Suppoſe the Diameter of the *Circle* or Side of the Square to be equal to 14 Feet, and the Circumference to 44 Feet; for as 7 is to 22, ſo is 14, the Diameter given, to 44, the Circumference given.

Demonstration. Multiply half the Diameter, 7 Feet, by half the Circumference, 22 Feet; and the Product is 154 Feet, the Area of the *Circle*.

Multiply one of the Sides of the Square, as into itself, *viz.* 14 by 14, and the Product will be 196, the Area of the Square.

Now by the Rule of Proportion:

As 154, the Area of the *Circle*, is to 196, the Area of the Square, so is 11 to 14.

```
154 : 196 :: 11 : 14
       11
      ---
      196
     196
     ----
154)2156(14
     154
     ----
      616
      616
      ---
      000 Remains.
```

Therefore the Proportion that a *Circle* has to a Square, whose Diameter and Sides are equal, is as 11 is to 14.

Corollary. Hence arises the stated Numbers and Rules, by which the Area of *Circles* are found, when the Diameters only are given, *viz.*

Square the Diameter given, multiply the Product by 11, and divide the last Product by 14, the Quotient is the Area required.

As for Example:

Let the Diameter of a given *Circle* be 14, as before.

```
              14
              14
              --
              56
              14
              ---
Product      196   The Diameter squared.
              11
             ---
Multiplied by 196
             196
            ----
        14) 2156 (154   The Area of the Circle as before,
            14          when half the Circumference
            --          was multiplied into the Semi-
            75          Diameter.
            70
            --
             56
             56
             --
             00
```

Or as 1 to 3.141539; so is the Diameter to the Circumference.

Let the Diameter (as in the former *Circle*) be 22.6; this being multiplied by 22, the Product will be 497.2; which being divided by 7, will give 71.028 for the Circumference.

Or (by the second Proportion) if 22.6 be multiplied by 355, the Product will be 802.3; this being divided by 113, the Quotient is 71, the Circumference.

Or (by the third Proportion) if 22.6 be multiplied into 3.141593, the Product is 71.0000018, the Circumference.

```
      22.6
        22
    ------
       452
      452
    ------
7) 4972 (71.028

  3.1415.93
        226
  ----------
  18849558
  6283186
  6283186
  ----------
  71.0000018
```

```
       355
      22.6
    ------
      2130
      710
     710
    ------
113) 8023.0 (71
    ------
     791
    ------
     113
     113
    ------
     ...
    ------
```

By Scale and Compasses.

Extend the Compasses from 7 to 22, or from 113 to 355, or from 1 to 3.14159; that Extent will reach from 22.6 to 71.

The Proportion of the Diameter of a *Circle* to the Circumference, was never yet exactly found, notwithstanding many eminent learned Men have laboured very far therein; amongst which the excellent *Van Cullen* hath hitherto out done all, in his having calculated the said Proportion to 36 Places of Decimals, which are engraved on his Tombstone in St. *Peter*'s Church at *Leyden*; which Numbers are these:

Diameter.

1.000 000.000.000.000.000.000.000.000.000.000.00.

Circumference.

3.14159.26535.89793.23846.26433 83279.50288.

Of which large Number, these 6 Places 3.14159, answering to the Diameter 1.00000, may be sufficient for these latter Proportions; as 7 to 22, 113 to 355, and 1 to 3.14159, the Reader may be at Liberty to use which of them he pleases. I shall only add, that the last two are the most exact, tho' the first is the most in use.

Problem III.

Having the Circumference of a Circle, *to find the Diameter.*

As 1 is to 318309, so is the Circumference to the Diameter.

Or, as 355 is to 113, so is the Circumference to the Diameter.

Or, as 22 is to 7, so is the Circumference to the Diameter.

Let the Circumference be 71, (as in the former *Circle*) if .318309 be multiplied by 71, (as by the firſt Proportion) the Product will be 22.599939 for the Diameter.

Or, by the ſecond Proportion, 113 multiplied by 71, the Product will be 8023; which divided by 355, the Quotient will be 22.6, the Diameter.

Or, by the third Proportion, 71 multiplied by 7, the Product will be 497; this being divided by 22, the Quotient will be 22.5909, the Diameter.

```
.318309          113                  71
     71           71                   7
-------          ---                 ---
 318309          113            22)497(22.59
2228163          791                  44
-------          ---                 ---
22.59939    355)8023(22.6             57
                 710                  44
                 ---                 ---
                 923                 130
                 710                 110
                 ---                 ---
                2130                 200
                2130                 198
                ----                 ---
                                       2
                                     ---
```

Thus by both the firſt Proportions, the Diameter is 22.6; but by the laſt it falls ſomething ſhort.

By Scale and Compaſſes.

Extend the Compaſſes from 314159 to 1, that Extent will reach 71 to 22.6; which is the Diameter ſought.

Or you may extend from 1 to 318309.

Or from 22 to 7.

Or from 355 to 113; the ſame will reach from 71 to 32.6, as fore.

Note, That if the Circumference be 1, the Diameter will be 318309.

PROBLEM IV.

Having the Diameter of a Circle, *to find the Area.*

All *Circles* are in Proportion one to another, as are the Squares of their Diameters, (by *Euclid* XII. 2.)

Now the Area of a *Circle*, whoſe Diameter is 1, will be 785.398, according to *Van Cullen*'s Proportion before mentioned; but for Practice 7854 will be ſufficient: Therefore

As 1 (the Square of the Diameter 1) is to 7.854, ſo is 510.76 (the Square of 226 the Diameter of the given *Circle*) to 401.15, (the Area of the given *Circle*:) But,

According

According to *Metius*'s Proportion,

As 452 : 355 :: 510.76 : 401.15, the ſame as before.

But if you uſe *Archimedes*'s Proportion, ſay,

As 14 : 11 :: 510.76 : 401.31; which Area is greater, than by the two former Proportions; tho' in ſmall *Circles* this is near enough the Truth.

See Operation of all theſe.

```
22.6                                  As 1 : .7854 :: 510.76
22.6                                                  .7854
----                                               --------
1356                                                204304
 452                                               255380
452                                               408608
----                                             357532
510.76  The Square of the Diameter.              ---------
------                                           401.150904 The Area.
                                                 ---------
```

By Scale and Compaſſes.

The Extent from 1 to 22.6 being twice turned over from .7854, will fall at laſt upon 401.15, the Area.

```
113          As 552 : 355 :: 510.76
  4                             355
---                          ------
452                          255380
---                         255380
                           153228
                          --------
                     452)181319.80(401.15
                          1808
                          ----
                            519
                            452
                            ---
                             678
                             652
                            ----
                            2260
                            2260
                            ....
                            ----
```

As 14 : 11 :: 510.76
11

510.76
51076

14) 561686 (402
56....

016
14

28
28

06

PROBLEM V.

Having the Circumference of a Circle, to find the Area.

Becauſe the Diameters of *Circles* are proportional to their Circumferences; that is, as the Diameter of one *Circle* is to its Circumference, ſo is the Diameter of another *Circle* to its Circumference.

Therefore the Area's of *Circles* are to one another, as the Squares of the Circumferences.

And if the Circumference of a *Circle* be 1, the Area of that *Circle* will be .07958; then the Square of 1 is 1, and the Square of 71 (the Circumference of the former Circles) is 5041; therefore it will be

Sq. Cir. *Area.* *Sq. Circumf.*
As 1 : .07958 :: 5041
5041

7958
31832
397900

401.16278 Area.

Or thus:

As 88 : 7 :: 5041
7

88) 35287 (400.98 Area.
352

0870
704

76

Or as 1420 : 113 :: 5041 : 401.15.

PROBLEM

Problem VI.

By having the Diameter, to find the Side of a Square that is equal in Area to that Circle.

If the Diameter of a *Circle* be 1, the Side of a Square equal thereto will be 8862: Therefore

As 1 : 8862 :: 22.6, the Diameter.

```
  22.6
 ------
  53172
 17724
17724
 ---------
 20.02812  The Side of the Square AC.
 ---------
```

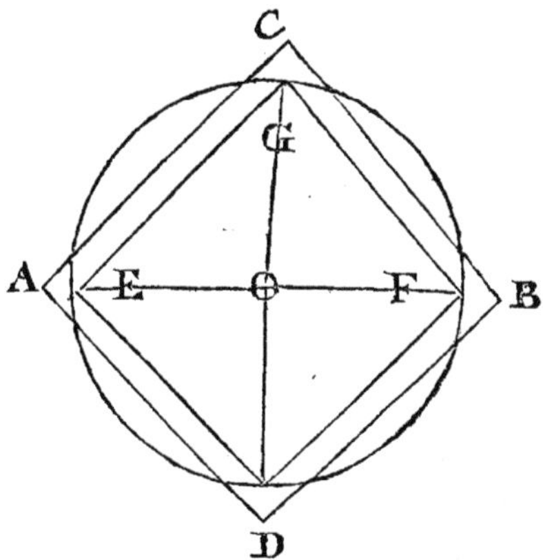

Let the Diameter of the *Circle* EF, or GH, be 22.6 (as before) to find the Side of the Square of the Side AC, AD, &c.

If .8862 be multiplied by 22.6, the Product will be 20.02812, which is the Side of a Square equal in Area to the *Circle* given; for if 20.02812 be multiplied square-wise, that is, by itself, it will produce 401.1255907344; which is nearly equal to the Area found in the last Problem.

You may find the Side of the Square equal, by extracting the Square Root out of the Area of the given *Circle*.

```
  401.15)20.0287295   Side of the Square.
  4
  ----------
4002)01.1500
  ----------
40048)349600
      320384
  ----------
       29216
       28034
  ----------
        1182
         801
  ----------
         381
         360
  ----------
          21
          20
  ----------
           1
  ----------
```

By this Method of extracting the Square Root of the Area, you may find the Side of a Square equal to any plain Figure, regular or irregular.

PROBLEM VII.

By having the Circumference, to find the Side of the Square equal.

If the Circumference of a *Circle* be 1, the Side of the Square equal will be .2821 : Therefore

```
As 1 : .2881 :: 71  The Circumf.
          71
      -------
        2821
       19747
      -------
     20.0291   The Side of the Square.
      -------
```

PROBLEM VIII.

Having the Diameter, to find the Side of a Square, which may be inscribed in that Circle.

If the Diameter of a *Circle* be 1, the Side of the Square inscribed will be .7071 : Therefore

```
As 1 : .7071 :: 22.6
          226
      -------
        42426
       14142
      14142
      -------
     15.98046  The Side EG inscrib'd.
      -------
```

Or if you square the Semi-Diameter, and double that Square, the Square Root of the doubled Square will be the Side of the Square inscribed: For (by *Euclid* 1. 47.) the Square of the Hypothenuse EG is equal to the Sum of the other two Legs EO and OG.

11.3

```
        11.3
        11.3
       ------
         339
        113
       113
       -------
       127.69 The Square of EO, which is double, because
            2                  EO=OG.
      --------
      255.38 (15.98 Root; which is the Side of the Square.
      1
      ----
  25)155
     125
     -----
 309)3038
     2781
     -----
3188)25700
     25504
     ------
       196
     ----
```

PROBLEM IX.

Having the Circumference, to find the Side of a Square, which may be inscribed.

If the Circumference be 1, the Side of the Square inscribed will be .2251: Therefore

```
As : 1 .2251 :: 71
              71
        ---------
          2251
        15757
        ---------
        15.9821  The Side of the
                 Square EG.
```

Because that in each of the four last Problems, *viz.* VIth, VIIth, VIIIth, and IXth, there is a Proportion laid down, it will be easy to work them by Scale and Compasses; for if you extend the Compasses from the first to the second, that Extent will reach from the third to the fourth: As in the last Problem, where the Proportion is as 1 to .2251, so is 71 to the Side of the Square 15.9821.

Here extend the Compasses from 1 to .2251; that Extent will reach from 71 to 15.98; and so of the rest.

But the fifth must be wrought like the fourth, thus: Extend the Compasses from 1 to 71; and that Extent turn'd over the same Way from .07958, will fall at the last upon 401.15.

PROBLEM X.

Having the Area, to find the Diameter.

If the Area of a *Circle* be 1, the Square of the Diameter thereof is 1.2732: Therefore

```
  Area. Sq.Diam.   Area.
As 1  : 1.2732 :: 401.15
          401.15
          ------
          63660
         12732
         12732
        50928
        ----------
        510 744180(22.599  The Diameter.
        4
        ---
     42)110
         84
        -----
    445)2674
        2225
        -----
   4509)44941
         40581
         ------
  45189)436080
         406701
         ------
          29379
```

By Scale and Compasses.

Extend the Compasses from 1 to 1.2732; that Extent will reach from 401.15 to 510.74, &c.

Then divide the Space between 1 and 510.74 into two equal Parts, and you'll find the middle Point at 22.6.

Or you may divide the Space upon the Line of Numbers, between 401.55 and .7854 into two equal Parts; and one of those Parts will reach from 1 to 22.6, the Diameter sought.

PROBLEM XI.

Having the Area, to find the Circumference.

If the Area of a Circle be 1, the Square of the Circumference will be 12.56637: Therefore

Area.

Area. *Sq.Circumf.* *Area.*
As 1 : 1.256637 :: 401.15.
401.15

6283185
1256637
1256637
5026548

Circumf.
5040.99932550(70.9990 Root.
49

109)14099
12681

14189)141893
127701

141989)1419225
1277901

14132450
12779901

1352549

By Scale and Compasses.

Divide the Space between 401.15 and .07958 upon the Line, into two Parts; one of those Parts will reach from 1 to 71, the Circumference sought.

Problem XII.

Having the Area, to find the Side of a Square inscribed.

If the Area of a *Circle* be 1, the Area of a Square inscribed within that *Circle* will be .6366: Therefore

As 1 : 401.15 :: .6366
.6366

240690
240690
120345
240690

255.372090(15.98 Roots; which is the Side of the Square sought.
1

25) 155
125

309) 3037
2781

3188) 25620
25504

11690

The same Reason may be given for the last Proportion, that was given before for the Proportion of *Circles* to the Squares of their Diameters and Circumferences; for not only the Squares of the Diameters and Circumferences are in Proportion to the Circles they belong to; but also Figures inscrib'd or circumscrib'd, have the Squares of their like Sides proportional to the *Circles* they are inscribed in, or circumscribed about, and also to the Figures themselves: The Square of any Side of one Figure, as the Square of the like Side of another similar Figure is to the Area thereof; as may be found prov'd at large in *Euclid*, *Sturmius*, *Mathesis Enucleata*, and other Authors.

By Scale and Compasses.

Extend the Compasses from 1 to 401.15; that Extent will reach from .6366 to 255.37; the half Space between that and 1, is at 15.98, the Side of the Square.

PROBLEM XIII.

Having the Side of a Square, to find the Diameter of the circumscribing Circle.

If the Side of a Square be 1, the Diameter of a *Circle*, that will circumscribe that Square, will be 1.4142: Therefore

As 1 : 1.4142 :: 15.98
1598

113136
127278
70710
14142

22.598916 The Diameter sought.

By

By Scale and Compasses:

Extend the Compasses from 1 to 1.4142, and that Extent will reach from 15.98 to 22.6, the Diameter sought.

Problem XIV.

Having the Side of a Square, to find the Diameter of a Circle *equal.*

If the Side of a Square be 1, the Diameter of a *Circle*, equal thereto, will be 1.128: Therefore

Side. Diam. Side of a Square.
As 1 : 1.128 :: 20.0291
1128

1602328
400582
200291
200291

22.5928248 Diamet.

By Scale and Compasses.

Extend the Compasses from 1 to 1.128, and that Extent will reach from 20.0291 (the Side of the Square given) to 22.6, the Diameter of the *Circle* sought.

Problem XV.

Having the Side of a Square, to find the Circumference of the circumscribing Circle.

If the Side of a Square be 1, the Circumference of a *Circle*, that will encompass that Square, will be 4.443: Therefore

Side Sq. Circumf. Side Sq.
As 1 : 4.443 :: 15.98
15.98

35544
39987
22217
4443

70.99914 Circumference.

By Scale and Compasses.

Extend the Compasses from 1 to 4.443, and that Extent will reach from 15.98 to 71, the Circumference.

Problem XVI.

Having the Side of a Square, to find the Circumference of a Circle *that will be equal thereto.*

If the Side of the Square be 1, the Circumference of a *Circle* that will be equal thereto, shall be 3.545: Then

As 1 : 3.545 :: 20.0291
3.545

1001455
801164
1001455
600873

71.0031595 *Circumf.*

By Scale and Compasses.

Extend the Compasses from 1 to 3.545, and that Extent will reach from 20.0291 to 71, the Circumference.

In

In ſeveral of the foregoing *Problems*, where the Diameter and Circumference is required, the Anſwers are not exactly the ſame as the Diameter and Circumference of the given *Circle*; but are ſometimes too much, and ſometimes too little, as in the two laſt *Problems*; where the Anſwers in each ſhould be 71, the one being too much, and the other too little.

The Reaſon of this is, the ſmall Defect that happens to be in the Decimal Fractions, they being ſometimes too great, and ſometimes too little; yet the Defect is ſo ſmall, that it is needleſs to calculate them to more Exactneſs.

Every *Circle* is ſuppoſed to be divided into 360 Degrees.

The Area of a *Circle* is found by multiplying the Periphery by the fourth Part of the Diameter, or half the Periphery by half the Diameter.

The Area is alſo found by finding a fourth Proportional to 1000.785, and the Square of the Diameter; or to 452.355, and the Square of the Diameter.

Circles, and ſimilar Figures inſcribed in them, are always as the Squares of the Diameters: So that they are in a duplicate Ratio of their Diameters, and therefore of their Radii.

A *Circle* is equal to a Triangle, whoſe Baſe is equal to the Periphery, and its Altitude to the Radius; therefore *Circles* are in a Ratio compounded of the Peripheries and the Radii.

CIRCULAR, any Thing that is deſcribed or moved in a Round; as the Circumference of a *Circle*, or the Surface of a Globe.

Circular Lines are ſuch ſtraight Lines as are divided from the Diviſions made in the Arch of a *Circle*; as Sines, Tangents, Secants, &c.

Circular Numbers are ſuch, whoſe Powers end in the Roots themſelves; as 5, whoſe Square is 25, and Cube 125.

CIRCUMVOLUTIONS. The Torus of the Spiral Line of the *Ionic* Volute.

CIRCUS, a large Building, either round or oval, uſed for the exhibiting Shews to the People.

The *Roman Circus* was a large Place or Square, arched at one End, encompaſſed with Porticoes, and furniſhed with Rows of Seats, placed aſcending over each other.

CISTERN, is properly uſed for a ſubterraneous Reſervoir of Rain-Warer; or a Veſſel made to ſerve as a Receptacle for Rain or other Water, for the neceſſary Uſes of a Family.

If you would make your *Ciſterns* under the Houſe, as a Cellar, which is the beſt Way to preſerve Water for culinary Uſes; then lay the Brick or Stone with Terras, and it will keep Water very well.

Or you may make a Cement to join your Brick or Stone withal, with a Compoſition made of ſlacked ſifted Lime and Linſeed Oil, tempered together with Tow or Cotton-Wooll. The Bottom ſhould be covered with Sand, to ſweeten and preſerve it.

Or

Or you may lay a Bed of good Clay, and on that lay the Bricks for the Floor; then raise the Wall round about, leaving a convenient Space behind the Wall to ram in Clay, which may be done as fast as the Wall is raised: So that when it is finished, it will be a *Cistern* of Clay walled with Brick; and being in a Cellar, the Brick will keep the Clay moist, (although it shall sometimes be empty of Water,) that it will never crack.

Mr. *Worlidge* says, he has known this to hold Water perfectly in a shady Place, though not in a Cellar.

Thus in a Garden, or other Place, may such a *Cistern* be made, and covered over, the Rain-Water being conveyed thereto by declining Channels running to it. Also in or near Houses, may the Water that falls from them, be conducted thereto.

Authors speak of a *Cistern* at *Constantinople*, the Vaults of which are supported by two Rows of Pillars 212 in each Row, each Pillar being two Feet in Diameter. They are planted circularly, and in Radii tending to that in the Centre.

There are some Persons very scrupulous about these Waters, which are received in *Cisterns*; for they pretend that they are not all good, without Distinction; that Rain which falls in a small Quantity during Heats, and the great Rains which fall presently after great Droughts, are reckoned in the Number of those that are bad: And thence it is, they say, that the Water which is sometimes taken out of *Cisterns*, has a very disagreeable Taste, and very often stinks. As for those Rains that fall during the Autumn, Spring and Winter, when the Weather is not violent; these, say they, will do. And in all fine Weather, they esteem the small Rains that fall in the Month of *May*, which should carefully be saved, to be the best, as being the purest and lightest, and even purify the Water they found in the *Cistern*.

As to the Way of making *Cisterns*, that is left to the Artist skilled that way; only it may be observed, the Walls should be good, and built to advantage, for fear the Water should be lost; that the Inside should be well cemented, especially in the Angles, without any Necessity of doing the same by the Arch or Roof, through which the Water cannot pass. As to the Bigness of the *Cistern*, that depends upon the Fancy of the Person.

The Manner of bringing together Rain-Water, is of Channels made of different Materials, fixed to the Edge of the Roofs of Houses, which convey the Water into a small Basou made of Lead or Tin, in the Midst of which, there is a Hole through which the Water passes into a Pipe that is there; and which, before it enters into the *Cistern*, helps it to fall into a Stone Trough made on purpose near the *Cistern*.

This Trough is placed to receive the Rain that falls from the Roofs of Houses, from whence it runs into the *Cistern*; but, as it has been observed before, that there is a Difference to be made between the Rains that

that fall, and which are received into these Conveyances, without Distinction, it is necessary you should know how to save those that are good and wholesome, and get rid of the rest; it must be by the Means of this Trough, which has a Hole in the Bottom, in a Corner, on that Side where the most Declivity appears. This Hole must, at the Time you judge it convenient to save the Water, be stopped, to the End, that the Trough coming to be filled up to a certain Place, where there is a Grate on the Side of the *Cistern*, it may supply a Passage for the inclosed Water to fall into the *Cistern*; and when, on the contrary, they do not value the Rains that fall, they only leave that Hole open, so that so fast as the Water comes into the Trough, so fast it runs out.

There are those who do not use any such Trough as this, but suffer the Rain to fall without any Distinction into a subterraneous Place built higher than the *Cistern*, in which they put some River Sand, pretending that the Water which passes thro' is purged of all ill qualities it may have; and that consequently the Water they take out of these *Cisterns* to drink, ought to be extreamly good.

CLAIR OBSCURE } in
CHIARO SCURO } Painting, is the Art of distributing to Advantage the Lights and Shades of a Piece; both with Regard to the easing the Eye, and the Effect of the whole Piece.

Thus, when a Painter gives his Figures a strong Relievo, loosens them from the Ground, and sets them free from each other by the Management of his Lights and Shades, he is said to understand the *Clair Obscure*.

The *Clair Obscure* makes one of the greatest Divisions, or Branches of Painting; the Whole of a Picture being resolvable into Light and Shadow.

The Doctrine of the *Clair Obscure*, may be brought under the following Rules:

Light may be considered either, *1st*, in regard to itself; *2dly*, in regard to its Effects; *3dly*, in regard to the Place wherein it is diffused; *4thly*, in regard to its Use.

For the first Light is either natural or artificial.

Natural Light comes either immediately from the Sun, which is brisk, and its Colour various, according to the Time of the Day; or it is that of a clear Air, through which Light is spread, and whose Colour is a little bluish; or a cloudy Air, which is darker, yet represents Objects in their genuine Colours, with Ease to the Sight.

Artificial Light proceeds from Fire or Flame, and tinges the Object with its own Colour; but the Light it projects is very narrow, and confined.

For the second: The Effects of Light are either principal, as when the Rays fall perpendicularly on the Top of a Body, without any Interruption; or glancing, as when it slides along Bodies; or secondary, which is for Things at a distance.

3. For the Place: It is either the open Campaign, which makes Objects appear with great Softness;

neſs; or an incloſ'd Place, where the Brightneſs is more vivid, its Diminution more haſty, and its Extremes more abrupt.

4. For the Uſe or Application: The Light of the Sun is always to be ſuppoſ'd without, and over-againſt the Picture, that it may heighten the foremoſt Figures; the Luminaries themſelves never appearing, becauſe the beſt Colours, can't expreſs them.

The chief Light to meet on the chief Group, and as much as poſſible, on the chief Figure of the Subject; the Light to be purſued over the great Parts, without being croſs'd or interrupted with little Shadows.

The full Force of the principal Light to be only in one Part of the Piece; taking care never to make two contrary Lights: Not to be ſcrupulouſly confined to one univerſal Light, but to ſuppoſe other acceſſary ones; as the opening of Clouds, *&c.* to looſen ſome Things, and produce other agreeable Effects.

Laſtly, The Light to be different, according to the Quality of Things whence it proceeds, and the Nature of the Subjects which receive it.

As for Shadows, they are diſtinguiſhed,

1. Into thoſe form'd on the Bodies themſelves by their proper Relievo's.

2. Thoſe made by adjacent Bodies; thoſe that make Parts of any Whole; and the different Effects, according to the Difference of Places.

For the firſt: Since the different Effects of Lights only appear by Shadows, their Degrees muſt be well managed.

The Place which admits no Light, and where the Colours are loſt, muſt be darker than any Part that has Relievo, and diſpoſed in the Front.

The Reflex or Return of Light, brings with it a Colour borrowed from the Subject that reflects it; and flies off at a greater or leſs Angle, according to the Situation of the reflecting Body, with regard to the luminous one. Hence its Effects muſt be different in Colour, and in Force, according to the Diſpoſition of Bodies.

Deepenings which admit not of any Light, or reflex, muſt never meet on the Relievo of any Member of any great elevated Part, but in the Cavities or Joints of Bodies, the Folds of Draperies, *&c.*

And to find Occaſions for introducing great Shadows, to ſerve for the Repoſe of the Sight, and the looſening of Things, inſtead of many little Shadows, which have a pitiful Effect.

For the ſecond: The Shadows made by Bodies, are either in plain or ſmooth Places, or on the Earth; wherein they are deeper than the Bodies that occaſion them, as receiving leſs reflex Light; yet ſtill diminiſh, as they depart further from their Cauſe, or on the neighbouring Bodies, where they are to follow the Form of the ſaid Bodies, according to its Magnitude and its Poſition, with regard to the Light.

For the third: In Shadows that have Parts, the Painter muſt obſerve to take for a Light in a ſhadow'd Place, the Teint or Luſtre of the light Part; and, on on the contrary, for the Shadow in the lightened Part, the Teint or Lustre in the Shadow: To make an agreeable Assemblage of Colour, Shadow and Reflex in the shadowed Part, but without interrupting the great Masses of Shadows; to avoid forming little Things in the Shadow, as not being perceiv'd, without closely look'd at; and to work, as it were, in the general, and at one Sight; never to set the strong Shadows against the Light, without softening the harsh Contrast, by the help of some intermediate Colour; though the Mass of Light may be placed either before or behind that of the Shadow, yet ought it to be so disposed, as to illumine the principal Parts of the Subject.

For the fourth: The Effects of Shadows are different, as the Place is either wide and spacious; as in those coming immediately from the Sun, which are very sensible, and their Extremes pretty abrupt; from the serene Air, which are fainter and more sweet; from the dark Air, which appear more diffused, and almost imperceptible; and those from an artificial Light, which makes the Shadows deep, and their Edges abrupt: Or as it is more narrow and confined, where the Light's coming from the same Place make the Shadow more strong, and the Reflex less sensible.

Clair-Obscure, or *Chiaro-Scuro*, is also used for a Design consisting only of two Colours, ordinarily black and white, sometimes black and yellow.

Or, it is a Design only wash'd with one Colour, the Shadows being of a dusky brown Colour; and the Lights heightened up with white.

A CLAMP, is a kind of Kiln, built above Ground (of Bricks unburnt,) for the burning of Bricks.

These *Clamps* are built much after the Method that the Arches are built in Kilns, *viz.* with a Vacuity betwixt each Brick's Breadth, for the Fire to ascend by, but with this Difference, that instead of arching, they *truss over*, or *over-span*, as they term it, *i.e.* they lay the End of one Brick about half way over the End of another, and so till both Sides meet within half a Brick's Length, and then a binding Brick at the Top finishes the Arch.

The Mouth (at which the Fire is to be put in) is left open about two Feet and a half wide, and about three Feet in Height; and then they begin to truss over, which they do for three Bricks in Height, and which, with a binding Brick at the Top, will close up the Arch.

But after they have begun to make the Place to receive the Fuel (before it is closed at the Top,) they fill it almost full with Wood, and upon that, lay Sea-Coal; then it being over-spanned like an Arch, they strew Sea-Coal on all the Surfaces, and then lay another Course of Bricks the other Way, laying them at a little Distance from one another, and strowing Sea-Coal upon them: And thus they continue laying one Course one Way, and another the other, and strowing Sea-Coal betwixt each Course, 'till they come to eight or ten Feet high, accord-ing

ing as the *Clamp* is to be for Bigneſs: When they have done this, they ſet the Wood on Fire, and that fires the Coals, which being all burnt out, the whole *Clamp* of Bricks is burnt.

CLAMP-NAILS, are ſuch Nails as are uſed to faſten on Clamps in the building or repairing of Ships.

CLAMPING, in Joinery, *&c.* is when a Piece of Board, *&c.* is fitted with the Grain to the End of another Piece of Board croſs the Grain, the firſt Board is ſaid to be clamped. Thus the Ends of Tables are commonly clamped, to prevent them from warping.

CLAY, a ſoft viſcous Earth, found in various Places, and uſed for divers Purpoſes, of ſeveral Kinds and Properties.

Dr. *Liſter*, in the *Philoſophical Tranſactions*, gives a Catalogue of twenty-two ſeveral *Clays* found in the ſeveral Counties of *England*; five of which he calls *Pure*, *i. e.* ſuch as are ſoft like Butter to the Teeth, with little or no Grittineſs in them, *viz.* Fullers Earth, which he diſtinguiſhes by its Colours, into yellowiſh brown, and white. 2. Boles. 3. Pale-yellow *Clay*, Cowſhot *Clay*, Dark-blue *Clay* or Marle.

Seventeen *Impure*; whereof eight are harſh and duſty when dry, as *Creta*, or Milk-white *Clay*. Two Potters Pale-yellow *Clay*. Potters Blue *Clay*; Blue *Clay*, wherein the Aſtroites is found; Yellow *Clay*, and fine Red *Clay*. Soft Chalky blue *Clay*; ſoft Chalky Red *Clay*.

Three are ſtony, when dry; *viz.* a Red Stony *Clay*, a Blue Stony *Clay*, and a White Stony *Clay*.

Three are mix'd Sand, or Pebbles, *viz.* one a Yellow Loam, a Red Sandy *Clay*. Three of a ſecond Species of the ſame *Clay*.

Laſtly, Three are mix'd with flat or thin Sand, glittering with *Mica*, *viz.* one Crouch White *Clay*. Two Grey, or Bluiſh Tobacco-Pipe *Clays*. Three Red *Clays*.

CLEAR [in Building] is ſometimes uſed by the Workmen for the Inſide Work of a Houſe.

CLEAVING of Laths, Pales, Shingles, and Timber.

CLINKERS, thoſe Bricks, which having naturally much Nitre, or Saltpetre in them, and lying next the Fire in the Clamp or Kiln, by the Violence of the Fire, are run and glazed over.

CLOISTER, a Habitation ſurrounded with Walls, and inhabited by Canons, and the Religious.

In a more general Senſe, *Cloiſter* is uſed for a Monaſtery of the Religious of either Sex; where Friars, Monks, and Nuns, live retir'd from the World. Alſo a long Place covered with a Floor or Plat-Fond, ſupported by Pillars.

CLOSET, a general Name for any very ſmall Room. The Contrivance of *Cloſets* in moſt Rooms, now ſo much practiſed in *England*, is one great Improvement in Modern Architecture.

COCHLEA,

COCHLEA, in Mechanicks, one of the five Mechanical Powers, otherwise called the Screw.

COCKLE-STAIRS. See WINDING-STAIRS.

COCK-PIT, a Sort of Theatre, where Game-Cocks fight their Battles. It is commonly a House, or Hovel, cover'd over.

COENOTAPH, an empty Tomb or Monument, erected in Memory of Some illustrious Defunct, who perishing by Shipwreck, in Battle, &c. his Body could not be found to be interred or deposited in the same.

COINS. See QUOINS.

COLARIN, the little Frize of the Capital of the *Tuscan* and *Doric* Column, placed between the Astragal and the Annulets; called by *Vitruvius*, Hypotrachelium. Called also Cincture.

Colarin is also used for the Orlo or Ring on the Top of the Shaft of the Column next the Capital.

COLLAR-BEAM, a Beam fram'd cross betwixt two principal Rafters.

COLLEGE, a Place set apart for the Society and Cohabitation of Students.

COLLER. See CINCTURE.

COLONNADE, a Peristyle, of a circular Figure, or a Series of Columns disposed in a Circle, and insulated withinside. Such is that of the little Park at *Versailles*; which consists of thirty-two *Ionic* Columns, all of solid Marble, and without Incrustation.

A *Polystyle Colonnade* is that whose Number of Columns is too great to be taken in by the Eye at a single View. Such is the *Colonnade* of the Palace at *St. Peter* at *Rome*, which consists of two hundred eighty-four Columns of the *Doric* Order, each above four Foot and a half Diameter, all in *Tiburtine* Marble.

COLUMN, a round Pillar, made to support or adorn a Building.

The *Column* is the principal or reigning Part of an Order.

The principal Laws and Properties of this eminent Member of Architecture, are thus deduced.

Every Fulcrum or Support is so much the more perfect, as it is the firmer, or carries the greater Appearance of Firmness. And hence all *Columns* or Pillars ought to have their Base or Foot broader than themselves.

Again, as a Cylinder, and a quadrangular Prism are more easily removed out of their Place than a truncated Cone or Pyramid, on the same Base, and of the same Altitude. The Figure of *Columns* ought not to be cylindrical, nor that of a Pilaster, pyramidical; but both the one and the other to be contracted or diminish'd, *i. e.* grow less and less like a truncated Cone, and a truncated Pyramid.

For the same Reasons, the lowest Parts of the *Columns* are to be cylindrical, and that of Pilasters, pyramidical. Hence, again, as *Columns* are more firm, if their Diameter bears greater Proportion to their Height, than if it bore a less, the greater Ratio is to be chosen where a large Weight,

Weight is to be ſuſtain'd, and a leſs, where a leſs.

Further, as the Deſign of a *Column* is to ſupport a Weight, it muſt never be ſuppos'd without an Entablature: Though a *Column* raiſed on an eminent Place, ſo as to leave no Room to fear its being thruſt out of its Place, needs no Pedeſtal.

The entire *Column* in each Order is compoſed of three principal Parts; the Baſe, the Shaft, and the Capital.

Each of theſe Parts is again ſubdivided into a great Number of leſſer, called Members or Mouldings: Some whereof are eſſential, and found in all *Columns*; others are only accidental, and found in all particular Orders.

Columns are made different, according to the ſeveral Orders they are uſed in; and likewiſe not only with Regard to their Order, but alſo to the Matter, Conſtruction, Form, Diſpoſition, and Uſe.

COLUMNS, *with Regard to Order.*

Tuſcan Column is the ſhorteſt and moſt ſimpleſt of all the *Columns*.

Its Height, according to *Vitruvius*, *Palladio*, and *Vignola*, is ſeven Diameters or fourteen Modules; according to *Scamozzi*, fifteen Modules; to *De Lorme*, twelve Modules; to *Trajan's Column*, ſixteen Modules.

Its Diminution, according to *Vitruvius*, is one Fourth of the Diameter; according to *Vignola*, a Fifth; and according to *Trajan's Column*, a Ninth.

The whole Height of this Column, *and the Height of each principal Part thereof, according to ſeveral Authors, is as in the following Table.*

Auth. Nam.	Whole Height		Pedeſt.		Baſe		Body		Capital		Archit.		Frieze		Cornic.	
	mo.	mi.	mo.	mi.	mo.	mi.	mo.	mi.	mo.	mi.	mo.	mi.	mo.	mi.	mo.	mi.
Vitr.	11	5	2	20	0	30	6	0	0	30	0	30	0	30	0	30
Vign.	11	5	2	20	0	30	6	0	0	30	0	30	0	35	0	40
Pall.	10	15	1	0	0	30	6	30	0	30	0	35	0	26	0	40
Sca.	11	15	1	52½	0	30	6	30	0	30	0	31½	0	41	0	41

The *Doric Column* is ſomething more delicate: Its Shaft is adorn'd with Flutings. Its Height, according to *Vitruvius*, is from fourteen to fifteen Modules; to *Scamozzi*, ſeventeen; to *Vignola*, ſixteen; in the *Coliſeum*, nineteen; in the Theatre of *Marcellus*, fifteen two Thirds. Its Diminutions, according to the Theatre of *Marcellus*, twelve Minutes; to the *Coliſeum*, four Minutes and a half.

The whole Height of this Column, *and the Height of each principal Part thereof, according to ſeveral Authors, is as in the following Table.*

Auth. Names	Whole Height		Pedeſt.		Baſe		Body.		Capital		Archit.		Frieze		Cornic.	
	mo.	mi.	mo.	mi	mo.	mi.	mo.	mi.	mo.	mi.	mo.	mi.	mo.	mi.	mo.	mi.
Vitr.	12	40	2	40	0	30	7	0	0	30	0	30	0	45	0	40
Vign	12	40	2	40	0	30	7	0	0	30	0	30	0	45	0	45
Pall.	13	0	2	20	0	30	7	45	0	30	0	30	0	45	0	35
Sca.	12	58	2	26	0	30	7	30	0	30	0	35	0	45	0	42

The *Ionic Column* is more delicate ſtill, it is diſtinguiſh'd from the reſt by the Volutes in its Capital, and by its Baſe.

Its Height, according to *Palladio*, is ſeventeen Modules and one Third; according to *Vignola*, eighteen.

Its Diminution, in the Temple of *Concord*, ten one half; of *Fortuna Virilis*, ſeven one half; of the *Coliſeum*, ten Minutes.

The whole Height of this Column, *and the Height of each principal Part thereof, according to ſeveral Authors, is as in this following Table.*

Auth. Nam.	Whole Height		Pedeſt.		Baſe		Body		Capital		Archit.		Frieze		Cornic.	
	mo.	mi.	mo.	mi	mo.	mi.	mo.	mi.	mo.	mi.	mo.	mi.	mo.	mi.	mo.	mi.
Vitr.	14	15	3	6	0	30	8	10	0	20	0	37½	0	30	0	52½
Vign	14	15	3	0	0	30	8	10	0	20	0	37½	0	45	0	52½
Pall.	13	28	2	40	0	52½	7	40	0	27½	0	34½	0	27	0	46½
Sca.	12	33¾	2	30	0	30	7	30	0	18¾	0	35	0	28	0	42

The

The *Corinthian Column* is the richeſt and moſt delicate of all the *Columns*.

Its Capital is adorn'd with two Rows of Leaves, and with Caulicoles, whence ſpring out little Volutes.

Its Height, according to *Vitruvius*, and many Remains of Porticoes, Temples, *&c.* is nineteen Modules; according to *Serlio*, eighteen; to the *Coliſeum*, Seventeen; to the three *Columns* in the *Campo Vaccino*, twenty; to the *Baſilick of Antoninus*, twenty.

Its Diminutions, according to the Temple of *Peace*, ſix Minutes and a half; the *Pantheon*, ſix and a half; the Temples of *Sybil*, and *Fauſtina*, eight; *Conſtantine's-Arch*, ſeven; Porticoe of *Septimius*, ſeven and a half.

The whole Height of this Column, *and the Height of each principal Part thereof, according to ſeveral Authors, is as in this following Table.*

Auth. Nam.	Whole Height mo. mi	Pedeſt. mo. mi.	Baſe mo. mi.	Body mo. mi.	Capital mo. mi.	Archit. mo. mi.	Frieze mo. mi.	Cornic. mo. mi.
Vitr.	16 0	3 30	0 30	8 20	1 10	0 30	0 37½	1 0
Vign.	16 0	3 30	0 30	8 20	1 10	0 45	04 5	1 0
Pall.	13 54	2 30	0 30	7 55	1 5	0 36	02 8	0 50
Sca.	14 42½	2 30	0 30	8 5	1 10	0 39	0 31¾	0 46¾

The *Compoſite Column* has two Rows of Leaves in its Capital, like the *Corinthian*; and angular Volutes, like the *Ionic*.

Its Height, according to *Vignola*, and *Titus's-Arch*, is twenty Modules; to *Scammozi*, and the Temple of *Bacchus*, nineteen and a half; to *Septimius's-Arch*, 19.09.

Its Diminution, according to *Titus's*, and *Septimius's-Arch*, ſeven Minutes; to the Baths of *Diocleſian*, eleven and one Third; to the Temple of *Bacchus*, ſixty and a half.

The whole Height of this Column, *and the Height of each principal Part, according to the several Authors, is, as in this following Table.*

Auth. Nam.	Whole Height		Pedest.		Base		Body		Capital		Archit.		Frieze		Cornic	
	mo.	mi.	mo.	mi	mo.	m	no.	mi	mo.	mi.	mo.	mi.	mo.	mi.	mo.	mi.
Vitr.	16	6[illegible]	3	3[illegible]	0	30	8	20	1	10	0	52½	0	52½	0	52½
Vign.	16		3	30	0	3[illegible]	8	20	1	10	0	45	0	45	1	0
Pall.	15	20	3	20	0	30	8	25	1	5	0	45	0	30	0	45
Sca.	15	20	3	2[illegible]	0	30	8	25	1	5	0	40	0	32	0	48

It may be here observed, that there seems to be more of Caprice, than Reason, in that Diversity found in the Heights of *Columns* of the same Order in different Authors; each of which frequently takes the Liberty of dispensing with his own Rules.

As for Instance, *Vitruvius* makes the *Doric Columns* of Temples shorter than those of Porches behind Theatres: *Palladio* gives a greater Height to *Columns* standing on Pedestals, than to those which have none: And *Serlio* makes his *Column* a Third shorter, when insulate or detached, than when contiguous to the Wall.

But notwithstanding the Diversity of Height in the *Columns* of the same Order, in different Authors, they still bear a true Proportion in the several Orders compared with each other; by which they go increasing, as the Orders are less massive.

But this Augmentation is greater in some Ordonnances than in others; for in the Antique, it is but five Modules, or Semidiameters, for the Five Orders; the shortest *Column*, *viz.* the *Tuscan*, being fifteen Modules; and the longest the *Composite*, twenty.

In *Vitruvius*, this Increase is also of five Modules, but commences from fourteen Modules, and ends at nineteen.

The Moderns usually make it greater: *Scamozzi* makes it five Modules and a half: *Palladio*, and *Serlio*, six.

From the several Proportions of *Columns* assign'd by several Authors. M. *Perrault* has formed a new one, which is a Mean betwixt the Extreams of the rest.

Thus he makes the *Tuscan Column* fourteen Modules and two Thirds; which is a Mean between the *Tuscan* of *Vitruvius*, fourteen; and that of *Trajan's-Column*, eighteen.

The Height of the *Doric Column* he makes sixteen Modules, which is a Mean between the fourteen of *Vitruvius*, and the nineteen of the *Coliseum.*

He makes the *Ionic* in Height seventeen Modules one Third; which is a Mean between the sixteen of *Serlio*, and the nineteen of the *Coliseum.*

He makes the Height of the *Corinthian Column* eighteen Modules two Thirds, as being a Medium between the sixteen Modules six Minutes of the Temple of the *Sybil*, and the twenty Modules six Minutes of the three *Columns* of the *Roman Forum.*

Lastly, he, by the same Rule, makes the *Composite Column* twenty Modules, that Height being a Mean between *Titus's-Arch*, and the Temple of *Bacchus.*

Indeed, the Rule he goes by, seems very rational, *viz.* the progressional Advance of each *Column* in the different Orders, be equal; so that having settled the whole Progression from the *Tuscan* to the *Composite*, at five Modules ten Minutes: This being a Mean between the Modules of the Antique and the five one half of the Moderns; he divides this Sum, which is a hundred and sixty Minutes, into four equal Parts, giving forty Minutes to the Progression of each Order: This makes the *Tuscan Column* fourteen Modules twenty Minutes. The *Doric* becomes sixteen; the *Ionic* seventeen, ten Minutes; the *Corinthian*, eighteen, twenty Min. and the *Composite*, twenty Mod.

The Method of Drawing a Column.

Example.

To draw the Base of a Column, *by which the Nature of describing any other Part may be known, the Rule being the same.*

Let the Base A B C D E F G K be given to be described in Profile.

1. Let L N be the Diameter of the *Column* given, divided into sixty Minutes as aforesaid.

2. Draw a Right Line at Pleasure, as *wk*; and in the Midst of that, raise the Perpendicular K L, which let let represent the Central Line of the *Column.*

3. Take from your Module ten Minutes, and set it on the Central Line from K to G; also take seven Minutes and a half, and set it from G to F; also take one Minute, and set it from F to E; and so in like manner, the Distance E D five half Minutes; the Distance D C one Minute; the Distance G B five Minutes; and the Distance B A one Minute.

4. Through all these several Points, draw Right Lines parallel to the Base *wk*.

5. Make K*k*, K*w*, G*i*, and G*v* each = 40 Minutes, and join *vw* and *ik*, and so will the Plinth be compleated.

6. Make F*r*, F*g*, E*q*, and E*f*, each = 35 Minutes, and join *qr* and *fg*, so will the lower be compleated; and if from the said

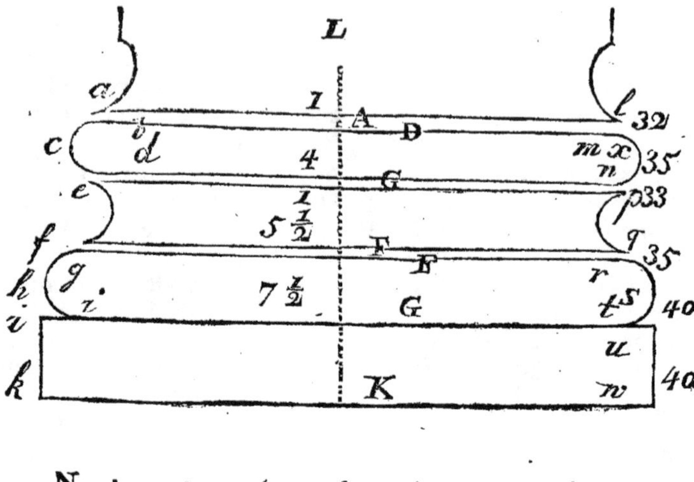

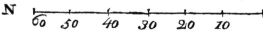

ſaid Fillet, you deſcribe Semicircles *r s t*, and *g h i*, they will compleat the lower Torus.

Laſtly, Proceed to ſet off the other Fillets *n d p e*, and *l m a b*, with the upper Torus *m b n d*, and Scotia *p e g f*, the whole Baſe will be compleated, as required.

Thus may the Baſe of any Order be eaſily delineated by the Proportions fixed thereto; and in the ſame Manner, you may likewiſe compleat an entire Order.

Note, That the Diameter of the *Column* at the Baſe, being divided into ſixty equal Parts, is called a Module; and the equal Parts are called Minutes; by which, the Heights and Projectures of every Member are ſet off from the Central Line of the *Column*.

There are ſome Occaſions, wherein an Architect cannot give his Building a ſufficient Projecture, patricularly where the Entablement would hinder the Sight of the Windows above, or intercept the Sight of the Apartments below.

In theſe Caſes, the *Columns* are to have one third Part of their Diameter inſerted or let into the Wall behind them.

But Recourſe ſhould never be had to this Shift, excepting in Caſes of Neceſſity; for the *Columns* here, loſe an infinite deal of that Beauty and Grace which they have when they ſtand alone.

It frequently happens too, that the *Columns* are let within the Wall, for the greater Solidity, and the further Strengthening of the Building.

This, however, ought to be obſerv'd, that they never loſe above one Third of their Diameter; the Reaſon of which will appear in the Article IMPOSTS.

When

When the *Columns* ſtand alone, they have uſually a Pilaſter placed behind them, joined to the Wall, or the Pillar of the Porticoe.

Sometimes, inſtead of a Pilaſter, we have a *Column* let within the Wall, in order to make the Symmetry more compleat.

Though we allow of a *Column* let within the Pillar of the Porticoe, yet we can never approve of the letting a *Column* within a Pilaſter.

Of Diminiſhing.

Columns of every Order muſt be ſo form'd, that the upper Part of the Body be leſs than the lower.

Which Diminiſhing muſt be more or leſs, according to the Proportion of their Heights; and is to be begun from one third Part of the whole Shaft upwards, (*i.e.* the lower third Part is to be of an equal Bigneſs) which *Philander* preſcribes by his own preciſe meaſuring of antient *Columns*,) as the moſt graceful Diminution. And as to the Quantity to be diminiſh'd, Architects lay down the following Rule:

That the	*Tuſcan*	*Column* be	$\frac{1}{4}$	Part
	Doric		$\frac{1}{5}$	
	Ionic		$\frac{1}{6}$	
	Corinthian		$\frac{1}{7}$	
	Compoſite		$\frac{1}{8}$	

ſmaller at the Top juſt under the Capital, than below juſt above the Baſe, *i. e.* the Diameter of

The Top of the	*Tuſcan*	*Column* is	$\frac{3}{4}$
	Doric		$\frac{4}{5}$
	Ionic		$\frac{5}{6}$
	Corinthian		$\frac{6}{7}$
	Compoſite.		$\frac{7}{8}$

of the Diameter of the *Column* below.

COLOURS, in Painting, is uſed both as to the Drugs themſelves, and to thoſe Teints produced by thoſe Drugs variouſly mix'd and apply'd.

The principal Colours uſed by Painters, are Red and White Lead, or Ceruſe, Yellow and Red Ocres: Several Kinds of Earth, as Umber, Orpiment, Black Lead, Cinnabar, or Vermilion, Gumbouge, Blue and Green Aſhes, Indigo, Biſtre, Lamp-Black, Smalt, Ultramarine, and Carmine.

Of theſe *Colours*, ſome are ground in Oil, others only in Freſco, and others in Water, and others for Miniature.

Painters reduce all *Colours* under theſe two Claſſes of Dark and Light *Colours*. Under Dark *Colours*, are comprehended Black, and all thoſe which are obſcure and earthy, as Umber, Biſter, &c. And under Light *Colours*, are

are comprehended White, and all those which approach nearest to it.

COMMENSURABLE QUANTITIES, in Geometry, are such as have some common Aliquot Part, or which may be measured by some common Measure, so as to leave no Remainder in either. Thus a Foot and a Yard are *Commensurables*, there being a third Quantity which will measure both, *viz.* an Inch, which taken twelve Times, makes a Foot, and thirty-six Times, a Yard.

Incommensurable is otherwise. The Ratio of *Commensurables* is rational; that of *Incommensurables* is irrational.

Commensurable Numbers, whether Integers, or Fractions, are such as have some other Number which will measure or divide them without any Remainder: Thus 6 and 8 $\frac{8}{12}$ $\frac{3}{4}$ are respectively by *Commensurable Numbers*.

COMMISSURE, in Architecture, &c. the Joint of two Stones; or the Application of the one to that of the other.

COMMON, in Geometry, is apply'd to an Angle, Line, or the like, which belongs equally to two Figures, or makes a necessary Part of both.

COMMON DIVISOR, in Arithmetick, is a Quantity or Number which exactly divides two or more other Quantities or Numbers, without leaving any Remainder.

COMPARTITION, in Architecture, signifies the useful and graceful Disposition of the whole Ground-Plot of an Edifice into Rooms of Office and Reception or Entertainment.

COMPARTIMENT, } a Design
COMPARTMENT, } composed of several different Figures, disposed with Symmetry, to adorn a Platfond, Parterre, Panes of Glass, or Pannels of Joinery, the Squares of a Ceiling, &c.

A *Compartiment* of *Tiles* is an Arrangement of white and red Tiles varnish'd for the Decoration of the Covering of a Roof.

COMPASSES, or } is
A Pair of COMPASSES } a Mathematical Instrument used for the describing of *Circles*, measuring the Distances of Points, Lines, &c.

The common *Compasses* consist of two Branches or Legs of Iron, Brass, or other Metal, pointed at Botrom, and at the Top joined by a River, whereon they move as on a Centre

Hair-Compasses are so contrived on the Inside, as to take an Extent to a Hair's Breadth.

Geman Compasses are those whose Legs are a little bent outwards towards the Top; so that when shut, only the Points meet.

Spring-Compasses, or *Dividers*, are made of harden'd Steel, the Head arch'd, which, by its Spring, opens the *Compasses*, the Opening being directed by a circular Screw, fastened to one Leg, and let through the other work'd with a Nut.

Turn-up-Compasses, a late Contrivance to save the Trouble of changing the Points.

Compasses of three Branches. Their Use is to take three Points at once, and so to form Triangles, &c.

Trisecting Compasses. Whose Use is for trisecting of Angles Geometrically.

Cylindriack Compasses.

Proportional Compasses.

Beam Compasses consist of a long Branch or Beam carrying two Brass Cursors; the one fixed at one End, the other sliding along the Beam, with a Screw to fasten it, on Occasion: To the Cursors may be skrew'd Points of any kind, whether Steel for Pencils, or the like. It is used to draw large Circles, take great Extents, &c.

Elliptical Compasses. Their Use is to draw *Ellipses* or Ovals of any kind.

COMPLEMENT, in Geometry, is what remains of a Quadrant of a Circle, or of ninety Degrees, after a certain Arch has been retrenched from it. Thus if an Arch or Angle be twenty-five Degrees, they say, its Complement is sixty-five, since 65 and 25, = to 90.

COMPOSITE ORDER, in Architecture, is the last of the five Orders of Columns, so called, because its Capital is composed out of those of the other Columns.

It borrows a Quarter-Round from the *Tuscan* and *Doric*; a Row of Leaves from the *Corinthian*; and Volutes from the *Ionic*. Its Cornice has simple Modillions or Dentils.

The *Composite* is also called the *Roman*, or *Italick Order*, as having been invented by the *Romans*, conformably to the rest which are denominated from the People, among whom they had their Rise.

This by most Authors is rank-after the *Corinthian*, either as being the richest, or the last that was invented. *Scamozzi* alone places it between the *Ionic* and *Corinthian*, out of Regard to its Delicacy and Richness, which he esteems inferior to that of the *Corinthian*, and therefore makes no Scruple to place it under it. In which he is follow'd by M. *Le Clerc*.

The Proportions of this Order are not fixed by *Vitruvius*; he only marks the general Character of it, by observing that its Capital is composed of several Parts taken from the *Doric*, *Ionic*, and *Corinthian*.

He does not seem to look upon it as a particular Order, nor does he vary it at all from the *Corinthian*, except in its Capital.

In Fact, it was *Serlio*, who first added the *Composite* Order to the Four of *Vitruvius*, forming it from the Remains of the Temple of *Bacchus*, the Arches of *Titus*, *Septimius*, and the *Goldsmiths*: Till that Time, this Order was esteem'd a Species of the *Corinthian*, only differing in its Capital.

This Order being thus left undetermin'd by the Antients, the Moderns have a kind of a Right to differ about its Proportions, &c.

Scamozzi, and after him, M. *Le Clerc*, makes its Column nineteen Modules and a half; which is less by half a Module, than that of the *Corinthian*, as in Effect,

fect, the Order is less delicate than the *Corinthian.*

Vignola makes it twenty, which is the same with that of his *Corinthian*. But *Serlio*, who first formed it into an Order, by giv- it a proper Entablature and Base, and, after him, M. *Perrault*, raise it still higher than the *Corinthian*.

This last does not think different Ornaments and Characters sufficient to constitute a different Order, unless it have a different Height too. Therefore conformably to his Rule of augmenting the Heights of the several Columns by a Series of two Modules in each, he makes the *Composite* twenty Modules, and the *Corinthian* eighteen; which, it seems, is a Medium between the Porch of *Titus* and the Temple of *Bacchus*.

M. *Perrault*, in his *Vitruvius*, makes a Distinction between the *Composite*, and *Composed* Order.

The *latter*, he says, is any Composition whose Parts and Ornaments are extraordinary and unusual; but have withal somewhat of Beauty, both on Account of their Novelty, and in respect to the Manner or Genius of the Architect: So that a Composed Order is an arbitrary humorous Composition, whether regular or irregular.

He likewise adds, That the *Corinthian* Order is the first *Composite* Order, as being compos'd of the *Doric* and *Ionic*, as *Vitruvius* himself observes.

The Composite Order, *by Equal Parts.*

The Height of the Pedestal being three Diameters and one Third, is divided into four, giving one to the Base, whose Plinth is two Thirds thereof: The other Part is divided into ten; three for the Torus, one for the Astragal, one half Part for the Fillet, three and a half for the Cymase, one and a half for the Astragal, one half Part for the Fillet which finishes to the Naked with a Hollow; the Breadth of the Naked is a Diameter and two Fifths.

The Projection is equal to its Height; the upper Astragal hath three of these Parts, and the lower Astragal eight.

The Height of the Cornice is half the Base, being one Eighth of the whole Height, and is divided into twelve Parts, giving a half Part to the Fillet, one and a half to the Astragal, three and a half to the Cymase, a half Part to the Fillet, three to the Corona, two to the Ogee, and one to the Fillet.

For the Projections, the Astragal hath two and a half of these Parts, the Fillet six, the Corona seven, and the Whole, nine.

The Base of the Column.

Make use of the same Base as in the *Corinthian* Order; though the Attic Base in the *Doric*, may very

very properly be made uſe of, for this, and alſo for the *Ionic*, and *Corinthian* Orders, eſpecially in Work expoſed to the Weather: Therefore inſtead of placing it here again, is ſhewn the Plane of either oftheCapitals as mentioned in the *Corinthian* Order.

The Diminiſhing of this Columu is the ſame as the laſt.

The Height of the Capital is a Diameter and one Sixth; and being divided, give two to each Height of Leaves whoſe Heads turn down a half Part, two Thirds of a Part to the Space between the Leaves and Fillet, one Third to the Aſtragal and Fillet, which is one Third of that, two Thirds to the Ovolo, one Third to the Space between the Ovolo and Abacus, a half Part to the Hollow, and a half Part to the Ovolo, whoſe Fillet hath one Third thereof.

The Projection is the ſame as the *Corinthian*.

The Height of the Entablature being two Diameters, is divided into ſix; giving two to the Architrave, one and a half to the Frieze, and two and a half to the Cornice.

The Height of the Architrave is divided into nine, giving two and a half to the firſt Face, a half Part to the Ogee, three and a half to the ſecond, one Fourth to the Aſtragal, three Fourths to the Ovolo, one to the Hollow, and a half Part to the Fillet

The Projection of the ſecond Face hath half a Part, the Ovolo one and one Fourth, and the Whole, two.

The Frieze is form'd after the ſame Method as the *Ionic*, being Part of a Circle that anſwers to the Naked, and Projection of the Architrave.

The Cornice is divided into ten Parts, giving one Fourth of a Part to the Fillet, as much to the Aſtragal; one Part to the Ogee, another to the firſt Face of the Modillions, a half Part to the Ogee, one and one Fourth to the ſecond Face, one Fourth to the Fillet, a half Part to the Ovolo, two to the Corona, three Fourths to the Cima Reverſa, one Fourth to the Fillet, one and a half to the Cima Recta, and a half Part to the Fillet.

For the Projections, the Ogee hath one and half of theſe Parts, the upper Face of the Modillions in Front two, but is two and half in Breadth, *viz.* a half Part within the Naked, the Cap thereof a half Part, the Return'd Modillion four and a half, and its Cap 5, the Corona ſeven and a half, the Cima Reverſa eight and a half, and the Whole ten being equal to the Height.

COM-

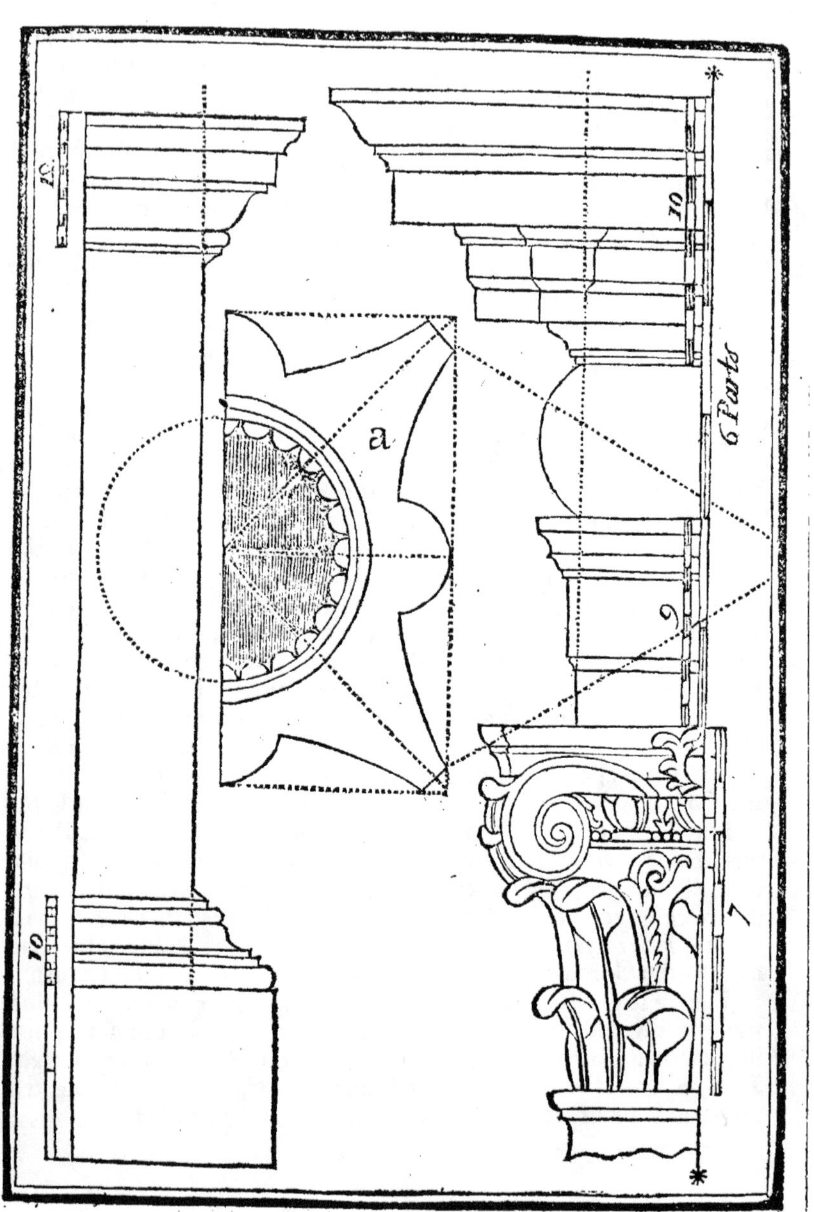
a
6 Parts
10
10
9
7

COMPOSITION, in Painting, includes the Invention and Disposition of the Figures, the Choice of Attitudes, *&c.*

Therefore *Composition* consists of two Parts; of which the one finds out by the Means of History, proper Objects for a Picture; and the other disposes them to Advantage.

Composition of Proportion signifies the comparing of the Sums of the Antecedent and Consequent; with the Consequent in two equal Ratio's: As suppose 4 : 8 : 3 : 6, by *Composition* of Proportion, we say 12 is to 8, as 9 is to 6.

Composition of Motion, in Mechanicks, is an Assemblage of several Directions of Motion, resulting from Powers acting in different, though not opposite Lines.

If a Point more or less, flow according to one and the same Direction, whether that be equable or not, yet it will still keep the same Right Line; the Celerity alone being changed, *i. e.* increased or diminish'd, according to the Forces with which it is impelled.

If the Directions be opposite, as one *e.gr.* directly downwards, the other upward, *&c.* yet still the Line of Motion will be the same.

But if the compounding Motion be not according to the same Line of Direction, the compound Motion will not be according to the Line of Direction of either of them, but a different one from them both; and this, either straight, or crooked, according as the Directions or Celerities shall require.

If two compounding Motions be each of them equable, the Line of the compound Motion will still be a straight Line: And this, though the Motions be neither at Right Angles one to another, nor equally swift, nor each to itself equable, provided they be but similar, that is, *both accelerated and retarded alike.*

COMPRESSION, the Act of pressing or squeezing something together, so as to set its Parts nearer to each other, and make it possess less Space.

To CONCAMERATE. To make an arched Roof, as in Vaults, *&c.* To arch over.

CONCAVE, is said to be the inner Surface of a hollow Body, especially if it be circular.

Concave is particularly understood of Mirrors and Lenses.

Concave Lenses are either *concave* on both Sides, called *Concavo-concave*, or *concave* on one Side, and plain on the other, called *Plano-concave*, or *concave* on the one Side, and *convex* on the other, called *concavo-convex*, or *convex-concave*, as the one or the other Surface is a Portion of the less Sphere.

The Properties of all *concave* Lenses are, that the Rays of Light, in passing through them, are deflected, or made to recede from one another; as in convex Glasses, they are inflected towards each other, and that the more, as the *Concavity*, or Convexity, are Portions of less Circles.

Hence parallel Rays, as those of the Sun, which by passing thro' a *concave* Lens, become diverging: Diverging Rays are made to diverge

diverge the more, and converging Rays, either made to converge less, or become parallel, or go out diverging.

Hence Objects view'd through *concave* Lenses, appear diminished; and the more so, as they are Portions of less Spheres: And this, in oblique, as well as in direct Rays.

Concave Mirrors have the contrary Effect to Lenses: They reflect the Rays which fall on them, so as to make them approach more to, or recede less from each other, than before; and that the more, as the *Concavity* is greater, or the Spheres, of which they are Segments, less.

Hence *concave* Mirrors magnify Objects presented to them; and that, in a greater Proportion of greater Spheres.

CONCENTRICK, that which has the same common Centre with another. The Word is principally used in speaking of round Bodies, and Figures, *viz.* Circular and Elliptical ones, *&c.* But may be used likewise in Polygons, drawn parallel to each other upon the same Centre.

CONCLAVE, in Architecture, is a Closet or inner Chamber.

CONDENSER, in Pneumaticks, is a Machine or Engine whereby an unusual Quantity of Air may be crowded into a given Space.

They can throw in three, four, five, or ten Atmospheres into the *Condenser*, *i. e.* three, four, *&c.* Times as much Air as there is in the same Compass without the Engine.

CONDUCTS, } Suits or
CONDUITS, } Gutters, to convey away the Suillage of an House. Also Canals or Pipes for the Conveyance of Water or other fluid Matter.

Sir *Henry Wotton* says, that in the first Place, Art should imitate Nature, in separating those ignoble Conveyances from the Sight; and (where a Running Water is wanting,) they should be placed in the most remote, and lowest Part of the Foundation, with secret Vents passing up through the Walls (like a Tunnel) to the wide Air: Which all *Italian* Artists commend for the Discharge of noisome Vapours.

A CONE, is a Solid, having a circular Base, and growing smaller and smaller, 'till it ends in a Point, which is called the Vertex, and may be nearly represented by a Sugar-Loaf.

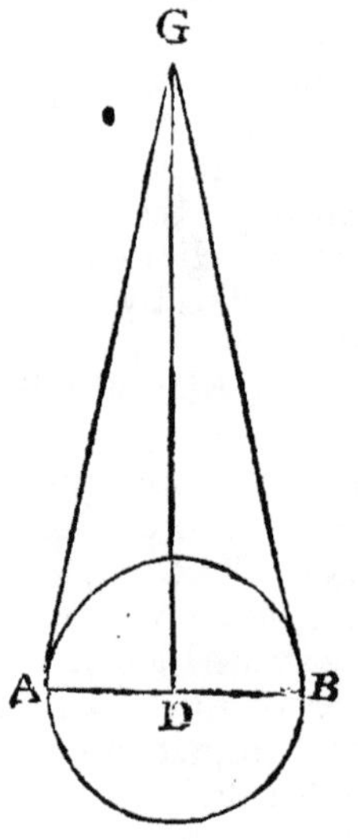

The

The RULE *to find out its Solidity.*

Multiply the Area of the Base by a third Part of the perpendicular Height, and the Product is the solid Content.

Let ABC be the *Cone*, the Diameter of whose Base AB is 26.5 Inches; and the Height of the *Cone* DC is 15.5 Feet.

First, Square the Diameter 26.5, and it will make 702 25, which multiply by 7854, and the Product will be 551.54715; which being multiply'd by 5.5. the Product will be 3033.407825; which being divided by 144, the Quotient will be .2107 *fere*, the solid Content of the *Cone*.

The Operation.

```
                                  26.5 the Diameter
                                  26.5
                                 -----
                                  1325
144)3033:503(21.07 Feet          1590
    ---                          530
     353 Content               -------
     947                        702.25 Square
                                 .7854
                               -------
                                280900
                               351125
                              561800
                             491575
                             -----------
                             551.54|715 Area of Base
                                 55 1/3 Pt. of the Height
                             ---------
                             275773
                            275773
                            --------
                            3033503
```

By Scale and Compasses.

Extend the Compasses from 13.54 to 26.5. (the Diameter) that extend, turn'd twice over from 5.5 (a third Part of the Height) will at last fall upon 21.07 Feet the Content.

To find the Superficial Content.

Multiply half the Circumference 41.626 by the slant Height AC 198 46, and the Product will be 8261.09596; which being divided by 144, the Quotient is 57.37 *fere* the Curve Surface; to which add the Base, and the Sum will be 61.2 Base, the Superficial Content.

```
 41.626
 198.46
--------
 249756
166504
333008
374634
41626
----------
8261.09596
```

```
144) 8261.09596 (57.37 Feet fere
     ------          3 83
     1061          -------
      530          61.20 Whole
       989               (Cont.
     -----
144) 551.54 (3.83
     1195
      434
     ----
       2
```

By Scale and Compasses.

Extend the Compasses from 144 to 198 46; and that Extent will reach from 41 626 to 57.37 Feet, the Curve Surface.

Then, extend the Compasses from 12 to 26.5 (the Diameter) and that Extent turn'd twice over from 7854 will at last fall upon 3.83 Feet, the Base; which being added to 57.37, the Sum will be 61.2 Feet, the Superficial Content.

DEMONSTRATION.

Every *Cone* is the third Part of a Cylinder of equal Base and Altitude.

The Truth of this may easily be conceived, by only considering that a *Cone* is but a round Pyramid; and therefore it must needs have the same Ratio to its circumscribing Cylinder, as the square Pyramid hath to its circumscribing Parallelopipedon, *viz.* as 1 to 3. However, to make it the clearer, let it be consider'd, That

Every right *Cone* is constituted of an infinite Series of Circles, whose Diameters do continually increase, in Arithmetical Progression, beginning at the Vertex or Point C, the Area of its Base AB being the greatest Term; and its perpendicular Height DC, the Number of all Terms; therefore the Area of the Circle of the Base multiply'd by a third Part of the Altitude DC, will be the Sum of all the Series, equal to the Solidity of the *Cone*.

The Curve Superficies of every Right *Cone* is equal to half the Rectangle of the Circumference of its Base into the Length of its Side.

For the Curve Surface of every Right *Cone* is equal to the Sector of a Circle, whose Arch BC

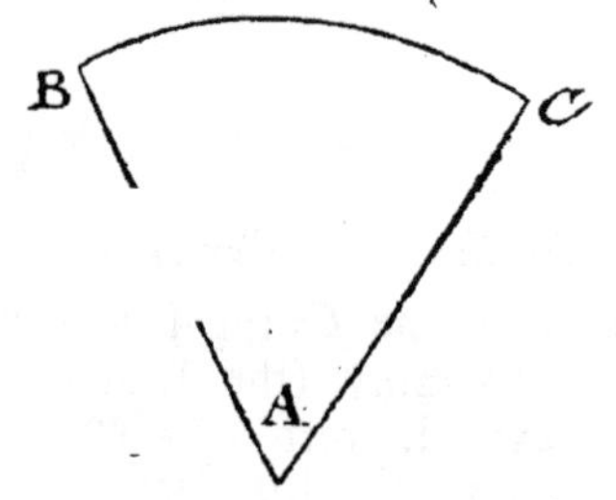

is equal to the Periphery of the Base of the *Cone*, and the Radius AB equal to the slant Side of the *Cone*.

Which

Which will appear very evident, if you cut a Piece of Paper in the Form of a Sector of a Circle ABC, and bend both the Sides AB and AC together till they meet, and you will find it to form a Right *Cone*.

CONGE [in Architecture, a Moulding, either in Form of a Quarter-Round, or a Cavetto, which serves to separate two Members from one another.

Such is that which joins the Shaft of the Column to the Cincture, called also *Apophyge*, which in *Greek*, signifies Flight, the Column seeming to arise. Hence, by the *Latins*, it is called *Scapus*, the Shaft of the Column.

CONGES, are Rings or Ferrils, heretofore used in the Extremities of Wooden Pillars, to keep them from splitting, afterwards imitated in Stone-Work.

CONICK SECTION, a curve Line arising from the Section of a Cone.

CONICKS, that Part of the Higher Geometry, or Geometry of Curves, which considers the *Cone*, and the several Curve Lines arising from the Sections.

CONOIDES, } CONOID, } in Geometry, a solid Body resembling a Cone, except in this, that instead of a perfect Circle for its Base it has an Ellipsis, or some other Curve approaching thereto.

CONSECTARY is a Proposition that follows, or is deduced from some preceding Definitions, Lemmata, Axioms, or the like; whence some chuse rather to call it a Consequence, and others a Corollary.

CONSOLE [of *consolider*, *Fr.* to re-unite, join, *&c.*] is an Ornament cut upon the Key of an Arch, which has a Projecture or Jetting, and on Occasion serves to support little Cornices, Figures, Busts, and Vases.

These are also upon Occasion called Mutules, Modillions, *&c.* according to their Form: Some of them are striated, or fluted; others in Form of Cartouches; others have Drops, in the Manner of Triglyphs.

Those made at the End of a Plank of Wood, cut Triangularwise, are called Ancones.

Consoles are frequently used as Keys of Arches, projecting out to support a Vase, or other Ornament.

M. *Le Clerc* is of Opinion, that a *Console* should always have something exceedingly massive, to sustain and serve it as a Rest.

CONSPIRING POWERS, in Mechanicks, are all such as act in Directions not opposite to one another.

CONTACT, the relative State of two Things that touch each other, or whose Surfaces join to each other without any Interstice.

Hence because very few Surfaces are capable of touching in all Points, and the Cohesion of Bodies is in Proportion to their *Contacts*, those Bodies will stick fastest together, which are capable of the most *Contact*.

CONTENT, the Capacity or Area of a Space, or the Quality of any Matter or Space included in certain Bounds.

The *Content* of a Ton of round Timber is forty-three solid Feet. A Load of hewn Timber contains fifty Cubick Feet. In a Foot of Timber are contained seventeen hundred and twenty-eight cubick or square Inches; and as often as seventeen hundred and twenty-eight Inches are contained in a Piece of Timber, be it round or square, so many Foot of Timber are contained in the Piece.

CONTIGNATION, in the antient Architecture, the Art of laying Rafters (in *Latin*, *Tigna*) together, and particularly Flooring.

CONTIGUOUS, signifies that two or more Things are dispos'd so near each other, as they join their Surfaces, or touch.

Contiguous Angles, in Geometry, such as have one Leg common to each Angle, otherwise called Adjoining Angles, in Contradiction to those produced by continuing their Legs through the Point of Contact; which are called Opposite or Vertical Angles.

CONTOUR, the Out-Line, or that which terminates or defines a Figure. In Architecture, it is the Out-Line of any Member, as that of a Base, a Cornice, or the like.

CONTRACTILE *Force*, in Mechanicks, is that Power or Property inherent in certain Bodies, whereby, when extended, they are enabled to draw themselves up again to their former Dimensions.

CONTRAMURE, in Architecture, an Out-Wall built about the Wall of a City.

TO CONTRAST, in Architecture, is to avoid the Repetition of the same Thing, in order to please by Variety; as is done in the Great Gallery in the *Louvre*, where the Pediments are alternately arched and angular.

CONTRAST, in Painting, and Sculpture, signifies an Opposition or Difference of Position, Attitude, &c. of two or more Figures, contriv'd to make Variety in Painting: Thus in a Group of three Figures, when one is shewn before, another behind, and a third sideways; there is said to be a *Contrast*.

The *Contrast* is not only to be observ'd in the Position of several Figures, but also in that of the several Members of the same Figure: Thus if the Right Arm advance the farthest, the Right Leg is to be hindmost: If the Eye be directed one Way, the Right Arm is to go the contrary Way.

Nay, the *Contrast* is to be persu'd even in the Drapery.

CONVERGING *Lines*, } CONVERGENT *Lines*, } in Geometry, are those which approximate, or whose Distance becomes continually less and less. These are oppos'd to divergent Lines, whose Distance becomes continually greater. Lines which converge one Way diverge another.

Converging Rays, in Dioptricks, are those Rays, which in their Passage out of one Medium into another, of a different Density, are refracted towards one another, so as if far enough continued, they will meet in a Point or Focus.

Thus

Thus all Convex Lenſes make the Rays *converge*, and Concave ones make them diverge, *i. e.* the one inflects them towards a Centre, and the other deflects them from it; and the more, as ſuch Lenſes are Portions of ſmaller Spheres.

On which Properties, all the Effects of Lenſes, Microſcopes, Teleſcopes, &c. depend.

Rays coming *converging* out of a denſer Medium into a rarer, *v. g.* from a Glaſs, into Air, become more *convergent*, and concur ſooner, than if they were to continue their Motion through the firſt.

Rays coming *converging* out of a rarer into a denſer Medium, *converge* leſs, and concur later, than if they had continu'd their Motion through the firſt.

Parallel Rays paſſing from a denſer, into a rater Medium, *v. g.* from Glaſs into Air, will become *convergent*, and concur in a Focus.

Converging Series, in the Mathematicks, is a Method of Approximation, or coming ſtill nearer and nearer towards the true Root of any Number or Equation; even though it be impoſſible to find any ſuch true Roots in Numbers.

CONVERSE, in Geometry, &c. A Propoſition is ſaid to be the *Converſe* of another, when, after drawing a Concluſion from ſomething firſt ſuppoſed, we proceed to ſuppoſe what had been before concluded, and to draw from it what had been ſuppoſed. As for Example;

Thus 'tis demonſtrated in Geometry, that if the two Sides of a Triangle be equal, the two Angles oppoſite to theſe Sides are equal alſo: The *Converſe* of the Propoſition is, that if the two Angles of a Triangle be equal, the two Sides oppoſite to theſe Angles are equal alſo.

CONVEXITY is the exterior Surface of a *Convex*, *i. e.* a gibbous and globular Thing, in Oppoſition to Concavity or the inner Surface, which is hollow or depreſs'd.

A *Convex Lens* is either *convex* on both Sides, called a *Convexo-Convex*, or it is plain on one Side, and *convex* on the other, called a *Convexo-Concave*, or *Concavo-Convex*, as the one or the other Surface prevails, *i. e.* as this or that is a ſmaller Portion of a Sphere.

All *Convex Lenſes* inflect the Rays of Light in their Paſſage, *i. e.* ſend them out from their *convex* Surface, converging ſo as that they concur in a Point or Focus.

Hence all *Convex Lenſes* magnify, *i. e.* repreſent their Images larger than their Objects; and this the more, as they are Portions of ſmaller Spheres.

A *Convex Mirror* repreſents the Images ſmaller than the Objects, as a Concave one repreſents them larger: A *Convex Mirror* reflects the Rays from it, diverging; and therefore diſperſes and weakens their Effect, as a *Convex* one reflects them converging, ſo as to concur in a Point, and have their Effect increas'd; and by how much the Mirror is the Portion of a ſmaller Sphere, by ſo much does it diminiſh the Objects, and diſperſe the Rays the more.

COPING *of a Wall*, the Top or Cover of a Wall, made floping, to carry off the Wet.

Coping over, in Carpentry, a fort of Hanging over, not fquare to its Upright, but bevelling on its Underfide till it end in an Edge,

The Price: Brick-Walls (of a Brick and half thick) have been cop'd with Stone for 4*d.* a Foot, lineal or running Meafure, the Workman drawing the Stones, into this Price.

Drawing of Stones for *Coping:* A Penny a Foot has been given for drawing them, for this Ufe.

CORBEILS [of the *Latin Corbis* a Basket] is a Piece of Carved Work in the Form of a Basket, full of Flowers or Fruits, ferving in Architecture to finifh fome Ornament.

CORBELS, in Architecture, the Reprefentation of a Basket fometimes feen on the Heads of the Caryatides.

Sometimes *Corbel* is ufed to fignify the Vafe or Tambour of the *Corinthian* Column, fo called on Account of the Refemblance it bears to a Basket; or becaufe it was firft form'd on the Model of a Basket.

CORBIL, } is alfo ufed in CORBEL, } Building, for a fhort Piece of Timber, placed in a Wall with its End fticking out fix or eight Inches, as Occafion ferves, in the Manner of a Shouldering-Piece.

The under Part of the End thus fticking out, is fometimes cut in the Form of a Boultin, fometimes of an Ogee, and fometimes of a Face, &c. according as the Workman's Fancy is, the upper Side being plain and flat.

Thefe *Corbels* are ufually placed for Strength, immediately under the Semi-Girders of a Platform, and fometimes under the Ends of Camber-Beams: In which latter Cafe, they are ufually placed a Foot or two below the Beam, and have a Piece of Timber ftanding upright clofe to the Wall from the *Corbel* to the Beam.

Corbel is alfo ufed by fome Architects, for the hollow Niches or Hollows left in the Walls for Images, Figures, or Statues to ftand in.

CORINTHIAN ORDER, the fourth, or, as *Scamozzi* and M. *Le Clerc* make it, the fifth and laft of the Orders of Architecture; being the nobleft, richeft, and moft delicate of all others.

The Invention of this Order is afcrib'd to *Callimachus*, an *Athenian* Sculptor, by moft of the Moderns, after *Vitruvius*, who paffing by the Tomb of a certain young Lady, over which her Nurfe had placed a Basket with fome of her Play-Things, and cover'd it up from the Wet with a Tile, the Whole having been placed on a Root of Acanthus, as it fprung up, the Branches encompafs'd the Basket, and bending down at the Top under the Corners of the Tiles, form'd a Sort of Volutes.

Hence *Callimachus* took his Hint: The Basket he imitated in the Bafe of his Column, the Leaves in the Volutes, and the Tile in the Abacus of this Order.

The *Corinthian Order* has feveral Characters by which it is diftinguifh'd from the reft. Its Capital is adorn'd with two

Rows

Rows of Leaves, between which arise little Stalks or Caulicoles, of which the Volutes are formed, which support the Abacus, and are in Number sixteen.

It has no Ovolo, nor even Abacus, properly speaking; for the Member which goes by that Name, is quite different from the Abacus of the other Orders, being cut with a Sweep, in the Middle of which is carved a Rose, or other Ornament.

Vitruvius observes that the *Corinthian* Order has no particular Ordonnance for its Cornice, or any of the other Ornaments of its Entablature; nor does he give it any other Proportions, than those of the *Ionic* Order: So that if it appears higher than the *Ionic*, 'tis purely owing to the Excess of the Height of its Capital.

He also makes the rest of the Entablature the same, and also uses the *Attic* Base indifferently for the one and the other.

But *Vitruvius* differs widely in this Order from all the Examples of Antiquity now remaining: The most beautiful of which have a particular Base, and the whole Order twenty Modules high; whereas the *Ionic* has but eighteen.

Again, its Capital is higher than that of *Vitruvius* by one Third of a Module; and its Entablature, which has Modillions, and sometimes Dentils together with the Modillions, is very different from the *Ionic* Entablature.

Most modern Architects pass by *Vitruvius*'s *Corinthian* Ordonnance, and follow that of the ancient Buildings; and select from them according to their several Tastes. So that the modern *Corinthian* is a kind of *Composite*, differing from many of the antient Buildings, and much more from *Vitruvius*.

Vignola and M. *Le Clerc* made the *Corinthian* Order twenty Modules in Height; yet *Serlio* makes it but eighteen; and M. *Perrault*, eighteen two Thirds, retrenching something from the nineteen of *Vitruvius*.

M. *Perrault* makes the Height of the Shaft less than that of the *Ionic*, by reason of the Excess of its Capital.

The Corinthian Column *by equal Parts.*

Corinthian Pedestal, being in Height three Diameters, and is divided into four, allowing one to the Base, whose Plinth is two Thirds of it: The other Part is divided into nine, allowing two and a half to the Torus, a half Part to the Fillet, three to the Cymase, a half Part to the Fillet, and two and a half to the Ogee; and the Breadth of the Die is a Diameter, and two Thirds.

The *Projection of its Base* is equal to its Height, the upper Fillet has three of these Parts, and the lower Fillet seven. The Height of the Cornice is half the Base being one Eighth of the whole Height; and is divided into eleven, by allowing one and a half to the Ogee, a half Part to the Fillet, three to the Cymase, three to the Corona, two to the Ogee, and one to Fillet. The Projection of the Fillet has two of these Parts, the Cymase four and a half, the

Corona six and a half, and the Whole eight and a half.

The Height of the *Base of the Column* is half a Diameter, which is divided into six, allowing three Fourths to the Plinth, one to the lower Torus, one Fourth to the Fillet, a half Part to the Scotia, one to the Astragals and Fillets (which are to be parted into six, allowing one to each Fillet, and two to each Astragal,) a half Part to the Scotia, one Fourth to the Fillet, and the other three Fourths to the Torus; the Fillet above the said Torus, is equal to the others, and Part of the Column.

The Projection is one Fifth of the Diameter; and the upper Fillet has one of the said six Parts, the upper Torus, and the lesser Fillets have one and a half, and one three Fourths are allowed to the Astragals, and lower Fillet.

Diminishing of this Column is one Eighth of the Diameter.

The *Corinthian Capital.* Divide the Diameter into six Parts, and take seven such Parts for the Height, allowing two to each Height of the Leaves (whose Heads turn down half a Part of it) allow another Part of the Stalks whose Heads turn down one Third of it; three Fourths to the small Volutes, and one Fourth to the Fillet; the large Volute is as high as the said Fillet; a half Part to the Hollow, and a half Part to the Ovolo, whose Fillet has one Third of it.

For the *Projection of the Capital,* make a Square, each Side being a Diameter and half, and draw the Diagonals (See the COMPOSITE *Order,*) and towards each Angle, mark a Diameter from the Centre, and draw the Cants at Right Angles with the said Diagonals: Then, for the Curvature of the Abacus, make an equilateral Triangle, (the Part of the Square cut off by the Cants being the Base,) and the opposite Angle the Centre. In the Circumference of the Column are eight Leaves, each Leaf having four Plants, and each Plant five Ruffles. The Projection of their Heads is found by a straight Line from the Abacus to the Colarino. The Rose is high as the Volute, and projects to the Side of the before-mentioned Square.

The *Architrave* is divided into nine Parts, allowing one and a half to the first Face, one and one Fourth to the small Bead, two to the second Face, three Fourths to the small Ogee, two and half to the third Face, a half Part to the Bead, one to the Ogee, and a half Part to the Fillet.

The *Projection* of the second Face has one Fourth of a Part, the third Face, one of those Parts, and the Whole, two.

The Height of the *Entablature* is two Diameters, and is divided into six Parts, two of which go to the Architrave, one and a half to the Frize, and two and a half to the Cornice.

The *Cornice* is divided into twelve Parts, allowing one and one Fourth to the Ogee, one

Fourth

Fourth to the Fillet, one and one Fourth to the Dentils, one Fourth to the Fillet, one Fourth to the Ovolo, one Fourth to the Fillet, two to the Modillions, a half Part to the Ogee, and one Fourth to the Fillet; one and three Fourths to the Corona, three Fourths to the Cima Reversa, one Fourth to the Fillet, one and a half to the Cima Recta, and a half Part to the Fillet.

As for the *Projections* of the *Cornice:* The Ogee is one half of these Parts, and the Dentils two and a half. The Dentils are in Breadth two Thirds of their Height, and the Spaces two Thirds of their Breadth.

The *Modillions* project three and three Fourths; and its Breadth is one Fifth of the Diameter; and one being in the Centre, gives the Spaces. The return'd Modillions eight and a half, the Cap nine, the Corona nine and a half, the Cima Reversa ten and a half, and the Whole twelve, being equal to the Height. See the Figure.

6 Parts
a
12
11
9
6
7

CORK, a Tree like the Holm ree, reſembling the ſame in its eaves, Catkins, and Fruit, the ark of it is light, ſpungy, and a grey Colour, inclining to ellow.

There are indeed ſeveral Sorts f this Tree; but two more re- iarkable, *viz.* One of a nar- wer, leſs jagged Leaf, and pe- nnial; the other of a broader eaf, and falling in the Winter.

It grows in the coldeſt Part of *iſcay*, in the North of *England*, *Italy*, *Provence*, and South- veſt Parts of *France*, eſpecially he ſecond Species, which are he fitteſt for our climate.

It grows in all ſorts of round, dry Heath, ſtoney and ocky Mountains, ſo that its oots run above the Earth, where they have little to cover hem; and therefore we have no reaſon to deſpair of their grow- ing with us.

There were none of them in *France* in *Pliny*'s Time; but there are large Woods of them in *Italy*, from whence it is probable they were tranſplanted hither.

The Manner of decorticating or taking off the Bark of the *Cork-Tree*, is as follows: They once in two or three Years ſtrip t in a dry Seaſon, otherwiſe the ntercutaneous Branches endan- ger the Tree, and therefore a ainy Seaſon is very pernicious o them when the Bark is off; hey unwrap it before the Fire, and preſs it even, and that with Weights, on the convex Part, and ſo it continues, being cold.

The beſt Bark ſhould be light, even; of a middling Thickneſs, without Cracks, having a few Knobs, and eaſy to be cut.

There are Cups made of one ſort of *Cork*, good for hectical perſons to drink out of.

The *Egyptians* made Coffins of *Cork*, which being lined with a reſinous Compoſition, preſer- ved their Dead uncorrupted.

They ſometimes in *Spain* line their Stone Walls with it, which renders them very warm, and corrects the Moiſture of the Air.

Beneath the *Cork*, or Bark of this Tree, there are two other Coats; one of them reddiſh, which they ſtrip from the Bole when fell'd, and is valued by Turners; the reſt of the Wood is not only good Firing, but alſo applicable to many other Uſes in Building.

CORNER *Stones*, are two Stones commonly of *Rigate* or Fire-Stone; of which there ſtands one in each Jaumb of a Chim- ney. Their Faces are hollow in Breadth, being a certain Sweep of a Circle. The Breadth of each Stone is equal to the Breadth of the Jaumb; and their Height reaches from the Hearth to the Mantletree.

As to the Price: Theſe Stones have been bought in *London* for 20*s.* *per* Pair.

CORNICE } [The Word CORNICHE } is formed from the *Latin*, *Coronis* a Crown- ing] is the uppermoſt Member of the Entablature of a Column, or that which crowns the Or- der.

The *Cornice* is the third grand Diviſion of the Trabeation com- mencing with the Frieze, and ending with the Cymatium.

The *Cornice* is different in the different Orders, there being as many

many Kinds of *Cornices*, as there are different Orders of Columns.

It is most plain in the *Tuscan* Order.

Vignola makes it to consist of an Ovum or Quarter-Round, an Astragal or Baguette, the Reglet or Fillet, the Larmier, and the Talon.

In the *Ionic*, the Members are in most Respects the same as in the *Doric*; except that they are frequently inriched with Carving, and have always Dentils.

In the *Doric*, *Vignola* makes the Capitals of the Triglyphs of the Frieze, with their Bandelettes, a Talon, Mutules, or Dentils; a Larmier, with its Guttæ underneath, a Talon, Fillet, Cavetto, and Reglet.

The *Corinthian Cornice* is the richest, and is distinguished by having both Modilions and Dentils, contrary to the Opinion of *Vitruvius*, who looks upon these two Ornaments as incompatible; and of M. *Le Clerc*, who accounts the Dentils as peculiar to the *Ionic*.

In the *Composite* there are Dentils, its Mouldings carved, and there are Channels under the Soffit.

The *Tuscan*, according to *Vitruvius*, the whole Height of this *Cornice* is one Module and a half; which Height being divided into four grand Divisions, the uppermost of which goes to the Boultin and Fillet under it; and this Division being subdivided into four Parts, three of them go to the Boultin, and one to the Fillet.

The two next grand Divisions go to the Corona, or Crown, (which is flat and plain,) and the lowermost grand Division goes to the Cymatium; which being again divided into three Parts, the uppermost of them goes to the Fillet, and the other two to the Cyma or Ogee.

The Projecture of the whole *Cornice*, as also of each Member thereof, he makes to be equal to its Height; and the under Side of the Corona he divides into 11 Parts, of which he allots two to the Fillet, and one to the Denticle, and so alternately; for, says he, it is fitting to have three as deep as they are large.

Scamozzi makes the whole Height of this *Cornice* 39 Minutes, and the Height of each particular Member of it (beginning at the Top, and descending orderly) is as follows: The upper List or Plinth of the *Cornice*, three M. the Supercilium, List, Tinea, or Eye-Brow, one M. and a half; the upper Cima or Ogee eight M. the List under it one Minute and a half; the Corona or Crown 9 $\frac{3}{4}$ M. the List 1 M. $\frac{1}{2}$; the Cima or greater Ogee, six M. (here is one and a half M. left betwixt for the Depth of the Dentils;) the Supercilium or List one and a half M. the Cymatium or little Ogee five M. the List, two M.

Palladio makes the whole Height of this *Cornice* forty-four M. of which the List at the Top is three and a half. M. the *Scima Recta* ten M. the List under it two and a half M. the Corona ten M. the Boultin nine, the List one and a half, and the Cavetto or Hollow seven and a half M.

The *Doric Cornice* is made by *Vitruvius* after two different Fashions; the whole Height of the one

one is half a Module, which being divided into two grand Divisions, one of them (*viz.* the upper one) is subdivided into eight Parts, of which one Part goes to the Lift at Top, and the other seven to the Ogee.

The other grand Division is again divided into four Parts, the uppermost and lowermost of which Parts go to the two Cymatiums, and the two middle Parts go to the Corona.

The Lift of each of these Cymatiums, is one Third of the whole Cymatium.

The whole Height of the other fashioned *Cornice* is forty M. which being divided into nine Parts, two are to go to the two Fascia's, one to the Thorus or Boultin above them, two to the Modilions above that, two to the Crown, and two to the Cima or Ogee at the Top.

The Modilions, as also the Crown, being divided each into three Parts, one of them shall go to their respective Cymatiums, of which their Lifts are each one Third of the Whole.

Scamozzi makes the whole Height of this *Cornice* forty-two M. of which the Lift at the Top is two M. the great Ogee seven M. the Lift one M. the little Ogee three M. the Corona eight M. the Lift one M. the Casement two M. the Boultin five M. the Lift one M. the Square seven M. the Lift one M. and the Boultin four M.

Palladio, in his verbal Description of this *Cornice*, makes the whole Height of it to be thirty-five M. But in his Figure it is but thirty-three M. and a quarter, of which the Lift at Top is two M. and a quarter, the *Cima Recta*, or Ogee, six M. three Fourths; the Lift one M. the *Cima Reversa* three M. one Fourth; the Corona eight M. the Ovolo or Boultin six M. the Lift one M. and the Casement at the Bottom five M.

The *Ionic*: *Vitruvius* makes the whole Height of this *Cornice* about fifty-two M. and a half. He describes two *Cornices* of different Fashions in this Order; in one of which he divides the whole Height into eleven Parts; the two uppermost of which go to the Cymatium, and the Boultin under it: And this Space is again subdivided into six Parts; two of which go to the Fillet of the Cymatium, three to the Ogee, and one to the Boultin.

The next two grand Divisions go to the Corona; and the next three to the Cartouses, and the Cymatium over them: And this Space being subdivided into five Parts, one of them makes the Cymatium, of which the Fillet is one Third of the Whole; then one and a half of the next grand Division goes to the Boultin and Fillet over it, of which the Fillet is one seventh Part of the Whole. And again, one and a half of the next grand Division goes to the Casement and Fillet over it, of which the Fillet is one Fourth of the Whole.

And the last grand Division goes to the Cymatium, of which the Fillet is one third Part of the Whole.

In the *Cornice* of the other Fashion, he divides the whole Height into six Parts; the uppermost of which he allows to the Ogee,

Ogee, the Fillet of which is one ſixth Part; the next grand Diviſion being ſubdivided into three Parts, the uppermoſt of them goes to the Cymatium, (the Fillet of which is one third Part,) and the other two to the Corona.

The next two grand Diviſions are ſubdivided into five Parts; the uppermoſt of which goes to the Cymatium (the Fillet of which is one third Part,) and the other four to the Cartouſes.

The next grand Diviſion is ſubdivided into four Parts; three of which go to the Boultin, and one to the Fillet under it.

And the laſt grand Diviſions being ſubdivided into four Parts, three of them are for the Caſement, and one for the Cymatium, of which its Fillet is one third Part.

According to *Scamozzi*, the whole Height of this *Cornice* is forty-two M. whereof the Liſt at the Top is two M. the Cima Recta five and a half M. the Liſt one M. the Cima Reverſa two and a half M. the Corona ſix and a half M the Cima Reverſa two and a half M. the Cartouſes ſeven M. the Boultin four M. the Liſt one M. the Square five M. the Liſt one M. and the Boultin four M.

According to *Palladio*, the whole Height of this *Cornice* is forty-ſix and a half M. of which the Liſt at the Top is two and a half M. the Cima Recta ſeven M. the Liſt one and one Fourth M. the Cima Reverſa three and a half M. the Corona eight M. the Cima Recta over the Modilions three and one Fourth M. the Modilions ſeven and a half M. the Liſt one M. the Ovolo or Boultin ſix M. the Liſt one and one half, and the Cavetto or Hollow five M.

The whole Height of the *Corinthian Cornice*, according to *Vitruvius*, is about one Mod. And he deſcribes two Forms of *Cornices*, in this Order: In one of which, he divides the whole Height into five Parts; the uppermoſt of which goes to the Ogee, of which its Fillet is one ſixth Part.

Then one and one Fourth goes to the Corona and Cymatium over it, of which Space the Cymatium is one third Part, and its Fillet one Third of that.

Then one and three Fourths of the next grand Diviſions go to the Modillions and Cymatium over them, of which Space, the Cymatium is one ſeventh Part.

And the laſt grand Diviſion goes to the Boultin and Fillets over and under it; and this being divided into three Parts, the lowermoſt goes to the Fillet; and the other two being ſubdivided into ſix Parts, five of them go to the Boultin, and the other to the Fillet over it.

In the *Cornice* of the other Faſhion, the whole Height is divided into nine Parts; of which the two uppermoſt being divided into four Parts, three of them go to the Ogee, (whoſe Fillet is one Sixth of the Whole,) and the other to the Cymatium over the Corona, (whoſe Fillet is one Third of the Whole,) and the next two grand Diviſions go to Corona.

The

The next two grand Divisions go to the Modillions, and the Cymatium over them: One Fourth of this Space goes to the Cymatium (whose Fillet is one Third of the whole Cymatium) and the rest of the Modillions.

The next two grand Divisions go to the Boultin, and the Fillet over and under it, which Fillets are each one Seventh of the Whole. And

The last grand Division goes to the Cima at the Foot of the *Cornice.*

Scamozzi makes the whole Height of this *Cornice* forty-six and three Fourths M. of which the List of the Cima Recta is two M. the Cima Recta six and a half M. the List of the Cima Reversa one M. the Cima Reversa three and one Fourth M. the Half-Round one and one half M. the Corona seven and one half M. the Cymatium three and one half M. the Modillions eight and a half M. the List one M. the Boultin five M. the List one M. and the Cima five M.

Palladio makes the whole Height of this *Cornice* fifty M. of which two and one Third M. go to the List of the Cima Recta; the Cima Recta is eight and one Third M. the List two Thirds M. the Cima Reversa three M. the Corona seven one Third M. the List of the Ogee over the Modillions two Thirds M. the Ogee 2 and two Thirds M. the Modillions eight and a half M. the Boultin four and one Third M. the List one M. the Boultin five and a half M. the List one M. and Ogee four and a half M.

The *Composite Cornice*: *Vitruvius* makes the whole Height of it equal to the Diameter of the Column above, which is about fifty-two and a half M.

He also describes two *Cornices* of this Order of a different Fashion; one of which he divides into two Parts, the uppermost of which goes to the Ogee (whose Fillet is one Seventh of the Whole,) and the undermost to the Corona and Cymatium over it; and this Space being divided into four Parts, three of them go the Corona, and one to the Cymatium, whose Fillet is one Third of the whole Cymatium.

Scamozzi makes the whole Height of this *Cornice* forty-eight M. and *Palladio* forty-five M. but for the Height of each particular Member, they leave us very much in the Dark.

Goldman makes the Height of the *Tuscan Cornice* one and one Third, and its Projecture two and two Fifths M. the Height of the *Doric* one and one Third, and its Projecture two and two Fifths. The Height of the *Ionic* one and three Fifths; its Projecture two and two Fifths. The Height of the *Composite* one and three Fifths; its Projecture two and thirteen Thirties. The Height of the *Corinthian* one and two Fifths; its Projecture two and thirteen Thirties.

The Projecture of the CORNICE.

'Tis an established Rule in Architecture, that the *Cornice* of the Entablement have its Projecture nearly equal to its Height:

Height; and yet the Projecture may may be safely made a little larger on Occasion, particularly where a beautiful Profile is required.

Cornice is used in general for all little Projectures of Masonry or Joinery, even where there are no Columns, as the *Cornice* of a Chimney, a Buffet, *&c.*

Architrave Cornice is one that is immediately contiguous to the Architrave, the Frieze being retrenched.

Mutilated Cornice is one whose Projecture is cut or interrupted to the Right of the Larmier, or reduced into a Platband with a Cymase.

Cantaliver Cornice, a Term used by Workmen for a *Cornice* that has Cantalivers underneath.

Coving Cornice, a *Cornice* which has a great Casement or Hollow in it, ordinarily lathed and plaistered upon Compass-Sprechets, or Brackets.

Modillion Cornice, a *Cornice* with Modillions under it.

Cornice is also used for the Crownings of Pedestals.

The *Cornice*, too, is different in the different Orders: In the *Tuscan*, according to M. *Perrault*, it has a Platband, which serves as a Corona and a Cavetto with its Fillet.

In the *Doric*, it has a Cavetto with a Fillet, which bears with a Drip crown'd with a Square.

In the *Ionic*, a Cavetto with its Fillet above, and a Drip, or hanging Square, crown'd with an Ogee, and its Fillet.

In the *Corinthian*, an Ogee with its Fillet, a Cymatium under the Ogee; Corona, and an Ogee with its Fillet.

Lastly, In the *Composite*, a Fillet with the Sweep over the Die, an Astragal, a Cima with its Fillet, Corona, and Ogee with its Fillet.

The Price: Mr. *Leybourn* tells us, some *Cornices* are valued by the Piece, dearer or cheaper, according to their Largeness, Goodness of the Stuff, and Curiosity of the Workmanship: Others are measured and rated by the Foot, running Measure, *i. e.* by the Number of Feet in Length only.

Some say, the common Rate for making of *Plain Cornices* (without any Carving) under the Eaves of a House, they commonly have 1*s.* *per* Foot running Measure.

Mr. *Wing* says, that *Cornices* are valued according to their Nature and Bigness. A *Modillion Cornice* of Freestone, of eighteen or twenty Inches thick, is worth 5*d.* or 6*d.* a Foor running Measure. And as to Joiners Work, a *Modillion Cornice*, with its carved Work, is worth 7*s.* a Foot, and a plain *Modillion Cornice* of twelve or fourteen Inches, will be worth 3*s.* 6*d.* or 4*s.* a Yard running Measure.

Some Workmen say a *Brick Cornice* is worth 2*s.* 6*d.* *per* Foot.

CORONA, CORONES, CROWN, CROWNING, } in Architecture, is a large flat strong Member of the *Cornice*, so called, because it crowns not only the Cornice, but the Entablature, and the whole Order.

The *French* call it the Larmier, our Workmen the Drip, as

as ſerving, by its great Projecture, to skreen the reſt of the Building from the Rain.

Some call it abſolutely the *Cornice*, as being the principal Member thereof *Vitruvius* frequently uſes the Word *Corona* for the whole Cornice.

The *Corona* is itſelf crown'd or finiſh'd with a Riglet or Fillet.

The *Corona*, M. *Le Clerc* ſays, is that large ſquare Moulding immediately under the Cymaſe.

It projects very much, both for the greater Beauty of the Entablature, and for the better ſheltering even of the whole Order

He alſo ſays, he makes this Part ſtronger than the Cymaſe, as being the ruling Member of the Entablature, and even of the Order.

Underneath this he uſually digs a Channel, for three Reaſons; the firſt is to give it more Grace and Ornament; the ſecond is to render it leſs heavy; and the third is to prevent Rain, or other Moiſture, from trickling down along the Order.

For the Water falling from the Top of the Cornice, not being able to aſcend into the Channel, is forced to fall Drop-by-drop on the Ground, by the Means of a little Ledge; and 'tis on this Account, that the Bottom of the *Corona* is called Larmier, or Drip.

CORNUCOPIA, in Architecture, Sculpture, &c. or Horn of Plenty, is repreſented under the Figure of a large Horn, out of which iſſue Fruits, Flowers, &c.

CORRIDOR, in Architecture, a Gallery, or long Iſle, around a Building, leading to ſeveral Chambers, at a Diſtance from each other.

CORSA. This Word, as, Faſcia and Tænia, in *Vitruvius*, ſignifies what is by us called Platband.

COUCH, in Painting, is uſed for each Lay or Impreſſion of Colour, whether in Oil, or in Water, wherewith the Painter covers the Wall, Wainſcot, or other Matter to be painted.

So they ſay, a Painting has had its laſt *Couch*, or Lay. A Ceiling has had two *Couches*.

The Word *Couch* is alſo uſed for a Lay or Impreſſion on any Thing, to make it more firm and conſiſtent, and to skreen it from the Weather; as Painting is coper'd with a *Couch* of Varniſh.

COVING, in Building, when Houſes are built projecting over the Ground-Plot, and the turn'd Projecture arch'd with Timber, (turn'd with a Quadrant of a Circle or Semi-Arch,) lathed and plaiſtered, under which People may walk dry, (as is much uſed at *Tunbridge-Wells*) on the Upper Walks,) the Work is commonly called *Coving*.

Mr. *Wing* ſays, that the Carpenters Work of *Coving* is worth 4 *s. per* Square.

COUNTER *Drawing*, in Painting, &c. is the Copying of a Deſign or Painting, by Means of a fine linnen Cloth, an oiled Paper, or other tranſparent Matter, whereon the the Strokes appearing through, are followed and traced with a Pencil, with or without Colour.

COUNTER FORTS, Buttreſſes, Spurs, or Pillars of Maſonry, ſerving to prop or ſuſtain Walls or Terraſſes ſubject to bulge, or be thrown down.

Theſe Works are uſually bent into Arches, and placed at a Diſtance from each other.

When any Thing is built on the Deſcent of a Mountain, it muſt be ſtrengthened with *Counterforts* well bound to the Wall, and at the Diſtance of about twelve Yards from each other.

Counter Gage, in Carpentry, a Method uſed in meaſuring the Joints, by transferring, *v. g.* the Breadth of a Mortoiſe to the Place in the Timber where the Tenon is to be, in order to make them fit each other.

Counter Light, in Painting, a Window or Light oppoſite to any Thing which makes it appear to a Diſadvantage.

Counter Mure } a little *Wall*
Counter Wall } built cloſe to another, to fortify and ſecure it, that it may not receive any Damages from Buildings made contiguous to it.

COURSE, in Architecture, a continued Range of Stones, level, or of the ſame Height throughout the whole Length of the Building, without being interrupted with any Aperture.

A *Courſe of Plinths*, is the Continuity of a *Plinth* of Stone or Plaſter in the Face of a Building, to mark the Separation of the Stones.

COUSSINET [*q. d.* a *Cuſhion*] in Architecture, ſignifies the Stone which crowns a Piedroit or Peer, or that lies immediately over the Capital of the Impoſt, and under the Sweep. The Bed of it is level below, and curved above, receiving the firſt

it be well grown together, it will ſcarce be perceived in ſome Stuffs, but only in working.

CROSS *Multiplication.* See MULTIPLICATION.

CROWN POST, in Architecture, a Poſt which in ſome Buildings ſtands upright in the Middle, between two principal Rafters.

Crown, in Architecture, the uppermoſt Member of the Cornice, called alſo Corona, and Larmier.

CROWNING, in Architecture, is generally underſtood when any Thing terminates or finiſhes a Decoration of Architecture: Thus a Cornice, a Pediment, Acroteres, &c. are called *Crownings*.

And thus alſo the Abacus is ſaid to crown the Capital: And any Member or Moulding is ſaid to be crowned, when it has a Fillet over it: And a Niche is crown'd, when it is cover'd with a Capital.

CRYPTA [of κρύπτω, *Gr.* to hide] a ſubterraneous Place or Vault, eſpecially under a Church, for the Interment of particular Families or Perſons.

Vitruvius uſes the Word for Part of a Building, anſwering nearly to our Cellar. Hence,

CRYPTO PORTICO, a ſubterraneous Place, arched or vaulted, uſed as an Underwork or Paſſage into old Walls The ſame Words are alſo uſed for the Decoration at the Entry of a Grotto.

CUBATURE, } is the Cu-
CUBATION, } bing of a Solid, or the meaſuring of the Space comprehended in a Solid, as in a Cone, Pyramid, Cylinder, &c.

The *Cubature* has Reſpect to the Content of a Solid as the Quadrature has to the Superficies of a Figure: So that the *Cubature* of the Sphere turns on the ſame Thing as the Quadrature of the Circle.

CUBE, in Geometry, a regular or ſolid Body, conſiſting of ſix ſquare and equal Faces and Sides, and its Angles all right, and therefore equal.

The *Cube* is ſuppoſed to be generated by the Motion of a ſquare Plane along a Line equal to one of its Sides, and at Right Angles thereto: Whence it follows, that the Planes of all Sections parallel to the Baſe, are Squares equal thereto, and conſequently to one another.

Cube, } in Arithme-
Cubick Numbers, } tick, is a Number ariſing from the Multiplication of a ſquare Number by its Root: Thus if the ſquare Number 4 be multiplied by its Root 2, the Factum 8 is a *Cube* or *Cube Root*.

Cube is a ſquare Solid, comprehended under ſix Geometrical Squares, being in the Form of a Dye, to find the ſolid Content. This is,

The RULE,

Multiply the Side of the *Cube* into itſelf, and that Product again by the Side, the laſt Product will be the Solidity or ſolid Content of the *Cube*.

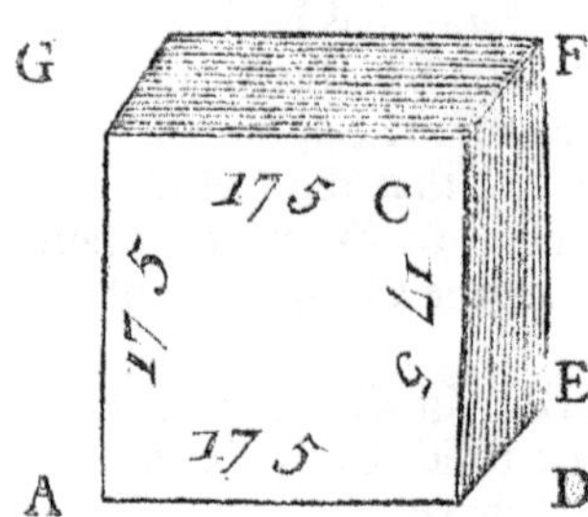

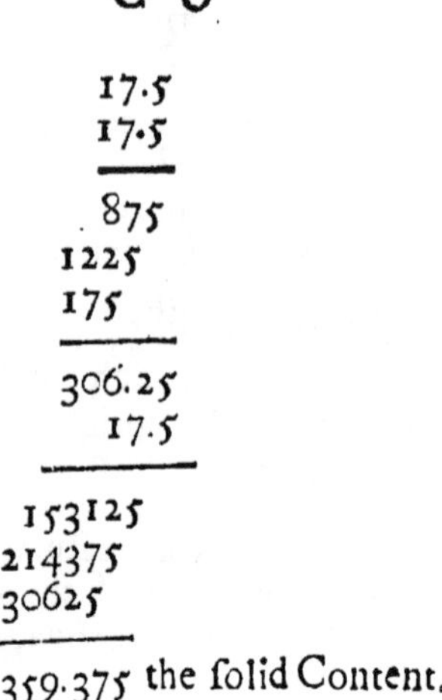

Suppoſe A B C D E F G a Cubical Piece of Stone or Wood, each Side thereof being ſeventeen Inches and a half, multiply 17.5 by 17.5, and the Product is 306.25 ; which being multiply'd by 17.5, the laſt Product will be 5359 ſolid Inches and 375 Parts.

To reduce the ſolid Inches to Feet, divide by 1728, (becauſe there are ſo many Cubical Inches in a Foot,) and the ſolid Feet in the *Cube* will be 3, and 175 Cubical Inches remaining.

By Scale and Compaſſes.

Extend the Compaſſes from 1 to 17.5, and that Extent turn'd over twice from 17.5, will reach to 5359 the ſolid Content in Inches: Then extend the Compaſſes to 1, turn'd the ſame Way from 5359, and they will reach to 3.1 Feet.

DEMONSTRATION.

If the Square A B C D be conceived to be mov'd down the Plane A D E F always remaining parallel to itſelf, there will be generated by ſuch a Motion, a Solid having ſix Planes ; the two oppoſite of which will be equal and parallel to each other ; whence it is called a Parallelopipedon, or ſquare Priſm.

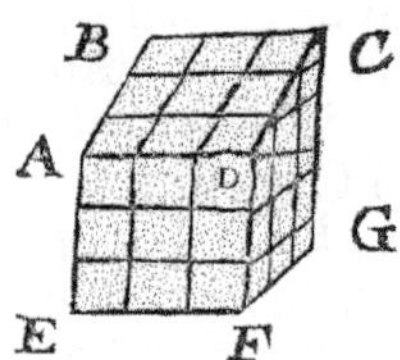

And if the Plane A D E F be a Square equal to the generating Plane A B C D, then will the generated Solid be a *Cube*.

From hence ſuch Solids may be conceived to be conſtituted of an infinite Series of equal Squares, each equal to the Square A B C D, and A E or D F will be the Number of Terms.

Therefore, if the Area of A B C D be multiply'd into the Number of Terms A E, the Product is the Sum of all that Series, (*per* Lemma 1.) and conſequently the

the Solidity of the Parallelopipedon, or *Cube*.

Or if the Base ABCD, being divided into little square Areas, be multiply'd into the Height AE, divided into a like Measure for Length: After this Way, you may conceive as many little *Cubes* to be generated in the whole Solid, as is the Number of the little Areas of the Base, multiply'd by the Number of Divisions the Side AE contains.

Thus if the Side of the Base AB be 3, that multiply'd into itself is 9, which is the Area of the square Base ABCD; then if AE be likewise 3, multiply 9 by 3, and the Product will be 27, and so many little *Cubes* will this Solid be cut into.

From this Demonstration, it is very plain, that if you multiply the Area of the Base of any Parallelogram into its Length, or Height, that Product will be the solid Content of such a Solid.

Extraction of the CUBE ROOT.

To extract the *Cube Root*, is nothing else but to find such a Number, as being first multiply'd into itself, and then into that Product, produceth the given Number. Which to perform, observe these following Directions:

First, You must point your given Number, beginning with the Unit's Place, and make a Point or Dot over every third Figure towards the Left Hand.

Secondly, Seek the greatest *Cube Number* in the first Point, towards the Left Hand, putting the Root thereof in the Quotient, and the said *Cube Number* under the first Point, and subtract it therefrom, and to the Remainder bring down the next Point, and call that the Resolvend.

Thirdly, Triple the Quotient, and place it under the Resolvend, the Unit's Place of this, under the Ten's Place, and call this the triple Quotient.

Fourthly, Square the Quotient, and triple the Square, and place it under the triple Quotient, the Units of this under the Ten's Place of the triple Quotient, and call this the triple Square.

Fifthly, Add these two together, in the same Order as they stand, and the Sum shall be the Divisor.

Sixthly, Seek how often the Divisor is contain'd in the Resolvend, rejecting the Unit's Place of the Resolvend, (as in the square Roor,) and put the Answer in the Quotient.

Seventhly, Cube the Figure last put into the Quotient, and put the Unit's Place thereof under the Unit's Place of the Resolvend.

Eighthly, Multiply the Square of the Figure last put in the Quotient, and place the Product under the last, one Place more to the Left Hand.

Ninthly, Multiply the triple Square by the Figure last put in the Quotient, and place it under the last, one Place more to the Left Hand.

Tenthly, Add the three last Numbers together, in the same Order as they stand, and call that the Subtrahend.

Lastly, Subtract the Subtrahend from the Resolvend, and if there

there be another Point, bring it down in the Remainder, and call that a new Refolvend; and proceed in all Refpects as before.

Example 1. Let 314432 be a *Cubick Number*, whofe Root is required.

314432(68 Root.
216

98432 Refolvend.

18 Triple Quotient of 6.
108 Triple Square of the Quotient 6.

1098 Divifor.

512 Cube of 8, the laft Figure of the Root.
1152 The Square of 8, by the triple Quotient.
864 The triple Square of the Quotient 6 by 8.

98432 The Subtrahend.

.....

After you have pointed the given Number, feek what is the greateft *Cube Number* in 314, you will find the firft Point to be 216, which is the neareft that is lefs than 314, and its Root is 6; which put in the Quotient, and 216 under 314, and fubtract it therefrom, and there remains 98; to which bring down the next Point, 432, and annex to 98; fo will it make 98432 for the Refolvend. Then triple the Quotient 6, it makes 18, which write down, the Unit's Place, 8, under 3, the Ten's Place of the Refolvend. Then fquare the Quotient 6, and triple that Square, and it makes 108, which write under the triple Quotient, one Place on the Left Hand; then add thofe two Numbers together, and they make 1098 for the Divifor. Then feek how often the Divifor is contained in the Refolvend, (rejecting the Unit's Place thereof) that is, how often 1098 in 9843, which is 8 Times; put 8 in the Quotient, and the *Cube* thereof below the Divifor, the Unit's Place under the Unit's Place of the the Refolvend. Then fquare the 8 laft put in the Quotient, and multiply 64, the Square thereof, by the triple Quotient 18, the Product is 1152; fet this under the *Cube* of 8, the Units of this under the Tens of that. Then multiply the triple Square of the Quotient by 8, the Figure laft put in the Quotient, the Product is 864; fet this down under the laft Product, a Place more to the Left Hand. Then draw a Line under thofe three

three, and add them together, and the Sum is 98432, which is called the Subtrahend; which being ſubtracted from the Reſolvend, the Remainder is nothing; which ſhews the Number to be a true *Cubic Number*, whoſe Root is 68, that is, if 68 be cubed, it will make 314432.

For, if 68 be multiply'd by 68, the Product will be 4624; and this Product, multiply'd again by 68, the laſt Product is 314432, which ſhews the Work to be right.

	68
	68
	544
The Work	408
	4624
	68
	36992
	27744
The Proof	314432

Example 2. Let the *Cube Root* of 5735339 be required.

After you have pointed the given Number, ſeek what is the greateſt *Cube Number* in 5, the firſt Point, which you will find to be one; which place under 5, and 1, the Root thereof, in the Quotient; and ſubtract 1 from 5, and there remains 4; to which bring down the next Point, it makes 4735 for the Reſolvend. Then triple the 1, and it makes 3; and the Square of 1 is 1, and the Triple thereof is 3; which ſet one under another, in their Order, and added, makes 33 for the Diviſor. Seek how often the Diviſor is in the Reſolvend, and proceed as in the laſt Example.

5735339 (179 Root.
1

4735 Resolvend.

3 The Triple of the Quotient 1, the first Figure.
3 The triple Square of the Quotient 1.

33 The Divisor.

343 The *Cube* of 7, the second Figure of the Root.
147 The Square of 7 multiply'd in the triple Quotient 3.
21 The triple Square of the Quotient multiply'd by 7.

3913 The Subtrahend.

822339 The new Resolvend.

51 The Triple Quotient 17, the two first Figures.
867 The triple Square of the Quotient 17.

8721 The Divisor.

729 The *Cube* of 9, the last Figure of the Root.
4131 The Square of 9, multiply'd by the triple Quotient 51.
7803 The triple Square of the Quotient 867 by 9.

822339 The Subtrahend.

:.....

In this Example, 33, the first Divisors seems to be contain'd more than seven times in 4735, the Resolvend; but if you work with 9 or 8, you will find that the Subtrahend will be greater than the Resolvend.

Some

Some more Examples for Practice.

32461759 (319 The Root.
27

5461 Resolvend.

9 The Triple of 3.
27 The triple Square of 3.

279 The Divisor.

1 The *Cube* of 1, the second Figure.
9 The triple Quotient by the Square of 1.
27 The triple Square multiply'd by 1, the second Figure.

2791 The Subtrahend.

2670759 A new Resolvend.

93 The Triple of 31.
2883 The triple Square of 31.

28923 The Divisor.

729 The *Cube* of 9, the last Figure.
7533 The Square of 9 by 93, the triple Quotient.
25947 The triple Square 2883 by 9.

2670759 The Subtrahend.

....

84604519 (439 The Root.
64

20604 Resolvend.

12 The Triple of 4.
48 The triple Square of 4.

492 The Divisor.

27 The *Cube* of 3.
108 The Square of 3 by the triple Quotient.
144 The triple Square of 3.

15507 The Subtrahend.

5097519 The Resolvend.

129 The Triple of 43.
5547 The triple Square of 43.

55599 The Divisor.

729 The *Cube* of 9.
10449 The Square of 9 by 129.
49923 The triple Square by 9.

5097519 The Subtrahend.

....

259697989 (638
216

43697 Resolvend.

18 The Triple of 6.
108 The triple Square of 6.

1098 The Divisor.

27 The *Cube* of 3, the second Figure.
162 The Square of 3 by 18.
324 The triple Square of 108 by 3.

34047 The Subtrahend.

9650989 Resolvend.

189 The Triple of 63.
11907 The triple Square of 63.

119259

512 The *Cube* of 8.
12096 The Square of 8 by 189.
95256 The triple Square of 11907 by 8.

9647072 The Subtrahend.

3917 The Remainder.

25917056 (295.9
8

17917 The Resolvend.

The Triple of 2.
The triple Square of 2.

The Divisor.

729 The *Cube* of 9, the second Figure.
486 The Square of 9 by 6.
108 The triple Square by 9.

16389 The Subtrahend.

1528056 The Resolvend.

87 The Triple of 29.
2523 The triple Square of 29.

25317 The Divisor.

125 The *Cube* of 5, the third Figure.
2175 The Square of 5 by 87.
12615 The triple Square by 5.

1283375 The Subtrahend.

244681000 The Resolvend.

885 The Triple of 295.
261075 The triple Square of 295.

2611635 The Divisor.

729 The *Cube* of 9, the last Figure.
71685 The Square of 9 by 885.
2349675 The triple Square by 9.

235685079 The Subtrahend.

8995921 The Remainder.

In this Example I annex 3 Cyphers to the Remainder, which makes the third Resolvend; by which Means I bring one to the Place of Decimals. And so you may proceed to more decimal Places at Pleasure, by annexing three Cyphers to the next Remainder, and carrying on the Work, as before.

2206981012ſ (280ſ
8

Reſolvend.

126	The Triple of 2.
	The triple Square of 2.
126	The Diviſor.
ſ12	The *Cube* of 8.
384	The Square of 8 by 6.
96	The triple Square by 8.
139ſ2	The Subtrahend.
11781012ſ	New Reſolvend.
84	The Triple of 28.
23ſ2	The triple Square of 28.
23604	The Diviſor.
840	The Triple of 280.
23ſ200	The triple Square of 280.
23ſ2840	New Diviſor.
12ſ	The *Cube* of ſ.
21000	The Square of ſ by 840.
1176000	The triple Square by ſ.
11781012ſ	The Subtrahend.
....	

In this Example, 139ſ2 being ſubtracted from the Reſolv. 14069, the Remainder is 117; to which bring down 810, the third Point, and it makes 117810, for a new Reſolvend; and the next Diviſor is 23604, which you cannot have in the ſaid Reſolv. (the Unit's Place being rejected,) ſo you muſt put 0 in the Quotient, and ſeek a new Diviſor, (after you have brought down your laſt Point to the Reſolvend;) which new Diviſor is 23ſ2840; which you'll find to be contained ſ times. So proceed to finiſh the reſt of the Work.

93759.

93759.575070 (45.42
64

29759 The Resolvend.

12 The Triple of 4, the first Figure.
48 The triple Square of 4.

492 The Divisor.

125 The *Cube* of 5, the second Figure.
300 The Square of 5 by 12, the triple Quotient.
240 The triple Square by 5.

27125 The Subtrahend.

2634575 The Resolvend.

135 The Triple of 45.
6075 The triple Square of 45.

60885 The Divisor.

64 The *Cube* of 4.
2160 The Square of 4 by 135.
24300 The triple Square by 4.

2451664 The Subtrahend.

182911070 The Resolvend.

1362 The Triple of 45.4.
618348 The triple Square of 45.4.

6184842 The Divisor.

8 The *Cube* of 2.
5448 The Square of 2 by 1362.
1236696 The triple Square by 2.

123724088 The Subtrahend.

59186982 The Remainder.

In extracting the *Cube Root* of a mix'd Number, always observe to make the decimal Part to consist of either three, six, nine, &c. Places, that is, always to consist of even Points, as in the last Example, where the decimal Places were five, to which

I

I annexed a Cypher to make up ſix, and ſo I proceed to point it; and by that Means I have a Point falls upon the Unit's Place of whole Numbers, which you muſt always obſerve.

To extract the Cube Root *out of a Fraction.*

This is the ſame to do as in whole Numbers, obſerve but the foregoing Directions for the true pointing thereof; for, as was before directed, the Decimal muſt always conſiſt of three, ſix, nine, *&c.* Places; and if it be not ſo, it muſt be made ſo, by annexing of Cyphers, as is aboveſaid.

If the *Cube Root* of a vulgar Fraction be required, you muſt firſt reduce it to a Decimal, and then extract the Root thereof.

Examples of each follow.

Example 1. Let the *Cube Root* of .401719179 be required.

.401719179 (.737 Root.
343
58719 Reſolvend.

21 Triple of 7.
147 Triple Square of 7.
1491 Diviſor.

27 *Cube* of 3.
189 Square of 3 by 21.
441 Triple Square by 3.
46017 Subtrahend.

12702179 Reſolvend.

219 Triple of 73.
15987 Triple Square of 73.
160089 Diviſor.

343 *Cube* of 7.
10731 Square of 7 by 219.
111909 Triple Square by 7.
11298553 Subtrahend.

1403626 Remainder.

Example 2. Let the *Cube Root* of .0001416 be required.

.000141600 (.052 Root.
125

16600 Resolvend.

15 The Triple of 5.
75 Triple Square of 5.

765 Divisor.

8 *Cube* of 2.
60 Square of 2 by 15.
150 Triple Square by 2.

15608 Subtrahend.

992 Remainder.

Example 3. Let $\frac{005}{276}$ be a vulgar Fraction, whose *Cube Root* is required.

By this Rule, reduce the vulgar Fraction to a Decimal.

276) 5.000000000(.018115942
....

2240
320
440
1640
2600
1160
560

8

.018115942 (.262 Root.
8

10115 Resolvend.

6 Triple of 2.
12 Triple Square of 2.

126 Divisor.

216 *Cube* of 6.
216 Square of 6 by the Triple (of 2.
72 The triple Square by 6.

9576 Subtrahend.

539942 Resolvend.

78 Triple of 26.
2028 Triple Square of 26.

20358 Divisor.

8 *Cube* of 2.
312 Square of 2 by 78.
4056 Trip. Square of 2028 (by 2.

408728 Subtrahend.

131214 Remainder.

You may prove the Truth of the Work, by cubing the Root found, as was shewn in the first Example; and if any Thing remains, add it to the said *Cube*, and the Sum will be the given Number, if the Work is rightly perform'd.

I will

I will ſhew the Proof of the fifth Example, the given Number being 259697989, whoſe Root is 638, it being a ſurd Number, there remains 3917.

	638
	638
	5104
	1914
	3828
The Square	407044
	638
	3256352
	1221132
	2442264
The *Cube*	253694072
The Remainder add	4917
Proof equal to the given Number	259697989

CUBICLE, a Bed-Chamber.

CULINARY, of or belonging to a Kitchin.

CULVERTAIL, the ſame as Dovetail.

CUPOLA, a ſpherical Vault, or the Round-Top of the Dome of a Church, in Form of a Cup inverted. Some call it a Lanthorn.

CURLING STUFF, in Joinery. See CROSS-GRAIN'D.

CURTICONE, in Geometry, a *Cone* whoſe Top is cut off by a Plane parallel to its Baſis.

CURVATURE of a Line, is its Bending, or Flexure, whereby it becomes a Curve of ſuch peculiar Properties.

CURVE, in Geometry, a Line, wherein the ſeveral Points it conſiſts of tend ſeveral Ways, or are poſited towards different Quarters.

In this Senſe the Word is uſed in Oppoſition to a ſtreight Line, whoſe ſeveral Points are pointed towards the ſame Quarter.

CURVILINEAL, crooked-lined, or conſiſting of crooked Lines.

Curvilineal Figures, in Geometry, are Spaces bounded by crooked Lines; as Circles, Ellipſis, ſpherical Triangles, *&c.*

CYCLOID, in Geometry, one of the mechanical; or, as others term them, the tranſcendental Curves, called alſo the *Trochoid.*

Cycloidal Space, the Space contained between the *Cycloid* and the Subtenſe thereof.

CYCLOMETRY, the Art of meaſuring Cycles, or Circles.

CYLINDER,

CYLINDER, is a round Solid, having its Bases circular, equal and parallel, in the Form of a Rolling-Stone.

To find the solid Content, this is the Rule.

Multiply the Area of the Base by the Length, and the Product is the solid Content.

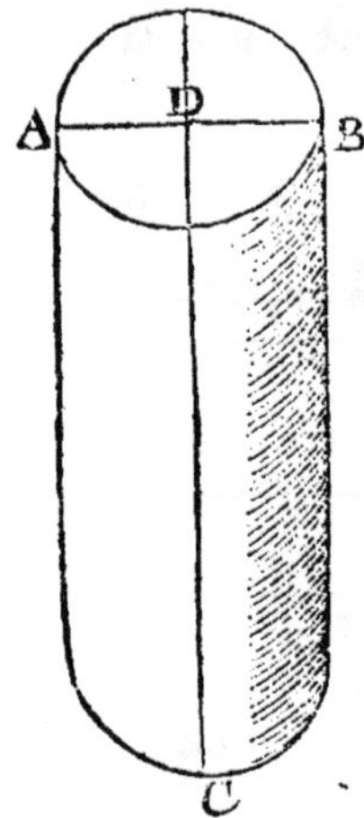

Let A B C be a *Cylinder*, whose Diameter A B is 21.5 Inches, and the Length C D is 16 Feet; what is the solid Content?

First square the Diameter 21.5, and make it 462.25; which multiplied by .7854, and the Product will be 363.05115.

Then multiply this by 16, and the Product will be 5808.8164. Divide this last Product by 144, and the Quotient will be 40.34 Feet, the solid Content.

By Scale and Compasses.

Extend the Compasses from 13.54 to 21.5, the Diameter, and that Extent (turned twice over from 16, the Length,) will at last fall upon 40.34, the solid Content.

To find the superficial Content.

First find the Circumference of the Base 67.54; which being divided by 12, the Quotient will be 90.05 Feet, the curved Surface: To which add 5.04 Feet, the Sum of the two Bases, and the Sum will be 95.09 Feet, the whole superficial Content.

```
      67.54
         16
    -------
      40524
      6754
    -------
12) 1080.64
    -------
      90.05
```

```
 90.05 } add
  5.04 }
 -----
 95.09
```

```
        363.05
             2
       -------
144) 726.10 (5.04
       -------
          610
       -------
           34
```

By Scale and Compasses.

Extend the Compasses from 12 to 67.54, (the Circumference) that Extent will reach from 16 (the Length) to 90.05 Feet, the curve Surface.

And

And extend the Compaſſes from 12 to 21.5 (the Diameter,) that Extent turn'd twice from .7854, will at laſt fall upon 2.52 the Area of one Baſe, which doubled is 5.04: This being added to the curve Surface, makes 95.09 Feet, the whole ſuperficial Content.

Demonstration.

The ſolid Content of every *Cylinder* is found by multiplying the Area of its Baſe into its Height, as aforeſaid: For every Right *Cylinder* is only a round Priſm, being conſtituted of an infinite Series of equal Circles; that of its Baſe or End being one of the Terms; and its Height CD (in the Figure) is the Number of all the Terms.

Therefore the Area of its Baſe AB being multiply'd into CD, will be its Solidity. Let D=AB, and H=CD.

CYLINDROID is a Fruſtum of a Cone, having its Baſes parallel to each other, but unlike.

The RULE.

To the longeſt Diameter of the greater Baſe, add half the longeſt Diameter of the leſſer Baſe, and multiply the Sum by the ſhorteſt Diameter of the greater Baſe, and reſerve the Product.

Then to the longeſt Diameter of the leſſer Baſe add half the longeſt Diameter of the leſſer Baſe, and multiply the Sum by the ſhorteſt Diameter of the leſſer Baſe, and add the Product to the former reſerved Sum, and that Sum will be the triple Square of a mean Diameter; which multiply'd by .7854, and that Product multiply'd by a third Part of the Height, the Product will be the ſolid Content.

Example.

Let ABCD be a *Cylindroid*, whoſe Bottom Baſe is an Oval, the tranſverſe Diameter being fourty-four Inches; and the upper Baſe is a Circle, whoſe Diameter is twenty-ſix Inches, and the Height of the Fruſtum is nine Feet; what is the Solidity?

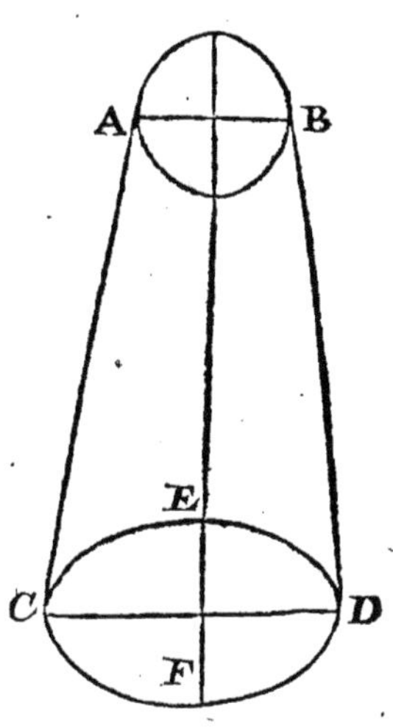

To 44 (the greater Diameter of the lower Baſe,) add 13, (half the Diameter of the leſſer Baſe,) and the Sum will be 57; which being multiply'd by 14, (the conjugate Diameter of the greater Baſe,) the Product will be 798; which reſerve.

Then, to 26, (the Diameter of the leſſer Baſe) add 22, (half the tranſverſe Diameter of the greater Baſe,) and the Sum will be 48; which being multiply'd by by 26, (the Diameter of the leffer Bafe,) the Product will be 1248; to which add the former referv'd Product, and the Sum will be 2046, which being multiply'd by 7854, the Product will be 1606.9284; which being multiply'd by 3, (a third Part of the Height,) the Product will be 4820.7852; which divided by 144, the Quotient will be 33.47 Feet, the folid Content.

```
44=CD           26=AB           1606.9284
13=half AB      22=half CD              3
——              ——              ————————
57 Sum          48 Sum      144) 4820.78|52(33.47
14=EF           26=AB            ————————
——              ——               500
228             228               687
57               96               1118
——              ————              ————
798 Prod. ref.  1248               110
                 798 add           ————
                ————
                 2046
                 7854
                ————
                 8184
                10230
               16368
              14322
              —————————
              1606-9284
              —————————
```

The Rule being the fame as that of the Fruftum of a Rectangled Pyramid or Prifmoid; the Proof of that may ferve as a fufficient Proof of this.

To find the fuperficial Content.

Add the Periphery of the Circle 81.68 to the Periphery of the Ellipfis 97.41, and the Sum will be 179 09; the half of which, 89.545, being multiply'd by 9, the Product will be 805.905; which being divided by 12, the Quotient will be 67.16 Feet, the curve Surface: Then the Area of the Ellipfis is 3.36 Feet; and the Area of the Circle is 3.68 Feet; both which added to the curve Surface, the Sum will be 74.2 Feet, the whole fuperficial Content.

CYMA in Architecture. See CIMA, SIMA, and CYMATIUM.

CYMATIUM, CIMATIUM, CIMA, } [of the *Greek* κυμάτιον, a little Wave] a Member or Moulding of the Cornice, the Profile of which is waved, *i. e.* concave at top, and convex at bottom. Which is oftentimes called alfo *Doucine Gorge*, or *Gula recta*

recta Goletta, by the *Italians*; but moſt uſually *Cymatium*, among us, as being the laſt, or uppermoſt Member, *q. d.* the *Cima* or Summit of the Cornice.

Some write it *Simaiſe*, from *Simus* an Ape, or *Camus* flat-noſed; but this Etymology ſeems not proper: The Beauty of the Moulding conſiſting in its having its projecture equal to its Height.

M. *Felibien* indeed will not allow of this Etymology; and contends, that the Moulding is not ſo denominated from its being the uppermoſt Member of the Cornice, but upon its being waved; which is the Opinion of *Vitruvius*.

Vitruvius does not confine *Cymatium* to the Cornice, but uſes it indifferently for any ſimilar Moulding, wherever he meets with it. In which he differs from the moſt accurate among the Moderns.

Felibien makes two Kinds of *Cymatiums*; the one right, and the other inverted: In the firſt, that Part which projects the fartheſt is concave; and is otherwiſe called *Gula recta*, and *Doucine*.

In the other the Part that projects fartheſt is convex, called *Gula inverſa*, or *Salon*.

The *Engliſh* Architects don't uſually give the Name of *Cymatium* to theſe Mouldings, except when they are found on the Tops of Cornices. But the Workmen uſe the Name indifferently, wherever they are found.

Tuſcan Cymatium conſiſts of an Ovolo or Quarter-Round. *Plulander* makes two *Doric Cymatiums*; of which this is one. *Baldus* calls this the *Lesbian* Aſtragal.

The *Doric Cymatium* is a Cavetto, or a Cavity leſs than a Semicircle, having its Projecture ſubduple to its Height.

Lesbian Cymatium, according to *Vitruvius*, is what our Architects otherwiſe call Talon, *viz.* a Concavo-convex Member, having its Projecture ſubduple its Height.

CYPRESS-TREE is of two Sorts, Wild, and the Sative, or Garden one, the moſt pyramidal and beautiful, and which is prepoſterouſly called the Male, and bears Cones.

The *Cypreſs* is a tall Tree, and ſhoots forth from its Roots a ſtraight Stalk, divided into ſeveral Branches that bear Leaves very much indented, thick, and of a browniſh-green Colour. At the Ends of theſe Branches grow Flowers like Cats Tails, compoſed of ſeveral little ſtrait Leaves or Scales, and barren.

Thoſe who would have *Cypreſs* in Standards, and grow wild, which may in Time come to be of large Subſtance, fit for the moſt immortal of Timber, and, indeed, are the leaſt obnoxious to the Rigour of Winter, provided they be never clipped or disbranched, muſt plant the Male Sort. It proſpers wonderfully where the Ground is hot and gravelly. The *Venetians* make great Profit of this Tree.

The Timber of *Cypreſs* is uſeful for Cheſts, Muſical Inſtruments, and other Utenſils; for it reſiſts the Worm, and Putre-

 faction;

faction, because of the Bitterness of its Juice. It never rifts nor cleaves, but with great Violence. And it may be worth observing, that the *Venetians* formerly made a considerable Revenue of it out of *Candia*; till the Forest there being set on Fire, either through Malice, or by Accident, in the Year 1400. It burnt seven Yeats together, because of the unctuous Nature of the Timber.

The Gates of *St. Peter's Church* at *Rome* were made of *Cypress* Wood, and lasted six hundred Yeats as fresh as if they had been new, till Pope *Eugenius* ordered Gates of Brass in their Stead.

The Chests of the *Egyptian* Mummies are, many of them, made of this Wood. The *Candiots* and *Maltese* use it in building.

The Root of the wilder Sort of *Cypress* is of an incomparable Beauty, by reason of its crisped Undulations. It was antiently made use of in building Ships, by *Alexander*, and others. And some will have it, that Gophir, of which *Noah's Ark* was built, was *Cypress*. *Plato* preferred it to Brass itself, for writing his Laws on.

D A

DADO, in Architecture, the Dye, or that Part in the Middle of the Pedestal of a Column which is between its Base and Cornice. It is of a cubick Form, and thence takes the Name of Dye.

DEALS. *Of Dressing them:* The Dressing of *Deals* is the rough Planing of them over with a Fore-Plane, in order to dry.

Mr. *Wing* says, this Work is worth 1 *s.* *per* Score: Though some say, they have had them done for 9 *d.*

Of the laying of *Deal-Floors*, *i. e.* the Planing and Joining them, is worth 5 *s.* *per* Square.

But if the Floors be laid with Dovetail or Key-Joints, without Plus or Nails, some Workmen say, they have 10 *s.* *per* Square for Workmanship only; but if the Workman finds Deals, and lays them the ordinary Way, it is worth from 24 *s.* to 30 *s.* *per* Square, according to the Goodness of the *Deals*.

But if the *Deals* are extraordinary, and laid either with Dovetail or Key-Joints (without Nails or Pins,) 'tis worth 35 *s.* or 40 *s.* *per* Square. See FLOORS.

DECAGON, in Geometry, is a plain Figure of ten Sides, and as many Angles; and if all the Sides are equal, and all the Angles, it is called a *Regular Decagon*, and may be inscribed in a Circle.

If the Side of a *Regular Decagon* be 1, the the Area thereof will be 8.69; whence as 1 to 8.69, so is the Side of the Square of any given *Decagon* to the Area of that *Decagon*.

DECASTYLE, in the antient Architecture, a Building with an Ordonance of ten Columns in Front, as the Temple of *Jupiter Olympius* was.

DECIMALS. A *Decimal Fraction* is an artificial Way of setting down and expressing of Natural or Vulgar Fractions, as whole

Numbers: And whereas the Denominators of Vulgar Fractions are divers, the Denominators of *Decimal Fractions* are always certain: For a *Decimal Fraction* hath always for its Denominator an Unit, with a Cypher or Cyphers annexed to it, and must therefore be either 10, 100, 1000, 10000, *&c.* And therefore, in writing down a *Decimal Fraction*, there is no Necessity of writing down the Denominator; for by bare Inspection it is certainly known, it consisting of an Unit, with as many Cyphers annexed to it, as there are Places (or Figures) in the Numerator.

Example. This *Decimal Fraction* $\frac{25}{100}$, may be written thus, .25, its Denominator being known to be an Unit with two Cyphers; because there are two Figures in the Numerator. In like Manner $\frac{125}{1000}$ may be thus written, .125; and $\frac{3575}{10000}$ thus, .3575; and $\frac{75}{1000}$ thus, .075; and $\frac{65}{10000}$ thus, .0065.

As whole Numbers increase in a Decuple, or tenfold Proportion, towards the Left Hand, so, on the contrary, *Decimals* decrease towards the Right Hand, in a decuple Proportion, as in the following Scheme.

Tens of Millions.	Millions.	Hundreds of Thousands.	Tens of Thousands.	Thousands.	Hundreds.	Tens.	Units.	Tenth Parts.	Hundredth Parts.	Thousandth Parts.	Ten Thousandth Parts.	Hundredth Thousandth Parts.	Millionth Parts, *&c.*
7	6	5	4	3	2	1 .	0	1	2	3	4	5	6

Hence it appears, that Cyphers put on the Right Hand of whole Numbers, do increase the Value of those Numbers in a decuple, or tenfold Proportion; but being annexed to the Right Hand of a *Decimal Fraction*, do neither increase nor decrease the Value thereof: So $\frac{2500}{10000}$ is equivalent to $\frac{025}{100}$ or .25. And, on the contrary, though in whole Numbers, Cyphers prefixed before them, do neither increase nor diminish the Value, yet Cyphers before a *Decimal Fraction* do diminish its Value in a decuple Proportion: For .25, if you prefix a Cypher before it, becomes $\frac{025}{1000}$, or .025: And .125 is $\frac{00125}{100000}$, by prefixing two Cyphers before it, thus, .00125. And therefore, when you are to write a *Decimal Fraction*, whose Denominator hath more Cyphers than there are Figures in the Numerator, they muſt be ſupply'd by prefixing ſo many Cyphers before the Figures of your Numerator; as ſuppoſe $\frac{19}{1000}$ were to be written down without its Denominator; here, becauſe there are three Cyphers in the Denominator, and but two Figures in the Numerator, therefore prefix a Cypher before 19, and ſet it down thus, .019.

The Integers are ſeparated from the *Decimals* ſeveral Ways, according to Mens Fancies; but the beſt and moſt uſual Way, is by a Point or Period; and if there be no whole Number, then a Point before the Fraction is ſufficient: Thus if you were to write down 317 $\frac{217}{1000}$, it may be thus expreſs'd, 317.217; and 59 $\frac{25}{10000}$, thus, 59.0025; and $\frac{75}{10000}$, thus, .0075, *&c.*

Reduction of Decimals.

In Reduction of *Decimals*, there are three Caſes. *Firſt*, To reduce a Vulgar Fraction to a *Decimal*: *Secondly*, To find the Value of a *Decimal* in the known Parts of Coin, Weights, Meaſures, *&c.* *Thirdly*, To reduce Coins, Weights, Meaſures, *&c.* to a *Decimal*. Of theſe, in their Order.

I. *To reduce a Vulgar Fraction to a* Decimal.

The RULE.

As the Denominator of the given Fraction is to its Numerator, ſo is an Unit (with a competent Number of Cyphers annex'd) to the *Decimal* requir'd.

Therefore, if to the Numerator given, you annex a competent Number of Cyphers, and divide the Reſult by the Denominator, the Quotient is the *Decimal* equivalent to the Vulgar Fraction given.

Example 1. Let $\frac{3}{4}$ be given to be reduced to a *Decimal* of two Places, or having 100 for its Denominator.

To 3 (the Numerator given) annex two Cyphers, and it makes 300; which divide by the Denominator 4, and the Quotient is .75, the *Decimal* required, and is equivalent to $\frac{3}{4}$ given.

Note, That ſo many Cyphers as you annex to the given Numerator, ſo many Places muſt be pricked off in the *Decimal* found; and if it ſhall happen that there are not ſo many Places of Figures in the Quotient, the Deficiency muſt be ſupply'd, by prefixing ſo many Cyphers before the Quotient-Figures, as in the next Example.

Example 2. Let $\frac{3}{573}$ be reduced to a *Decimal*, having ſix Places.

To the Numerator annex ſix Cyphers, and divide by the Denominator, and the Quotient is 5235: But it was required to have ſix Places, therefore you muſt prefix two Cyphers before it, and then it will be .005235, which is the *Decimal* required, and is equivalent to $\frac{3}{573}$.

See the Work of theſe two Examples.

4) 3.00(.75

```
4)3.00(.75     573)3.000000(5235
  28                      ....
  --              -------
  20               1350
  20                2040
  --                 3210
  ..                -------
  --                  345
                     ---
```

In the ſecond Example, there remains 345, which Remainder is very inſignificant, it being leſs than $\frac{1}{1000000}$ Part of an Unit, and therefore is rejected.

II. *To find the Value of a* Decimal *in the known Parts of Money, Weight, Meaſures,* &c.

The RULE.

Multiply the given *Decimal* by the Number of Parts in the next inferior Denomination, and from the Product prick off ſo many Places to the Right Hand as there were Places in the *Decimal* given; and multiply thoſe Figures prick'd off by the Number of Parts in the next inferior Denomination and prick off ſo many Places as before, and ſo continue to do, till you have brought it to the loweſt Denomination required.

Example 1. Let .7565 of a Pound Sterling be given to be reduced to Shillings, Pence, and Farthings.

Multiply by 20, by 12, and 4, as the Rule directs, and always prick off four Places to the Right Hand, and you will find it to make 15*s.* 1*d.* 2*q.* See the Work.

```
   .7565
      20
s. -----
 15.1300
      12
d. -----
  1.5600
       4
q. -----
  2.2400
  ------
```

A more compendious Way of finding the Value of a Decimal *of a Pound Sterling.*

Double the firſt Figure, (or Place of Primes,) and it makes ſo many Shillings; and if the next Figure (or Place of Seconds) be 5, or more than 5, for the 5 add another Shilling to the former Shillings; then for every Unit in the ſecond Place count ten, and to that add the Figure in the third Place, and reckon that ſo many Farthings; but if they make above 13, abate 1; and if it be above 38, abate 2, and add the remaining Farthings to the Shillings before found.

Example 1. Let .695 of a Pound be reduced to Shillings, Pence, and Farthings.

Firſt, Double your 6, and it makes 12*s.* then take 5 out of 9, and for that reckon another Shilling, and it makes 13*s.* and the 4 remaining is four Tens, and the 5 makes 45, which being above 38, you muſt therefore caſt away 2, and there reſts 43 Farthings, which is 10*d.* $\frac{3}{4}$. So the Anſwer is 13*s.* 10*d.* $\frac{3}{4}$.

	l.	s.	d.
So the Value of .725 =		14	6
And the Value of .878 =		17	$6\frac{3}{4}$
And the Value of .417 =		8	4

And ſo of any other.

Let .59755 of a Pound Troy be reduced to Ounces, Penny-Weights, and Grains.

Multiply by 12, by 20, and by 24, and always prick off five Places towards the Right Hand, and you will find the Anſwer to be 7 *oz.* 3 *pwt.* 10 *gr. ferè*. See the Work.

```
.59755
    12
--------
7.17060
     20
--------
3.41200    Facit  oz. pwt. gr.
     24            7   3   9.888
--------
 164800
 82400
--------
9.88800
```

Let .43569 of a Ton be reduced to Hundreds, Quarters, and Pounds.

Multiply by 20, by 4, and by 28, and the Anſwer will be 8 *C.* 2 *qrs.* 24 *lb. ferè*.

```
.43569
     20
--------
8.71380
      4          C. qrs. lb.
--------   Facit 8   2   23.9456
2.85520
     28
--------
23.94560
```

Let .9595 of a Foot be reduced into Inches, and Quarters.

```
 .9595
    12
--------
11.5140
      4    Facit 11 Inches 2 Qrs.
--------
 2.0560.
```

III. *To reduce the known Parts of Money, Meaſure, &c. to a Decimal.*

The RULE.

To the Number of Parts of the leſſer Denomination given, annex a competent Number of Cyphers, and divide by the Number of ſuch Parts that are contained in the greater Denomination, to which the *Decimal* is to be be brought; and the Quotient is the *Decimal* sought.

Example 1. Let 6*d.* be reduced to the *Decimal* of a Pound.

To 6 annex a competent Number of Cyphers, (suppose 3) and divide the Result by 240, (the Pence in a Pound,) and the Quotient is the *Decimal* requir'd.

```
240)6.000(.025
    ----
    1200        Facit .025
    ----
    ....
    ----
```

Example 2. Let 3*d.* ¾ be reduced to the *Decimal* of a Pound, having six Places.

In 3*d.* ¾ there are fifteen Farthings, therefore to 15 annex six Cyphers, (because there are to be six Places in the *Decimal* requir'd,) and divide by 960, (the Farthings in a Pound,) and the Quotient is .015625.

```
96|0)15 00000|0(.015625
        .....
        -----
       540
        600
         240
          480
          ---
          ...
          ---
```

Example 3. Let 3¼ Inches be reduced to the *Decimal* of a Foot, consisting of four Places.

In three ¼ Inches there are 13 Quarters; therefore to 13 annex four Cyphers, and divide by 48, (the Quarters in a Foot,) and the Quotient is .2708.

```
48)13.0000(.2708
   ----
    340
     400
     ---
      16
      ---
```

Example 4. Let 9*C.* 1 *qr.* 16*lb.* be reduced to the *Decimal* of a Ton, having six Places.

```
C. qr. lb.
9  1   16
4
-
37 qrs.
28
---
302     Facit .469642
75
-----
10552 Pounds.
-----
```

```
224|0)1052.00000|0(.469642
           ..... .
       ----------
       15600
        21600
         14400
          9600
           6400
           ----
           1920
           ----
```

Addition of Decimals.

Addition of *Decimals* is perform'd the ſame Way as Addition of whole Numbers, only you muſt obſerve to place your Numbers right, that is, Units under Units, Primes under Primes, Seconds under Seconds, &c.

Example. Let 317.25, 17.125, 275.5, 47 3579, and 12.75, be added together into one Sum.

	317.25
	17.125
	275.5
	47.3579
	12.75
Sum	669.9829

This is ſo plain, that more Examples I think needleſs.

Subtraction of Decimals.

Subtraction of *Decimals* is perform'd likewiſe the ſame Way as in whole Numbers, Reſpect being had to the right placing the Numbers, (as in Addition,) as in the following Examples.

	(1)		(2)
From	212.0137	From	201.1250
Subtr.	31.1275	Subtr.	5.5785
Reſts	180.8862	Reſts	195.5465
Proof	212.0137	Proof	201.1250

	(3)		(4)
From	2051.315	From	30.5
Subtr.	79.172	Subtr.	7.2597
Reſts	1972 143	Reſts	23.2403
Proof	2051 315	Proof	30.5

Note, If the Number of Places in the *Decimals* be more in that which is to be ſubtracted, than in that which you ſubtract from, you muſt ſuppoſe Cyphers to make up the Number of Places, as in the fourth Example.

Multiplication

Multiplication of Decimals.

Multiplication of *Decimals* is alſo perform'd the ſame Way as Multiplication of whole Numbers; but to know the Value of the Product, obſerve this Rule.

Cut off, or ſeparate by a Comma, or Prick, ſo many *Decimal Places* in the Product, as there are Places of *Decimals* in both Factors, *viz.* in the Multiplicand and Multiplier, which I ſhall farther explain in the following Examples.

Let 3.125 be multiplied by 2.75; multiply the Numbers together, as if they were whole Numbers, and the Product is 8.59375; and becauſe there were three Places of *Decimals* prick'd off in the Multiplicand, and two Places in the Multiplier, therefore you muſt prick off five Places of *Decimals* in the Product, as you may ſee by the Work.

```
   3.125
    2.75
  ------
   15625
  21875
  6250
  ------
 8.59375
```

Let 79.25 be multiplied by .459.

In this Example, becauſe two Places of *Decimals* are prick'd off in the Multiplicand, and three in the Multiplier, therefore there muſt be five prick'd off in the Product.

```
   79.25
    .459
  ------
   71325
  39625
 31700
 -------
36.37575
```

Let .135272 be multiplied by .00425.

In this Example, becauſe in the Multiplicand are ſix *Decimal* Places and in the Multiplier five Places; therefore in the Product there muſt be eleven Places of *Decimals*, but when the Multiplication is finiſhed, the Product is but 57490600, *viz.* only eight Places; therefore, in this Caſe, you muſt prefix three Cyphers before the Product-Figures, to make up the Number of eleven Places; ſo the true Product will be .00057490600.

```
      .135272
       .00425
      -------
       676360
      270544
     541088
 ------------
 .00057490600
```

More Examples for Practice.

```
    .001472        .017532
     .1045            347
    -------        -------
      7360         122724
     5888          70128
    1472          52596
 -----------     ---------
 .0001538240     6.083604

    279.25        32.0752
     .445           .0325
   --------      ---------
    139625        1603760
   111700         641504
  111700         962256
  ---------     ----------
  124.26625     1.04244400

    4.443         20.0291
    15.98           35.45
   -------       ---------
    35544        1001455
   39987         801164
  22215         1001455
  4443          600873
 ---------     -----------
 70.99914      710.031595

   7.3564         .75432
    .0126          .0356
   -------       --------
   441384         452592
  147128         377160
  73564         226296
 ----------    ------------
 .09269064     .026853792
```

Contracted Multiplication of Decimals.

Becaufe in Multiplication of *Decimal Parts* and mix'd Numbers, there is no need to exprefs all the Figures of the Product, but in moft Cafes two, three, or four Places of *Decimals* will be fufficient; therefore, to contract the Work, obferve this following

RULE.

Write the Unit's Place of the Multiplier under that Place of the Multiplicand, whofe Place you intend to keep in the Product; then invert the Order of all the other Figures, that is, write them all the contrary Way. Then in multiplying always begin at that Figure in the Multiplicand which ftands over the Figure you are then multiplying withal, and fet down the firft Figure of each particular Product directly one under the other; but yet a due Regard muft be had to the Increafe arifing from the Figures on the Right Hand of that Figure in the Multiplicand, which you begin to multiply at. This will appear more plain by Examples.

EXAMPLE I.

Let 2.38645 be multiply'd by 8.2175, and let there be only four Places retain'd in the *Decimals* of the Product.

First,

Firſt, according to the Directions, write down the Multiplicand, and under it write the Multiplier, thus; Place 8 (being the Unit's Place of the Multiplier) under 4, the fourth Place of *Decimals* in the Multiplicand, and write the reſt of the Figures quite contrary to the uſual Way, as in the following Work: Then begin to multiply, firſt the 5 which is left out, (only with Regard to the Increaſe, which muſt be carry'd from it,) ſaying, 8 times 5 is 40, carry 4 in your Mind, and ſay 8 times 4 is 32, and 4 I carry is 36; ſet down 6, and carry 3, and proceed through the reſt of the Figures, as in common Multiplication: Then begin to multiply with 2, ſaying, two times 4 is 8, for which I carry 1, (becauſe it is above 5,) and ſay, two times 6 is 12, and 1 that I carry is 13; ſet down 3, and carry 1, and proceed through the reſt of the Figures: Then multiply with 1, ſaying, once 6 is 6, for which carry 1, and ſay, once 8 is 8, and 1 is 9; ſet down 9, and proceed: Then multiply with 7, ſaying, ſeven times 8 is 56, for which carry 6, (becauſe it is above 55,) and ſay, ſeven times 3 is 21, and 6 that I carry is 27; ſet down 7, and carry 2, and proceed: Then multiply with 5, ſaying, five times 3 is 15, for which carry 2, and ſay, five times 2 is 10, and 2 I carry is 12, which ſet down, and add all the Products together; and the total Product will be 19.6107. See the Work.

```
2.38645
5712.8
-------
 190916
   4773
    239
    167
     12
-------
19.6107
-------
```

Note, That in multiplying the Figure left out every time next the Right Hand in the Multiplicand, if the Product be 5, or upwards to 10, you carry 1; and if it be 15, or upwards to 20, carry 2; and if 25, or upwards to 30, carry 3, &c.

I have here ſet down the Work of the laſt Example, wrought by the common Way, by which you may ſee both the Reaſon and Excellency of this Way, all the Figures on the right Hand the Line being wholly omitted.

```
     2.38645
      8.2175
------------
     11|93225
    167|0515
    238|645
   4772|90
 190916|0
------------
 19.6106|52875
------------
```

Example 2. Let 375.13758 be multiplied by 167324, ſo that the Product may have but four Places of *Decimals*.

Firſt,

First, set 6, the Unit's Place of the Multiplier, under 5, being the fourth Place of *Decimals* in the Multiplicand, (because four Places of *Decimals* were to be prick'd off,) and write all the rest of the Figures backward. Then multiply all the Figures of the Multiplicand by 1, after the common Way. Then begin with the second Figure of the Multiplier 6, saying six times 8 is 48, for which I carry 5, (in respect of the 8 left out,) and six times 5 is 30, and 5 that I carry is 35; set down 5 and carry 3, and proceed after the common Method. Then begin with 7, the third Figure of the Multiplier, and say, seven times 5 is 35, for which carry 4, and say, seven times 7 is 49, and 4 I carry is 53; set down 3 under the first, and carry 5, and proceed as before. Then beginning with 3, the fourth Figure of the Multiplier, and say, three times 7 is 21, carry 2, and say three times 3 is 9, and 2 I carry is 11; set down 1, and carry 1, and proceed as before Then begin with 2, the fifth Figure, and say, two times 3 is 6, for which I carry 1, and say, two times 1 is 2, and 1 carry is 3; set down 3; and two times 5 is 10; set down 0, and carry 1, and proceed as before. Then begin with 4, the last Figure of the Multiplier, and say, four times 1 is 4, for which I carry nothing, because 'tis less than 5; then say, four times 5 is 20; set down 0, and carry 2; and proceed through the rest of the Figures of the Multiplicand. Then add all up together, and the Product is 6276.9520. See the Work.

375.13758	the Multiplicand.
4237.61	the Multiplier revers'd.
37513758	the Product with 1.
22508255	the Product with 6 increas'd with 6 × 8.
2625963	the Product with 7 increas'd with 7 × 5.
112541	the Product with 3 increas'd with 3 × 7.
7503	the Product with 2 increased with 2 × 3.
1500	the Product with 4 increas'd with 0.
6276 9520	the Product requir'd.

Let the same Example be repeated, and let only one Place in the *Decimals* be prick'd off.

375·13758 the Multiplicand.
4237·61 the Multiplier inverted.

37514 the Product by 1 with the Increase of 1 × 7.
22508 the Product with 6 increased with 6 × 3.
2626 the Product of 7 increased with 7 × 1.
113 the Product with 3 increased with 3 × 5.
7 the Product with 2 increas'd with 2 × 7.
1 the Increase only of 4 × 3.

6276.9 the Product is the same as before.

More Examples for Practice.

Multiply 395·3756 by ·75642, and prick off four Places in *Decimals*.

395·3756 the Multiplicand.
24657· the Multiplier revers'd.

2767629 the Product by 7 increased with 7 × 6.
197688 the Product by 5 increased with 5 × 5.
23722 the Product by 6 increased with 6 × 7.
1581 the Product by 4 increased with 4 × 3.
79 the Product by 2 increased with 2 × 5.

299 0699 the Product required.

Let the same Example be repeated, and let there be only one Place of *Decimals*.

395·3756
24657·

1767 the Product by 7 increased with 7 ×
198 the Product by 5 increased with 5 × 5.
24 the Product by 6 increased with 6 × 9 + 6 × 5.
2 the Increase of 4 × 9 + 4 × 3.

299 1 the Product.

Characters

Characters, and their Signification,

Note, That this Mark + ſignifies Addition; as 8+5, that is, 8 more 5, or 8 added to 5; and 8+3 +7 denotes theſe Numbers are to be added into one Sum.

This — ſignifies Subtraction, as 9—4 ſignifies that 4 is to be taken from 9.

This Mark × ſignifies Multiplication, as 7×5 ſignifies that 7 is to be multiply'd into 5.

This Mark ÷ ſignifies Diviſion, as 12 ÷ 4 ſignifies 12 is to be divided by 4.

This Mark ſignifies Equality, or Equation; that is, when = is placed between Numbers, or Quantities, it denotes them to be equal, as 7+5=12, that is, 7 more 5 is equal to 12; and 15—7=8, that is, 15 leſs by 7, is equal to 8, or ſubtract 7 from 15, and there remains 8.

This Mark : : is the Sign of Proportion, or the Golden Rule, it being always placed betwixt the two middle Terms or Numbers in Proportion, thus, 4 : 20 : : 6 : 30, to be thus read, as 4 is to 20, ſo is 6 to 30.

Diviſion of Decimals.

Diviſion of *Decimals* is perform'd after the ſame Manner as Diviſion of whole Numbers; but to know the Value or Denomination of the Quotient, is the only Difficulty; for the reſolving of which, obſerve either of the following

RULES.

I. The firſt Figure in the Quotient muſt be of the ſame Denomination with that Figure in the Dividend which ſtands (or is to be ſuppoſed to ſtand) over the Unit's Place in the Diviſor, at the firſt ſeeking.

II. When the Work of Diviſion is ended, count how many Places of *Decimal* Parts there are in the Dividend more than in the Diviſor, for that Exceſs is the Number of Places which muſt be ſeparated in the Quotient for *Decimals:* But if there be not ſo many Figures in the Quotient, as is the ſaid Exceſs, that Deficiency muſt be ſupply'd with Cyphers in the Quotient, prefixed before the ſignificant Figures thereof, towards the Left Hand, with a Point before them; ſo ſhall you plainly diſcover the Value of the Quotient.

Theſe following Directions ought alſo to be carefully obſerv'd.

If the Diviſor conſiſts of more Places than the Dividend, there muſt be a competent Number of Cyphers annexed to the Dividend, to make it conſiſt of as many (at leaſt) or more Places of *Decimals* than the Diviſor; for the Cyphers added muſt be reckon'd as *Decimals.*

Conſider whether there be as many *Decimal* Parts in the Dividend as there are in the Diviſor; if there be not, make them ſo many, or more, by annexing of Cyphers.

In

In dividing of whole or mixed Numbers, if there be a Remainder, you may bring down more Cyphers, and by continuing your Diviſion, carry the Quotient to as many Places of *Decimals* as you pleaſe.

Theſe Things being conſidered, I ſhall proceed to the Practice of Diviſion of *Decimals*, which I ſhall endeavour to explain in as familiar and as eaſy a Method as poſſible.

Example 1. Let 48 be divided by 144.

In this Example the Diviſor 144 is greater than the Dividend 48; therefore, according to the Directions above, I annex a competent Number of Cyphers, (*viz.* four,) with a Point between them, and divide according to the uſual Way.

```
144) 48.0000(.3333
     -------
     480
      480
       480
      ----
        48
```

But, firſt, in ſeeking how often 144 in 48.0, (the firſt three Figures of the Dividend,) I find the Unit's Place of the Diviſor to fall under the firſt Place of the *Decimals*; therefore the firſt Figure in the Quotient is in the firſt Place of *Decimals*: Or, by the ſecond Rule, there being four Places of *Decimals* in the Dividend, and none in the Diviſor; ſo the Exceſs of *decimal* Places in the Dividend, above that in the Diviſor, is four; ſo that when the Diviſion is ended, there muſt be four Places of *Decimals* in the Quotient. See the Work.

Example 2. Let 217.75 be divided by 65.

Firſt, in ſeeking how oft 65 in 217, (the firſt three Figures of the Dividend,) I find the Unit's Place of the Diviſor to fall under the Unit's Place of the Dividend; therefore the firſt Figure in the Quotient will be Units, and all the reſt *Decimals*. Or, by the ſecond Rule, there being two Places of *Decimals* in the Dividend, and no *Decimals* in the Diviſor, therefore the Exceſs of *decimal* Places in the Dividend, above the Diviſor, is two; ſo when the Diviſion is ended, ſeparate two Places in the Quotient towards the Right Hand by a Point. See the Work.

```
65)217.75(3.35
   . . .
   -----
    227
     325
     ---
     . . .
     ---
```

Example 3. Let 267.15975 be divided by 13.25.

13.25)267.15975(20.163
. . . . .

2159
8347
3975

. . . .

In this Example, 3, the Unit's Place of the Divisor, falls under 6, the Ten's Place of the Dividend; therefore (by the first Rule) the first Figure in the Quotient is Tens. Or, by the second Rule, the Excess of *Decimal* Places in the Dividend, above the Divisor, is three; there being five Places of *Decimals* in the Dividend, and but two in the Divisor, so there must be three Places of *Decimals* in the Quotient.

Example 4. Let 15.675159 be divided by 375.89.

375.89)15.675159(.0417
. .

63955
263669

546

In this Example, 5, the Unit's Place of the Divisor, falls under 7, the second Place of *Decimals* in the Dividend; therefore (by the first Rule) the first Figure in the Quotient is in the second Place of *Decimals*; so that you must put a Cypher before the first Figure in the Quotient: And (by the second Rule) the Excess of *decimal* Places in the Dividend, above the Number of *decimal* Places in the Divisor, is 4; for the *decimal* Places in the Dividend is 6, and the Number of Places in the Divisor but two; therefore there must be four Places of *Decimals* in the Quotient. But the Division being finished after the common Way, the Figures in the Quotient are but three, therefore you must prefix a Cypher before the significant Figures.

Example 5. Let 72.1564 be divided by .1347.

.1347)72.1564(535.68
. . .

4806
7654
9190
11080

304

In this Example, the Divisor being a *Decimal*, the first Figure thereof falls under the Ten's Place in the Dividend; therefore the Units (if there had been any) should fall under the Hundred's Place in the Dividend; and so the first Figure in the Quotient is Hundreds. And, by the second Rule, there being four Places of *Decimals* in the Dividend, and as many in the Divisor, so the Excess is nothing; but in dividing, I put two Cyphers

phers to the Remainders, and continue the Divifion in two Places farther, fo I have two Places of *Decimals*. See the Work.

Example 6. Let .125 be divided by .0457.

```
.0457).1250000(2.735
        ....
     -------
     914
     -------
     3360
     3199
     -------
      1610
      1371
     -------
       2390
       2285
       -------
        105
       ----
```

In this Example, the Unit's Place of the Divifor (if there had been any) would fall under the Units Place of the Dividend; therefore the firft Figure of the Quotient is Units. And by the fecond Rule, there being feven Places of *Decimals* in the Dividend, and but four Places in the Divifor, fo the Excefs is three; therefore there muft be three Places of *Decimals* in the Quotient.

I fhall fet down only the Work of fome few Examples more, and fo proceed to *Contracted Division*.

```
.00456(.0000059791(.00131
                ..
         -----------
                1419
                 511
                 ----
                  55
                 ----
```

Let it be divided by 282.

```
282)1.0000000(.0035461 ferè.
        .....
     --------
     1540
      1300
       1720
        280
        ----
```

```
.325).400000(1.2307
          ...
       ---------
       750
       1000
        2500
        -----
         225
        -----
```

```
.042)495.00000(11785.71
       . .....
      -----------
      75
      330
       360
        240
         300
          60
          ----
          18
          ----
```

Diviſion of Decimals *contracted.*

In Diviſion of *Decimals* the common Way, when the Diviſor hath many Figures, and it is required to continue the Diviſion till the Value of the Remainder be but ſmall, the Operation will ſometimes be large and tedious, but may be excellently contracted by the following Method.

The RULE.

By the firſt Rule of this Chapter, find what is the Value of the firſt Figure in the Quotient; then, by knowing the firſt Figure's Denomination, you may have as many or as few Places of *Decimals* as you pleaſe, by taking as many of the left Hand Figures of the Diviſor as you think convenient for the firſt Diviſor; and then take as many Figures of the Dividend as will anſwer them; and in dividing, omit one Figure of the Diviſor at each following Operation. A few Examples will make it plain.

Example 1. Let 721 17562 be divided by 2 25743, and let there be three Places of *Decimals* in the Quotient.

```
2.25743) 721.175|62 (319.467
 .....   677229
         ------
          43946
          22574
          -----
          21372
          20317
          -----
           1055
            903
           ----
            152
            135
            ---
             17
             15
             --
              2
             --
```

In this Example the Unit's Place of the Diviſor falls under the Hundred's Place in the Dividend, and it is required that three Places of *Decimals* be in the Quotient; ſo there muſt be ſix Places in all, that is, three Places of the whole Numbers, and three Places of *Decimals*. Then, becauſe I can have the Diviſor in the firſt ſix Figures of the Dividend, I cut off the 62 with a Daſh of the Pen, as uſeleſs; then I ſeek how oft the Diviſor in the Dividend, and the Anſwer is three Times; put three

in

in the Quotient, and multiply and subtract as in the common Division, and the Remainder is 43946. Then prick off three in the Divisor, and seek how often the remaining Figures may be had in 43946, the Remainder, which can be but once; put 1 in the Quotient, and multiply and subtract, and the next Remainder is 21372. Then prick off the 4 in the Divisor, and seek how often the remaining Figures may be had in 21372, which will be 9 times; put 9 in the Quotient, multiply thus; saying nine times 4 is 36, for which I carry 4, (in respect of the 4 last prick'd off,) and nine times 7 is 63, and 4 is 67; set down 7, and carry 6, and so proceed till the Division be finished, always respecting the Increase made from the Figures prick'd off. Observe the Work, which will better inform you than many Words.

```
2.25743) 721.17562 (319.467
                 . . .
       677229|
       ------
        43946|6
        22574|3
       ------ --
        21372|32
        20316|87
       ------ ---
         1055|450
          902|972
          --- ---
          152|4780
          135|4458
          --- ----
           17|03220
           15|80201
           -- -----
            1|23019
           -- -----
```

I have set down the Work of this last Example at large, according to the common Way, that thereby the Learner may see the Reason of the Rule, all the Figures on the right Side the perpendicular Line being wholly omitted.

Example 2. Let 5171.59165 be divided by 8.758615, and let it be required that four Places of *Decimals* be prick'd off in the Quotient.

```
8.758615)5171.5916|5(590.4577
 . . . . . .        .
           43793075
           --------
            7922841
            7882754
            -------
              40087
              35034
              -----
               5053
               4379
               ----
                674
                613
                ---
                 61
                 61
                 --
                 ..
                 --
```

In this Example I can't have 8, the first Figure in the Divisor, in 5, the first Figure of the Dividend; so that the Unit's Place of the Divisor falls under the Hundred's Place of the Dividend; so that there will be seven Figures in the Quotient, that is, three of whole Numbers, and four of *Decimals*; therefore there must be seven Figures in the Divisor,

 (because

(because the Number of Places in the Divisor and Quotient will be equal,) and there must be eight Places in the Dividend; so that I cut off the Figure 5 with a Dash, as useless. Thus having proportion'd the Dividend to the Divisor, and both to the Number of Places or Figures desired in the Quotient, I proceed to divide as before, saying, how often 8 in 51, which will be five times; put 5 in the Quotient, and multiply and subtract, and the Remainder is 7922841. Then I prick off the first Figure in the Divisor 5, and seek how often the remaining Figures of the Divisor are in the aforesaid Remainder, which I find 9 times; put 9 in the Quotient, and multiply thereby, saying, nine times 5 (the Figure prick'd off) is 45, for which I carry 5; and say nine times 1 is 9, and 5 I carry is 14; set down 4, and carry 1, and proceed to multiply the rest of the Figures, and subtract, and the Remainder will be 40087. Then prick off the Figure 1, and seek how often 87586 in the Remainder 40087, the Answer will be 0; so put 0 in the Quotient, and prick off the Figure 6, and seek how often 8758 in 40087, which will be four times; put 4 in the Quotient, and multiply, saying, four times 6 (the Figure last prick'd off) is 24, for which I carry 2; and say, four times 8 is 32, and 2 I carry is 34; set down 4 and carry 3; multiply the rest of the Figures, and subtract as before, and so proceed after the same Manner, until all the Figures of the Divisor be prick'd off, to the last Figure. See the Work.

Example 3. Let 25.1367 be divided by 217.3543, and let there be five Places of *Decimals* in the Quotient.

In this Example, 7, the Unit's Place of the Divisor, falls under 1, the first Place of *Decimals*; therefore the first Figure of the Quotient is in the first Place of *Decimals*, so the Quotient will be all *Decimals*: Then because the Quotient-Figures, and the Figures of the Divisor will be of an equal Number, dash off the 43 in the Divisor, and the 7 in the Dividend, as useless, and divide as before.

```
217.35|43) 25.136|7(.11564
            21735
            -----
             3401
             2174
             ----
             1227
             1087
             ----
              140
              130
              ---
               10
                8
                -
                2
                -
```

Although I have hitherto given Directions for proportioning the Divisor and Dividend, so as to bring into the Quotient what Number of *Decimals* you please, yet there is no absolute Necessity for it; but you may carry on your Division to what Degree you please, before you begin to prick off the Figures of the Divisor, in order to contract the Work, as in the following Examples,

amples, where it is not required to prick off any determinate Number of *Decimals*, but it may be done according to Discretion.

```
2.756756) 741476717 (2689.67118
 ......       ...
          5513512
          --------
          19012551
          16540536
          --------
           24720157
           22054048
           --------
            2666109
            2481080
            -------
             185029
             165405
             ------
              19624
              19297
              -----
                327
                276
                ---
                 51
                 28
                 --
                 23
                 22
                 --
                  1
```

```
12.34254)514.75498(41.705757
     ... 4937016
         -------
         2105338
         1234254
         -------
          871084
          863978
         -------
            7106
            6171
            ----
             935
             864
             ---
              71
              62
              --
               9
               8
               -
               1
               -
```

DECK-NAILS. See NAILS.

DECORATION, in Architecture, an Ornament in a Church, or other publick Place; or what adorns and inriches a Building, Triumphal Arch, *&c.* either on the Inside or without.

The Orders of Architecture contribute a great deal to the *Decoration*; but then the several Parts of these Orders must have their just Proportion, Characters, and Ornaments; or otherwise the finest Order will bring Confusion rather than Richness.

Decorations in Churches are Paintings, Vases, Festoons, *&c.* occasionally placed on or against the Walls, but so discretionally, as not to take off any thing from the Form and Beauty of the Architecture, as is much practised in *Italy* at the solemn Feasts.

Decorations also signify the Scenes of Theatres. The *Decorations* in Opera's, and other theatrical Performances, must be often changed, in Conformity to the Subject.

The Antients had two Sorts of *Decorations* for their Theatres: The first, called *Versatiles*, having three Sides or Faces, which were turned successively to the Spectators. The other, called *Ductiles*, which were drawn or slidden before others.

The latter Sort of *Decoration* is still in use, and to greater Advantage among us than the Antients, who were under a Necessity of drawing a Curtain whenever a Change was made in the *Decoration*; whereas the Change is made in a Moment, and without scarce being perceiv'd upon our Stage.

DECORUM } i. e. *Decency*,
DECOR } is particularly used in Architecture for the Suitableness of a Building, and the several Parts and Ornaments thereof, to the Station and Occasion.

Vitruvius is very exact in this Point, and gives Rules expresly for the appropriating or suiting the several Orders to their natural Characters: So that, *e.g.* a *Corinthian* Column should not be set at the Entrance of a Prison or Gate-House, or a *Tuscan* in the Portico of a Church, as has been done by some of our Builders, who have offended even in the Disposition of the Offices in our ordinary Houses; we often finding the Kitchen set where the Parlour should be; and that in the first and the best Story, which should have been condemned to the lowest and worst.

Some interpret *Decorum* to signify the observing a due Respect between the Inhabitant and Habitation: Whence *Palladio* concludes, that the principal Entrance must never be regulated by any certain Dimensions, but according to the Dignity of the Person who is to live in it; yet to exceed rather in the more than in the less, is a Token of Generosity, and may be excused with some notable Emblem or Inscription, as that of the *Conti di Bavilaiqua* over his large Gate at *Verona*, (where probably there had been some Disproportion committed;) *Patet janua Cor magis*, i.e. My Gate is wide, but my Heart more wide.

DECUPLE, in Arithmetick, a Term of Relation or Proportion, implying a Thing to be ten times as much as another.

DECUSSATION, in Geometry, Opticks, &c. the Point in which two Lines, Rays, &c. cross or intersect each other.

DEFICIENT NUMBERS, in Arithmetick, are such, whose Parts added together, make less than the Integer, *e.g.* 8, whose quota Parts are 1, 2, and 4, which together make no more than 7.

DENTICLES } in Architecture, an Ornament in Corniches, bearing some Resemblance to Teeth, particularly affected in the *Ionic* and *Corinthian* Orders.
DENTILS }

They are cut on a little square Member, properly called *Denticulus*; and the Notches or Ornaments themselves *Dentes*, from *Dens*, *L.* as resembling a Row of Teeth.

In antient Times *Dentils* were never used in the *Ionic* Cornice; yet they are found in the Remains of the Theatre of *Marcellus*; which some take for an Argument, that *Vitruvius* had not the Direction of that Building.

Vitruvius prescribes the Breadth of each *Dentil*, or Tooth, to be its Height; and the Indenture or Interval between each two, he directs to be two Thirds of the Breadth of the *Dentil*.

He also in his fourth Book observes, that the *Greeks* never put *Dentils* under Modillons, because Modilions represent Purlins; whereas *Dentils* represent the Ends of Rafters, which can never be placed underneath Purlins.

The *Romans* were not so scrupulous as to this *Decorum*, except in the Pantheon, where there are no *Dentils* under the Modilions, neither in the Portico, nor the Inside of the Building.

DESCRIBENT, in Geometry, a Term expressing some Line or Surface, which by its Motion produces a Plane Figure, or a Solid.

DESIGN, in Architecture, &c. is the Draught, or the Thought, Plan, geometrical Representation, Distribution, and Construction of a Building, &c.

In Building, the Term Ichnography may be used, when by *Design* is only meant the Plan of a Building, or a flat Figure drawn on Paper. And when some Side or Face of the Building is raised from the Ground, we may use the Term Orthography; and when both Front and

and Sides are ſeen in Perſpective, we may call it Scenography.

DESIGNING is the Art of delineating or drawing the Appearance of natural Objects by Lines on a Plane.

DIAGONAL, in Geometry, is a Right Line drawn acroſs a Figure of ſeveral Sides, from the Vertex of one Angle. Some Authors call it Diameter, and others Diametral of the Figure.

Firſt, It is demonſtrated, that every *Diagonal* divides a Parallelogram into two equal Parts. *Secondly*, That two *Diagonals* drawn in any Parallelogram, biſect each other. *Thirdly*, That the *Diagonal* of a Square is incommenſurable with one of its Sides.

DIAGRAM, in Geometry, a Scheme for the Explanation or Demonſtration of any Figure or Proportion belonging to it.

DIAL, an Inſtrument whereby to know the Hour or Time of the Day, when the Sun ſhines. The firſt *Sun-Dial* that was ſet up in *Rome*, was erected by *Papyrius Curſor*, about the Year of the City, 447: For *Pliny* ſays, there was no Mention of any Account of Time, but of the Sun's Setting and Riſing. This Dial was ſet up in the Temple of *Quirinus*, but it went not right.

About thirty Years after this, M. *Valerius Meſſala*, ſays *Varro*, being Conſul, brought out of *Sicily*, from the taking of *Catana*, another *Dial* which he ſet up on a Pillar near the *Roſtrum*; but it not being made for that particular Latitude, it could not go true. Nevertheleſs they made uſe of it for eleven Years; and then *Marcius Philippus*, who was Cenſor with *Lucius Paulus*, ſet up another that was more exact.

The *Greeks* alſo were a long Time without Clocks and *Sun-Dials*. Some aſcribe the Invention of *Sun-Dials* to *Anaximenes Miletius*, and others to *Thales*.

There are many Kinds of *Dials* mention'd by *Vitruvius*; as one invented by *Beroſus* the *Chaldean*, which was on a reclining Plain, parallel almoſt to the Equinoctial, there was an half Circle upon it; and thence it was called *Hemicyclus*.

Ariſtarchus Samius found out the *Hemiſphere-Dial*. And there were ſome Spherical ones with a Needle for a Gnomon. The Diſcus of *Ariſtarchus* was an *Horizontal Dial*, with its Limb raiſed up all round to prevent the Shadow from extending itſelf too far off.

Dial-Planes are of two Sorts. *Firſt*, Such as are made on the Wall of a Building. Or, *Secondly*, ſuch as are drawn on Tables of Wood, commonly called *Dial-Boards*.

The firſt Sort, if they are made on Brick Work is done by Plaiſtering on the Wall with Lime, Sand, and Hair mix'd: This, if well drench'd with Linſeed Oil, after it is dry, or as long as it will drink in any, and afterwards with Oil and white Lead, may be durable enough.

But it will be a better Way to temper the Lime, Sand and Hair, with Ox Blood, which will be no great Charge, but of great Advantage; for this Mixture will equal in Time the Hardneſs

neſs of a Freeſtone, and keep the Surface as much from the Injuries of the Weather; but it muſt be afterwards painted white.

If you are to work on a Stone, the beſt Way is to drench the Stone with Linſeed Oil and White Lead, very thin, till it will drink in no more; then will the *Dial* you paint upon laſt the longer, and be the better prepar'd to reſiſt the Ruins of Time

Now for Tables or *Dial-Boards* of Wood, they being moſt common, I ſhall give ſuch Directions for the making of them, as have been always found moſt profitable and fit for the Purpoſe.

The beſt Wood for this Purpoſe is the cleareſt Oak, and the reddeſt Fir, provided it be not turpentiny. There is but little Difference between theſe two Woods as to their Alteration by the Weather, both being ſubject to ſplit, in caſe they are bound, and have not free Liberty to ſhrink with dry Weather, and to ſwell with Wet. But as to their laſting, Oak ſeems to be the better of the two: Though good Red Fir that is hard, will ordinarily laſt the Age of a Man, if it be ſecured as ſuch Things ought to be.

In working either of theſe Kinds of Woods, firſt cut the Boards to ſuch a Length as you intend the Deal Board ſhould be, and ſo many of them as may make up the Breadth deſigned; then let them be jointed on the Edges, and planed on both Sides, and afterwards ſet to dry: For it has been obſerv'd, that though Boards have lain in an Houſe ever ſo long, and are ever ſo dry, yet when they are thus ſhot and planed, they will ſhrink afterwards beyond Belief, if kept dry. When they have been thought to have been dry enough, and will ſhrink no more, let them be again ſhot with good Joints, and let every Joint be ſecured by two Wooden Dovetails, let in croſs the Joint on the Backſide; but let this be done when the Boards are glued together, and well dry'd.

After they have been thus glued, and the Joints are ſufficiently dry'd, then let the Face of the Board be well planed, and try'd every Way, that it may be both ſmooth and true, and all of a Thickneſs, as Pannels of Wainſcot are commonly wrought.

The Edges muſt be thus true and even, that they may fit into the Rabet of a Moulding put round it, juſt as a Pannel of Wainſcot does in its Frame. This will give Liberty to the Board to ſhrink, and ſwell without tearing; whereas Mouldings that are nailed round the Edge, as the common Way is, do ſo reſtrain the Motion of the Wood, that it cannot ſhrink without tearing; but Boards wrought after this Manner will laſt a long Time, without either parting in the Joints, or ſplitting in the Wood.

Dials are ſometimes drawn on Planes lin'd with Copper or Lead, that they may be free from ſplitting or tearing; but a Board (if it be made as above directed) is thought preferable to them in many Reſpects.

As,

As, *First*, It is much cheaper. *Secondly*, Lead (and Copper too, a little) will ſwell with the Heat of the Sun, and grow in Time hollow outwards or convex, inſtead of a perfect Flat, ſo that the Truth of its Shadow will be much injured. *Thirdly*, the Colours will be apt to peel from the Metal, and the *Dial* will by that Means be in danger to be ſooner defaced, than if it were painted on a Wooden Plane.

For Gluing the Joints of *Dial* Boards, ſee the Article GLUE.

Directions for Painting, &c. *a* Sun-Dial.

Four Colours are ſufficient for this Work, *viz. Spaniſh Brown*, for the Priming, or firſt Colour.

White Lead for the ſecond Colour, and finiſhing the Face of the Table.

Vermilion for drawing the Hour-Lines. And,

Lamp-Black for the Figures in the Margin, reſpecting the Lines of every Hour, if it be a plain *Dial*.

But if you would have the Figures gilded, then ſome others are required, as Gold, and the Size to lay it on, and Smalt, for a blue Ground, if you would have a rich Colour.

But ſome lay the Ground where the Figures are gilt, with Vermillion: And that ſhews well, if the Figures are liſted with black, and a black Moulding round the *Dial*.

The next Particular ſhould be the Practice of painting the *Dial*; but before that can be done, the Draught muſt be drawn; and therefore it will not be unſeaſonable to direct to the beſt Authors who have wrote on the Subject of *Dialling*. As,

First, *Stirrup*'s *Dialling*, as being of excellent Uſe to acquaint a young Learner with the Knowledge of the Sphere, that he may underſtand the Nature and Reaſon of *Dials*.

Secondly, *Collins*'s *Dialling*, a very uſeful Book.

Thirdly, *Leybourn*'s *Dialling*, in which you have the beſt Ways for drawing Eaſt and Weſt *Dials*, and far *Decliners*.

Fourthly, *Collins*'s *Sector on a Quadrant*: In which you have communicated the Cut of a Scale, that by knowing the Declination, gives all the reſt of the Requiſites of an Upright Decliner, by Inſpection only, with as great Exactneſs, as by the niceſt Calculation: Beſides, it teaches the Way of drawing the Hours of a *Dial* by a Tangent Line, and alſo by the Scale of Hours; two of the beſt and moſt expeditious Ways that ever were yet found out.

The Practice of painting Sun-Dials.

When according to the Rules given in the Books before mentioned, you have drawn on the Paper the Draught of your *Dial*, and your Board is ready, and alſo your Colours prepared, according to the Directions before given, you ſhould then in painting your *Dial*, proceed as follows:

Take *Spaniſh Brown*, that has been well ground, and mixed moderately

moderately thin, and with a large Briſtle-Bruſh dipped in it, colour the Board or Plane all over, on the Back as well as on the Fore-ſide, to preſerve it the better, ſo that no Part be left uncolour'd: This is called the Priming of a *Dial*. When this firſt Colour is dry, do it over again with more of the ſame Colour, temper'd ſomewhat thicker; and when this is alſo dry, you may, if you pleaſe, do it over again with the ſame Colour: The Work will be the ſubſtantialler, and laſt the longer.

When this laſt Time of Colouring with the Priming is done, then colour the Face of the *Dial-Plane* over with White Lead; and when that is dry, work it over again three or four Times more, ſucceſſively after each Drying; and ſo will the Face of the *Dial-Plane* be ſufficiently defended againſt the many Years Fury and Violence of the Weather.

When the laſt Colouring of the White is drawn, you muſt draw on the Plane, with a Black Lead Pencil, a Horizontal Line, ſo far diſtant from the uppermoſt Edge of the *Dial*, as your Diſcretion ſhall think fit, or your Experience finds ſhall be moſt becoming the Plane. Then ſet out the Margin of the *Dial* with Boundary Lines for the Hour, Half Hour, Quarters, and Quarter-Diviſions of the *Dial*, as you ſee done in moſt *Dials*.

When the Margin and Boundary Lines of the *Dial* are ſet out, then take your Paper-Draught, that has been fairly drawn, and place the Horizontal Line of that, on the Horizontal Line that you before drew on the Plane; in doing of which, obſerve to place the Centre according as the Situation of your Plane, for Convenience ſake, requires. Thus if your *Dial* be a Full-South *Dial*, then let the Centre be exactly in the Middle of the Plane: But if your *Dial* decline from the South, either Eaſt or Weſt, then place not the Centre of your Draught in the Centre of your Plane, but nearer to one Side or other of it, according as it declines, having alſo Regard to the Quantity of its Declination.

As for Example: If your *Dial* decline Eaſtwards, then let the Centre of your Draught be placed between the Centre and the Eaſtern Side of your Plane, the Quantity thereof muſt be according as your *Dial* declines: If it declines but a little, then place the Centre of your Draught but a little from the Centre of your Plane; and if it declines much, place the Centre of your Draught the more out of the Centre of your Plane.

The Reaſon of adviſing this, is, that by ſo doing you may gain a greater Diſtance for thoſe Hour-Lines, which in declining Planes, fall nearer together on one Side than they are on the other: For which Reaſon, it is uſual ſo to do in declining Planes, except they decline far, as between eighty and ninety Degrees: For in this Caſe, they are commonly drawn without Centres, to gain the greater Diſtance for the Hour-Lines.

When

When the Paper-Draught has been thus artificially placed on the Plane, and faſtened with Pins, or ſmall Tacks, then let the Draught of it be transferred on the Plane, by laying a Ruler over every Hour and Quarter-Diviſion, and where the Ruler cuts or interſects the Boundary Lines of the Margin, there make Marks, by drawing Lines with a Black Lead Pencil, of ſuch a Length as each Diviſion requires, (or is deſign'd by your Boundary Lines,) obſerving always to draw the Hour and and Half Hour Lines quite through your Margin, that they may be Guides for the right placing of the Figures, and for a ſmall Spot that is uſually placed in the Margin, right againſt the Half Hour.

When this *Dial* Draught has been thus transferred to the Plane itſelf, you muſt not forget to draw the Subſtilar Line according as it lies in your Draught, to be a Guide for the right placing the Stile or Cock; for you muſt be very exact in every Particular, or elſe the *Dial* will not be right.

Every Thing that is requir'd being taken from the Draught, and transferred to the Plane, then take the Draught off, and with Vermillion very well ground and prepar'd, as is before taught, let the Boundary Lines of your *Dial*, as alſo the Hour, Half Hour, and Quarter-Diviſion be drawn therewith: Let your Colour be as thick and as ſtiff as you poſſibly can work it, ſo as to draw a clear and ſmooth Line; becauſe this is to be done but once.

When your Vermillion Lines are drawn, then make the Figures with Lamp-Black, and a Spot in the Middle of the Margin, right againſt the Half Hour Line; and, if you pleaſe, in the Margin at the Top of your Plane, you may put the Date of the Year, your Name, and ſome Sentence as is uſual in Things of this Nature. Then fit in your Cock, ſo as to make right Angles with the Plane. So ſhall your *Dial* be drawn, and finiſh'd in all Reſpects as a plain *Dial* ought to be.

If you would have the Figures in Gold, ſee GILDING.

DIAMETER, in Geometry, is a Right Line, paſſing through the Centre of a Circle, and terminating on each Side at its Circumference.

The Properties of the *Diameter*, are,

Firſt, That it divides the Circumference into three equal Parts. And hence we have a Method of deſcribing a Semicircle upon any Line aſſuming a Point therein for the Centre.

Secondly, The *Diameter* is the greateſt of all the Chords.

Thirdly, To find the Ratio of the *Diameter* to the Circumference.

Archimides has found the Ratio of the *Diameter* to the Circumference, as 7 to 22.

Modern Practical Geometricians aſſume the *Diameter* to be to the Circumference, as 100 to 314.

Ad. Metius gives us the Ratio of the *Diameter* to the Circumference, as 113 to 355, which is the moſt accurate of all thoſe expreſs'd in ſmall Numbers, as not erring 3 in 100000000.

Diameter of a Column, is that taken juſt above the Baſe. From this the Module is taken, which meaſures all the other Parts of a Column.

Diameter of the Swelling, is that taken at the Height.

Diameter of the Diminution of Columns, is that taken from the Top of the Shafts.

DIAMOND-GLASS. See QUARRY.

Diamond-Pavement. See PAVING.

DIAPHANOUS, tranſparent, or pellucid, *i.e.* giving Paſſage to the Rays of Light, as Water, Air, Glaſs, Talk, fine Porcelane, *&c.*

DIAPHANEITY, the Quality of a tranſparent or pellucid Body.

DIASTYLE, in the antient Architecture, an Edifice where the Columns ſtand at ſuch a Diſtance one from another, that eight Modules, or four Diameters, are allowed for the Intercolumniation.

DIE, a Term apply'd to any ſquare Body, as the Trunk or Naked of a Pedeſtal, which is that Part included between its Baſe and Cornice.

DIGGING. The *Digging* of the Ground for Cellars, and for the Foundations of Buildings, is commonly done by the Yard ſolid, containing twenty-ſeven ſolid Feet, which is commouly counted a Load.

Therefore take the Dimenſion in Feet, multiply the Length by the Breadth, and the Product by the Depth, and then divide the laſt Product by 27, and the Quotient will give the Content in ſolid Yards.

DIMENSION is the Extenſion of a Body conſidered as capable of being meaſured.

Hence, as we conceive a Body extended, and capable of being meaſured in Length, Breadth, and Depth, we conceive a trine *Dimenſion*, *viz.* Length, Breadth, and Thickneſs. The firſt is called a Line, the ſecond a Surface, and the third a Solid

DIMINISHING *of Columns*. See COLUMN, *and* DIMINUTION.

DIMINUTION, in Architecture, is the Contraction of the upper Part of a Column whereby its Diameter is made leſs than that of the lower Part.

All Architects have made their Columns leſs above than below, with Deſign to attain thoſe two important Points in Architecture, Strength, and the Appearance.

Some again have made them a little bigger towards the Middle, than towards the Bottom, which is called the Swelling.

Indeed, neither *Diminution*, nor Swelling, are obſerved by the *Gothick* Architects, who make their Columns perfectly Cylindrical; for which Reaſon they are properly called Pillars, in Contradiſtinction to Columns.

The *Diminution of Columns* commences generally from one Third of the Height of the Column; though ſome begin it from the very Baſe of the Column, and ſo go on tapering to the Capital; but this is not eſteemed to have ſo good an Effect.

Vitruvius himſelf would have the *Diminution* of Columns be different, according to their Height, and not according to their Diameter: As for Example, he diminiſhes a Column of 15 Feet in Height a ſixth Part of its Diameter, and another of 50 only one eighth Part; but this Rule of *Diminution* is not found to have been obſerved in the Antique.

Mr. *Perrault* obſerves, that a Difference of Orders does not infer a Difference in *Diminutions*; there being in different Works of the ſame Order both ſmall and great *Diminutions*; but however, except the *Tuſcan* Order, which *Vitruvius* diminiſhes by a fourth Part; though *Vignola* does it only by a fifth, and the *Trajan* Column by a ninth.

Diminutions in Antique Buildings, are very differently adjuſted, as well as in different modern Authors.

M. *Le Clerc* ſays, all *Diminutions* of Columns begin to diminiſh in Thickneſs from one Third of their Height. But in Proportion, as their Orders are more delicate, their *Diminution* ought to be leſs ſenſible.

For inſtance, in the *Tuſcan* Order, where the Column is but 14 Modules high, its Semidiameter under the Aſtragal may be diminiſhed ſix Minutes.

In the *Tuſcan* Order, where the Column is 15 Modules high, its *Diminution* under the Aſtragal may be but 5 Minutes and a half.

In the *Doric* Order, where the Column is 16 Modules, the *Diminution* may be but five Minutes.

In the *Ionic*, where the Column is 18 Modules, the *Diminution* may be but four Minutes and a half.

And in the *Roman* and *Corinthian*, no more than four, that is, a *Diminution* of four Minutes on each Side the Axis, is the utmoſt that the Column will undergo, though it always increaſes in Height in Proportion to its Thickneſs.

Indeed, ſays he, according to ſome Authors, the *Diminution* of Columns, even of the ſame Order, ought to be greater or leſs, according as their Heights are greater or leſs.

For Inſtance, a *Doric* Column, ſay they, 20 Feet high, muſt have leſs *Diminution* than another of 15 Feet; and one of 30 leſs than one of 20. The Reaſon they give for this, is, that the Greatneſs of the Height eaſily impoſes on the Sight; and hence they conclude, that a very tall Column muſt of itſelf appear diminiſhed towards the Top.

Nor can it be denied, but that this holds true, where the Eye is placed near, and looks up from the Bottom to the Top of the Column; but then it is to be conſidered, that large Columns are never made with deſign to be viewed thus near; but always at a Diſtance ſuitable to their Height: And it would be ridiculous to ſpoil their Proportions, out of Complaiſance to ſuch as ſhould pleaſe to view them at an improper Diſtance.

Therefore, ſays he, in my Opinion, when any certain *Diminution* of a Column has been once eſtabliſhed, provided it does but pleaſe the Eye when viewed at a Diſtance, it ought never to be changed on Occaſion of any. Alteration in the Height of the Column, excepting it ſhould be found in ſome cloſe narrow Place; which yet can never happen, unleſs in the Inſide of a Building: For Inſtance, of a Dome, or the like; to which a prudent Architect will always have a particular Regard.

But, ſays he, here it muſt be remember'd, that I am here ſpeaking of Columns of the ſame Order; for in different Orders the *Diminution* muſt be different. But as to the *Doric* Column, for Inſtance, be its Height an hundred Feet, or be it but ten, its *Diminution* ſhould always be the ſame, at leaſt, this is my Opinion.

DIOPTRICKS is the Doctrine of refracted Viſion, which are alſo called *Diaclaſticks*. They are properly the third Branch of Opticks, whoſe Office is to conſider and explain the Effects of Light, refracted by paſſing through different Mediums of Air, Water, Glaſs, *&c.*

DIPTERE } in the antient
DIPTERON } Architecture, ſignified a Temple ſurrounded with two Rows of Columns, which form a ſort of Portico's, called *Wings* or *Iſles*.

Pſeudo Diptere is the ſame, excepting, that inſtead of the double Row of Columns, this was only encompaſſed with a ſingle one

DIRECT. VISION, in Opticks, is that which is perform'd by direct Rays, in Contradiſtinction to *Viſion* by refracted or reflected Rays.

Direct Viſion is the Subject of Opticks, which preſcribes the Laws and Rules thereof.

Direct Rays, are ſuch as paſs in direct Lines from the Luminary to the Eye; without being turned out of their rectilineary Direction by any intermediate Body, either opake or pellucid.

Direct, in Arithmetick. The *Rule of Three Direct*, is that oppoſite to the *Inverſe*. In the *Direct*, the fourth Number required increaſes the Proportion; and in the *Inverſe*, diminiſhes it.

DIRECTION, in Mechanicks, or *Line of Direction*, is particularly uſed for a Line paſſing from the Centre of the Earth through the Centre of Gravity of a Body, and the Support or *Fulcrum* that bears it. A Man muſt of Neceſſity fall down, as ſoon as the Centre of his Gravity is out of the *Line of Direction*.

Angle of Direction, in Mechanicks, is that which is comprehended between the *Lines of Direction* of two conſpiring Powers.

Line of Direction, in Mechanicks, is that *Line* in which a Body moves, or endeavours to proceed.

Magnetical Direction is uſed, in the general, to ſignify the Tendency or Turning of the Earth, and all *magnetical Bodies*, to certain Points.

The Situation of our Earth is known to be ſuch, that its Axis is the Axis of the Univerſe, and therefore its Poles and Cardinal Points correſpond exactly to thoſe of it.

This Situation ſome account for hence, that it is the moſt commodious, commodius, in respect to the Aspects and Influences of the heavenly Bodies, and renders it the fittest Habitation for Man.

Others maintain, that this Position of the Earth is the Effect of a magnetick Virtue; and suppose a celestial Pole, with a like magnetick Virtue, which extending as far as our Earth, draws the correspondent Part of it, the Pole, towards itself.

DIRECTLY, in Geometry, a Term used of two Lines, which are said to be *directly* against each other, when they are Parts of the same Right Line.

In Mechanicks, a Body is said to strike *directly* against another, if it strike in a Right Line perpendicular to the Point of Contact.

A Sphere is said to strike *directly* against another, when the Line of Direction passes through both their Centres.

DIRECTRIX } in Geome-
DIRIGENT } try, a Term which expresses the Line of Motion, along which the describent Line or Surface is carried in the *Genesis* of any Plane or solid Figure.

DISC } in Opticks, the Mag-
DISK } nitude of Telescope-Glasses, or the Width of their Apertures, whatever their Figure be, whether plain, convex, menisens, &c.

DISCREET PROPORTION }
DISJUNCT PROPORTION }
is when the Ratio of two or more Pairs of Numbers, or Quantities, is the same, but not continual, that is, when the Ratio of the Consequent of one Pair of Numbers, or Quantities, to the Antecedent of the next Pair, is not the same as of the Antecedent of one Pair to its Consequent; as 3 : 6 :: 8 : 16 are *discreet Proportionals*, because the Ratio of 3 to 6, is equal to the Ratio of 8 to 16; but the Ratio of 3 to 6, or 8 to 16, is not the same as of 6 to 8.

Discreet Quantity is such as is not continuous, and joined together; as Numbers whose Parts being distinct Units, cannot be united into one *Continuum*; for in a *Continuum* there are no actual determinate Parts before Division, but they are potentially infinite.

DISTANCE is properly the shortest Line between two Points.

Line of Distance, in Perspective, is a Right Line drawn from the Eye to the principal Point.

Point of Distance, in Perspective, is a Point in the Horizontal Line, at such Distance from the principal Point, as is that of the Eye from the same.

DISPOSITION *of Pictures and Paintings*; the Manner how and where Gentlemen, &c. who are possessed of several Sorts of them, should place them in their Houses, &c.

I. Antique Works, or Grotesco, may become a Wall, and the Borders and Friezes of other Works; but if there be any Draughts in Figures of Men and Women to the Life on the Wall, they will be best of Black and White, or of one Colour heightened: If they be naked, let them be as large as the Place will afford: If of Marble Columns, Aqueducts, Arches, Ruins,

and

and Cataracts, let them be bold, high, and of large Proportion.

II. Let the best Pieces be placed to be seen with single Lights; for so the Shadows fall naturally, being always barred to answer one Light; and the more under or below the Light, the better, especially in Men's Faces, and large Pieces.

III. Let the Porch or Entrance into the House be set out with rustick Figures, and Things rural.

IV. Let the Hall be adorn'd with Shepherds, Peasants, Neat-Herds, with Milk-Maids, Flocks of Sheep, and the like, in their respective Places, and with proper Attendants; as also Fowls, Fish, and the like.

V. Let the Stair-Case be set off with some admirable Monument or Building, either new or ruinous, to be seen and observed at a View passing up; and let the Ceiling over the Top-Stair be with Figures fore-shortened, looking down out of the Clouds, with Garlands, and Cornucopia's.

VI. Let Landskips, Hunting, Fishing, Fowling, Histories, and Antiquities, be put in the Great Chamber.

VII. Let the Pictutes of the King, Prince, &c. or their Coats of Arms, be placed in the Dining-Room, forbearing to put any other Pictures of the Life, as not being worthy to be their Companions, unless at the lower End, two or three of the chief Nobility, as Attendants on their Royal Persons; for want of which, you may place some few of the nearest Blood.

VIII. In the inward, or with-drawing Chambers, put other Draughts of the Life, of Persons of Honour, intimate, or special Friends, or Acquaintance; or of Artists only.

IX. In Banquetting-Rooms, place chearful and merry Paintings, as of *Bacchus*, Centaurs, Satyrs, Syrens, and the like, forbearing all obscene Pictures.

X. Histories, grave Stories, and the best Works become Galleries, where any one may walk, and exercise their Senses in viewing, examining, delighting, judging, and censuring.

XI. Place Castles, Churches, or some fair Buildings in Summer-Houses, and Stone-Walks. In Terrasses, put Boscage, and wild Works. Upon Chimney-Pieces, only Landskips; for they chiefly adorn.

XII. Place your own, your Wife and Childrens Pictures, in in your Bed-Chambers, as only becoming the most private Room, and your Modesty; least (says our Author,) if your Wife be a Beauty, some wanton Guest should gaze on't too long, and commend the Work for her sake.

XIII. In hanging Pictures, if they hang high above Reach, let them bend somewhat forward at the Top; because, otherwise, it is observed, that the visual Beams of your Eye which extend to the Top of the Picture, appear further off than those of the Foot.

DISTEMPER, in Painting, is the Working up of Colours with something else besides bare Water, or Oil; as if the Colours are mixed with Size, Whites of Eggs, or any ſuch proper glutinous or unctuous Subſtance, and not with Oil, then the Painting is ſaid to be done *in Diſtemper*, as the admirable Cartoons at *Hampton-Court* are.

DISTINCT BASE, in Opticks, is that Diſtance from the Pole of a convex Glaſs, in which Objects beheld through it appear *diſtinct*, and well defined, and is what is otherwiſe called the *Focus*.

DISTRIBUTION, in Architecture, as the *Diſtribution of the Plan*, is the dividing and diſpenſing the ſeveral Parts and Pieces, which compoſe the Plan of a Building.

Diſtribution of Ornaments, is an equal orderly placing of the Ornaments in any Member of Architecture.

DITRYGLYPH, in Architecture, the Space between two *Triglyphs*.

DIVERGENT, DIVERGING *Lines*, in Geometry, are ſuch Lines whoſe Diſtance is continually increaſing. Lines which converge one Way, diverge the oppoſite Way.

Divergent Rays, *Diverging Rays*, in Opticks, are thoſe Rays which iſſue from a Point of a viſible Object, are diſperſed, and continually depart from one another, according as they remove from the Object.

In this Senſe, the Word is oppoſed to *Convergent*; which implies that the Rays approach each other, or to tend to the Centre, where, when they are arrived, they interſect, and if continued further, they become *diverging*.

Concave Glaſſes render the Rays *diverging*, and Convex ones *converging*.

Concave Mirrors make the Rays *converge*, and Convex ones *diverge*.

It is demonſtrated in Opticks, that as the Diameter of a pretty large Pupil does not exceed two M. or one Fifth of a Digit. *Diverging Rays*, flowing from a radiant Point, will enter the Pupil, parallel to all Intents and Purpoſes, if the Diſtance of the Radiant from the Eye be four thouſand Feet.

Diverging Hyperbola, is one whoſe Legs turn their Convexities towards one another, and run contrary Ways.

DIVIDEND, in Arithmetick, is the Number that is to be divided into equal Parts by another Number.

DIVISIBILITY is that Diſpoſition of a Body whereby it is conceived to have Parts, into which it may actually or mentally be divided; or it is defined a paſſive Power, or Property in Quantity, whereby it becomes ſeparable into Parts, either actually, or at leaſt mentally.

Body is diviſible *in infinitum*, *i.e.* you cannot conceive any Part of the Extenſion ever ſo ſmall, but that ſtill there may be a ſmaller.

There are no ſuch Things as Parts infinitely ſmall; but yet the Subtilty of the Parts of ſeveral Bodies is ſuch, that they very much paſs our Conception. And there are innnmerable Inſtances in Nature of ſuch Parts that are actually ſeparated from one another.

Mr.

Mr. *Boyle* gives us ſeveral Inſtances: He mentions a Silken Thread that was three hundred Yards long, which weighed but two Grains and a half.

He alſo meaſured Leaf-Gold, and found that fifty ſquare Inches of Leaf-Gold weighed but one Grain. Now if the Length of an Inch be divided into two hundred Parts, the Eye may diſtinguiſh them all; therefore there are in one ſquare Inch forty thouſand viſible Parts; and in one Grain of Gold there are two Millions of ſuch Parts; which viſible Parts may be further divided.

DIVISION is one of the four great Rules of Arithmetick, being that whereby we find how often a leſs Quantity is contained in a greater, and the Overplus.

Diviſion of Numbers, is in Reality only a compendious Subtraction: The Effect of it being to take a leſs Number from a greater, as often as it is poſſible, that is, as oft as it is contain'd therein. There are three Numbers contained in *Diviſion*: *Firſt*, That given to be divided, called the Dividend. *Secondly*, That whereby the Dividend is to be divided, which is called the Diviſor. *Thirdly*, That which expreſſes how oft the Diviſor is contained in the Dividend, or the Number reſulting from the *Diviſion* of the Dividend by the Diviſor, called the Quotient.

DIVISOR, is the Dividing Number, or that which ſhews how many Parts the Dividend is to be divided into.

DODECAGON, a regular Polygon, conſiſting of 12 equal Sides and Angles.

DODECAEDRON, in Geometry, is one of the regular *Platonick* Bodies, comprehended under 12 equal Sides, each of which is a *Pentagon*. Or,

A *Dodecaedron* may be conceived to conſiſt of 12 quinquangular Pyramids, whoſe Vertexes or Tops meet in the Centre of a Sphere, conceived to circumſcribe the Solid, and of conſequence they may have their Baſes and Altitudes equal.

To find the Solidity of the Dodecaedron.

Firſt, Find that of one of the Pyramids, and multiply it by the Number of Baſes, *viz.* 12, and the Product will be the Solidity of the whole Body; or the Solidity of the whole Body may be found by multiplying the Baſe into one Third of its Diſtance from the Centre 12 Times; And to find this Diſtance, take the Diſtance of two parallel Faces, and the Half will be the Height: Or,

Multiply the Area of the Pentagonal Faces of it by 12, and then this latter Product of it by one Third of the Diſtance from the Diſtance from the Centre of the *Dodecaedron*, which is the ſame as the circumſcribing Sphere.

The Side of a *Dodecaedron* inſcribed in a Sphere, is the greater Part of the Side of a Cube inſcrib'd in that Sphere, cut into extream and mean Proportion.

If the Diameter of the Sphere be 1000, the Side of a *Dodecaedron* inſcribed in it, will be ·35682.

All *Dodecaedrons* are ſimilar, and are to one another as the Cubes of their Sides; and their Surfaces are alſo ſimilar, and are therefore as the Square of their Sides; whence, as .509282 is to 10.51462, ſo the Square of the Side of any *Dodecaedron* to its Superficies; and as .3637 to 2.78516, ſo is the Cube of the Side of any *Dodecaedron* to the Solidity of it.

Let ABCDEFGHIK be a *Dodecaedron*, each Side of which being 12 Inches, the ſolid Content, and ſuperficial Content is requir'd.

The Solidity of the *Dodecaedron* is compoſed of 12 Pentangled Pyramids, whoſe Vertexes all meet in the Centre.

Therefore if we find the Solidity of one of thoſe Pyramids, and multiply that by 12, that Product will be the Solidity of the *Dodecaedron*.

The Altitude of one of the Pentangled Pyramids will be found to be 13.36219.

The Perpendicular of the Pentagon will be 8.258292.

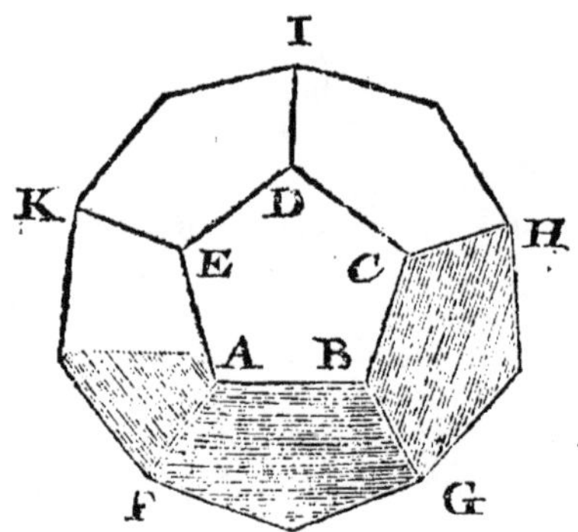


```
847.748760
        30 half Sum
----------
247.748760
   60454.4
----------
  99099504
   9909950
   1238744
     99099
      1486
----------
1103 48783 Content of one Pyramid.
        12
----------
13241.85396 the Solidity of the Dodecaedron.
```

If

If the Area of the Pentagon be multiplied by 12, the Product will be the superficial Content.

```
247.7487
      12
------------
2972.98512  the superficial Content.
```

Example 1. If the Side of a *Dodecaedron* be 12 Inches, what is the Content solid and superficial?

```
7.663119  the tabular Number.
    1728  the Cube of the Side.
-------------
 61304952
 15326238
 53641833
 7663119
-------------
13241.869632  the solid Content, nearly the same as before.
```

```
20.645729
      144
-------------
 82582916
 82582916
 20645729
-------------
2972.984976  the superficial Content.
```

By Scale and Compasses.

Extend the Compasses from 1 to 12, (the Side,) that Extent being turn'd three Times over from 7.63119, will at last fall upon 13241.86, &c. the solid Content.

And if you apply the same Extent twice from 20.645729, it will at last fall upon 2972.98, &c. the superficial Content.

Example 2. If the Side of an *Octaedron* be 20 Inches, what is the solid and superficial Content?

.4714045	the tabular Number.
8000	the Cube of the Side.
3771.2360000	the solid content.
3.464102	the tabular Number.
400	the Square of the Side.
1385.640800	the superficial Content.

By Scale and Compasses.

Extend the Compasses from 1 to 20, and that Extent turn'd three Times over from .4714045, will at last fall upon 3771.236, the solid Content.

The same Extent turn'd twice over from 3.464, &c. will at last fall upon 1385.64, the superficial Content.

DOME, in Architecture, a spherical Roof, or a Roof of a spherical Form raised over the Middle of a Building, as a Church, Hall, Pavilion, Vestible, Stair-Case, &c. by way of crowning.

Domes are the same that the *Italians* call *Couppola's*, and we *Cupola's*. *Vitruvius* calls them *Tholi*.

They are generally made round, or resembling the Bell of a great Clock; but there are some Instances of square ones, as those of the *Louvre*; and also some of them are in the Form of Polygons, as that of the Jesuit's Church in the *Rue St. Antboine* at *Paris*.

Domes have commonly Columns ranged around their Outsides, both for the sake of Ornament, and Support to the Vault.

DOORS, in Architecture, are Apertures in Walls, to give Entrance and Exit into and out of a Building, or an Apartment of it.

It is laid down as a Rule, that the *Doors* of an House be as few in Number, and as moderate in Dimensions as possible: For, in a Word, all Openings are Weaknings.

Secondly, That they do not approach too near the Angles of the Walls, it being a very great Solecism to weaken that Part which should strengthen all the rest.

A Precept well recorded, but illy practised by the *Italians*, particularly at *Venice*.

Thirdly, That the *Doors*, if possible, be placed over one another, that Void may be over Void, and Full over Full; which will be a great Strengthening to the whole Fabrick.

Fourthly, That, if possible, they may be opposite to each other, in such manner, that one may see from one End of the House to the other; which will not only be very graceful, but most convenient,

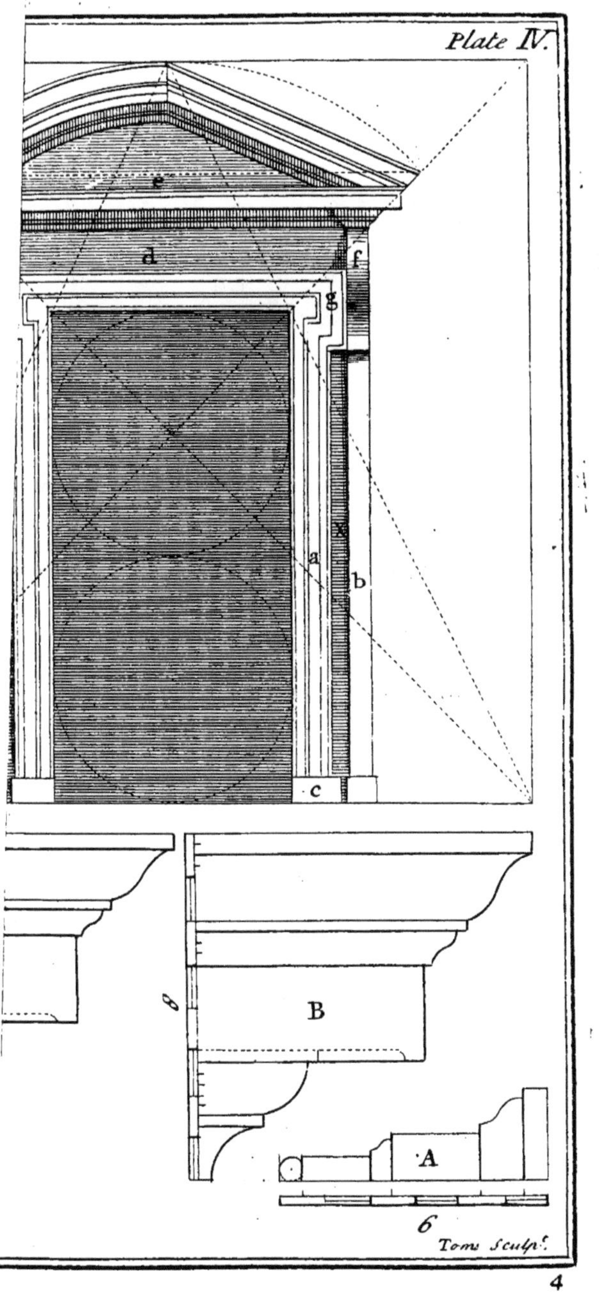
Plate IV.
e
d
f
g
a
b
c
B
8
A
6
Toms Sculp.

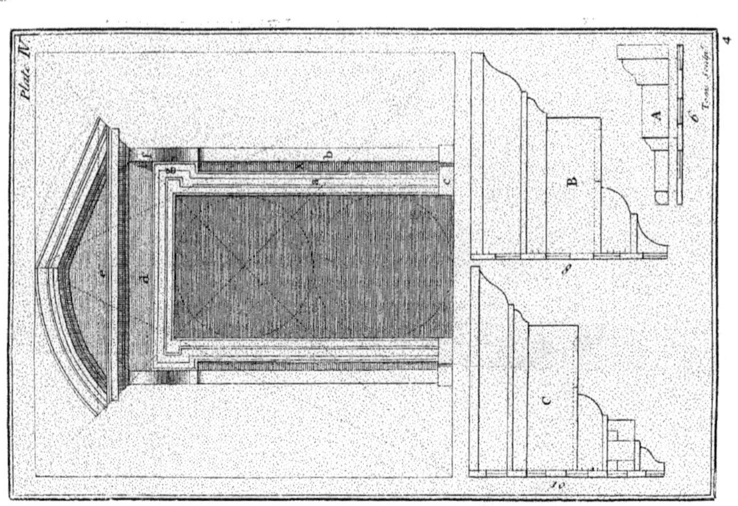

venient, in reſpect that it affords Means of cooling the Houſe in Summer, by letting the Air through the Houſe; and by keeping out the Wind in Winter, which Way ſoever it ſit.

Fifthly, 'Tis not only ornamental, but very ſecure, to turn Arches over *Doors*; which will diſcharge them, in great Meaſure, of the ſuperincumbent Weight.

The Proportion of *Doors* is adjuſted by that of a Man.

In large Buildings they muſt always be larger than in ſmall; but ſhould not be leſs than ſix Foot high in any, to admit a Man of juſt Statute erect And as the Breadth of a Man with his Arms placed a-kembo, is nearly ſubduple his Height, the Width ought never to be leſs than three Feet.

Some Architects give us theſe Dimenſions following:

In ſmall Buildings, the Breadth of the *Door*, four Feet, or four and a half; in middling Buildings, five or ſix; in large ones, ſeven or eight; in Chambers of the firſt Story, three and a half, three and three Fourths, or four: of the ſecond, four, or four and a half; and of the third, five or ſix; in Churches, ſeven or eight; in Gates, nine, ten, or twelve. Hence their Height is eaſily determin'd, except for the Gates of Cities, which ſhould only be four Fifths of their Breadth.

Palladio has an Obſervation, that the principal *Door*, or Enrance of an Houſe muſt never be regulated by any certain Dimenſions, but by the Dignity of the Perſon who is to live in it; yet to exceed rather in the more than the leſs is a Token of Generoſity, and may be excuſed with ſome noble Emblem, as that of the *Conte di Bevilacqua*, over his large Gate at *Verona*, where a little Diſproportion had been committed, *Patet janua cor magis*.

As to the Price of *Doors*: Thoſe that are made of plain, Whole Deal, and rabbeted, are for Stuff, Nails, and Workmanſhip, valued, as ſome Workmen ſay, at 3*d.* or 4*d.* the ſuperficial Foot; the Workmanſhip only at 2*s.* or 2*s.* 6*d.* *per Door.*

Double Doors batten'd, or made Wainſcot-faſhion, may be worth 7*d.* the Foot for Workmanſhip and Materials, and 4 *s.* or 5*s.* *per Door*, for Workmanſhip alone.

Folding Doors and Caſes are uſually valued at 20*s.* or 30*s.* *per* Pair; and *Balcony Doors* and Caſes at the ſame Rate.

Ordinary Doors, without plaining, are uſually valued at 1*s.* *per Door*, making and hanging up.

Architrave Door-Caſes, in Brick Buildings, are worth, according to their Mouldings, a Penny an Inch, *i. e.* if the Breadth of the Moulding (from the Outſide to the Inſide of the Frame) be nine Inches, it is worth 9*d.* *per* Foot running Meaſure; if ten Inches, 10*d.* *per* Foot; and ſo either more or leſs, in Proportion.

Frontiſh Doors, in large Buildings, with their uſual Ornaments, as Pilaſters, *&c.* are worth (according to their Largeneſs and Variety of Workmanſhip included) from 3*l.* to 5. 10, or 20*l.* or more, *per Door*. See BATTEN DOOR.

M. *Le Clerc* ſays, when a little Door is made in the Front of an ordinary, but regular Building, it ſhould be raiſed to the juſt Height of the Windows that accompany it; but its Breadth muſt a little exceed that of the Windows, leaſt while it is adjuſted to the reſt of the Building, it appear ill-proportion'd in itſelf.

If it is deſir'd to have the *Door* adorn'd with an Order of Columns, it muſt be raiſed higher.

A Geometrical Rule for a Door, *or Window.*

The Breadth being given, take it three Times for the Side of a a Square, and draw the Diagonals, whoſe Interſection will be the Centre of the Pediment's Arch; then from the Top of the Pediment, draw Lines to the oppoſite Angles of the Square, and their Interſection with the Diagonals, is the Height of the *Door*.

The Breadth of the *Door* being divided into ſix, one is for the Breadth of the Architrave [a], one Third to the Space [x], two Thirds to the Pilaſter [b]; the Plinth [c] is two Thirds high; the Height of the Kneel g of the Architrave is twice its Breadth; the Height of the Frize [d] is equal to the Breadth of the Architrave, and the Cornice [e] one Fourth higher; the Length of the Truſs [f] is from the Top of the Frize to the Bottom of the Kneel. For the ſeveral Members, divide the Breadth of the Architrave A into ſix Parts, giving half a Part to the Bead, one and a half to the firſt Face, half a Part to the ſmall Ogee, two to the ſecond Face, one to the Ogee, a half Part to the Fillet.

For the Projections, the firſt Face is a half Part, the ſecond Face one, and the Whole two.

For the plain Cornice B, divide the Height into eight Parts, giving one and one Fourth to the Cavetto, one Fourth to the Fillet, one and one Fourth to the Ovolo, one Fourth to the Fillet, two to the Corona, three Fourths to the Cima Reverſa, one Fourth to the Fillet, one and a half to the Cima Recta, and a half Part to the Fillet. For the Projections, the Cavetto hath one and a half, the Ovolo two and a half, the Corona five and a half, the Cima Reverſa ſix and a half, and the Whole eight.

For the Dentil Cornice C, divide the Height into ten Parts, giving one and one Fourth to the Ogee, one Fourth to the Fillet, one and a half to the Dentils, (whoſe Breadth is two Thirds of their Height,) one Fourth to the Fillet, one and one Fourth to the Ovolo, one Fourth to the Fillet, two and one Fourth to the Corona, three Fourths to the Cima Reverſa, one Fourth to the Fillet, one and a half to the Cima Recta, and a half Part to the Fillet. For the Projections, the Ogee hath one and a half, the Dentils two and a half, the Ovolo four, the Corona ſeven and a half, the Cima Reverſa eight and a half, and the Whole ten. See the Plate.

DORIC, in Architecture, is the ſecond of the Five Orders; and is that between the *Tuſcan* and the *Ionic*.

This Order ſeems the moſt natural and beſt-proportion'd of all the Orders, all the Parts of it being founded on the natural Poſition of ſolid Bodies.

Accordingly,

Accordingly, the *Doric* is the firſt and moſt antient of the Orders of Architecture, and is that which gave the firſt Idea or Notion of regular Building.

It was, indeed, more ſimple at its firſt Invention than it is at preſent; and when they came in After-Times to adorn and enrich it more, the Appellation of *Doric* was reſtrained to this richer Manner; and then they called the primitive ſimple Manner by the new Name of *Tuſcan.*

Tradition delivers that *Dorus* King of *Achaia* having built a Temple of this Order at *Argos*, dedicated to *Juno*, cauſed it to be called *Doric:* Though ſome derive its Name from its having been invented or uſed by the *Dorians.*

Some Time after its Invention, it was reduced to the Proportions, Strength, and Beauty of the Body of a Man.

Hence as the Foot of a Man was judged the ſixth Part of its Height, they made the *Doric* Column ſix Diameters high. After that they added another Diameter to it, and made it ſeven; which Augmentation ſeem'd to bring it nearer to the Proportion of a Man; the human Foot, at leaſt, in our Days, not being exactly a Sixth, but nearly the Seventh of the Body.

The Characters of the *Doric* Order, as they are now manag'd, are the Height of its Column, which is now eight Diameters; the Frize, which is adorn'd with Triglyphs, Drops, and Metopes; its Capital, which is without Volutes; and its admitting of Cymatiums.

It has been already obſerved, that the Antients had two *Dorics:* The firſt of which was the more ſimple and maſſive, and chiefly uſed in Temples; the ſecond, which was the more light and delicate, they uſed in Porticoes and Theatres.

Indeed, *Vitruvius* complains of the *Doric*, as being very troubleſome and perplexing on Account of the Triglyphs and Metopes, ſo as to be ſcarce capable of being uſed, except in the Pycnoſtyle, by placing a Triglyph between each two Columns; or in the Aræoſtyle, by placing three Triglyphs between each Column.

The *Doric* is uſed by the Moderns on Account of its Solidity in large ſtrong Buildings, as in the Gates of Cities and Citadels, the Outſides of Churches, and other maſſy Works, in which Delicacy of Ornaments would be unſuitable.

The moſt conſiderable antient Monument of this Order, is the the Theatre of *Marcellus* at *Rome*, the Capital, the Height of the Frize, and Projecture of which, are much ſmaller than in the modern Architecture.

Vignola adjuſts the Proportions of the *Doric* Order, as follows: He divides the whole Height of the Order, without the Pedeſtal, into twenty Parts, or Modules; one of which he allows to the Baſe, fourteen to the Shaft or Fuſt, one to the Capital, and four to the Entablature. The particular Proportions of the ſeveral Parts and Members may be ſeen under their reſpective Articles.

The

The DORIC ORDER *delineated by equal Parts, instead of Modules and Minutes.*

The Height of the Pedestal being two Diameters and one Third, is divided into four, giving one to the Base, whose Plinth is two Thirds thereof; the other Part is divided into seven, giving four to the Torus, one to the Fillet, and two to the Hollow. The Breadth of the Die is a Diameter and one Third. The Projection of the Base is equal to its Height, and the Fillet hath four of these Parts. The Height of the Cornice is half the Base, being one Eighth of the whole Height; and is divided into nine, giving two to the Hollow, one to the Fillet, five to the Corona, and one to the Fillet; the Projection of the Hollow is three of these Parts, the Corona six, and the Whole seven.

Base of the Column.

The Height is half a Diameter, and is divided into six, giving two to the Plinth, one and a half to the lower Torus, one Fourth to the Fillet, one to the Scotia, ¼ to the Fillet, and one to the upper Torus. The Fillet above the Torus is equal to the others, and is Part of the Column. The Projection is two of these Parts, and one Third thereof is for the upper Fillet, two Thirds to the upper Torus, and the Fillet under it, is perpendicular to the Centre.

For forming the Scotia, divide the Height into three, as in the Scheme, and on the Line that separates the one Part above from the two Parts below, and perpendicular to the Fillet, is the Centre for the first Quarter-Sweep, and the same Distance forwards, on the Line, is the Centre for the other Quarter; and is also the Projecture of the lower Fillet.

The Diminishing of this Column is one Eighth of the Diameter. The Height of the Capital is half a Diameter, and is divided into nine, giving three to the Frize of the Capital, one to the Fillets, which are three, and are equal, two to the Ovolo, two to the Abacus, and one to the Ogee b and Fillet, which is one Third.

For the Projections, the Fillets have one of these Parts, the Abacus three, and the Whole four.

The Height of the Architrave is half a Diameter, and is divided into six Parts, giving two to the first Face, two to the second, one to the Belis c and Fillet, which is one Third, and one to the Band at top. The Projection is equal to the Band.

The Frize is in Height three Fourths of the Diameter, and the Triglyphs d are in Breadth half a Diameter; which are divided into six, giving one to each of the Channels, half a Part to each Half-Channel, and one to the Spaces between the Channels.

The Projection from the Naked of the Frize is three Fourths of a Part, and the Spaces, or Metopes, between the Triglyphs, ought to be equal to the Height of the Frize.

The

The Height of the Cornice is three Fourths of the Diameter, and is divided into nine, giving one to the Cape of the Triglyph, one to the Hollow and Fillet, which is one Sixth, one to the Ovolo, one to the Mutules and Fillet under it, which is equal to the other, a half Part to the Cap of the Mutule and Fillet, which is one Third, one and three Fourths to the Corona, three Fourths to the Cima Reversa, one Fourth to the Fillet, one and one Fourth to the Cima Recta, and a half Part to the Fillet.

For the Projections, the Cap of the Triglyph hath one of these Parts, the Hollow one and three Fourths, the Ovolo two and three Fourths, the Mutule eight and three Fourths, the Corona nine and three Fourths, the Cima Reversa ten and three Fourths, and the Whole twelve Parts.

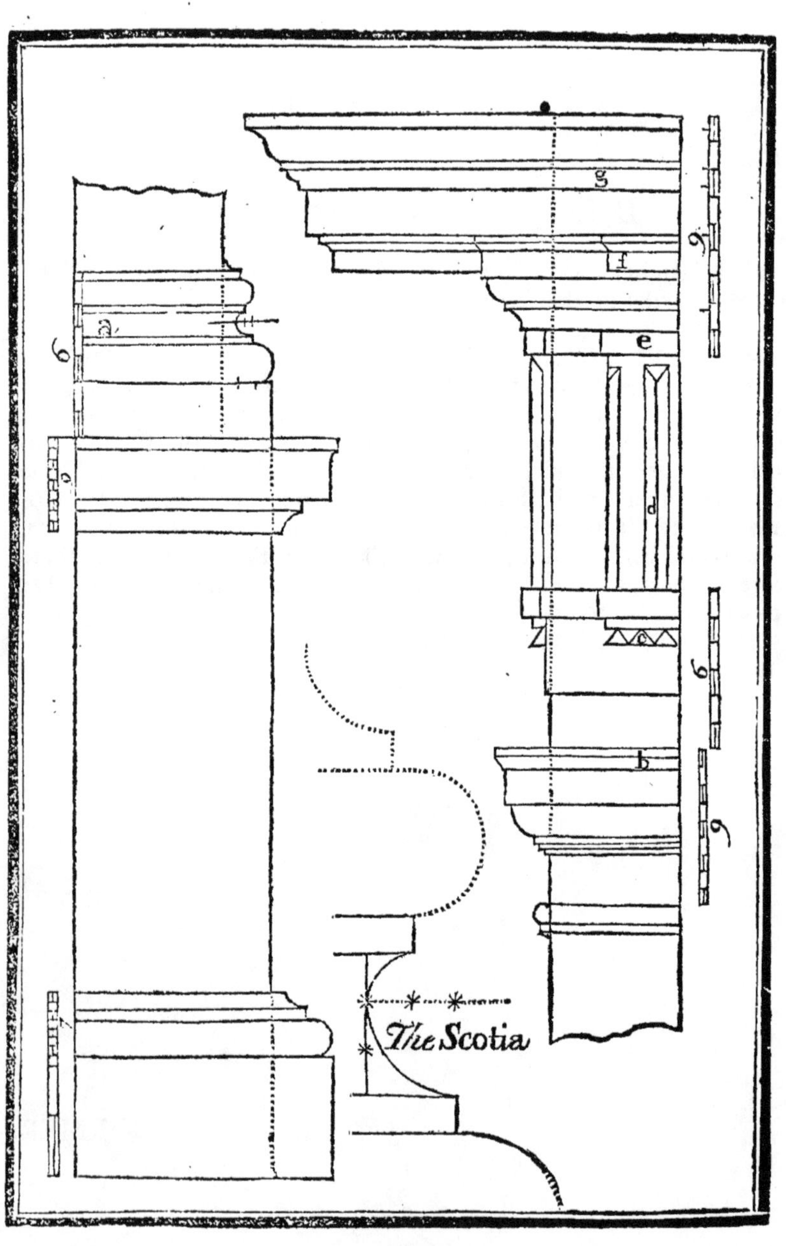
The Scotia

DORMER, } DORMANT, } in Architecture, is the Window made in the Roof of a Houſe, or above the Entablature, being raiſed upon the Rafters

Dormers are commonly rated at ſo much *per* Piece.

Dormant-Tree is a Name given by Workmen to a great Beam lying croſs a Houſe, commonly called, a *Summer*

Dormant-Tiles. See TILES.

DORMITORY, a Gallery in Convents or Religious Houſes, divided into ſeveral Cells, in which the Religious ſleep or lodge.

DOUCINE, in Architecture, is a Moulding or Ornament on the higheſt Part of the Cornice, in the Form of a Wave, half convex, and half concave.

The *Doucine* is the ſame as a Cymatium, or Gula.

DOVE-TAILING, in Carpentry, is a Manner of faſtening Boards (or other Timber) together, by letting one Piece into another, in the Form of the Tail of a Dove. It is the ſtrongeſt of the Kinds of Jointings or Aſſemblages, wherein the Tenon, or Piece of Wood which is put into the other, goes widening to the Extreme; ſo that it cannot be drawn out again by reaſon the Extreme or Tip is bigger than the Hole.

It is called by the *French*, *Queue d'Aronde*, *i. e. Swallow-Tail*; which Name is alſo uſed by the *Engliſh* themſelves in Fortification.

DRAG. A Door is ſaid to *drag*, when in opening or ſhutting, it hangs or grates upon the Floor, or Cell.

DRAGON BEAMS are two ſtrong Braces or Struts which ſtand under a Breſt-Summer, meeting in an Angle upon the Shoulder of the Kingpiece.

DRAPERY, in Sculpture and Painting, the Repreſentation of the Garments or Clothing of human Figures. It includes not only Garments, but Tapeſtry, Hangings, Curtains, *&c.*

DRAUGHT, } DRAFT, } in Architecture, is the Figure of an intended Building deſcribed on Paper; in which is laid down, by Scale and Compaſſes, the ſeveral Diviſions and Partitions of the Apartments, Rooms, Doors, Paſſages, Conveniencies, *&c.* in their due Proportion to the whole Building.

It is cuſtomary, and alſo exceedingly convenient for any Perſon, before he begins to erect a Building, to have Deſigns or *Draughts* drawn upon Paper or Vellum, wherein the Ichnography or Ground-Plot of every Floor or Story is delineated or repreſented, as alſo the Form or Faſhion of each Front, with the Windows, Doors, Ornaments. *&c.* in an Orthography, or Upright.

Sometimes the ſeveral Fronts, *&c.* are taken and repreſented in the ſame *Draught*, to ſhew the Effect of the whole Building, which is called a Scenography or Perſpective.

But this not being eaſily underſtood, except by thoſe who underſtand the Rules of Perſpective, therefore it will be more intelligible to the ſeveral Workmen, to have a *Draught* of each Front, in a particular Paper by itſelf;

itſelf; and alſo a *Draught* of the Ichnography or Grouud-Plot of each Floor or Story, in a Paper by itſelf; becauſe oftentimes the Contrivance and Conveniencies of one Story differ from thoſe of another, either as to the Largeneſs of the Chimneys or Diviſions of the Rooms, ſome being larger in one Story than another, &c.

All which Things being well conſidered and drawn on Paper before the Building is begun, theſe *Draughts* will be a great Guide to the Workmen, and ſave them a great deal of Time in contriving their Work; and, beſides, there will be no need of Alterations, or pulling the Building to Pieces after 'tis begun; which, beſides the Hinderance of the Procedure, makes the Building lame and deficient; nothing being ſo well done, when 'tis put up, and pull'd down, and ſet up again, as if it were done at firſt.

To draw any Object in its Out-Lines as exact as the Life, or Nature.

Take a Sheet of the thinneſt, or whiteſt brown Paper, and bruſh it over with Oil of Turpentine, which will immediately render it tranſparent and then put the Paper to dry in the Air; when it is dry, ſtrain it upon a Frame, and fix it againſt any Object you deſign to draw, as an Houſe, &c. then juſt before it place a Piece of Wood with a Hole in it, fit for one Eye to look through; and as you meet any Outlines of the Object you deſire upon the tranſparent Paper, trace them over with a Pencil, ſo will you be ſure you cannot err; for there will be nothing but juſt Proportion, and a true Repreſentation of Nature.

To make this ſtill of more Elegancy, obſerve the Tracing of your *Draughts*, where the Shades are, and mark them with your Pencil; for all the Art in the World can never diſpoſe the Shades ſo regularly, as one may touch by this Method: But the Shades muſt be done quickly after the Outlines are drawn, and not at different Times, becauſe every Inſtant the Sun changes them.

In this too, obſerve, that in certain Objects, you will have fainter, ſtronger, and yet more dark Shades; and in your Remarks of them, take ſuch Memorandums, as may direct you how to finiſh them, with *Indian* Ink, or other Colour, when you ſit down to compleat your Work.

The beſt Way is to prepare three Shells or Gallipots of *Indian* Ink mixed with common Water, before you attempt to trace out your Object, *viz.* one of a very faint Black, the next of a middling Black, and the other of an intenſe Black: Number them 1, 2, 3, from the lighteſt to the darkeſt; and as you make your Obſervation on the Shades of your Object, mark upon your *Draught* the ſame Numbers, as they happen to appear; ſo that afterwards you may finiſh with Certainty.

Again, it is neceſſary in the drawing of any Thing after this Manner, to obſerve that the

L'nes

Lines on the ſhady Side ſhould be thinner, in Proportion to the Light that falls upon them.

As for Example: In the darkeſt Part a Line may be of that Thickneſs, in the next Part ſomewhat thinner, and in the other thus; unleſs in Things of a great Diſtance hardly to be underſtood, or ſo faint as hardly to be perceived thus; a mere Shadow, as it were.

Some have been guilty of a great Fault, though they have taken the Outlines very exact, by making all their Lines of an equal Thickneſs.

If an Object be repreſented at a great Diſtance, as half a Mile, or two Miles off, and the Drawing be as ſtrong in that Part of the Picture, as if it was next the Eye, or not ten or twenty Feet from the *Draughts-Man*, it would not appear pleaſant or natural to the Eye.

A Man muſt not be expreſs'd with Buttons on his Coat at a Mile's, *&c.* Diſtance, no more than they muſt be omitted in a Perſon ſo near the Eye, as ten or twenty Feet: Though this has inadvertently been done by ſome that paſſed for great Men: Nor a Capital, *&c.* with Carvings and Mouldings, *&c.*

And the Shades, in thoſe diſtant Appearances, muſt be in Proportion to the Strength of the Objects as they appear to us, *i. e.* imperfect.

Three or four well-directed Touches of the Pencil on the ſhady Side, will repreſent a Figure at the Diſtance we can diſcern it, as lively as ſome Hundreds will near the Eye.

The tranſparent Paper, before mention'd, is alſo of another Uſe; for if it be laid upon any Picture or Print, all the Lines may be ſeen through it; and then you may draw or copy it with the greateſt Pleaſure.

You will then, if the Pictures be done by a good Maſter, ſee which Lines are ſtrong, and which tender or ſoft, and imitate them.

There is yet another Way to take Views and Landskips, which ſome prefer to the tranſparent Paper, that is, either with white, or black Tiffany or Lawn ſtrained upon a Frame, and uſed in the ſame Manner as the Paper, excepting that the Black Lead Pencil is uſed to the Paper, on the Tiffany and the Lawn, Charcoal, finely pointed, and very ſoft, is uſed; but on the black Tiffany white Chalk of the tendereſt Sort.

How to bring theſe Drawings *to Uſe, and to copy from Prints, Paintings*, &c.

If you draw upon tranſparent Paper, to take a *Drawing* from it regularly, get a Piece of Paper of the ſame Size, and rub on one Side of it ſome Powder of Black Lead, till it is well and equally black'd, and ſo well rubb'd, that a Touch of a Finger will hardly be tinged with it.

Then take the *Drawing* you have made, or Print, and lay the black'd Paper under it, with the black'd Side downwards, upon a Piece of white Paper, and pin the three together in two or three Places; take then a Pin or Needle a little blunt at the Point, and trace it over the Outlines of your Picture, which, with a little Preſſing, will direct the black Paper to impreſs the undermoſt white Paper, ſo as to receive every Stroke you draw.

When this is done, you muſt with a Black Lead Pencil, correct what Errors you find, and ſlightly clean the *Draught* new made, with ſome Crumbs of ſtale Bread.

Black Lead Pencils, that are tolerably good, are difficult to be got; if we find a good Piece of Lead in the Beginning, when it has been uſed an Inch or two, the reſt generally proves hard, gritty, and full of Knots.

'Tis a great Pleaſure to a *Draughts-Man* to work with a good Pencil, and as great a Plague to work with a bad one.

As for *Draughts* drawn on Tiffany, or Lawn, lay them only on one Paper, *i.e.* that which is drawn with Charcoal, on white Paper, and that drawn with Chalk on black or blue Paper; then giving each of them a Knock or two with an Hammer, the Charcoal, or the Chalk, will fall through them upon the Paper directly in the Lines they were drawn, and give you the true Repreſentation of the Object you drew from the Life: Upon the black Paper, you will ſee it in white Lines; and on the white in black.

Then you are to ſtrengthen theſe Shadows of *Drawings* with with your Black Lead Pencil, or Chalk, or Red Oker, on the Sheets of Paper where they have made their Marks, otherwiſe the Lines will be eaſily rubb'd out.

But you muſt take care that this Amendment be made ſuddenly, for theſe tender *Draughts* will ſoon vaniſh, if Care be not taken to ſtrengthen them immediately. Begin firſt at the Bottom of the *Drawing*, that you may not rub out the faint Lines by ſtrengthening the upper Part.

Another Method, is to take a thin Piece of Paper, and hold it againſt a Glaſs Window, eſpecially at ſuch a one as is ſaſh'd; for the Interruptions of the Lead, in the ſmaller glaz'd Windows, will hinder Part of the Proſpect: The Point is to draw what you ſee from the Glaſs, and then the Black Lead Pencil is to be uſed as before directed.

There is ſtill another Way, which may be more eaſy to the Hand or Arm of a Perſon not accuſtomed to drawing upon Paper or Lawn placed uptight; which is, by the Uſe of a portable *Camera Obſcura*: Though to help the firſt one may hold a Baguette, or ſuch a Stick, in the Left Hand, as the Oil-Painters do to reſt the Right Hand upon; or have ſome other Reſt made for the Right Hand, as may be ſcrew'd up and down at one's Pleaſure.

But there is this Difference ſtill by drawing a Piece of Perſpective, or View on a tranſparent Paper or Lawn, placed upright againſt any Object, that ſuch a Piece will take in more of the View or Object, and from

from a greater Diſtance than the portable *Camera Obſcura* will.

However, as the portable Camera will at firſt be moſt eaſy to the Arm of the Beginner, by reaſon that the Objects appear upon an Horizontal Plane, ſuch as a Table, the Hand will have a more proper Reſt, and more readily follow the Lines repreſented on the Plane with Exactneſs.

Indeed, ſuch a portable *Camera* is of ſome Expence; and ſuch as can afford it, may have them of any Price, from 30 *s.* to 5 *l.* apiece, as they pleaſe, of Mr. *John Fowler* Mathematical Inſtrument Maker, in *Swithen's-Alley*, near the *Royal Exchange.*

What will make the Differences in the Prices, will be the Largeneſs of the Sizes of the Glaſſes which lie horizontally, and receive the Objects which we are to trace out with the Pencil. The ſmaller of theſe Glaſſes may be perhaps four Inches ſquare, and the larger fifteen Inches.

On ſuch Glaſſes, you will have the exact Repreſentation (ſmaller or larger, according to the Bigneſs of the Machines) of the Objects they are pointed or directed to, each one adorn'd with the natural Colours agreeable to the Point of Diſtance, ſtronger near the Eye, and gradually declining, as the Objects are more remote from it: The Shades of the ſeveral Colours are in this Way expreſs'd in a very lively Manner.

A few Leſſons, with good Conſideration, will be of good Information, not only to a young Beginner, but to a Maſter of the Pencil.

But ſtill to advance the Knowledge and Uſe of this portable *Camera*, I ſuppoſe, that inſtead of the Glaſſes, which receive the Objects ſpoken of, there ſhould be placed Frames of tranſparent Paper, to receive the Objects that are to be taken, upon which the Pencil may ſtill be uſed with greater Freedom.

A Dozen or two may be had with each *Camera*; or one Frame will ſerve for as as many Papers as you pleaſe to ſtrain upon it, if a Perſon has Patience to paſte them on.

There is alſo another Way of drawing Objects in the *Camera Obſcura* Way; which is by making a Room as dark as may be, only leaving an Hole in one of the Window-Shutters, as low as poſſible, to receive an Ox-Eye Glaſs, as it is called; which is ſold at the Mathematical Inſtrument Makers.

This turns in a Socket, ſo as to direct every Object, within a certain Reach, to a Sheet of Paper; ſo that you may draw them in great Perfection; but they all appear revers'd, or the wrong End downwards: However, they are in as exact Proportion and Beauty as thoſe repreſented in the former.

In this Caſe, however, 'tis nor more difficult to draw, or rather copy the ſeveral Things that are ſeen upright on the Frames of tranſparent Paper, Lawn, or Tiffany; for to trace Lines, will be done as eaſily one Way as the other.

And though the Objects falling on the Sheet of Paper, will, while

while they are drawing, be reversed, 'tis but turning the Sheet of Paper upside-down when they are done, and the *Drawing* will be right to the Eye.

When this is shewn by Way of Curiosity, to those who are unacquainted with the Reasons why the Images represented on a Sheet of Paper appear upside-down, it would not have so desirable an Effect, as if they should be viewed in their natural Situation: But to obviate this Difficulty, let the Sheet of Paper, which is to receive the Objects, be placed against the Back of a Chair, and let them look on the several Objects represented on the Paper over the Back of the Chair, and it will set them upright to the Eye.

This Way of bringing them to rights, is thought on but by very few, tho' at the first Proof every one will wonder that he did not find it sooner.

Thus far is shewn how one may copy either a Print, *Drawing*, or Piece of Painting; or even make an exact Representation from the Life.

But I shall yet add, concerning the taking off of Prints or *Drawings*, a Method or two, which are easy or diverting, not before mentioned.

The one is, prick with a Pin any Out-Lines of a Print or *Drawing* you have a mind to copy, and then lay the said Paper on a Sheet of Paper; then take a Powder Puff, or a Tuft of Cotton, and dip it now and then in Charcoal Dust, or red Chalk Dust, and beat it over the prick'd Lines through the Picture, renewing it with Dust by frequent dipping; and then you will have full Directions marked on your Cloth or Paper, sufficient to finish a just *Drawing*.

Another Way there is to make an Impression from the Print or Picture, which shall give a just Copy of it. This is of great Use when we would carry every Stroke of the Engraver along with it.

It will indeed sully the Print, though not very much, if it be done with Care: Which may be perform'd in a few Minutes Time, when the *Drawing* of it with every Stroke the Engraver has made, would cost you whole Hours, nay, Days.

To do this, take some Soap, either of the white or green Sort; mix this with Water, till near the Consistence of a Gelly; wet the Paper you would have to receive the Impression from it with a wet Spunge, then lay it on the Print, and cover all with two or three Pieces of dry Paper, and rub it very hard all over with any Thing that is very smooth and pollish'd; and the wetted Paper will have the reverse of the Print you rubbed it upon, with every distinct Line in the Original, if it has been equally rubbed.

To take a Drawing *with fix'd Ink.*

Take a thin Sheet of Paper, and rub it all over with fresh Butter, as equally as possibly; then dry it well by the Fire, and rub the buttered Side with Carmine, till 'tis all equally coloured; or else rub it over with Lamp-Black, or with Black-Lead

Lead Powder, or with blue Bice finely ground; take Care in the rubbing on any of theſe, that the Colour will not come off by a very ſlight Touch of the Finger, and it will be fit for your Work.

When you have choſen a Print or Deſign that you would copy, lay the coloured Side of your buttered Paper upon a Piece of clean Paper, and your Print upon the buttered Paper; and then with a fine Pin or Needle, blunted a very little at the Point, trace the Out-Lines of your *Drawing* carefully, and you will have a good Copy of it upon your white Paper, which may be touch'd up afterwards with Crayons of the like Colour.

Short Rules for Drawing *in* Perſpective.

Many are deterred from applying themſelves to *Drawing*, by the Apprehenſion of the Time it will take up to render themſelves Maſters of Perſpective, and from being frightened at the Difficulties they conceive are in the Study; particularly imagining, that one muſt be firſt well grounded in the mathematical Sciences; but for the Encouragement of ſuch Perſons, I ſhall lay down in the following Leſſons how to lay any Plan in Perſpective, and raiſe Pillars or Buildings, according to their proper Diſtances.

Lesson I.
Of the Plan.

Suppoſe we have a ſquare Piece of Pavement, as in Fig. I. conſiſting of 25 Pieces of Marble, each a Foot ſquare, it muſt be meaſured exactly, and laid regular down upon Paper. You may likewiſe, for your better Obſervation, mark every other Stone or Marble black, which will better inform you how every particular Square will appear when we have a true perſpective View of them; or elſe you may number one, and when the following Leſſon is done, number thoſe in the perſpective Plan with the ſame Figures as are marked on the firſt Plan.

Lesson II.
Of laying Figure I. *in Perſpective.*

It is to be underſtood in Perſpective, that there are two Points to be conſidered; the firſt we call the *Point of Sight*, that is, which relates to every Thing in our View from the Place where we ſtand: And it matters little where we ſtand to take our View; for the Perſpective will ſtill be true, according to the Appearance of the Plan to our Eye, if we ſtand at a Corner, or in the Middle, or at any Point. The Method I ſhall preſcribe preſently, will lay our Plan juſtly before us as it will appear.

The other Point is called the *Point of Diſtance*, becauſe it governs the Diſtances and Proportions of every Thing we can truly ſee of the Plan, in whatever Poſition we happen to be.

At A you ſee the Plan of Fig. I. This is divided into Squares, as mentioned in that Figure; the three at the Bottom marked BCD in both *a* the Plan A, marked 1,2,3,4, are thoſe which are marked in Perſpective with the ſame Numbers.

Now to lay your Plan in Perſpective, fix your Point of Sight, as you obſerve in the Figure, or more or leſs to the Right or Left, as you think proper; then draw the Line K K, parallel to, and at what Diſtance you will from the Line L L; then raiſe a Line on each Side from L to K, to form the Figure you ſee as a Frame to you Picture; then draw a Line from the Corner of K, which is the Point of Diſtance to L, and this Line will regulate your Work. Then draw Lines from the Squares of your Plan to the Point of Sight, as exactly as poſſible; and wherever your Line of Diſtance cuts thoſe Lines, which are drawn from the Square of your Plan to the Point of Sight, that marks where your Square in Perſpective ought to be; then draw Lines parallel to the Line L L, where the Line of Diſtance cuts, and that will give you a true Figure of every Square. So D in the perſpective Plan, anſwers to D in the meaſured Plan, and 1, 2, 3, 4, anſwers to the others in the ſame.

When you have done this, the next Rule you are to know, is how to raiſe Pillars, Trees, Houſes, or any other Bodies, according to their reſpective Heights, at different Diſtances and Proportion on the Plan you have laid down.

How to raiſe Pillars, or any Bodies of a certain Proportion in Perſpective. See Fig. III.

You have now your Plan meaſured out in Perſpective into Squares of a Foot; one of theſe Squares in this Leſſon ſerves for the Baſe or Bottom of a Pillar a Foot thick.

This Figure III. is exactly of the ſame Dimenſions of the Plan laid in Perſpective at Fig. II.

First mark the Line L K in equal Proportions, by the ſame Scale of the ground Plan, Fig. II. as *a*, *b*, *c*, *d*, which are ſo many Feet in Height; and they ſtanding on the Baſe of the firſt Figure, are Uprights not in Perſpective: Then draw a Line parallel with L 1, from Number 4, which gives you the Front of the Body you are to raiſe; if it is to be only three Feet high, draw a Line croſs from Number 4. and that determines the Height, which you will then find to be a Foot wide, and three Feet high by Meaſure: Then from the Top of the Line 4, draw a Line with a Black-Lead Pencil to the Point of Sight; and raiſe another Line from 3 parallel to the Line 4, till it touches the penciled Line paſſing from 5 to the Point of Sight; which gives you the Side Appearance of the Column or Body, as you will ſee it from the Place where you ſtand, [the Line from Point 3 ſhould be drawn with a Pen, becauſe it is to remain;] then with a Pencil draw a Line from C to the Point of Sight, which will determine the other Line, to make the Shape on the Top of the Column: And then raiſe a Line parallel with L 1, with a Pencil from the Point, till it touches the Line from C to the Point of Sight; then draw a parallel Line to C 5, at 6, 7, and you

you will have the Square at the Top of the Pillar or Column, as you can obſerve it from the Place where you ſtand, which is ſuppoſed to be at A. [You muſt remember, that the Line drawn from 2 to 6, is only an imaginary Line, to be rubbed out; for it cannot be ſeen from the Place where you ſtand, and therefore muſt not appear in the *Drawing*; but you ſhould not leave it out, becauſe it ſhews you where to regulate the Top of the Column, and teaches you to place your Column upon its Baſe with Certainty.]

By this Means you may ſee Front, and one Side of your Column: And the Line from 1 to 2 muſt alſo be rubbed out, becauſe it can't be ſeen.

Then finiſh your Column only with the Lines

From 1 to C	From C to 5
From 4 to 5	From 6 to 7, and
From 3 to 7	From 1 to 4.

And it will be drawn without any Imperfection, and appear as follows in Fig. IV.

When this is done, you may place another Column on any one of the Squares erected in the ſame manner, obſerving to fling your Shades all on one Side, and then you cannot err: But eſpecially mind where the dotted Lines are in Fig. III.

DRAUGHT COMPASSES are *Compaſſes* with ſeveral moveable Points, to draw fine *Draughts* in Architecture.

DRAW-BRIDGE is a *Bridge* made to draw up, or let down, as Occaſion ſerves, before the Gate of a Town or Caſtle: And they are made after ſeveral Manners; but the moſt common are made with Plyers twice the Length of the Gate, and a Foot in Diameter.

The inner Square is traverſed with a Croſs, which ſerves for a Counterpoiſe; and the Chains which hang from the other Extremities of the Plyers, to lift up or let down the *Bridge*, are of Iron or Braſs.

DRIP, in Architecture. See LARIMER.

Drips are alſo uſed in Building for a certain kind of Steps, made on flat Roofs to walk upon; a Way of Building much uſed in *Italy*, where the Roof is not made quite flat, but a little raiſed in the Middle with *Drips*, or Steps, lying a little to the Horizon.

DROPS, in Architecture, an Ornament in the *Doric* Entablature, repreſenting *Drops*, or little Bells, immediately under the Triglyphs.

DUPLA RATIO } i. e. *Double*
DUPLE } *Ratio*, in Architecture, is where the antecedent Term is *double* the Conſequent; or where the Exponent of the Ratio is 2; thus 6 : 3 is in a *Duple Ratio*.

SUB-DUPLE, or *Double Sub-Duple Ratio*, is where the conſequent Term is *double* the Antecedent, or the Exponent of the Ratio is $\frac{1}{2}$; thus 3 : 6 is a *ſub-duple Ratio*.

DUPLICATE *Ratio* ought to be well diſtinguiſhed from *Duple*.

In a Series of Geometrical Proportions, the firſt Term to the third is ſaid to be in a *Duplicate Ratio* of the firſt to the ſecond, or as its Square is to the Square of the ſecond: Thus 2. 4. 8. 16. the Ratio of 2 to 8 is *Duplicate* of that of 2 to 4, or as the Square of 2 to the Square of 4; for which Reaſon, *Duplicate Ratio* is the Proportion of Squares, as *Triplicate Ratio* is of Cubes, *&c.* And the Ratio of 2 to 8, is ſaid to be compounded of that of 2 to 4, and of 4 to 8.

DUPLICATION, i. e. *Doubling*, in Arithmetick and Geometry, is the multiplying a Quantity diſcreet, or continued by two.

The Term is chiefly uſed of the Cube, as the *Duplication of the Cube*, which is a famous Propoſition that the Geometricians have ſought this 2000 Years.

The *Duplication of a Cubic*, is to find the Side of a Cube that ſhall be equal in Solidity to a Cube given.

This has been attempted by ſeveral geometrically; but it is in vain to pretend to it, for it cannot be done without the Solution of a cubick Equation; and ſo a Conick Section, or ſome higher Curve, muſt be uſed for determining the Problem.

DYE, in Architecture, is any ſquare Body, as the Trunk or notch'd Part of a Pedeſtal; or it is the Middle of the Pedeſtal, or that Part included between the Baſe and the Cornice; or is ſo called, becauſe it is often made in the Form of a Cube or *Dye*.

Dye is alſo uſed for a Cube of Stone, placed under the Feet of a Statue, and over its Pedeſtal, to raiſe it, and ſhew it the more.

DYPTERE } in the antient DIPTERE } Architecture, was a kind of Temple encompaſſed with a double Row of Columns; and the *Pſeudo Dyptere*, or *Falſe Diptere*, was the ſame, only that this was encompaſſed with a ſingle Row of Columns, inſtead of a double Row.

EA

EAGLE, in Architecture, a Figure of that Bird, antiently uſed as an Attribute or Cognizance of *Jupiter* in the Capitals and Friezes of the Columns of Temples conſecrated to that God.

EAVES, in Architecture, is the Margin or Edge of the Roof of an Houſe; being the loweſt Tiles, Slates, or the like, that hang over the Walls, to throw off Water to a Diſtance from the Wall.

Eaves-Lath, is a thick feather-edg'd Board, generally nailed round the *Eaves* of an Houſe for the lowermoſt Tiles, Slates, or Shingles to reſt upon.

Eaves-Laths are commonly ſold for three Half-pence or Two-pence *per* Foot, (running Meaſure,) according as they are in Goodneſs.

ECCENTRICK } in Geo- EXCENTRIC } metry, a Term apply'd where two Circles or Spheres, though contained in ſome Meaſure within each other, yet have not the ſame Centre, and of conſequence are not parallel in Oppoſition to Concentrick, where they have one and the ſame common Centre, and are parallel.

ECCENTRICITY

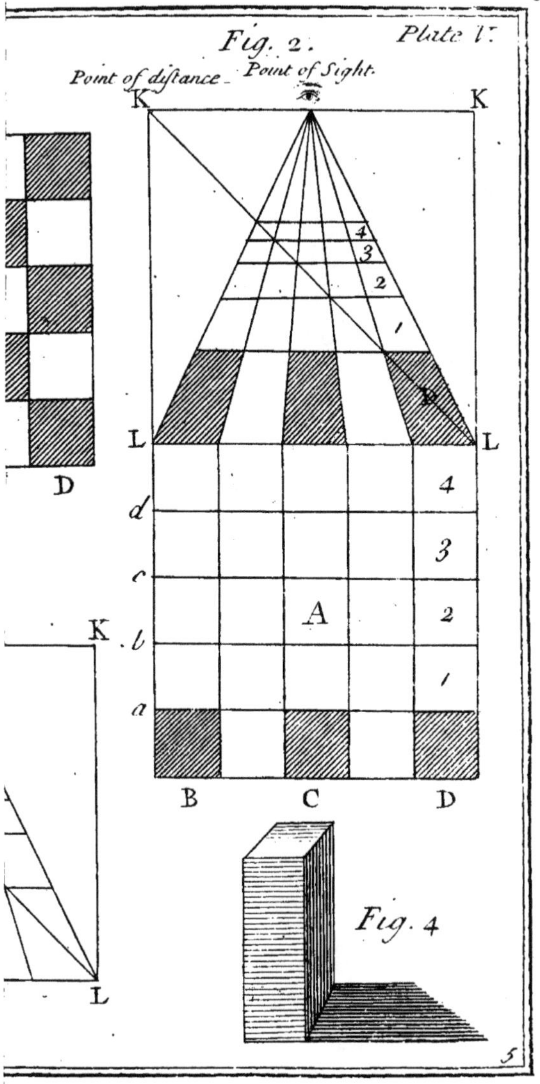
Fig. 2.
Plate V.
Point of distance
Point of Sight
K
K
4
3
2
1
L
L
D
d
4
c
3
A
2
K
b
1
a
B
C
D
Fig. 4
L
5

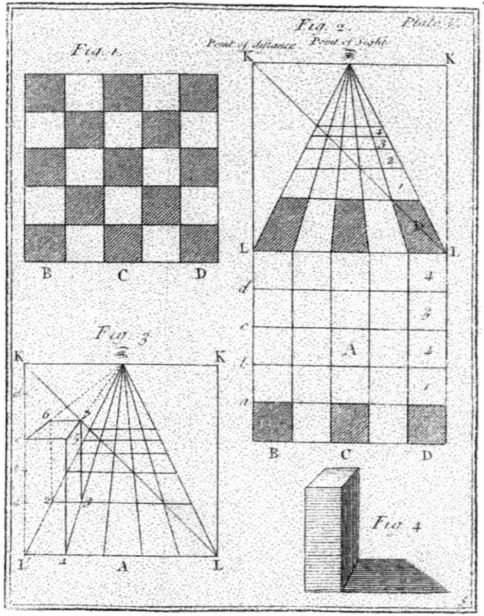
Plate I.
Fig. 1.
B
C
D
Fig. 2.
Point of distance
Point of sight
K
K
L
L
A
a
b
c
d
B
C
D
Fig. 3
K
K
L
L
A
Fig. 4

ECCENTRICITY } is the EXCENTRICITY } Distances between the Centres of two Circles or Spheres, which have not the same Centre.

ECHINUS, in Architecture, is a Member or Ornament near the Bottom of the *Ionic*, *Corinthian*, and *Composite* Capitals, which the *French* call *Quart de Rond*, from its circular Form or Contour; and the *English*, *Quarter Round*, or *Boultin*; the *Italians* call it *Ovolo*, from *Ovum*; and the *French*, *Ove*, from the *Latin*, *Ovum* an Egg; and thence the *English* call it *Eggs and Anchors*. See ANCHOR.

The *Greeks* call it Ἐχῖνος, a Chesnut, from the Egg's being encompassed with a Cover something resembling a Chesnut cut open.

ECHO, in Architecture, is a Term applied to certain Kinds of Vaults and Arches, most commonly of elliptical and parabolical Figures, used to redouble Sounds, and produce *artificial Echoes*.

The Jesuit *Blanc*, in his *Echometry*, at the End of his first Book of the *Sphere*, teaches the Method of making an *artificial Echo*.

Vitruvius relates, that in divers Parts of *Greece* and *Italy*, there were brazen Vessels artfully ranged under the Seats of the Theatres, to render the Sound of the Voices of the Actors more clear, and make a Kind of *Echo*; by which Means, every Person of that prodigious Multitude who assisted at the Spectacles, might hear with Ease and Pleasure.

Echo is a Sound reflected or reverberated from a solid concave Body, and so repeated to the Ear.

The *Peripateticks*, who imagined Sound to be I know not what Species, or Image of the sonorous Body impressed on the adjoining Air, account for *Echo* from a Resilition or leaping back of the Species, caused by its meeting some Obstacle in the Way.

But modern Naturalists, who know that Sounds consist in a certain Tremor or Vibration in the sonorous Body, communicated to the contiguous Air, and by that Means to the Ear, give us a more consistent Account of *Echo*.

For 'tis evident, that a tremulous Body, striking upon another solid Body, may be repelled without destroying or diminishing its Tremor; and of consequence, that a Sound may be redoubled by the Resilition of the tremulous Body or Air.

But a simple Reflection of the sonorous Air is not enough to solve the *Echo*; for then every plain Surface of a solid Body, as being fit to reflect a Voice or Sound, would redouble it; which, as is found by Experience, does not hold.

Therefore it should seem, that a kind of Concameration or Vaulting were necessary to produce an *Echo*, in order to collect; and by collecting, to heighten and increase, and afterwards to reflect the Sound; as it is found is the Case in reflecting the Rays of Light, where a concave Mirrour is required.

In

In Effect, as often as a Sound strikes on a Wall perpendicularly, behind which, is any Thing of an Arch, or even another parallel Wall, so often it will be reverberated, either in the same Line, or other adjacent ones.

Therefore it is necessary, in order that an *Echo* be heard, that the Ear be in the Line of Reflection; and in order that the same Person may hear it echo, who made the Sound, it is necssary, that he be perpendicular to the Place which reflects it. And as for a tautological or manifold *Echo*, it is necessary, that there be a Number of Walls and Vaults, or Cavities, either placed behind each other, or fronting each.

A single Arch or Concavity, *&c.* can scarce ever stop, and reflect the Sound; but if there be a convenient Disposition behind it, Part of the Sound, that is propagated thither, being collected and reflected as before, will present another *Echo*; or if there be another Concavity opposed at a due Distance to the former, the Sound reflected from the one upon the other, will be toss'd back again upon this latter, *&c.*

The Bishop of *Leighs*, &c. has well consider'd many of the Phænomena of *Echo*.

He remarks, that any Sound falling either directly or obliquely on any dense Body of a smooth Superficies, whether arched or plain, is reflected more or less.

He says, the Surface must be smooth, or else the Air by Reverberation will be put out of its regular Motion, and the Sound thereby broke or extinguish'd.

He likewise adds, that it echoes more or less, to shew that when all Things are, as before described, there is still an *Echoing*, though it be not always heard, either because the direct Sound is too weak to be beat quite back again to him that made it, or that it does return to him, but so weak, that it cannot be discern'd; or else that he stands in a wrong Place to receive the reflected Sound, which passes either over his Head, or under his Feet, or on one Side of him, and which therefore may be heard by a Man who stands in the Place where the reflecting Sound will come, provided no interposed Body intercept it, but not by him that first made it.

Echoes may be produced with different Circumstances:

First, A Plane Obstacle reflects the Sound back in its due Tone and Loudness, Allowance being made for the proportionable Decrease of the Sound according to its Distance.

Secondly, A Convex Obstacle reflects the Sound somewhat smaller, and somewhat quicker, though weaker than it otherwise would be.

Thirdly, A Concave Obstacle echoes back the Sound bigger, slower, and also inverted, but never according to the Order of Words.

Nor does it seem possible to contrive any single *Echo* that shall invert the Sound, and repeat backwards; because in such case, the Word which was last spoken,

that

that is, which laſt occurs to the Obſtacle, muſt be repelled firſt, which cannot be: For, where, in the mean Time, ſhould the firſt Words hang and be concealed? or, how after ſuch a Pauſe, be revived and animated again into Motion?

From the determinate Concavity or Archedneſs of the reflecting Bodies, it may happen that ſome of them ſhall only echo back one determinate Note, and only from one Place.

Fourthly, The echoing Body being removed further off, it reflects more of the Sound than when nearer; which is the Reaſon why ſome *Echoes* repeat but one Syllable, ſome one Word, and ſome many.

Fifthly, Echoing Bodies may be ſo contriv'd and placed, as that reflecting the Sound from one to the other, either directly and mutually, or obliquely and by Succeſſion, out of one Sound ſhall a *Multiple Echo*, or many *Echoes* ariſe.

To this may be added, that a *Multiple Echo* may be made by ſo placing the echoing Bodies at unequal Diſtances, as that they may reflect all one Way. and not one on the other; by which Means a manifold ſucceſſive Sound will be heard: One Clap of the Hands will be heard like many; one *Hah*, like a Laughter; one Word, like many of the ſame Tone and Accent; and ſo one Vial like many of the ſame Kind, imitating each other.

Laſtly, Echoing Bodies may be ſo order'd, that from any one Sound given, they ſhall produce many *Echoes*, different both as to Tone and Intention.

By which Means a Muſical Room may be ſo contriv'd, that not only Playing on an Inſtrument in it, ſhall ſeem many of the ſame Sort and Size, but even a Conſort of different ones, only placing certain echoing Bodies ſo, as that any Note play'd ſhall be return'd by them, in Thirds, Fifths, and Eighths.

Echoes are diſtinguiſh'd into divers Kinds, *viz.*

I. *Single Echoes*, which return the Voice but once; of which, ſome are Tonical, which only return a Voice when modulated in ſome particular Tone.

Others Polyſyllabical, which may return many Syllables, Words, and Sentences.

II. *Multiple* or *Tautological Echoes*, which may return Syllables the ſame oftentimes repeated.

In *Echoes*, the Place where the Speaker ſtands, is called the *Centrum Phonicum*, and the Object or Place which returns the Voice, *Centrum Phonicamp-ticum*.

At the Sepulchre of *Metella*, Wife of *Craſſus*, was an *Echo* which repeated what a Man ſaid five Times. And Authors mention a Tower at *Cyzicus*, where the *Echo* was repeated ſeven Times.

EFFECTION, in Geometry, is uſed in the ſame Senſe with the Geometrical Conſtruction of Propoſitions, and often of Problems and Practices; which, when they are reducible from, or founded upon ſome general Propoſition, are called the *Geometrical*

mettical Effections thereunto belonging.

EFFICIENTS, in Arithmetical Progression, are the Numbers given for an Operation of Multiplication called Factors. These *Efficients* are the Multiplicand, and the Multiplicator

EFFIGIES, } a Portrait Fi-
EFFIGY, } gure or Representation of a Person to the Life.

EGGS, in Architecture, an Ornament in that Form, cut in the Echinus or Quarter-Round of the *Ionic* and *Composite* Capitals. The Profile or Contour of an Echinus is enrich'd with *Eggs* and Anchors placed alternately.

ELABORATORY. See LABORATORY.

ELASTICITY is that Property of Bodies whereby they return to their former Figure, when it has been altered by any Force: For if a compact Body be dented in, without the Parts falling into that Dent, the Body will return to its former Figure, from the mutual Attraction of its Parts.

All Bodies, in which we observe *Elasticity*, consist of small Threads or Filaments, or at least may be conceived as consisting of such Threads; and it may be suppos'd that those Threads laid together make up one Body: Therefore that we may examine *Elasticity* in the Case which is the least complex, we must consider Strings of Musical Instruments, and such as are of Metal; for Cat-Gut Strings have a spiral Twist, and cannot be consider'd in the same Manner as those Fibres with which Bodies were form'd.

The *Elasticity of Fibres* consists in this, that they can be extended, and taking away the Force by which they are lengthened, they will return to the Length they had at first.

Fibres have no *Elasticity*, unless they are extended with a certain Force; as it appears in Strings which have their Ends fix'd without being stretch'd; for if you remove them a little from their Position, they do not return to it; but what the Degree of Tension is, which gives beginning to their *Elasticity*, is not yet determin'd by Experiments.

When a Fibre is extended with too much Force, and this Degree of Tension is also unknown: This we do know, that the Degree of Tension in Fibres, which constitutes *Elasticity*, is confin'd to certain Limits.

Hence appears the Difference of Bodies that are Elastick, and such as are not so; why a Body loses its *Elasticity*, and how a Body destitute of *Elasticity*, acquires that Property. A Plate of Metal, by repeated Blows of an Hammer, becomes Elastick, and by being heated, does again lose that Virtue.

Between the Limits of Tension that terminate *Elasticity*, there is a different Force requir'd for different Degrees of Tension in, or to stretch Cords to certain Lengths. What this Proportion is, must be determin'd by Experiments; which must be made with Chords of Metal.

ELBOW, in Architecture, a Term used for an Obtuse Angle of

of a Wall, Building, Road, &c. which divides it from its Right Line.

ELEMENTS, by Geometricians, Natural Philosophers, &c. are usually taken to signify the same as Principles or Rudiments of any Science. So when Natural Philosophers say, *the Elementary Principles of mixed Bodies*, they mean the simple Particles out of which the mixed Body is compos'd, and into which it is ultimately resolvable.

ELLIPSIS, in Geometry, is one of the Conick Sections, properly call'd an Oval or Oblong.

Ellipsis, or *Oval*, is a Figure bounded by a regular Curve Line, returning into itself; but of its two Diameters cutting each other in the Centre, one is longer than the other, in which it differs from the Circle.

To find the Area thereof, this is the RULE.

Multiply the Transverse Diameter by the Conjugate, and multiply that Product by .7854, and this last Product will be the Area of the *Ellipsis*.

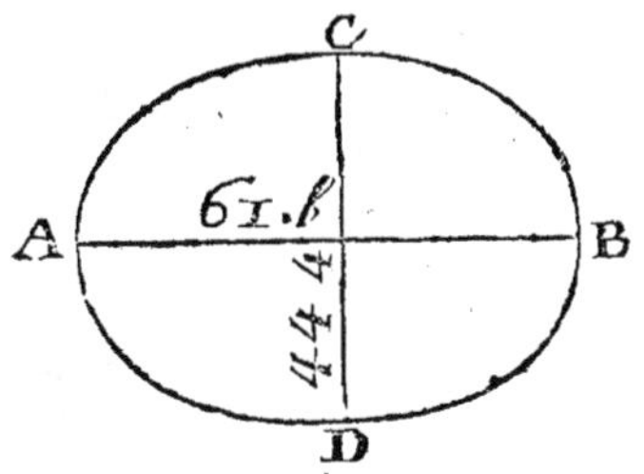


```
     61.6. the Transverse Diameter.
     44.4
    ------
     2464
    2464
   2464
  -------
   2735.04
     .7854
  --------
   1094016
  1267520
 2188032
1914528
-----------
2148.100416 the Area of the Oval.
```

DEMONSTRATION.

If you circumscribe any *Ellipsis* with a Circle, and suppose an infinite Number of Chord Lines drawn therein, all parallel to the Conjugate Diameter, as those in the Figure above, then it will be,

As DA the Diameter of the Circle is to N*n* the Conjugate Diameter of the *Ellipsis*; so is B *a* B any Chord in the Circle to

b a b

bab, its respective Ordinate in the *Ellipsis*.

For, according to the Property of the Circle,

It is	1	$aS \times Ta = \square\, Ba$, by the Property of the *Ellipsis*.
and	2	$\square\, TC : NC :: S \times Ta : \square\, ba$
it is	3	$\square\, TC : NC :: \square\, Ba : \square\, ba$
1. 2	4	$TC : NC :: Ba : ba$.

3 Hence Consequ.	5	$2\, TC : 2\, NC :: Ba : ba$
That is	6	$DA : Nn :: BaB : bab$.

But the Sum of an infinite Series of such Chords as B*ab* do constitute the Area of the Circle; and the Sum of the like Series of their respective Ordinates, as *bab*, do constitute the Area of the *Ellipsis*.

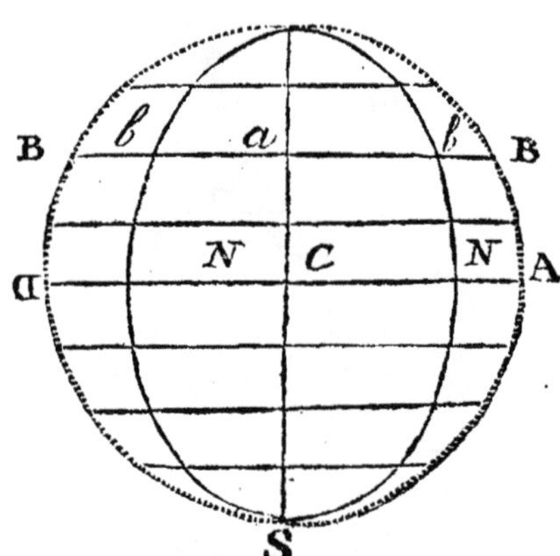

Therefore $TS : Nn ::$ Circle's Area : the *Ellipsis Area* but $TS : Nn :: \square\, TS \cdot TS \times Nn$; whence it follows that

$\square\, TS$: Circle's Area $:: TS \times Nn$: *Ellipsis* Area.

Consequently, as 1 is to .7854, so is the Rectangle or Product of the Tranverse and Conjugate Diameter of any *Ellipsis* to its Area.

Hence it is easy to conceive that the square Root of the Product of the Transverse and Conjugate Diameters will be the Diameter of a Circle equal to the *Ellipsis*.

ELLIPTICK, } of or per-
ELLIPTICAL, } taining to an Ellipsis. *Serlio*, *Hartman*, &c. demonstrate, that the best Form for Arches or Vaults, is *Elliptical*.

Elliptick Space is the Area contain'd within the Circumference or Curve of the Ellipsis.

Elliptical Compasses, an Instrument usually made of Brass for drawing an Ellipsis or Oval at one Revolution of an Index.

ELM

ELM is of ſingular Uſes, where it may lie continually wet or dry in Extreams, therefore proper for Water-Works, Mills, Ladies, and Soles of Wheel-Pipes, Aqueducts, Pales, Ship-Planks, beneath the Water-Line. Some of it found in Bogs has turn'd like the moſt poliſh'd and hardeſt Ebony.

It is alſo of Uſe for Wheelwrights, Handles for ſingle Saws, the knotty Parts for Naves, and Hubbs; the ſtraight and ſmooth for Axletrees; and the very Roots for curiouſly dappled Works, Kerbs of Coppers, Featheredge, and Weather-Boards, Trunks, Coffins, and Shovel-Board Tables. The Tenor of the Grain makes it alſo fit for all Kinds of Carved Work, and moſt Ornaments belonging to Architecture.

Vitruvius commends it for Tenons and Mortoiſes.

EMBOSSING, } in Architec-
IMBOSSING, } ture, Sculpture, &c. is the forming, or faſhioning of Works in Relievo, whether cut with a Chiſſel, or otherwiſe; it is a kind of Sculpture or Engraving, wherein the Figures ſtick out from the Plane whereon it is engraven, and according as they are more or leſs protuberant.

It is called by the *Italians*, Baſſo, Mezzo, or Alto Relievo, and by the *Engliſh*, Baſs-Relief, Mean Relief, or High Relief.

EMBRASURE, in Architecture, is the Enlargement made of a Gap or Aperture of a Door or Window on the Inſide of a Wall.

Its Uſe is to give the greater Play for the opening of the Door, Wicket, Caſement, &c. or to take in the more Light.

The *Embraſure* coming ſloping inwards, renders the inner Angles obtuſe. When the Wall is very thick, they ſometimes make *Embraſures* on the Outſide.

EMPASTING, in Painting, is the laying on Colours thick and bold; or the applying ſeveral Lays of Colours, to the End that they may appear thick.

ENDECAGON, } in Geo-
HENDECAGON, } metry, a Figure having eleven Angles, and conſequently as many Sides.

ENGINE, in general, is any Mechanick Inſtrument compos'd of Wheels, Screws, Pullies, &c. by the Help of which, a Body is either mov'd or hindred from moving.

Firſt, When the Quantities of Motion, in the Weight and Power, are equal, the *Engine* ſhall ſtand *in æquilibrio*; but when they are unequal, the greater Quantity of Motion ſhall overcome and work the *Engine*.

Secondly, Of Forces in themſelves equal, that which is neareſt to that Point of the *Engine*, about which the Weight and Power move, or upon which they ſuſtain each other, is relatively the weakeſt upon the *Engine*; for as the *Engine* works, the neareſt Force moves the floweſt, and therefore has the leaſt Quantity of Motion.

Thirdly, The Effect of any Force upon the *Engine*, will not be changed, if, without changing the Line of Direction, it is only placed in ſome other Point of the ſame Line.

The

The Nature of any *Engine* is explain'd, when it is known in what Circumstances the Weight and Power will be *in æquilibrio* upon that *Engine*.

Fourthly, In all *Engines* whatsoever, the Weight and Power will be *in æquilibrio*, when their Quantities are in the reciprocal Proportion of the Velocities, which the Working of the *Engine* will give them.

If an *Engine* be compounded of several simple *Engines*, the Power is to the Resistance, when it counterbalances it in a Ratio, compounded of all the Ratio's which the Powers in each simple *Engine* would have to the Resistance, if they were separately apply'd.

ENNEAGON, a Figure of nine Angles and nine Sides.

ENTABLATURE, } in Architecture,
ENTABLEMENT, }
which *Vitruvius* and *Vignola* call Ornament, is that Part of an Order of a Column, which is over the Capital, and comprehends the Architrave, Frize, and Cornice.

The *Entablature* is also called the Trabeation, and seems borrow'd from the *Latin Trabs*, a Beam. But some derive it from *Tabulatum*, *L.* a Ceiling; because the Frize is suppos'd to be form'd by the Ends of the Joists, which bear upon the Architrave.

It is different in different Orders: Indeed, it does consist of the three grand Parts or Divisions above mentioned in all; but those Parts consist of a great or less Number of particular Members or Subdivisions, according as the Order is more or less rich.

Vignola makes the *Entablature* a quarter of the Height of the whole Column in all the Orders.

In the *Tuscan*, and *Doric*, the Architrave, Frize, and Cornice, are all of the same Height.

In the *Ionic*, *Corinthian*, and *Composite*, the whole *Entablature* being fifteen Parts, five of these go to the Architrave, four to the Frize, and six to the Cornice.

Entablature, } in Masonry, is
Entablement, }
used for the last Row of Stones on the Top of the Wall of a Building, on which the Timber and the Covering rest.

It is often made to project beyond the Naked of a Wall, to carry off the Rain.

The *Entablement* of the *Tuscan* Order, says M. *Le Clerc*, consists of three principal Parts; a Cornice, a Frize, and an Architrave.

To the first, that is, the Cornice, he gives about two Fifths of the Height of the *Entablement*.

The Frize he makes somewhat higher than the Architrave, to the End that those two Members may appear to have nearly the same Height; the Overplus given to the Frize, being intended to supply the Place of that Part hidden from the Eye by the Tænia, which finishes the Architrave. And this same Rule, he says, he uses in all his *Entablements*.

Of the *Entablement* of the Column raised on a Pedestal, he tells

tells us, he always makes two Deſigns of an *Entablement*; the one a ſmall Matter higher than the other; the firſt for Columns that have no Pedeſtals, and the ſecond for thoſe that have. This Difference of *Entablement* is a Thing highly reaſonable, in regard Columns that have Pedeſtals, are in a more ſtately Ordonnance than thoſe which have none, provided the Columns be but equal in other Reſpects. Whence 'tis apparent, the *Entablement* of the firſt ſhould be ſtronger than that of the laſt: Accordingly, making one Entablement three Modules and fifteen Minutes, which is the common Height, he ſays, he could not think it adviſeable to make the other, which is for Columns without Pedeſtals, above three Modules, ten Minutes; which comes five Minutes ſhort of the former.

I am ſenſible, ſays he, that were we only to have Regard to the Laws of Strength and Weakneſs, we ſhould diminiſh the *Entablements* of Columns that have Pedeſtals, rather than thoſe which have none.

But we are here, ſays he, conſulting Beauty, not Strength; and it may be obſerved, I don't augment the Strength of this *Entablement*, but diminiſh that of the former, where the Porticoes are leſs grand, and the Columns leſt diſtant.

Of *Entablatures* which have Breaks, and project unequally, M. *Le Clerc* ſays, the *Entablature* is ſometimes made to give back or retreat a little between the Columns; but on etraordinary Occaſions, and for ſpecial Reaſons, as where there are not large Stones ſufficient to carry out the whole *Entablature* to its due Pitch; or where a great Projecture between the Columns might intercept the Light neceſſary underneath, or prevent the View of any Thing above. But, however, it muſt not be forgot that the principal End of the *Entablature*, is to ſhelter what is underneath; which, in this Caſe, it only does by Halves, as having nothing but the bare Projecture of the Cornice for that Purpoſe.

ENTERSOLE, in Architecture, a kind of little Story, ſometimes called a Mezanzine, contriv'd occaſionally at the Top of the firſt Story, for the Conveniency of a Wardrobe, *&c.*

ENTRY, a Door, Gate, Paſſage, *&c.* through which we arrive at any Place.

EPICYCLOID, in Geometry, a Curve, generated by the Revolution of a Point of the Periphery of a Circle along the convex or concave Part of another Circle.

EPISTYLE, in the antient Architecture, a Term uſed by the *Greeks*, for what we call Architrave. *viz.* a maſſive Piece of Stone or Wood laid immediately over the Capital of a Column.

EQUAL is a Term of Relation between two or more Things of the ſame Magnitude, Quantity, or Quality.

Equal Circles are thoſe whoſe Diameters are *equal.*

Equal Angles are thoſe whoſe Sides are inclin'd alike to each other, or that are meaſured by ſimilar Parts of their Circles.

Equal Figures are those whose Areas are *equal*, whether the Figures be similar or not.

Equal Solids are such as comprehend or contain each as much as the other, or whose Solidities and Capacities are *equal*.

Equal Geometrical Ratio's are those whose least Terms are similar aliquot or aliquant Parts of the greater.

Equal Arithmetical Ratio's are those wherein the Difference of the two less Terms is equal to the Difference of the two greater.

Proportion of EQUALITY, evenly ranged, or *Ex æquo ordinata*, is that wherein two Terms in a Rank or Series are proportional to as many Terms of another Rank. *i. e.* the first of one Rank to the first of another, and the second to the second, *&c.*

Proportion of Equality evenly disturb'd, called also, *Ex æquo perturbata*, is that wherein more than two Terms of a Rank are proportional to as many Terms of another Rank, compared to each other in a different and interrupted Order.

EQUIANGULAR, in Geometry, is apply'd to Figures whose Angles are all equal, as a Square is an *equiangular* Figure. All equilateral Triangles are also *equiangular*.

When the three Angles of one Triangle are severally equal to the three Angles of another Triangle, the Triangles are also said to be *equiangular*.

EQUICRURAL TRIANGLE, *i. e.* having equal Legs, is what we more usually call an Isosceles.

EQUIDIFFERENT, in Arithmetick. If in a Series of three Quantities, there be the same Difference between the first and second, as there is between the second and third, they are said to be *continually equidifferent*.

But if in a Series of four Quantities, there be the same Difference between the first and second, as there is between the third and fourth, they are said to be *discreetly indifferent*. Thus 3, 6, 7, and 10, are *discreetly equidifferent*, and 3, 6, and 9, *continually equidifferent*.

EQUIDISTANT, in Geometry, is a Term of Relation between two Things which are every where at one Equal, or the same Distance from each other: Thus parallel Lines are said to be *equidistant*; as they neither approach nor recede: And Parallel Walls are *equidistant* from each other.

EQUILATERAL is a Term apply'd to any Thing, the Sides of which are all equal. Thus an *Equilateral Triangle* is one whose Sides are all of an equal Length.

In an *Equilateral Triangle*, all the Angles are likewise equal.

All Regular Polygons, and Regular Bodies, are *equilateral*.

EQUILIBRIUM, in Mechanicks, a Term that implies an exact Equality of Weight between two Bodies.

EQUIMULTIPLE, in Arithmetick, is apply'd to simple Magnitudes when multiply'd equally, *viz.* by equal Quantities or Multipliers.

In Arithmetick, we generally use the Term *Equimultiples* for

for Numbers which contain equally, or an equal Number of Times, their Submultiples.

Thus 12 and 6 are *Equimultiples*, of their Submultiples 4 and 2, in as much as each of them contains its Submultiple three Times.

EVEN NUMBER is that which can be divided into two equal Parts, as 4, 6, 8, *&c.*

EVENLY-EVEN *Number*, is that which an even Number measures by an even one; as 16 is an *Evenly-even Number*, because 8 an even Number measures it by 2, an even Number.

Evenly-odd Number is that which an even Number measures by an odd one, as 20, which the even Number 4 measures by the odd one 5.

EVOLVENT, in Geometry, a Term used by some Writers for the Curve which results from the Evolution of a Curve, in Contradistinction to the Evolute, which is the first Curve supposed to be opened or evolved.

EVOLUTE, in the Higher Geometry, a Curve supposed to be evolv'd or open'd, and which in opening describes other Curves.

EVOLUTION, in Geometry, is the Unfolding or Opening of a Curve, and making it describe a Volute.

Evolution is also used for the Extraction of Roots out of Powers; in which Sense it is directly contrary to Involution.

EURITHMY, in Architecture, Sculpture, and Painting, is a certain Majesty, Elegance, and Easiness, appearing in the Composition of divers Members or Parts of a Body, Painting, or Sculpture, and resulting from the fine Proportion of it.

Vitruvius ranks the *Eurithmia* among the essential Parts of Architecture. He describes it as consisting in the Beauty of the Construction or Assemblage of the several Parts of the Work, which render its Aspect, or its whole Appearance grateful; *e.g.* when the Height corresponds to the Breadth, and the Breadth to the Length, *&c.*

From these three Ideas, or Designs, *viz.* Orthography, Scenography, and Profile, it is, that the same *Eurithmia*, majestick and beautiful Appearance of an Edifice, does result; which creates that agreeable Harmony between the several Dimensions, *i. e.* between the Length, Breadth, and Height of each Room in a Fabrick, so that nothing seems disproportional, too long for this, or too broad for that, but corresponds in a just and regular Symmetry and Consent of all the Parts with the Whole. *Evelyn's Account of Architects.*

EUSTYLE, in Architecture, a sort of Building in which the Pillars are placed at the most convenient Distance one from another; the Intercolumniations being all just two Diameters, and a quarter of the Column; except those in the Middle of the Face, before and behind, which are three Diameters distant.

EXAGGERATION, in Painting, is a Method of representing Things wherein they are charged too much, or mark'd too

 strong;

ftrong either in refpect to the Defign or the Colouring.

EXAGON. See HEXAGON.

EXPERIMENTUM *Crucis* is a capital leading or decifive Experiment, thus called, as either like a Crofs, or Poft of Direction placed in the Meeting of feveral Roads: It guides and directs Men to the true Knowledge of the Nature of the Thing, as it were, extorted by Violence.

EXPONENT, in Arithmetick, or *Exponent of a Power*, the Number which expreffes the Degree of the Power; or which fhews how often a given Power is to be divided by its Root, before it be brought down to Unity.

EYE, in Architecture, is ufed to fignify any round Window made in a Pediment, an Attick, the Reins of a Vault, or the like.

Eye of a Dome is an Aperture at the Top of the Dome; as that of the *Pantheon* at *Rome*, or of *St. Paul's* at *London*. It is ufually cover'd with a Lanthorn.

Eye of the Volute, in Architecture, is the Centre of the Volutes, or that Point in which the Helix or Spiral, of which it is form'd, commences: Or it is the little Circle in the Middle of the Volutes, in which are found the thirteen Centres for the defcribing the Circumvolutions of it.

Eye-Brow, in Architecture, is ufed in the fame Senfe as Lift or Fillet.

F A

FABRICK, the Structure or Conftruction of any Thing, particularly a Building, as an Houfe, Hall, Church, *&c.*

FACADE, } in Architecture,
FACE, } the Front of a Building, or the Side on which the chief Entrance is. Alfo it is fometimes ufed for the Side that it prefents to a Street, Garden, Court, *&c.* And fometimes for any Side oppofite to the Eye.

FACE, } in Architecture, is
FACIA, } a flat Member, ha-
FASCIA, } ving a confiderable Breadth, and but a fmall Projecture; as the Bands of an Architrave, Larmier, *&c.*

Face of a Stone, in Mafonry, is the Superficies or Plane of the Stone that is to lie in the Front of the Work; which is very eafily known, when the *Face* is fcapted, the *Face* being always oppofite to the Back; and the Back going rough as it comes from the Quarry.

But in rough Stones, Workmen generally chufe to make one of thofe Sides the *Face*, which, when in the Quarry, lay perpendicular to the Horizon, and confequently the breaking, and not the cleaving Way of the Stone.

For a better underftanding of which, fee STONE.

FACEING,

FACEING *of Timber Building with Brick.* The Manner of this is as follows: All betwixt the Timber and the Wall is a Brick's Length thick, (or a nine Inch Wall,) and against the Timber but half a Brick, or a four and a half Inch Wall.

But this Way of *facing* Timber Buildings is not approved of, by reason that the Mortar does so extremely burn the Timber.

FACIA, FASCIA, FACIO, FACE, in Architecture, according to M. *Perrault*, signifies any flat Member, as the Band of an Architrave, &c. Some write it *Fasce*, as though from the *Latin*, *Fascia*, a Swathe, or large Turban, which *Vitruvius* uses on the like Occasion. In effect it is no more than a broad List, or Fillet.

They are commonly made in Architraves, and in the Cornices of Pedestals.

Fascia's, in Brick Buildings, are certain Juttings-out of the Bricks over the Windows of each Story, except the upper one. And these are sometimes plain, like those of Columns; but sometimes they are moulded, which make a very handsome Appearance: And this Moulding is usually a Scima Reversa at the Bottom, above which are two plain Courses of Bricks, then an Astragal, and lastly a Boultin.

It is the same in Stone Buildings as it is in Brick, and they are also sometimes plain, and sometimes moulded with a Scima Reversa, or Ogee.

The Price of *Fascia's* is, if the Workman finds materials, usually about 10 *d.* *per* Foot running Measure; but Workmanship only, is about 6 *d.* or 8 *d.* *per* Foot.

Vitruvius means by the Term *Fascia*, (as also *Tænia* and *Corsa*,) what we call *Plat-Band.*

FACTORS, in Arithmetick, is a Name given to the Multiplicand and Multiplicator, because they do *facere productum*, *i. e.* make or constitute the Product.

FACTUM, in Arithmetick, the Product of two Quantities multiplied by each other.

FEATHER-EDG'D *Boards* or *Planks*, are such as are thicker on one Side than the other.

FELLING *of Timber.* See TIMBER.

FENCING *with Pales:* As Paling with three Rails, cleft Pales, Rails, and Posts, cleaving and setting up, is usually done at 3 *s.* 6 *d.* the Rod, reckoning the felling of the Timber into the Bargain; but then their Materials are laid down to their Hands.

Fencing with single Rails and Posts, together with felling, cleaving, and setting up, is usually done at 8 *d.* or 10 *d.* the Rod; but then also their Materials must be laid down to their Hands, that they may have no carrying.

Some say, that they have known it done for 4 *d.* 5 *d.* or 6 *d.* *per* Rod, felling, cleaving, and setting up; but then it is when the *Fence* is cross a Field, or the like, where the Post Holes may be easily dug, (and when pretty many Rods are to be done together, and the Materials are also laid down to their Hands,)

 and

and not in Gaps in Hedges, and the like, where the Digging is hard Work, and but a little at a Place; for in such Work it is worth 8 *d.* 10 *d.* or the Rod.

FENCE WALLS, are Walls of Brick or Stone made about Gardens, &c.

FESTOON, an Ornament or Garland of Flowers, Fruits, and Leaves, intermix'd or twisted to-

gether. They were antiently much used at the Gates of Temples, where Feasts or solemn Rejoycings were held; or at any other Places where marks of publick Joy and Gaiety were desired; as on Triumphal Arches, Tournaments, &c.

FESTOONS, in Architecture, &c. a Decoration used by Architects, Painters, Joiners, &c. to enrich their Works. It consists of a String or Column of Flowers, Fruits, and Leaves tyed together, somewhat biggest in the Middle, and extended by the two Extremes; besides which, the main Part which falls down in an Arch, two lesser Parts hang perpendicularly. See the Figure.

Festoons are now chiefly used in Friezes, and other vacant Places, which require to be fill'd up and adorn'd.

This Ornament is made in Imitation of the *Festoons*, or long Clusters of Flowers, which the Antients placed on the Doors of their Temples, &c. on festival Occasions.

FIGURE is the Surface or terminating Extremes of a Body.

Figure also signifies all Representations or Images of Things in Sculpture, Prints, &c.

Figure, in Geometry, signifies a Surface inclos'd, or circumscribed with one or more Lines; as Circles, Ellipses, Triangles, Squares, Polygons, &c.

Figures are either rectilineal or curvilineal, or mix'd, according as the Perimeter consists of Right Lines, Curve Lines, or both.

Rectilineal Figures are those which have their Extremities all Right Lines, as Triangles, Quadrilaterals, &c. Polygons Regular, Irregular, &c.

Curvilineal Figures are such as have their Extremities crooked; as Circles Ellipses, &c.

Mix'd Figures are such as are bounded partly by Right Lines, and partly by crooked ones; as a Semi-Circle, Segment of a Circle, &c.

Plane Figures, or Plane Surfaces, are such as are terminated and bounded by Right Lines only.

A *Regular Figure*, is that which is equilateral and equiangular.

An *Irregular Figure*, is that which is not both.

Figure, in Conic Sections, according to *Apollonius*, is the Rectangle under the *Latus Rectum* & *Transversum* in the Hyberbola and Ellipsis.

Figure of the Diameter, the Rectangle under any Diameter, and

and its proper Parameter, is in the Ellipsis and Hyperbola, call'd the *Figure of that Diameter*.

Figure, in Painting and Designing, is the Lines and Colours, that Form the Representation of a Man, or other Animal.

Figure, in Architecture, *&c.* signifies the Representations of Things made in solid Matter, as Statues, *&c.*

Figures, in Arithmetick, are the nine Digits, or numerical Characters, 1, 2, 3, 4, 5, 6, 7, 8, 9, 0; or those by which Numbers are expressed or written.

FILLET } in Architecture, a
FILET } little square Member, or Ornament, or Moulding, used in divers Places, and upon divers Occasions; but generally as a Corona for Crowning over a greater Moulding.

The *Fillet* is the same that is by the *Italians* called *Lista*, or *Listella*; by the *French*, *Reglet*; and by others *Band*, and *Bandelette*.

Fillet, in Painting, Gilding, *&c.* is a little Rule, or Reglet of Leaf Gold drawn over some Mouldings; or on the Edges of Frames, Pannels, *&c.* especially when painted white, by Way of Enrichment.

FINISHING, with Architects, *&c.* is frequently used of a Crowning, Acroter, *&c.* raised over a Piece of Building, to terminate and *finish*, or compleat it.

FIRE-STONE, a sort of Stone called also *Rygate-Stone*, of the Name of the Place from whence it is chiefly brought, being very good for Fire-Hearths, Ovens, Stoves, *&c.*

As to the Price of *Fire-Stone*, Hearths of it are usually sold at 1 *s. per* Foot, and Chimney-Corner Stones at 20 *s. per* Pair; and Blocks for setting up Coppers, each being about three Feet long, one and a half broad, and eight or nine Inches thick, at 6 *s.* 8 *d. per* Piece.

FLEMISH BRICKS, a neat strong Sort of Bricks, of a yellowish Colour, brought from *Flanders*, and much used in paving Yards, Stables, *&c.* being much neater and stronger, than common or Clay Bricks.

These Bricks are six Inches and a quarter in Length, two and a half in Breadth, and one and a quarter thick. Now, allowing one Fourth of an Inch for the Joint, 72 of them will pave a Yard square; but if they be set edge-ways, then a Yard square will require 100.

These *Bricks* are usually sold for 2 *s. per* Hundred.

FLIGHT of a Stair-Case. See STAIR-CASE.

FLINT WALLS. See WALLS.

FLOOR, in Architecture, is the Underside of the Room, or that Part whereon we walk.

Floors are of several Sorts; some of Earth, some of Brick, some of Stone, and some of Wood.

Carpenters by the Word *Floor*, understand as well the fram'd Work of Timber, as the Boarding over it.

Earthen Floors are commonly made of Loam, and sometimes (for *Floors* to make Malt on) of Lime and Brook Sand, and Gun-Dust,

Dust, or Anvil-Dust from the Forge.

FLOORING, a rural Sort of Work, by which, in this Place, are not meant *Floors* laid with Boards or Planks, but such as are used in plain Country Habitations, and the Manner of making them.

Take two Thirds of Lime, and one of Coal-Ashes well-sifted, with a small Quantity of loamy Clay; mix the Whole that you intend to use together, and temper it well with Water; making it up into a Heap, let it lie a Week or ten Days, in which Time it will mellow and digest: Then temper it well over again, and be sure that your Quantity of Water does not exceed, but rather that it may obtain a mellow Softness and Toughness from Labour: Then heap it up again for three or four Days, and repeat the Tempering very high, till it becomes smooth and yielding, tough and glewy.

Then the Ground being levelled, lay your *Floor* therewith about two and a half, or three Inches thick, making it smooth with a Trowel: The hotter the Season is, the better; and when it is thoroughly dry'd, it will continue Time out of Mind.

This makes the best *Floors* for Houses, especially for Malt-Houses; but as for those who cannot get these Materials, or go to the Charge of them, they may take of clayey Loam and new soft Horse-Dung one Third, with a small Quantity of Coal-Ashes, if they can be had, and temper these after the aforementioned Manner; and lay the *Floor* with the Stuff three or four Inches thick, smooth and even, which will cement, become hard, strong, and durable, being done in a hot and dry Season; good for Cottages, Barns, and other small Houses.

But if any would have more beautiful *Floors* than these, they must lay their *Floors* even, smooth, and fine, either with the first or last mentioned *Flooring*; then take Lime made of Rag-Stones, and temper it with a little Whites of Eggs, the more Eggs the better, to a very high pitch, with which cover your *Floor* about a quarter or half an Inch thick, before the under *Flooring* be too dry, that they may well incorporate together: This being well done, and thoroughly dry, if sometimes rubbed over with Mops or Cloths, with a little Oil thereon, it will look very beautiful and transparent, as if it were pollished Metal or Glass, provided the Eggs and Lime were thoroughly tempered, and otherwise well performed.

Sir *Hugh Plat* gives us a Receipt for making an artificial Composition wherewith to make smooth, glittering, and hard *Floors*; and which may also serve for plastering of Walls.

Ox Blood and fine Clay tempered together, he says, makes the finest *Floor* in the World; and that this Mixture laid in any *Floor* or Wall, will become a very strong and binding Substance.

For Brick and Stone *Floors*, see PAVING.

Concerning Boarded *Floors*, it is to be observed, that the Carpenters

penters never *Floor* their Rooms with Boards, till the Carcase of the House is set up, and also is inclosed with Walls, lest the Weather should wrong the *Flooring*; yet they generally rough-plane the Boards for *Flooring*, before they begin any thing else about the Building, that they may set them by to season, which is done as follows: They lean them one by one on end aslant, with the Edge of the Board against a Balk somewhat higher than half the Length of the Board, and then they set another Board in the same Posture on the other Side of the Balk, so that the Boards cross one another above the Balk; then on the first Side they set another Board in the same Posture, and on the second Side another, and so proceeding alternately, till the whole Number of Boards is thus set on end.

The Boards being set up in this Posture, there is left a Space of the Thickness of a Board all the Length of the Boards, but just where they cross one another, for the Air to pass through to dry and shrink them; but they are set under some covered Shed, that neither the Rain nor Sun may come at them: For if they should be wetted with Rain, that would swell them, instead of shrinking them; and if the Sun should shine very hot upon them, it would dry them so fast, that they will split or crack, which is what they call *tearing*, or *shaking*.

There is also another Way of drying and seasoning Boards for *Floors*, *viz.* by laying them flat upon three or four Balks, each Board about the Breadth of a Board asunder the whole Length of the Balks; then they lay another Lay of Boards athwart the last, and so till they have laid them all after this Manner; so that in this Position they also lie hollow, for the Air to play between them.

Of measuring *Floors*: Boarded *Floors* are usually measured by the Square, (of 100 superficial Feet,) by multiplying the Length of the Room in Feet by the Breadth in Feet, and the Product is the Content in Feet; then the Chimney-Ways and Well-Holes for Stairs are measured by themselves, and their Content in Feet is deducted from the whole Content; and afterwards cut off two Figures from the Remainder on the right Hand, and what remains on the left Hand is Squares, and what are cut off are odd Feet of the Content of the *Flooring* in that Room.

The Price: Mr. *Wing* says, the Framing of *Floors* in ordinary Buildings is worth seven or eight Shillings *per* Square, and in great Buildings ten or eleven Shillings.

Some *Sussex* Workmen say, that they usually have but four Shillings *per* Square for framing *Floors* in ordinary Buildings; and that if they frame the Joists the whole Depth of the Girder, and pay for sawing the Timber, they have nine or ten Shillings *per* Square.

Mr. *Leybourn* says the Price of laying (*i. e.* boarding) of *Floors*, is various, according to the Goodness of the Stuff, from twelve to twenty Shillings *per* Square; but

but if the Builder himself finds the Boards, the common Allowance for planing, jointing, and laying of Boards, is four or five Shillings *per* Square, besides Nails; of which 200 is a competent Allowance for one Square of *Flooring*.

Some *Sussex* Workmen say, they will lay Deal Boards braded, and plain Joints broken at every four or five Boards, for three Shillings *per* Square; and if they break Joint at every Board, then six Shillings, and others say six Shillings and eight Pence, or seven Shillings *per* Square.

Mr. *Wing* likewise says, Plaister *Floors* running, the Workman finding all, are worth one Shilling and four Pence *per* Yard; but the working Part only is worth four Pence, five Pence, or six Pence *per* Yard.

The Manner of Framing a Floor, *with the Names of each Member.*

1. The Thickness of the Wall and Lintel, or Wall-Plate, and if it be in Timber Work, then a Bress-Summer.
2. The Summer.
3. The Girders fram'd into the Summer.
4. Spaces between the Joists.
5. Joists.
6. Trimmers for the Chimney-Way.
7. Trimmers for the Stair-Case, or Well-Hole for the Stairs.

See the Plate.

FLUIDITY is the State or Affection of Bodies, which denotes or renders them fluid, and is directly opposite to Firmness, or Solidity.

It is distinguished from Liquidity and Humidity, in that the Idea of the first is absolute, and the Property contained in the Thing itself; whereas that of the latter is relative, and implies wetting or adhering; *i.e.* somewhat that gives it the Sensation of Wetness or Moisture, and which would have no Existence, but for our Senses.

Thus melted Metals, Air, and even Smoak, and Flame itself, are fluid Bodies, but not liquid ones, their Parts being actually dry, and not leaving any Sense of Moisture.

FLUIDS are Bodies whose Parts are but weakly connected, their mutual Cohesion being in great measure prevented from some external Cause; in which Sense a *Fluid* is opposite to a Solid.

Sir *Isaac Newton* defines a *fluid* Body to be that whose Parts yield to the smallest Force impress'd, and by yielding are easily moved among each other.

Fluids are either natural, as Water, Mercury, &c. or animal, as Blood, Milk, Urine, &c. or fictitious, as Wines, Drinks, Spirits, Oils, &c.

The Knowledge of the Properties of *Fluids* being of use in Hydraulicks, or the conducting of Water, I shall here give some Laws of *Fluids* in that Science.

Hydraulick Laws of Fluids.

1. The Velocity of a *Fluid*, as Water moved by the Pressure of a superincumbent *Fluid*; as Air, is equal at equal Depths.

2. The

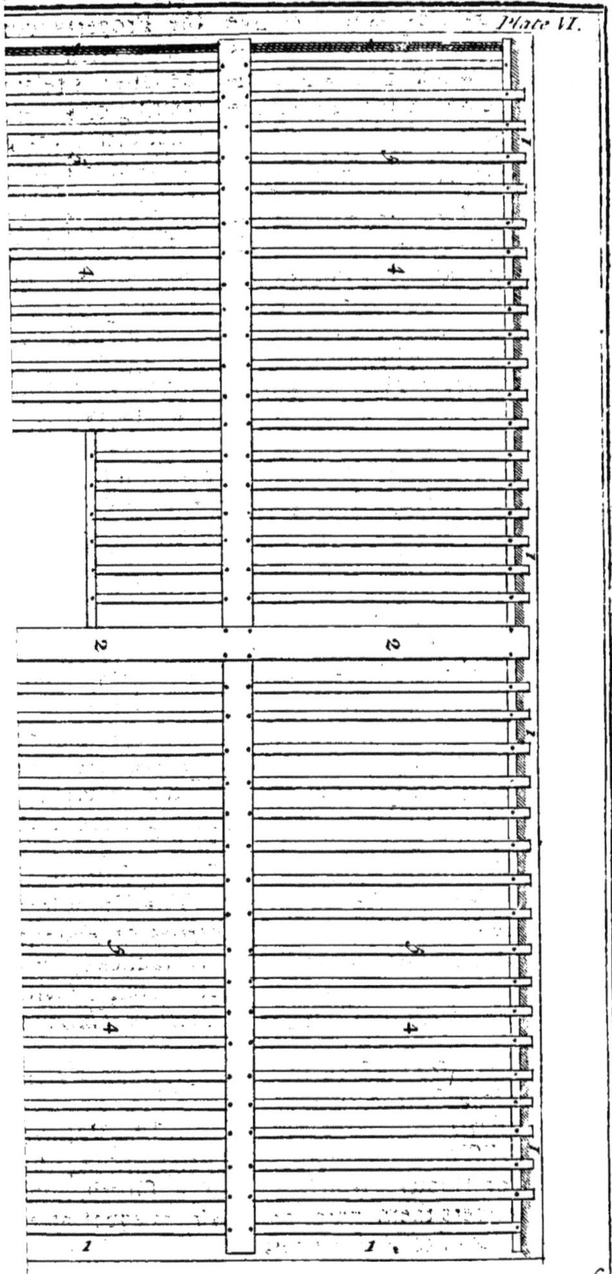
1
2
4
1
2
2
4
4
2
1
1
1
1
1
1
6

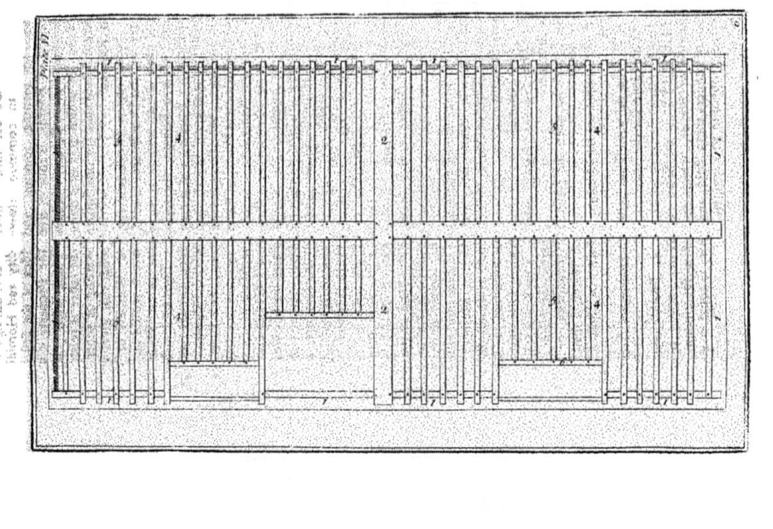

2. The Velocity of a *Fluid* arising from the Pressure of a superincumbent *Fluid* at any Depth, is the same as that which a Body would acquire in falling from a Height equal to the Depth.

3. If two Tubes, of equal Diameters, full of any *Fluid*, be of the same Altitude, they will discharge equal Quantities of the *Fluid* in equal Times.

4. If two Tubes of equal Altitudes, but unequal Apertures, or Diameters, be kept constantly full of Water, the Quantities of Water they yield in the same Time, will be as the Diameters; and this whether they be erect, or any way inclin'd.

5. If the Apertures of two Tubes be equal, the Quantities of Water discharged at the same Time, are as their Velocities.

6. If two Tubes have equal Apertures, and unequal Altitudes, the Quantity of Water discharged from the greater, will be to that discharged from the other in the same Time, in a subduplicate Ratio of the Altitudes.

7. If the Altitude of two Tubes be unequal, and the Apertures likewise unequal, the Quantities of Water discharged in the same Time, will be in a Ratio compounded of the simple Ratio of the Apertures, and the Subduplicate of one of the Altitudes.

8. If the Altitudes of two Tubes be equal, the Water will flow out with equal Velocity, however unequal the Apertures be.

9. If the Altitude of two Tubes, as also their Apertures, be unequal, the Velocities of the Water discharged are in a subduplicate Ratio of their Altitudes.

10. The Altitudes and Apertures of two Cylinders full of Water being the same, one of them will discharge double the Quantity of Water discharged in the same Time by the other, if the first be kept continually full, while the other runs itself empty.

11. If two Tubes have the same Altitudes, and equal Apertures, the Times wherein they will empty themselves, will be in the Ratio of their Bases.

12. Cylindrick and Prismatick Vessels empty themselves by this Law, That the Quantities of Water discharged in equal Times, decrease according to the uneven Numbers 1, 3, 5, 7, 9, *&c.* taken backwards.

13. If Water descending through a Tube Spout up at the Aperture, whose Direction is vertical, it will rise to the same Altitude at which the Level of the Water in the Vessel does stand.

14. The Length or Distances to which Water will Spout, either through an inclin'd or an horizontal Aperture, are in a subduplicate Ratio of the Altitudes in the Vessel or Tube. See HYDRAULICKS and WATER.

FLUTES } in Architecture, are per-
FLUTINGS } pendicular Channels or Cavities cut along the Shaft of a Column or Pilaster.

They

They are ſuppoſed, as *Vitruvius* ſays, to have been firſt introduced in Imitation of the Plaits of Women's Robes or Garments.

The *French* call them *Channellures*, *i. e.* Channellings, as being Excavations; and we call them *Flutes*, or *Flutings*, as bearing ſome Reſemblance to the muſical Inſtrument called a *Flute*.

They are chiefly affected in the *Ionic* Order, where they had their firſt Riſe; though indeed they are uſed in all the richer Orders, as the *Corinthian* and *Compoſite*, but ſeldom in the *Doric*, and ſcarce ever in the *Tuſcan*.

Each Column has 24 *Flutes*, and each *Flute* is hollow'd in exactly a Quadrant of a Circle; but the *Doric* has but 20.

Between the *Flutes* are little Spaces that ſeparate them, which *Vitruvius* calls *Striæ*, and we *Liſts*; though in the *Doric* the *Flutes* are frequently made to join to one another, without any intermediate Space at all, the Liſt being ſharpened off to a thin Edge, which forms a Part of each *Flute*.

Vignola determines the Depth of the *Flutes* by taking the Angle of the equilateral Triangle for the Centre.

Vitruvius deſcribes the Depth from the Middle of the Square, whoſe Side is the Breadth of the *Flute*: Which latter Method makes them deep.

Some Columns have *Flutes* that go winding round the Shaft ſpirally; but this is rather accounted an Abuſe.

The *Flutes* or *Striæ* are frequently filled up with a prominent or ſwelling Ornament; ſometimes plain, in Form of a Staff or Reed, and ſometimes a little carved or enriched, in Imitation of a Rope, or otherwiſe; and therefore called *Rudenture*, or *Cabling*, and the Columns that are thus enrich'd, *Cabled Columns*.

This is moſt often done in the *Corinthian* Order; the Cablings or Fillings up commence from about one Third of the Height of the Column, reckoning from the Baſe, and are continued to the Capital, that is, they begin and end with the Diminution of the Column.

Flutings of the *Doric* Column, M. *Le Clerc* ſays, ought not to exceed 20, which is the Number obſerved by *Vignola*.

Palladio indeed has 24; but they appear too ſlender for this Order.

Theſe ſhould always be ſo diſpoſed, as that there may be one to ſtand full in the Middle of the Column.

Vignola determines their Depth by an equilateral Triangle, having one of its Angles in the Middle of the *Fluting*.

Vitruvius will have the Depth to be the Middle of a Square, one of whoſe Sides is the Width of the *Fluting*; which laſt muſt indeed be the deeper of the two.

Sometimes the *Flutings* are made flat, and are called *Facettes*; but theſe never have ſo good an Effect as the others, and for that Reaſon are not ſo much in uſe; though it can't be deny'd, but that

that they are more ſuitable to the Solidity of the Order.

He likewiſe adds, that the *Flutings* ought always to begin and end in the Shaft, near the Extremity of the Apophyges.

When there are *Flutings* in the Column, there ought alſo to be Eggs and Anchors in the Quarter-Round of the Capital, and even Pearls and Olives in a Baguette, to be made underneath in lieu of Anulets.

Theſe Eggs and Olives ought to be in the ſame Number with the *Flutings*, and to be regularly diſtributed.

The *Flutings* of the *Roman* Order he makes alſo 24, as in the *Ionic*; but to diſtinguiſh them, he makes thoſe of the *Compoſite* flat at Bottom, and only a Minute and a quarter deep, but twice as much in Width.

As to the *Flutings* of the *Corinthian* Order, he ſays, if we were only to have regard to this Order, the *Flutings* of the *Ionic* would ſuit it very well; but when the two Orders may be compared together, as will be the Caſe if they are placed over one another, then theſe *Flutings* may eaſily have the additional Ornaments of a little Fillet running quite round.

When *Flutings*, ſays he, are uſed in Pilaſters, their Number ſhould be ſeven on each Side: The firſt and laſt of which may be a little further from the Angle than the reſt are from each other, that the Extremities of the Pilaſters may not be too much weakened.

He adds, in ſome old Monuments we find Pilaſters which have only five *Flutings* on a Side; but then theſe are too large, and make the Pilaſters appear little and pitiful; and if there were nine, they would be too fine and ſlender, even for the moſt delicate Orders.

We never, ſays he, make *Flutings* in the *Tuſcan* Pilaſter; and if by chance we make any in the *Doric*, (which however is very rare,) we leave pretty large Spaces next to the Extremities, in order to fortify the Angles.

One may either add a ſingle *Fluting* in the Projecture or Thickneſs of the Pilaſter, or leave it quite plain, provided it don't exceed ten Minutes in Breadth.

FLYERS, in Architecture, are ſuch Stairs as go ſtraight, and do not wind round, nor are its Steps made tapering, but the fore and back Part of each Stair, and the Ends, reſpectively parallel to one another; or the ſecond of theſe *Flyers* ſtanding parallel behind the firſt, and the third behind the ſecond, and ſo of the reſt.

So that if one Flight do not carry to your intended Height, then there is a broad half Space; from thence you begin to fly again, with Steps every where of the ſame Breadth and Length, as before.

FOCUS, in Geometry and Conic Sections, is a Term applied to certain Points in the Parabola, Ellipſis, and Hyperbola, wherein the Rays reflected from all Parts of thoſe Curves do concur and meet.

FOLIAGE, a Cluſter or Aſſemblage of Flowers, Leaves, Branches, &c.

Foliage is particularly used for the Representations of such Flowers, Leaves, Branches, Rinds, &c. whether natural or artificial, used as Inrichments on Capitals, Friezes, Pediments, &c.

FOOT is a Measure consisting of 12 Inches; supposed to be the ordinary Length of the *Foot* of a Man.

The *Foot* long is divided into 12 Inches; and the Inch is supposed to be the Length of three Corns of Barley laid lengthwise.

The *Foot* is divided by Geometricians into 10 Digits, and the Digit into 10 Lines, &c.

The *French* divide their *Foot*, like us, into 12 Inches, and the Inch into 12 Lines, &c.

The *Foot square* is the same Measure both in Length and Breadth, containing 144 square or superficial Inches.

The *Cubic* or *Solid Foot* is the same Measure in all the three Dimensions, containing 1728 Cubic Inches.

The *Foot* is of different Lengths in different Countries.

The *Paris Royal Foot* exceeds the *English* by 7 Lines and a half.

The antient *Roman Foot* of the Capital consisted of four Palms, equal to four Inches and seven Tenths *English*.

The *Rhinland* or *Leyden Foot*, by which all the Northern Nations reckon, is to the *Roman Foot* as 950 to 1000.

The Proportion of the principal *Feet* of several Nations, compared with the *English* and *French*, are as follows:

The *English Foot* being divided into 1000 Parts, or into 12 Inches, the other *Feet* will be as follows:

			Ft.	Inc.	Lin.
London	Foot	1000	0	12	0
Paris, the Royal,	Foot	1068	1	0	8
Amsterdam	Foot	942	0	11	3
Antwerp	Foot	946	0	11	2
Dort	Foot	1184	1	2	2
Rhinland, or *Leyden*	Foot	1033	1	0	4
Mechlin	Foot	919	0	11	0
Middleburg	Foot	991	0	11	9
Strasburg	Foot	920	0	11	0
Lorrain	Foot	958	0	11	4
Cologn	Foot	954	0	11	4
Bremen	Foot	964	0	11	6
Frankfort on the Mayn	Foot	948	0	11	4
Spanish	Foot	1001	0	11	0
Toledo	Foot	899	0	10	7
Roman	Foot	967	0	11	6
Bolonia	Foot	1204	1	2	4
Mantua	Foot	1569	1	6	8
Venice	Foot	1162	1	[illegible]	9
Dantzick	Foot	944	0	11	3

Copenhagen

Copenhagen	Foot	965			
Riga	Foot	1831			
The *Greek*	Foot	1007			
The *Paris* (by Dr. *Bernard*)	Foot	1066	1	0	1
Old Roman	Foot	970	0	11	0

The *Paris Foot* being supposed to contain 1440 Parts, the rest will be as follows:

Paris	Foot	1440
Rhinland	Foot	1391 $\frac{3}{10}$
Roman	Foot	1320
London	Foot	1350
Swedish	Foot	1320
Danish	Foot	1403 $\frac{1}{2}$
Venetian	Foot	1540
Constantinopolitan	Foot	1320
Bononian	Foot	1682 $\frac{2}{5}$
Strasburg	Foot	1282 $\frac{1}{4}$
Norimberg	Foot	1346 $\frac{3}{4}$
Dantzick	Foot	1721 $\frac{1}{2}$
Italy	Foot	1320

FOOT-PACE, } is a Part of HALF-PACE, } a Pair of Stairs, whereon, after four or six Steps, you arrive at a broad Place, where you may take two or three Paces before you ascend another Step, by that Means to ease the Legs in ascending the rest of the Steps.

FORNICATION is an Arching or Vaulting.

FOUNDATION is that Part of a Building, which is under Ground, or the Mass of Stone, Brick, &c. which supports a Building, or upon which the Walls of a Superstructure are raised Or it is the Coffer or Bed dug below the Level of the Ground, to raise a Building upon; in which Sense, the Foundation either goes to the whole Area or Extent of the Building; as when there are to be Vaults, Cellars, or the like; or it is drawn in Cuts or Trenches, as when only Walls are to be raised.

The *Foundation* is properly so much of the Masoury, or Bricklayers Work, as reaches as high as the Surface of the Ground, and ought always to be proportion'd to the Load or Weight of the Building that it is to bear.

Sometimes the *Foundation* is massive, and continu'd under the whole Building, as in the antique Arches and Aqueducts, and some Amphitheatres: But it is more usually in Spaces or Intervals, either to avoid Expence, or because the Vacuities are at too great

great a Distance, in which latter Case, they make use of insulated Pillars, bound together by Arches.

Of Digging for, and Laying of Foundations.

There are several Things to be well consider'd in laying the *Foundations* of a Building, the most material of which are here extracted from the best Architects Antient and Modern.

That we may *found* our Habitation firmly, requires the exactest Care: For, says Sir *Henry Wooton*, if the *Foundation dance*, 'twill marr all the Mirth in the House.

Therefore, says that excellent Architect, we must first examine the Bed of Earth upon which we are to build, and then the under Fillings or Substruction, as the Antients called it.

For the former, we have a general Precept in *Vitruvius*, twice repeated by him as a Point indeed of main Consequence; *Substructionis* Fundationes *fodiantur, si queant inveniri ad solidum & in solido*: By which, he recommends not only a diligent, but even jealous Examination what the Soil will bear; advising us not to rest upon any appearing Solidity, unless the whole Mould through which we cut, have likewise been solid.

But he has no where determin'd how far we should go in this Search, as perhaps depending more upon Discretion than Regularity, according to the Weight of the Work.

But yet *Palladio* has ventured to reduce it to a Rule; and allows a sixth Part of the Height of the whole Building for the *Cavasione*, *i. e.* a hollowing or under-digging, unless there be Cellars under Ground; in which Case, he would have it somewhat lower. See Sir *Henry Wooton's Elements of Architecture*.

Palladio also lays down several Rules to know if the Earth be firm enough for the *Foundation*, (without artificial Helps,) by Observations from the digging of Wells, Cisterns, and the like, (which he would have to be done in the first Place) and from Herbs growing there, if there be such as usually spring up in firm Ground; also if a great Weight be thrown on the Ground, it neither sounds nor shakes; or if a Drum being set on the Ground, or lightly touch'd, it does not resound again, nor shake the Water in a Vessel set near it. These, says he, are Signs of firm Ground But the best Way to discover the Nature of the Soil, is to try it with an Iron Crow, or else with a Borer, such as is used by Well-Diggers.

Architects ought to use the utmost Diligence in this Point; for, of all the Errors that may happen in building, those are the most pernicious, which are committed in the *Foundation*; because they bring with them the Ruin of the whole Building; nor can they be amended without very great Difficulty.

Foundations are either natural, or artificial: Natural, as when we build on a Rock, or very solid Earth; in which Case, we need not seek for any further Strengthening: For these, without Digging, or other artificial Helps, are of themselves excellent *Foundations*, and most fit to uphold the greatest Buildings.

But if the Ground be sandy or marshy, or have lately been dug, in such Case, Recourse must be had to Art. In the former Case, the Architect must adjust the Depth of the *Foundation* by the Height, Weight, &c. of the Building: A sixth Part of the whole Height is looked upon as a Medium; and as to Thickness, double that of the Width of a Wall is a good Rule.

If you build upon mossy and loose Earth, then you must dig till you find sound Ground.

This sound Ground (fit to uphold a Building) is of divers Kinds; (as *Alberti* well observes,) in some Places 'tis so hard, as hardly to be cut with Iron, in other Places very stiff, in other Places blackish, in others whitish, (which is accounted the the weakest,) in others like Chalk, and in others sandy; but of all these, that is the best that requires most Labour in cutting or digging, and when wet, does not disolve into Dirt.

If the Earth to be built on is very soft, as in Moorish Grounds, or such that the natural *Foundation* cannot be trusted, then you must get good Pieces of Oak, whose Length must be the Breadth of the Trench, or about two Foot longer than the Breadth of the Wall: These must be laid cross the *Foundation*, about two Feet asunder, and being well ramm'd down, lay long Planks upon them; which Planks need not lie so broad as the Pieces are long, but only about four Inches of a Side wider than the Basis or Foot of the Wall is to be, and pinn'd or spik'd down to the Pieces of Oak on which they lie.

But if the Ground be so very bad, that this will not do, then you must provide good Piles of Oak of such a Length as will reach the good Ground, and whose Diameter must be about one twelfth Part of their Length. These Piles must be forced or drove down with a Commander, or a Machine or Engine for that Purpose, and must be placed as close as one can stand by another; then lay long Planks upon them, and spike or pin them down fast.

But if the Ground be faulty, but here and there a Place, and the rest of the Ground be good, you may turn Arches over these loose Places, which will discharge them of the Weight.

You must not forget to place the Piles not only under the outer Walls, but also under the inner Walls that divide the Building; for if these should sink, it would be a Means to make the outer Wall crack, and so ruin the whole Building.

Having thus far consider'd the Bed of Earth on which the Building is to be erected, we shall next consider the Substruction, as it was called by the Antients, but the Moderns generally call it the *Foundation*.

This is the Ground-Work of the whole Edifice which must sustain the Walls, and is a kind of Artificial, as the other was Natural: As to which, these Things

Things that follow, are the most necessary to be observ'd.

First, That the Bottom may be exactly level; therefore lay a Platform of good Board.

Secondly, That the lowest Ledge or Row be all of Stone, (the broader the better,) laid close, without Mortar; which is a general Caution for all Parts of a Building that are contiguous to Board or Timber; because Lime and Wood are utter Enemies to one another; and if unfit Confiners any where, then they are more especially so in the *Foundation*.

Thirdly, That the Breadth of the Substruction be at least double the Breadth of the Wall that is to be raised upon it.

But even in this Case, Art ought to give way to Discretion; and the Substruction may be made either broader or narrower, according as the Goodness of the Ground, and the Ponderosity of the Edifice requires.

Fourthly, That the *Foundation* be made to diminish as it rises, but yet so, that there may be as much left on the one Side as the other; so that the Middle of that above, may be perpendicularly over the Middle of that below: Which ought in like manner to be observed in diminishing the Walls above Ground; for by this Means the Building will become much stronger, than it would be, if the Diminution were made any other Way.

Fifthly, That you ought never to build upon the Ruins of an old *Foundation*, unless you are well assured of its Depth, and that its Strength is sufficient to bear the Building.

Lastly, There is a curious Precept in the Writings of some antient Architects, *That the Stones in the Foundation should be laid as they naturally lay in the Quarry*; they supposing them to have most Strength in their natural Position.

This Precept is generally observ'd by all good modern Artists, not only in the *Foundation*, but also in all the Parts of the Superstructure; and that for a better Reason than that bare Conjecture, *viz.* because they find the Stones to have a cleaving Grain, or that they are subject to cleave that Way of the Stone that lay horizontally in the Quarry: And for that Reason, if the horizontal Position of the Stones in the Quarry should be placed vertically in the Building, the super-incumbent Weight would be apt to cleave them, and so render the Building ruinous: For, as it has been observ'd by *Philip D'Orme*, the Breaking or Yielding of a Stone in the *Foundation*, although it should be but the Breadth of the Back of a Knife, it will make a Cleft of more than half a Foot in the Fabrick aloft. See *Stone-Bed*, and *Face of a Stone*.

In some Places they found the Peers of Bridges, and other Buildings near the Water, on Sacks of Wool laid like Matrasses, which being well press'd and greasy, will never give way nor rot in Water.

How to value Foundations.

Mr. *Phillips* tells us there are several Ways of valuing *Foundations*, or Ground-Plots of Houses.

First,

First, Some value Buildings by their Length or Breadth towards the Street, reckoning every Foot in Front to be worth 4, 5, 6, 8, or 10*s.* yearly, according to the Street or Place they stand in; and this yearly Value they reckon at twenty Years Purchase; and at those Rates, every Foot in Front is worth 4, 5, 6, 8, or 10*l.*

But this is a very uncertain Way, by reason of the great Difference in the Depth of Houses, &c.

Secondly, Others value *Foundations* by their Length and Breadth, and measure them by the Foot, reckoning every Foot to be worth 3, or 4*s.* But this Method is as uncertain and fallible as the other.

For Ground being scarce and dear in the City of *London*, &c. each Foot of it there, may in some Places, be worth 8 or 10*s.* which in other Places of it shall not be worth half so much, and in the Country not half a Farthing, though you reckon Land at 20*s. per* Acre, and twenty Years Purchase; for at that Rate, it is worth but one Penny *per* Yard, and every Yard contains nine Feet.

Thirdly, But a certain Author (*viz.* Mr. *Wing*) prescribes a more certain and general Way of valuing these *Foundations*; which is by getting a true and indifferent Estimate of the yearly Rent the Houses formerly went at, at a moderate Rack-Rent, without any Abatement or Diminution of it by Fines, or any other Considerations; which being known, you may reckon the true Value of these *Foundations* to be four, five, or six Years Purchase, according to the said yearly Rent, that is, about the third Part of the whole Worth or Purchase of the Fee-Simple of the House.

But in order to make a more exact Judgment and Valuation of the true Worth of *Foundations*, it will be best to distribute them into three Sorts, reckoning the first and lowest Sort of Houses, which yield least Rent, at four Years Purchase; the second Sort, which yield a moderate Rent, at five Years Purchase; and the third Sort, which yield the largest Rent, at six Years Purchase.

Mr. *Wing* attempts to demonstrate, that this Way of valuing *Foundations* is better, and to be preferred to any other.

The Foundations of Bridges.

Of all the Antients (says M. *Gautier*) in Architecture, who have given us any Rules for the founding of Bridges, *Scamozzi* is the only one that has said any Thing to the Purpose.

He tells us, that the *Foundations* are laid after different Mannets.

The first is by enclosing all round the Space of Ground you would build upon, by Dams made with Piles set deep in the Ground in double Rows, well strengthened and bound together with cross Pieces and Cords, and filling the vacant Spaces between them with Chalk or other earthy Matter.

This being done, the Water must be empty'd out, and the *Foundation* digged according to the Quality of the Ground; driving down Piles, if it be necescessary; upon which the Walls of the *Foundation* must be laid.

But this Method is only practicable in building on such Rivers where the Water is neither very rapid, nor very deep.

The second is done by laying the *Foundation* or Grate-Work, Rafts of stout Oak well bound together and made fast at the Surface of the Water with Cables or Machines, and building upon them large Quarters of Stone cramp'd together, and joined with good Mortar or Cement, and afterwards letting them descend softly by those Cables and Machines perpendicularly to the Bottom of the Water; as he says, was done in the Time of the Emperor *Claudius*, at the Port of *Ostia*; and as *Draguet Reys* did in the last Age at *Constantinople*, in the fine Mosque that he built upon the Sea. This Manner requires a good Bottom, equal and very even.

The third is by drawing off all or the greatest Part of the Water of the River into some other Place, or, by digging it another Bed, or letting it out into deep Ditches; in which, says he, great Diligence must be used, to have all the Materials ready, and to have Workmen enough ready, sufficient to compleat it in a short Time, to the End that the Masonry may be well confirm'd and settled before there is a Necessity to let the River into its former Bed.

The last Method, which is that which *Scamozzi* says, he believes *Trajan* made use of in building a Bridge over the *Danube*, is to dig a new Bed for the River, in a Place which seems, as it were, to meet itself, after having made a great Elbow or Compass about; which being done, the Bridge may be built with Ease, and that dry-shod, in that Place. And when the Bridge has been well settled, to open the Passage of the Current at the two Ends, stopping the Bed up with strong Banks or Moles, and so the River will take to its old Course again; this, says he, is the surest Method of all.

To lay the *Foundation* of the Piers of a Bridge, if the Earth be soft; it must be piled, after as much of it has been carried away as can well be. The same is to be done, if it be Sand or Gravel; which must be dug out as deep as can be, all round about, to a reasonable Distance; which must be surrounded with pointed Piles or Stakes well fastened to one another, filling the Spaces between Pile and Pile with Chalk or solid Earth well rammed in; which will for a Time, hinder the Current from washing away the Piles and Sand, and ruining the Work.

The Piles ought to be made tapering from Top to Bottom; the Arches unequal in Number, and carried up higher than the highest Inundation. The Architecture of Bridges ought to be plain and rustick.

Scamozzi afterwards gives the Design of his fine Bridge of Stone, and another of Carpentry. The Profile of this last, may be seen in Monsieur *De La Hire's Treatise of Carpentry*.

M. *Blondel* relates the Method he made use of in laying the

Foundation

Foundation of the Bridge of *Xaintes*, which he caused to be built over the *Charante*.

The ancient Bridge had been born down, it having been built on Potters Earth or Clay, and piled, so that the Swelling of the *Foundation* had raised the Piles, and threw down the Bridge. The Piles, by the Swelling of the Clay, started out above a Foot higher than the Level of the rest.

The Plummets went into this Clay, to the Depth of sixty Feet, made with a large Borer, the Arms of which were of Iron, each three Feet in Length, and well jointed one into another with good Pins.

After they had caused it to be dug seven Feet below the Bottem of the Water, all the Work being counterguarded and encompassed with a good Dam, and the Hollowness rak'd level, then a Grate-Work of good Oaken Wood was laid all over the *Foundation*, of twelve or fourteen Inches in Thickness; and square the whole Length and Breadth of the Building, not only that Part that was piled, but also the Opening or Space between the Piers and Abutments of the Bridge, or the void Spaces of the Arches.

The Chambers of the *Grille*, or Grate-Work, filled with good Quarters of hewn Stone, and the upper Part covered with Planks of five or six Inches thick, well fastened on with Pins all over the Grate-Work. Afterwards upon this Work of Carpentry is laid a *Foundation* of Masonry five Foot thick, all level, with good hewn Stones well fastened together with Cramp-Irons.

Upon this flat *Foundation* of five Feet thick, the Piers are erected; which for the first Year are brought no higher than the Imposts, to the End that they may settle well during the Winter.

M. *Blondel* makes it appear, that whatever Precautions Architects take to secure their Works by good *Foundations*, yet they are very conjectural and uncertain.

He in this compares an Architect to a Physician, who proceeds only upon Conjectures.

For who can venture to say, says he, that building upon a *Foundation* of Consistence, as it appears to him to be, that he shall not meet with soft or bad Ground underneath; which the Weight of the Building will press down and sink into, and by that Means be overturn'd.

Upon this Occasion, says M. *Gautier*, I can give an Example that happen'd in one of the Isles of *Oleron*, or *Rhe*, where the King causing Fortifications to be built, one Face of the Wall sunk, or fell down, notwithstanding it was built on a Bank of Rock; because it had a Hollow underneath, which could not be, or was not discovered.

M. *Blondel* relates in Confirmation of what he has said, that the vast Walls of the Church of *Val de Grace*, sunk in one Side, though built upon a good *Foundation*; because there were underneath large Hollows which had been made in former Times, for taking out of Stones some Fa-

 thoms

thoms lower, there having been a Quarry there.

Michael Ange Bonarote caused the *Foundation* of the Dome of *St. Peter's* at *Rome* to be laid with all the Precautions imaginable. But for all that, this Work did gap or split, which they cured by binding it about with a Hoop of Iron of an extraordinary Breadth and Thickness, which cost above 100000 Crowns.

It is supposed, that this Fracture in the Dome is an Effect of the Waters of a subterraneous Source, from a Spring which runs down from the high Mountains of the *Vatican* and *Janiculus*, which have washed the *Foundations* of this huge Edifice. So that according to these Examples, no Body can be answerable for the *Foundations* of a Building.

The *Corderie* of *Rochefort*, the Design of M. *Blondel*, is in Length two hundred and sixteen Toises, not comprising the Pavillions that are at the two Ends, and four Toises, the Breadth between the Walls, to two Stories, built upon a *Grillage*, or Grate-Work, as well in the Full, as in the Void of ten or twelve Inches thick, laid upon a Bottom of Potters Clay.

Upon this *Grillage*, are laid Platforms well fastened together with Pins, and upon them a Couch or Course of hewn Stones, and good rough Stones, the Building being raised every where, level continually, that it may be every where equal, that there may be no more Weight on the one Side than the other, that all the Parts of the Work may be *in æquilibrio*. This Building thus raised, has succeeded perfectly well.

M. *Blondel* remarks further, that the Materials at *Paris* not being of the same Solidity as those of *Italy*, as perhaps Marble, and harder, will not permit to make Bridges at *Paris* with so much Delicacy, and so disengaged as those which are made in *Italy*; which have a great deal less Thickness at the Place of the Keys of the Arcades.

Artificial FOUNTAIN, a Machine whereby the Water is violently spouted or darted out, called by the *French* a *Jet d'Eau*.

Of these there are divers Kinds, some founded on the Pressure or Weight of the Water, and others on the Pressure or Weight of the Air.

The Construction of an artificial Fountain *playing by the Pressure or Weight of the Water.*

The Fund or Reservoir of Water must be placed considerably higher than the *Fountain* itself, (whether the Fund or Reservoir has been placed there by Nature, or whether it has been raised by Art for that Purpose by any Kind of Machine, as Pump, Siphon, or the like;) and from the Reservoir vertical Tubes must be laid for the Water to descend through; to which vertical Tubes horizontal ones are to be fitted under Ground, to convey the Water to the Place where the *Fountain* is to play.

Lastly, vertical Tubes must be erected from these horizontal ones,

ones, by way of Ajutages, Spouts, or Jets: Their Altitudes must be much less than that of the Tubes, whereby the Water was brought from the Reservoir to the horizontal ones.

If this be done, the Water will, by the Pressure of the superincumbent Column, be spouted up at these Jets, and that to the Height or Level of the Water in the Reservoir; and this will be, though the Tubes be bent and incurvated after any Manner whatsoever.

By this Means Water may be spouted to any given Height at Pleasure; and the Tubes may be so proportioned, as to yield any given Quantity of Water, in a given Time; or several Tubes of the same *Fountain* may be made to yield Water in any given Ratio; or different Tubes may may be made to project the Water to different Altitudes.

As for Instance: In a

Fountain *that shall spout the Water in various Directions.*

Suppose the vertical Tube or Spout, in which the Water rises, to be B A, [see the Plate, Fig. I.] into this fit several other Tubes, some horizontal, others oblique, some inclining, others reclining, as P O, N M, L F, *&c.* Then as all Water retains the Direction of the Aperture through which it is spouted, that issuing through B will rise perpendicularly; and that through H, L, P, N, E, will describe Arches of different Magnitudes, and tend different Ways.

Or thus: Suppose the vertical Tube B A, (see the Plate, Fig. II.) through which the Water rises, to be stopped at the Top, as in B, and instead of Pipes or Jets, let it be only perforated with little Holes all around, or only half its Surface; then will the Water spirt or spin forth in all Directions through the little Apertures, and to a Distance proportional to the Height of the Fall of the Water.

And hence, if the Tube BA be supposed to be the Height of a Man, and be furnished with a Cock at D, upon the opening the Cock, the Standers-by not being apprehensive of any such Thing, will be covered and surrounded with a Shower.

But here this is to be observed, that the Diameters of the Apertures out of which the Water is emitted, must be a great deal less than that of the Tube in which the Water is conducted, lest the Resistance of the Air and other Impediments break the Force of the Water.

A Fountain *which spouts the Water in Manner of a Shower, is thus constructed.*

To the Tube wherein the Water is to rise, fit a Head that is in the Form of a Sphere or Lens, as in the Plate, Fig. III. made of a Plate of Metal, and perforated at the Top with a great many small Apertures.

And then the Water rising with a great Force towards B A, will be there parted into innumerable little Threads, which will afterwards break and disperse into the finest Drops.

The Construction of a Fountain, *the Stream of which raises and plays a Brass Ball.* See the Plate, Fig. IV.

Procure a hollow Globe or Ball made of thin Plate, whose Weight shall not be too heavy for the Force of the Water. Let the Tube CB, through which the Water rises, be exactly perpendicular to the Horizon.

Then let the Ball be laid in the Bottom of the Cup or Bason C, and it will be forced up by the Stream, and sustain'd at a considerable Height, as at A, vibrating alternately, or playing up and down.

And as the Figure of the Ball does not contribute any Thing to its rising and falling reciprocally, any other Body, in several other Forms, that are not too heavy, may be used instead of it, *e. g.* a Bird with its Wings stretched forth, &c.

But this Sort of *Fountain* should only be play'd in a Place that is not exposed to the Wind.

And it is also necessary, that the Ball, when on the Descent, should keep the same precise Perpendicular in which it rose, since otherwise it would miss the Stream, and fall downright.

The Construction of Fountains, *which spout Water out of the Figures of Men and other Animals.*

Since Water may be deriv'd or convey'd by Tubes in any Situation, and always retains the Direction of the Apertures or Holes, all that is required in making such *Fountains*, is to inclose Tubes within the Figures of those Animals, having their Orifices in those Parts out of which the Water is to spout.

Therefore, from the Principles before laid down, it will be very easy to deduce whatsoever relates to the Furniture of *Fountains*, and the various Forms that Water may be put into by their Means: All which depend on the Magnitude, Figure, and the Direction of the Ajutages or Apertures.

The Construction of a Fountain *which spreads the Water in the Form of a Cloth.*

Solder two spherical Segments D and C to the Tube BA, Figure V. almost touching each other, with a Screw F, that so you may either contract or enlarge the Interstice or Chink between them at Pleasure.

Or you may make a smooth even Nitch in a spherical or lenticular Head, fitted upon the Tube; and then the Water spouting through the Chink or Nitch, will expand itself in manner of a Cloth.

So much may suffice, as to such *Fountains* as play by Means of the Pressure or Weight of the Water.

The

The Construction of an artificial Fountain, *which plays by the Spring or Elasticity of the Air.* See the Plate, Fig. VI.

Provide a Vessel of Metal, Glass, or the like, proper for a Reservoir, ending in a small Neck *d* at the Top.

Put the Tube *db* through this Neck, traversing the Middle of the Vessel till its lower Orifice *e* does almost, but not quite, reach the Bottom of the Vessel, the Vessel being first half fill'd with Water.

The Neck is to be so contriv'd, as that a Syringe or condensing Pipe may be screw'd upon the Tube; by Means of which a large Quantity of Air may be intruded through the Tube into the Water, out of which it will disengage itself, and emerge into the vacant Part of the Vessel, and lie over the Surface of the Water DC.

Now the Water here contain'd being thus press'd by the Air, which is, *e.g.* twice as dense as the external Air, and the elastick Force of Air being equal to its gravitating Force, the Effect will be the same as if the Weight, or the Column of Air over the Surface of the Water, were double that of the Column pressing in the Tube; so that the Water must of necessity spout up through the Tube, with a Force equal to the Excess of Pressure of the included, above that of the external Air.

The Construction of a Fountain *which will play by the Draught of the Breath.* See the Plate, Fig. VII.

Suppose a Sphere be made of Glass, or Metal, into which a Tube is fitted DC, having a little Orifice in C, and reaching almost to D, the Bottom of the Sphere; if now the Air be suck'd out of the Tube DC, and the Orifice C be immediately immersed under cold Water, the Water will ascend through the Tube into the Sphere.

Thus proceeding by repeated Exsuctions, till the Vessel be above half full of Water, and then applying the Mouth to C, and blowing Air into the Tube, upon removing the Mouth, the Water will spout forth.

Or, if the Sphere be put in hot Water, the Air being thereby rarefied, will make the Water spout out as before.

This Kind of *Fountain* is call'd *Pila Heronis*, i. e. *Hero's Ball*, from the Name of its Inventor.

The Construction of a Fountain, *which, when it has done spouting, may be turned like an Hour-Glass.* See the Plate, Fig. VIII.

Be provided with two Vessels, ML and ON, to be so much the bigger, as the *Fountain* is to play the longer; and placed at so much the greater Distance from each other NP, as the Water is desired to spout the higher.

Let

Let B A C be a crooked Tube, furnished in C with a Cock; and F E D, another crooked Tube, furnished with a Cock in D: In I and K let there be other lesser Tubes, open at both Ends, and reaching near to the Bottoms of the Vessels N O and L M, to which the Tubes Q R and S T are likewise to reach.

If now the Vessel L M be filled with Water, it will descend through the Tube B A, and upon opening the Cock C, will spout up near to the Height of K; and after it is full again, will sink through the little Tube I into the Vessel N O, and expel the Air through the Tube Q O.

At length, when all the Water is emptied out of the Vessel L M, by turning the Machine upside down, the Vessel N O will be the Reservoir, and make the Water spout up through the Cock D.

Hence, if the Vessels L M and N O contain just as much Water as will be spouted up in an Hour's Time, there will be a spouting Clepsydra, or Water Clock, which may be graduated or divided into Quarters, Minutes, &c. See the Figure.

The Construction of a Fountain *which begins to play upon the lighting of Candles, and ceases as they go out.* See Fig. IX.

Take two cylindrical Vessels A B and C D, and connect them by Tubes open at both Ends K L, B F, &c. so that the Air may descend out of the higher into the lower.

To the Tubes solder Candlesticks H, &c. and to the hollow Cover of the lower Vessel C E fit a little Tube or Jet F E, furnished with a Cock G, and reaching almost to the Bottom of the Vessels. In G let there be an Aperture furnished with a Screw, whereby Water may be poured into C D.

Then upon lighting the Candle, H, &c. the Air, in the contiguous Tubes, becoming rarefied thereby, the Water will begin to spout through E F.

By the same Contrivance may a Statue be made to shed Tears upon the Presence of the Sun, or the lighting of a Candle, &c. all that is here required being to lay Tubes from the Cavity wherein the Air is rarefied to some Cavities near the Eyes, and full of Water.

Vulgar FRACTIONS, *q. d.* a broken Number, in Arithmetick, is a Part or Division of an Unite or Integer; or it may be defined to be a Number that stands in Relation to an Unite, as a Part to its Whole.

Vulgar Fractions, also called *simple Fractions*, are always expressed by two Numbers, the one written over the other, with a short Line drawn between them.

The upper Number, which is called the *Numerator of the Fraction*, expresses the Parts given of the Denominator; and the lower Number, which is called the *Denominator of the Fraction*, denotes the Unite or Whole, which is divided into Parts.

Thus

Thus three Fourths of a Line, or other Thing, are written $\frac{3}{4}$, where the Denominator 4 shews that the whole Line is suppos'd to be divided into four equal Parts, and the Numerator 3 indicates, or assigns three of such Parts.

Again, seven Twelfths is written $\frac{7}{12}$, where the Numerator 7 expresses seven Parts of an Integer divided into twelve; and the Denominator 12 gives the Denomination of these Parts, which are called *Twelfths*.

In all *Fractions*, as the Numerator is to the Denominator, so is the *Fraction* itself to the Whole of which it is a *Fraction*.

Thus, supposing $\frac{3}{4}$ of a Pound equal to 15*s*. it is evident, that 3 : 4 :: 15 : 20; whence it follows, that there may be infinite *Fractions* of the same Value with one another, inasmuch as there may be infinite Numbers found, which shall have the Ratio of 3 to 4.

Fractions are distinguished into *proper* and *improper*: A *proper Fraction* is one whose Numerator is less than its Denominator, and consequently the *Fraction* less than the Whole, or Integer, as $\frac{2}{12}$.

An *improper Fraction* is one whose Denominator is either equal to, or bigger than the Denominator; and consequently the *Fraction* equal to, or greater than the Whole, as $\frac{12}{12}$ or $\frac{14}{12}$.

Fractions are also distinguish'd into *simple* and *compound*: *Simple Fractions* are such as consist of one Numerator and one Denominator, as $\frac{3}{12}$, $\frac{26}{144}$, &c.

Compound Fractions are such as consist of several Numerators and Denominators, as $\frac{1}{2}$ of $\frac{3}{4}$ of $\frac{5}{12}$, &c. These are also called *Fractions of Fractions*.

The Arithmetick of *Fractions* consists in *Reduction*, *Addition*, *Subtraction*, *Multiplication*, and *Division*.

I. *Reduction of Fractions*.

First, To reduce a given whole Number into a *Fraction* of any given Denominator;

Multiply the given Integer by the given Denominator, and the Factum will be the Numerator.

Thus we shall find $3=\frac{72}{24}$, and $5=\frac{30}{06}$, and $7=\frac{28}{04}$.

If no Denominator be given, the Number is reduced to a *Fraction* by writing 1 under it as a Denominator, as $\frac{2}{1}$, $\frac{3}{1}$, $\frac{11}{01}$.

Secondly, To reduce a given *Fraction* to its lowest Terms, *i. e.* to find a *Fraction* equivalent to a given *Fraction*, as $\frac{12}{48}$, divide both the Numerator 12, and the Denominator 48 by some one Number that will divide them both without any Remainder, as here by 4, the Quotients 3 and 12 make a new *Fraction*, as $\frac{3}{12}$ are equal to $\frac{12}{48}$.

And if the Division be performed by the greatest Number that will divide them both, then will the *Fraction* be reduced into its lowest Terms.

Now to find the greatest common Divisor of two Quantities, divide the greater by the less; then divide the Divisor of the Division by the Remainder thereof. Again, divide the Divisor of of the ſecond Diviſion by the Remainder of the ſecond, and ſo on till there remain nothing; and the laſt Diviſor will be the greateſt common Meaſure of the given Numbers.

If it happen that Unity is the only common Meaſure of the Numerator and Denominator, then the *Fraction* cannot be reduced any lower.

Thirdly, To reduce two or more *Fractions* to the ſame Denomination, *i. e.* to find *Fractions* equal to the given ones, and with the ſame Denominators;

If only two *Fractions* be given, multiply the Numerator and Denominator of each into the Product of the Denominator of the other, and the Products given will be the new *Fractions* required.

Thus 5) $\frac{2}{3}$, and 3) $\frac{4}{5}$, make $\frac{10}{15}$ and $\frac{12}{15}$: If more than two be given, multiply both the Numerator and Denominator of each into the Product of the Denominators of the reſt: Thus 24) $\frac{2}{3}$, 12) $\frac{1}{6}$, 18) $\frac{3}{4}$, are $= \frac{48}{72}, \frac{12}{72}, \frac{54}{72}$.

Fourthly, To find the Value of a *Fraction* in the known Parts of its Integer; *e.gr.* As ſuppoſe it were required to know what is the Value of $\frac{9}{16}$ of a Pound, multiply the Numerator 9 by 20, and divide the Product by the Denominator 16, and the Quotient will be 11 *s.*

Then multiply the Remainder 4 by 12, the Number of known Parts in the next inferior Denomination; and dividing the Product by 16, as before, the Quotient will be 3 *d.* ſo that $\frac{2}{16}$ of a Pound is $=$ 11 *s.* 3 *d.*

Fifthly, To reduce a mix'd Number, as $6\frac{11}{12}$, into an improper *Fraction* of the ſame Value;

Multiply the Integer 6 by 12, the Denominator of the *Fraction*, and to the Product 72 add the Numerator: The Sum 83 ſet over the former Denominator $\frac{83}{12}$, conſtitutes the *Fraction* required:

Sixthly, To reduce an *improper Fraction* into its equivalent mix'd Number.

Suppoſe the given *Fraction* to be $\frac{41}{12}$, divide the Numerator by the Denominator, and the Quotient $3\frac{5}{12}$ is the Number ſought.

Seventhly, To reduce a compound *Fraction* into a ſimple one, multiply all the Numerators into each other, for a new Numerator; and all the Denominators for a new Denominator thus, $\frac{1}{2}$ of $\frac{3}{4}$ of $\frac{5}{6}$ will be $\frac{14}{45}$.

II. *Addition of* Vugar Fractions.

Firſt, If the given *Fractions* have different Denominators, reduce them into the ſame; then add the Numerators together, and under the Sum write the common Denominator, thus, *e. g.* $\frac{2}{3} + \frac{4}{5} = \frac{10}{15} + \frac{12}{15} = 1\frac{7}{15}$. and $\frac{2}{3} + \frac{1}{6} + \frac{3}{4} = \frac{48}{72} + \frac{54}{72} = \frac{114}{72} = 1\frac{42}{72} = 1\frac{07}{12}$.

Secondly, If compound *Fractions* are given to be added, they muſt firſt be reduced to ſimple ones: and if the *Fractions* be of different Denominations, as $\frac{5}{8}$ of a Pound, and $\frac{3}{4}$ of a Shilling, they muſt be firſt reduced to *Fractions* of the ſame Denomination of Pounds.

Thirdly, To add mix'd Numbers, the Integers are firſt to be added

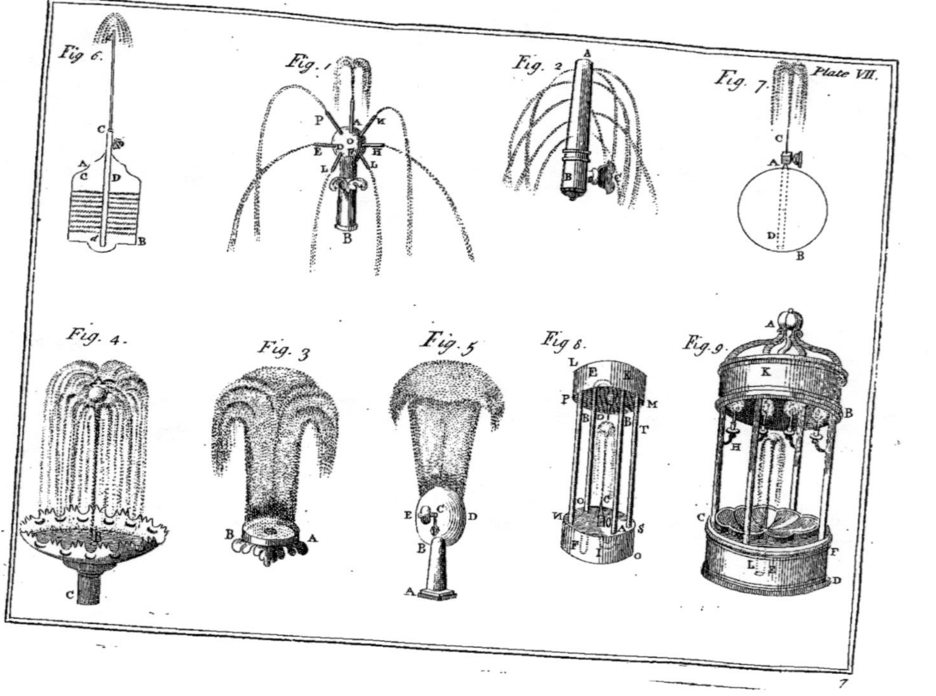
Plate VII.
Fig. 6.
Fig. 1
Fig. 2
Fig. 7.
Fig. 4.
Fig. 3
Fig. 5
Fig. 8.
Fig. 9.
7

added, then the Fractional Parts; and if their Sum be a proper Fraction, only annex it to the Sum of the Integers.

If it be an improper *Fraction*, reduce it to a mix'd Number, adding the Integral Part thereof to the Sum of the Integers, and the *Fractional* Part after it, as $5\frac{2}{3} + 4\frac{2}{3} = 10\frac{1}{3}$.

III. *Subtraction of* Fractions.

First, If they have the same common Denominator, subtract the lesser Numerator from the greater, and set the Remainder over the common Denominator.

Thus from $\frac{9}{12}$ take $\frac{5}{12}$, and there will remain $\frac{4}{12}$.

Secondly, if they have not a common Denominator, they must be reduced to *Fractions* of the same Value, having a common Denominator; and then as in the first Rule,

Thus $\frac{6}{7}\ \frac{2}{5} = \frac{30}{35} - \frac{14}{35} = \frac{16}{35}$.

Thirdly, To subtract a whole Number from a mix'd Number, or one mix'd Number for another, or reduce the whole, or mix'd Numbers to improper *Fractions*, then proceed, as in the first and second Rule.

IV. *Multiplication of* Fractions.

First, If the *Fractions* propos'd be both single, multiply the Numerators one by another for a new Numerator, and the Denominators for a new Denominator.

Thus $\frac{3}{4}$ being multiply'd by $\frac{7}{8}$, produces $\frac{21}{32}$.

Secondly, If one of them be a mix'd, or whole Number, it must be reduced to an improper *Fraction*; and then you are to proceed as in the last Rule,

Thus $\frac{4}{9}$ being multiply'd into $\frac{10}{1} = \frac{40}{9}$.

In Multiplication of *Fractions*, it is to be observ'd, that the Product is less in Value, than either the Multiplicand and Multiplicator; because in all Multiplications, as Unity is to the Multiplicator, so is the Multiplicand to the Product.

But Unity is bigger than either Factor, if the *Fractions* be proper; and therefore either of them must be greater than the Product.

Thus, in whole Numbers, if 5 be multiply'd by 8, it will be as 1 : 5 :: 8 : 40; or 1 : 8 :: 5 : 40.

Wherefore in *Fractions* also, as $1 : \frac{3}{4} :: \frac{5}{8} : \frac{15}{32}$; or as $1 : \frac{5}{8} :: \frac{3}{4} : \frac{15}{32}$.

But 1 is greater than either $\frac{3}{4}$ or $\frac{5}{8}$; wherefore either of them must be bigger than $\frac{15}{32}$.

V. *Division of* Fractions.

First, If the *Fractions* propos'd be both simple, multiply the Denominator of the Divisor by the Dividend, and the Product will be the Numerator of the Quotient.

Then multiply the Numerator of the Divisor by the Denominator of the Dividend, and the Product will be the Denominator of the Quotient,

Thus $\frac{3}{5})\frac{4}{9}(\frac{20}{27}$.

Secondly, If either Dividend, Divisor, or both, be whole, or mixed

mixed Numbers, reduce them to improper *Fractions*; and if they be compound *Fractions*, reduce them to simple ones; and then proceed as in the first Rule.

In Division of *Fractions* it is to be observ'd, that the Quotient is always greater than the Dividend; because in all Division, as the Divisor is to Unity, so is the Dividend to the Quotient; as if 3 divide 12, it will be as 3 : 1 :: 12 : 4.

Now 3 is greater than 1; wherefore 12 must be greater than 4; but in *Fractions*, as $\frac{1}{3}$: 1 :: $\frac{4}{9}$: $\frac{20}{27}$, where $\frac{1}{3}$ is less than 1; wherefore $\frac{4}{9}$ must be also less than $\frac{20}{27}$.

FRAMING *of an House*, is the Carcase, Flooring, Partitioning, Roofing, Ceiling, Beams, Ashlering, *&c.* all together.

As to the Carcase of a House, Mr. *Leybourn* says, that Carpenters commonly work by the Square of ten Feet in erecting the Carcase, that is, (as he says) the *framing* and setting up, with the Partitions, Floors, Rafters, and such like; for which, he says, they have in running Buildings from 15 to 20*s.* the Square, and some may deserve 30*s.* or more. And he adds, that to a Square of good Carcase, 20 Feet of Ground-Roof Timber may be allowed: But it does not appear, whether he means that the Carpenter fells, hews, and saws the Timber into that Price.

Some *Sussex* Workmen have said, that they have but 8*s.* *per* Square for *framing* the Carcase of a House, and sawing the Timber; and but 4*s.* 6*d.* without sawing the Timber.

As to the Carcase of a Barn: Some Workmen say, that they have for *framing* of Barns 3*s.* 6*d.* *per* Square; and that the Charge of the Carcase of a Barn may be thus computed, *viz.* 4*s.* *per* Square for sawing the Boards, considering the Slabbing and the Boards lying over one another; 2*s.* *per* Square for sawing the Timber; 3*s.* 6*d.* *per* Square for *framing*; and 4*s.* *per* Square for the Timber, reckoning at 12*s.* *per* Ton, and one Ton to make three Square of *Framing*: So that the whole Charge of the Carcase will be at least 13*s.* 6*d.* *per* Square; but if the Timber be more than 12*s.* *per* Ton, then the whole Charge will be more than what has been before mentioned.

Partitions: Though indeed some Workmen reckon Partitions into the Carcase, yet others reckon for them by themselves; for which, and sawing the Timber, some say they have 6 or 7*s.* *per* Square, and for the Workmanship only, but 4*s.* 6*d.* *per* Square.

Roofs: Mr. *Leybourn* says, that Carpenters commonly reckon 4 or 5*s.* in the Square more for *framing* of Roofs, than for the rest of the Building.

But some *Sussex* Workmen say, that for *framing* Roofs, and sawing the Timber, they have no more than 8 or 9*s.* *per* Square; and for the Workmanship, only 4*s.* 6*d.* *per* Square.

Thorough Framing, is *framing* and making all Doors and Windows. This, some Workmen say, they have 5*s.* a Square for the Workmanship only.

By

By the Great Square: A Term used by Carpenters, who sometimes work by the Great Square in Brick Buildings; and then besides *framing* the Floors, Partitions, Roof, *&c.* they also make Doors, Windows, Cornices, Stair-Cases, and (in general) all that is Carpenters Work, and sawing of Timber, (except Modilions or Cantalivers,) for which Work they have 6*l. per* Square.

But this is to be noted, that in this Way of Working they measure only the Ichnography or Ground-Plot, only to the Dimensions they add one of the Projectures in Front, and not in Flank, and so cast it up.

This Sort of Work is measured by the Square, as *Floors*.

FREEZE } in Architecture,
FRIEZE } is properly a large flat Face or Member, separating the Architrave from the Cornice, and that Part of the Entablature between the Architrave and Cornice.

This Member was called by the Antients *Zoophorus*, because it was commonly inriched with the Figures of Animals.

Frieze is said to be formed from the *Latin*, *Phrygio* an Embroiderer, because usually adorned with Sculptures in *Basso-Relievo*, imitating Embroidery.

The *Frieze* is supposed to be designed to represent the Heads of the transverse Beams which sustain the Roof or Covering.

This Member is quite plain in the *Tuscan* Order; but is inrich'd with Triglyphs in the *Doric*: Is sometimes made arch'd or swelling in the *Ionic*, in which Case particularly, *Vitruvius* calls it *Pulvinatus*, *q. d.* pillowed. In the *Corinthian* and *Composite*, it is frequently joined to the Architrave by a little Sweep, and sometimes to the Cornice: And in these richer Orders, it is commonly adorned with Sculpture, Figures, Compartiments, Histories, Foliages, Festoons, *&c.*

As to the Height of the *Frieze*, it is in the general much the same as that of the Architrave.

Vitruvius makes the *Tuscan Frieze* flat and plain, and in Height 30 Minutes.

Palladio, who makes it convex or swelling, gives it but 26 Minutes: And *Scamozzi* makes it plain, and in Height 42 Minutes.

The *Doric*: *Vitruvius* and *Vignola* make this *Frieze* flat, only carved with Triglyphs and Metopes, and its Height 30 or 40 Minutes. And *Palladio* and *Scamozzi* make it like *Vitruvius*, and in Height 45 Min.

The *Ionic*: *Vitruvius* makes this *Frieze* flat, only carved with Acanthus Leaves, Lions, Men, *&c.* and in Height 30 Min. *Vignola*, he also makes it flat, and allows it 45 Min. And *Palladio*, who makes it convex or swelling, 27, and *Scamozzi* 28.

The *Corinthian Vitruvius* inriches with Acanthus, human Figures, *&c.* and makes its Height 37 Min. *Vignola* 45, *Palladio* 28, and *Scamozzi* 31 $\frac{1}{2}$.

Lastly, the *Composite*: This in *Vitruvius* is set with Cartouses, and carved between them, is 52$\frac{1}{2}$ Min. *Vignola* makes it like *Vitruvius*, but gives it but 45 Min. *Palladio*, who makes it swelling, has but 30, and *Scamozzi* 32.

From

From these Variety of Inrichments which adorn the *Freezes*, they obtain various Denominations.

Convex Freezes } are such
Pulvinated Freezes } whose Profile is a Curve; the best Proportion of which, is when drawn on the Base of an equilateral Triangle.

In some the Swelling is only at Top, as in a Console; in others at Bottom, as in a Baluster.

Flourish'd Freezes are such as are inrich'd with Rinds of imaginary Foliages; as the *Corinthian Freeze* of the Frontispiece of *Nero*; or with natural Leaves, either in Clusters or Garlands; or continued as in the *Ionic* of the Gallery of *Apollo* in the *Louvre*.

Historical Freezes are such as are adorn'd with Bass Relievo's, representing History, Sacrifices, *&c.* as the Arch of *Titus* at *Rome*.

Marine Freezes are such as represent Sea-Horses, Tritons, and other Attributes of the Sea; or Shells, Baths, Grotto's, *&c.*

Rustic Freezes are such whose Courses are rusticated or imboss'd, as in the *Tuscan Freeze* of *Palladio*.

Symbolical Freezes are those adorned with the Attributes of Religion, as the *Corinthian* of the Temple behind the Capital at *Rome*, whereon are represented the Instruments and Apparatus of Sacrifice.

We sometimes, says M. *Le Clerc*, make the *Frieze* of the Entablature convex; but then this should never be done without some good Reason, mere Caprice being not sufficient to warrant such an Alteration.

When one Order is raised over another, and the upper Column has its due Bigness, its Pedestal nevertheless goes beyond the Naked of the under Column, which to some Persons has a disagreeable Effect.

This, says he, inclines me to think, that the first Architect who made a convex *Frieze*, did it with a Design to extenuate this Appearance of a Defect

This is evident, that as the Naked of a *Frieze* is hidden by this Swelling, the Pedestal of the upper Order appears less, to exceed the Naked of the under Order, as may be easily observed where the two Orders are seen over each other. Were it not on this Account, the convex *Frieze*, just mentioned, ought not, says he, in my Opinion, to be imitated: The *Frieze* may be made convex in all the Orders except in the *Doric*, where this Swelling can't be allow'd, by reason of the Triglyphs.

FRESCO, a Method of Painting, or rather Plastering on Walls to endure the Weather, and representing Birds, Beasts, Herbs, Fruits, *&c.* in Relief.

It is performed on fresh Plaster, or on a Wall laid with Mortar, not yet dry, and with Water Colours.

This Sort of Painting has a great Adantage by its incorporating with the Mortar, and drying along with it, it is rendered extreme durable, and never fails or falls but along with it.

Of the Method of this Painting: To make the Compost or Plaster of old rubbish Stones, and

and mix it with well-burnt Flint (or Lime) and Water; but wash out the Saltness of the Lime, by often pouring the Water, and putting fresh to it. This should not be done in moist Weather, because that has a great Influence on the Walls.

And in order to render the Plaister the more durable, they strike into the Joints of the Brick or Stone Wall Stumps of Horse Nails, at about six Inches Distance, to prevent the Plaister from peeling off.

With this Plaister the Wall is first to be plaistered a good Thickness, and left for some Time to dry; and the Design and Colours being first ready prepared.

This Painting is chiefly performed on Walls and Vaults newly plaistered with Lime and Sand; but the Plaister is only to be laid in Proportion as the Painting goes on, no more being to be done at once than the Painter can dispatch in a Day, while it is dry.

Before the Painting is begun, there is usually a Cartoon or Design made on Paper, to be calk'd and transferred to the Wall, about half an Hour after the Plaister has been laid on.

The Colour being prepared and mingled, the Wall is to be plaistered over again a second Time about the Thickness of Half-a-Crown, but only so much as you intend presently to Work upon; and while it is wet, you must work the Colours therein, which will mix and incorporate with the Plaister, so as never to wash out.

The Painting must be, for the Work to come out in all its Beauty, wrought quick, and with a free Hand; for there can be no Alteration after the first Painting, and therefore make your Colour high enough at first; you may deepen, but not easily heighten.

Nor must they ever be retouch'd dry, with Colours mix'd up with the White of an Egg, or Size or Gum, as some Workmen do, by reason such Colours grow blackish; nor do any preserve themselves, but such as were laid on hastily at first.

In this Painting all the compound and artificial Colours, and almost all the Minerals are set aside, and scarce any Thing used but Earths; which are capable of preserving their Colour, defending it from the burning of the Lime, and resisting its Salt, which *Vitruvius* calls its Bitterness.

The Colours used are White made of Lime slack'd long ago, and white Marble Dust; Oker, both red and yellow, Violet Red, Verditer, Lapis Lazuli, Smalt, Earth, black Spanish Brown, Spanish White, &c. All which are only ground and work'd up with Water; and most of them grow brighter and brighter as the *Fresco* dries.

The Brushes and Pencils for this Work must be long and soft, or else they will rake and raze the Painting: The Colours must be full, and flowing from the Brush; the Design perfect in the Image or Paper Copy; for in this Work you cannot alter or add upon any Colour.

This Sort of Painting has a great Advantage, by its incorporating with the Mortar, and drying

ing along with it, is rendered extreamly durable, and never fails nor falls but along with it.

The Antients painted on Stuck; And it is worthy Obſervation in *Vitruvius*, what infinite Care they took in making the Incruſtation or Plaiſtering of their Buildings, to render them beautiful and laſting: Though the modern Painters find a Plaiſter made of Lime and Sand preferrable to Stuck, both becauſe it does not dry too haſtily, and as being a little browniſh, it is fitter to lay Colours on, than a Ground ſo white as Stuck.

This Kind of Painting was the antient *Grecian* Way of Painting, and ſince much uſed by the *Romans*.

Plutarch informs us, that *Aratas*, the great Commander under *Ptolemy* King of *Egypt*, (in a Compliment to the Emperor's Affections that Way,) forbore to ſack a wealthy City, merely for the Excellency of the *Freſco* Painting upon the Walls of the Houſes.

There have been ſeveral whole Towns of this Work in *Germany*, excellently well done, but now ruin'd by Wars.

At *Rome* there are three Chambers (in the *Popes* Palace) of *Freſco*, done by *Raphael Urbin*, and *Julio Romano*, (his Diſciple) who finiſhed his Maſter's Work, which is yet called *Raphael*'s Deſign.

There are other Places done by *Andrea del Sexto* and *Michael Angelo*, and ſome other Artiſts.

There is an excellent *Freſco* Work at *Fountainbleau* in *France*. It is the continued Travels of *Ulyſſes*, in ſixty Pieces, done by *Bollmneo Martin Rouſe*, a *Florentine*, and others.

FRET, FRETTE, in Architecture, is a Kind of Knot or Ornament, conſiſting of two Liſts or ſmall Fillets va-

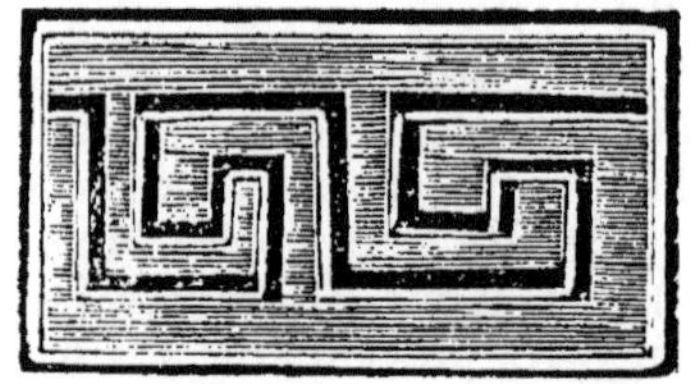

riouſly interlaced or interwove, and running at parallel Diſtances equal to their Breadth.

Every Return and Interſection of theſe *Frets* muſt be at Right Angles. This is ſo indiſpenſibly neceſſary, that they have no Beauty without it, but become perfectly *Gothic*.

Sometimes the *Fret* conſiſts but of a ſingle Fillet, which however may be ſo ordered, as to fill its Space exceedingly well, if well managed.

Theſe *Frets* were very much in uſe among the Antients, who apply'd them chiefly on even flat Members, or Parts of Buildings; as the Faces of the Corona, and Eves of Cornices; under the Roofs, Soffits, &c. on the Plinths of Baſes, &c.

The Name *Frette* was hence occaſion'd, that *Frette* literally ſignifies the Timber-Work of a Roof; which conſiſts chiefly of Beams, Rafters, &c. laid acroſs each other, and, as it were, *fretted*.

FRET-WORK, an Enrichment of *Frette*, or a Place adorned with ſomething in the Manner thereof.

Fret-Work is sometimes used for the filling up and enriching flat empty Spaces, but it is principally practised in Roofs which are *fretted* over with Plaister-Work.

The *Italians* also use *Fret-Work* in the Mantlings of Chimneys with great Figures, a cheap Piece of Magnificence, and as durable almost within Doors, as harder Matters in the Weather.

FRICTION, in Mechanicks, is the Resistance which a moving Body meets withal from the Surface on which it moves.

Friction is produced by the Asperity or Roughness of the Surface of the Body mov'd on, and that of the Body moving.

For such Surfaces consisting alternately of Eminencies, and Cavities, either the Eminencies of the one must be raised over those of the other, or they must be both broken or worn off: But neither of these can happen without Motion; nor can Motion be produced without a Force impress'd.

Hence the Force apply'd to move the Body is either wholly or partly spent on this Effect, and of Consequence there arises a Resistance or *Friction*, which will be be greater, *cæteris paribus*, as the Eminencies are the greater, and the Substance the harder.

And as the Body by continual *Friction* grows more and more polish'd, the *Friction* diminishes.

Hence it follows, that the Surfaces of the Parts of Machines, which touch each other, ought to be as smooth and polish'd as possible.

But as no Body can be so much polish'd, as to take all Inequality quite away; witness those numerous Ridges which may be discover'd by the Help of a Microscope on the smoothest Surfaces; hence arises the Necessity of anointing the Parts that touch with Oil, or other fatty Matter.

The Laws of *Friction* are,

First, As the Weight of a Body moving on another, is increased, so is its *Friction*.

This is experimentally seen in a Balance; which when only charged with a small Weight, easily turns; but with a greater, a greater Force is required.

Hence if the Line of Direction of a moving Body be oblique to the Surface moved on, the *Friction* is the greater; this having the Effect as the Increase of Weight.

And hence again, as a perpendicular, Stroke or Impression is to an oblique one; as the whole Sine is to the Sine of the Angle of Incidence; and the Sine of the greater Angle is greater, and that of the lesser less; the *Friction* is greater, as the Line of Direction approaches nearer to a Perpendicular.

This is easily observable; and especially in the Teeth of Wheels, which are frequently broke on this very Account.

The *Friction* therefore is taken away, if the Line of Direction of the moving Body be parallel to the Surface.

Secondly, The *Friction* is less in a Body that rolls, than it would be, were the same Body to ſlide; as is eaſily demonſtrated.

For ſuppoſe a dented Ruler, and ſuppoſe a dented Wheel to move along it, with its Teeth perpendicular to the Circumference:

If now the Body were to ſlide, the Tooth, when it touch'd the Ruler, would deſcribe a Right Line on the Surface thereof; and as the Tooth of the Ruler reſiſts the ſame, it could not proceed without either removing or breaking either the Tooth of the Wheel, or its own. And the ſame will hold of the ſliding of any rough Surface upon another, where all the *Friction* will take Place, that can any way ariſe from the Roughneſs of the Surface.

But if the Wheel roll along the Ruler, then the Tooth will no longer reſiſt its Motion, only as it is to be hoiſted out of the Cavity over the Eminence of the Tooth; and the ſame holds in the rubbing of any rough Body over the Surface of another.

Hence in Machines, leaſt the *Friction* ſhould employ a good Part of the Power, Care is to be taken that no Part of the Machine ſlide along another, if it can be avoided, but rather that they roll or turn upon each other.

With this View, it may be proper to lay the Axes of Cylinders, not as is uſually done in a Groove, or concave Matrix, but between little Wheels moveable on their reſpective Axes.

This was long ago recomended by *P. Caſatus*; and it is confirm'd by Experience, that a great deal of Force is ſaved by it.

Hence alſo it is, that a Pully moveable on its Axis, reſiſts leſs than if it was fixed; and the ſame may be obſerv'd of the Wheels of Coaches, and other Carriages.

From theſe Principles, with a little further Help from the Higher Geometry, *Olaus Roemer* determin'd the Figure of the Teeth of Wheels, that ſhould make the leaſt Reſiſtance poſſible, and which would be Epicycloidal; and the ſame was afterwards demonſtrated by M. *De La Hire*. Though the Thing is not yet taken into Practice.

Hence, in Sawing-Mills, the Sides of the Wooden Rectangle, the Saws are fitted into, ſhould be furniſh'd with Rotules or little Wheels, which would greatly leſſen the *Friction*; and the like in other Caſes. Add, that as Winches, or curved Axes prevent all *Friction*, thoſe ſhould be uſed inſtead of Wheels, as often as poſſible.

The Calculation of the Quantity or Value of Friction.

Tho' *Friction* is a Point of the utmoſt Importance in Machines, and by all Means to be conſidered in calculating the Force thereof; yet it is generally overlook'd in ſuch Calculations: But this, principally, by reaſon the preciſe Value is not known.

It is not yet reduced to certain and infallible Rules. The common Method is barely to compute what the Advantage is which a moving Power has from the

the Machine, either on account of its Diſtance from a fixed Point, or of the Direction in which it acts. And in all the Demonſtrations, it is ſuppoſed, that the Surfaces of Bodies are perfectly ſmooth and poliſh'd.

Indeed, the Engineers expect that in the Practice they ſhould loſe Part of the Advantage of their Force by their *Friction*; but how much it is ſuppoſed, nothing but the Practice can determine.

M. *Amontons*, indeed, has made an Attempt to ſettle by Experiment a Foundation for a preciſe Calculation of the Quantity of *Friction*; and M. *Parent* has confirm'd it by Reaſoning and Geometry: But their Theory, however warranted, is not generally and fully received.

M. *Amontons*'s Principle is, that the *Friction* of two Bodies depend on the Weight or Force wherewith they bear on each other; and only increaſes as the Bodies are more ſtrongly preſs'd and apply'd againſt each other; or are charged with a greater Weight; and that it is a vulgar Error, that the Quantity of *Friction* has any Dependance upon the Bigneſs of the Surfaces rubbed againſt each other; or that the *Friction* increaſes as the Surfaces do.

Upon the firſt Propoſal of this Paradox, M. *De La Hire* had Recourſe to Experiments, which ſucceeded much in Favour of the new Syſtem.

He laid ſeveral Pieces of rough Wood on a rough Table: Their Sizes were unequal; but he laid Weights upou them, ſo as to render them all equally heavy, and he found that the ſame preciſe Force or Weight apply'd to them, by a little Pully, was required to put each in Motion, notwithſtanding all the Inequality of the Surfaces. The Experiment ſucceeded in the ſame Manner, in Pieces of Marble laid on a Marble Table.

Upon this M. *De La Hire* apply'd himſelf to conſider the Rationale of the Thing; and has given us a Phyſical Solution of the Effect. And M. *Amontons* ſettled a Calculus of the Value of *Friction*, and the Loſs ſuſtained thereby in Machines, upon the Footing of the new Principle.

In Wood, Iron, Lead, and Braſs, which are the principal Materials uſed in Machines, he finds the Reſiſtance cauſed by *Friction* to be nearly the ſame, when thoſe Materials are anointed with Oil, or other fatty Matter: And this Reſiſtance, independant of the Quantity of the Surface, he makes to be nearly equal to a third Part of the Force wherewith the Bodies are preſſed againſt each other.

Beſide the Preſſion, the Magnitude whereof determines that of the *Friction*, there is another Circumſtance to be conſidered, *viz.* Velocity.

The *Friction* is the greater, and the more difficult to ſurmount, as the Parts are rubb'd together with the greater Swiftneſs: So that this Velocity muſt be compared with that of the Power neceſſary to move the Machine, and overcome the *Friction*.

If the Velocity of the Power be double that of the Parts rubb'd, it acquires by that Means an an Advantage, that makes it double, or, which amounts to the ſame, diminiſhes the contrary Force of *Friction* by one half, and reduces it to a ſixth Part of the Weight or Preſſion.

But this Velocity, M. *Amontons* only conſiders as a Circumſtance that only augments or diminiſhes the Effect of the Preſſion, *i.e.* the Difficulty of the Motion; ſo that the *Friction* ſtill follows the Proportion of the Weight.

Only we are hereby directed to diſpoſe the Parts of Machines that rub againſt each other in ſuch Manner, as that they may have the leaſt Velocity poſſible. And thus the Diameter of the Axis of a Wheel ſhould be as ſmall as poſſible, with Regard to that of the Wheel, in that the leſſer the Axis, the ſlower will be Motion of the Surfaces rubbing againſt each other, ſince the Velocity of a circular Motion always goes diminiſhing from the Circumference to the Centre.

And for the ſame Reaſon, the Teeth of dented Wheels ſhould be as ſmall and as thin as poſſible; for a Tooth catching on a Notch, &c. rubs one of its Sides againſt a Surface equal to its own; and is to diſengage itſelf in a certain Time by paſſing over a Space equal to the Surface; conſequently the leſs the Surface, the leſs Space it has to move, the Littleneſs of the Surface diminiſhing the Reſiſtance of the *Friction*; not that it is a leſs Surface that rubs, but as there is a leſs Space to move.

But notwithſtanding all the Confirmations and Illuſtrations of the Theory of *Friction*, the Publick, nor even the Academy itſelf, where it was propoſed, could not be brought fully to acquieſce in it.

It is granted, the Preſſion has a great Effect, and is, in many Caſes, the only Thing to be conſidered in *Frictions*. But it will be hard to perſwade us abſolutely to exclude the Conſideration of the Surface. And in Effect, the contrary ſeems capable of a Metaphyſical Demonſtration.

If two Bodies with plain Surfaces, ſuppos'd infinitely hard and poliſh'd, be moved along each other, the *Friction* will either be none, or infinitely ſmall: But if inſtead of ſuch Suppoſition, which has no Place in Nature, we ſuppoſe two Bodies with rough uneven Surfaces, the Difficulty of moving one of them upon the other muſt ariſe either from this, that the firſt muſt be raiſed, in order to diſengage the Parts, catch'd or lock'd into the ſecond; or that the Parts muſt be broke or worn off, or both.

In the firſt Caſe, the Difficulty of raiſing one of the Bodies, makes that of the Motion; and of conſequence the *Friction* ariſes wholly from the Weight or Preſſion, and the Surface has nothing to do.

In the ſecond Caſe, the Magnitude of the Surface would be all, were it poſſible this ſecond Caſe could be abſolutely extracted from the firſt, *i.e.* could the Parts of one Body be rubb'd and worn againſt thoſe of the other, without raiſing one of them; it being viſible, that a greater Number

ber of Parts to be broke would make a greater Refiſtance than a leſs.

But as in Practice, we never rub or grind without railing the Body, the Reſiſtance ariſing from the Greatneſs of the Surface, is always combin'd in the ſecond Caſe with that of the Preſſion; whereas in the former Caſe, that ariſing from the Preſſion, may be alone, and uncompounded.

Add to this, that what is wore off a Body, is ordinarily very little, with Regard to the great Number of Times the Body muſt have been raiſed during the *Friction*, and all the little Heights added together, which the Body muſt have been raiſed to.

Hence as the Reſiſtance from Preſſion may be ſingle, and as the ſame always accompanies that ariſing from the Magnitude of the Surfaces, and is uſually the much more conſiderable of the two, when it does accompany it; for these Reaſons, in moſt of the Experiments that are made, it is the only one perceived, and the only one that needs to be conſidered.

But then, as 'tis poſſible, in certain Caſes for the Preſſion to be very ſlender, and the Number of Parts to be rubb'd, very great, it muſt needs be own'd, there are Caſes wherein the *Friction* follows very ſenſibly the Proportion of the Surfaces.

FRIEZE, FRIZE, FREEZE, } in Architecture, is a Member or Diviſion of the Entablature of Columns, by the Antients called *Zoophorus*.

FRIGERATORY, a Place to make or keep Things cool in Summer.

FRONT, in Architecture, is the principal Face or Side of a Building, or that which is preſented to the chief Aſpect or View.

Of Setting of Fronts.

The *Setting* (that is making) of the *Fronts* of great Buildings, *viz.* Aſhlar (or Stones) Architrave Windows, or Doors, with the Ground-Table, Faſcias, and other Members, (Mr. *Wing* ſays) are worth from 3 *l.* 10 *s.* to 5 *l.* *per* Rod, according to the Goodneſs of the Work.

Front, in Perſpective, is a Projection or Repreſentation of the Face or Forepart of an Object, or of that Part directly oppoſite to the Eye; which is more uſually called the *Orthography*.

FRONTAL, a little Fronton or Pediment, ſometimes placed over a little Door or Window.

FRONTISPIECE, in Architecture, the Portrait, or principal Face of a Building.

FRONTON, in Architecture, an Ornament which is more uſually called among us, *Pediment*.

FROWEY. Workmen ſay Timber is *frowey*, when it is evenly tempered all the Way, and works freely without tearing.

FRUSTUM, in Mathematicks, a Piece cut off, and ſeparated from a Body. Thus the *Fruſtum of a Pyramid*, or *Cone*, is a Part or Piece of it cut off uſually by a Plane parallel to the Baſe.

Frustrum of a Pyramid, is the remaining Part, when the Top is cut off by a Plane parallel to the Base.

To find the solid Content, there are several Rules:

RULE I.

To the Rectangle, (or Product) of the Sides of the two Bases, add the Sum of their Squares; that Sum being multiply'd into one Third of the Height of the *Frustum*, will give its Solidity, if the Bases be square.

Or thus, which is the same in Effect:

Multiply the Areas of the two Bases together, and add the two Areas to the square Root; and that Sum multiply'd by one Third of the Height, gives the Solidity of any *Frustum*, either square or multangled.

RULE II.

Add one third Part of the Square of their Difference to the Rectangle of the Sides of two Bases; and that Sum being multiply'd into the Height, will, if the Bases be square, produce the Solidity: But if they be triangular, or multangular, the said Rectangle of the Sides, with the third Part of the Square of their Difference, will be the Square of a mean Side; and the square Root thereof will be such a mean Side, as will reduce the tapering Solid to a Prism equal to it.

Example. Let A B C D be the *Frustum* of a square Pyramid, the Side of the greater Base 18 Inches, and the Side of the lesser

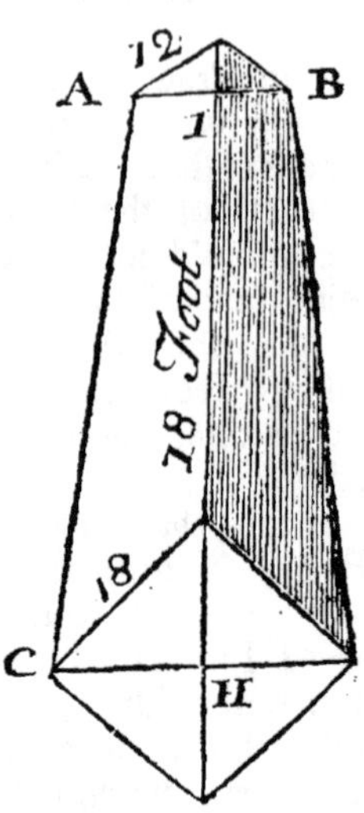

12 Inches, and the Height 18 Feet; what is the Solidity of it?

First, Multiply the two Sides together 18 by 12, and the Product will be 216; and the Difference of the Sides is 6, the Square of which is 36, a third Part of which is 12; which being added to 216, the Sum is 228 Inches, the Area of a mean Base; which being multiplied by 18 Feet, the Length, the Product will be 4104: This being divided by 144, the Quotient will be 28.5 Feet, the Content.

Or, by the first Rule, thus:

The Square of 18 is 324, and the Square of 12 is 144, and the Rectangle of 18 by 12 is 216: The Sum of these three is 684, which multiply'd by 6, the Product will be 4104; which divided by 144, the Quotient will be 28.5 Feet, the same as before.

The

The Operation both Ways.

```
 18      6 Diff.              18      12
 12      6                    18      12
---     --                   ----------
216    3)36(Square.         {324 Sq.  144 Squ.
 12 Add. --                 {144
---      12 a Third.        {216
228 the Sum.                 ----
 18 the Height.              684 the Sum.
----                           6 a Third of the Height.
1824                        -----
228                         144)4104(28.5 Feet.
144)4104(28.5                ----
----                         1224
1224                         ----
 720                           720
 ...                           ...
----                          ----
```

By Feet and Inches thus:

```
          F.  I.   I.
Multiply  1 : 6 : 6                 Or thus;
      by  1        6
         --       --             F.  I.
 Product  1   6 3)369             2  3 Square of the greater.
     Add  0   1                   1  6 the Rectangle.
         -------                  1  0 Square of the less.
Multiply  1 : 7                  -----
      by 18   0 Height.           4  9 Trip of a mean Area.
         -------                  6  0 a Third of the Height.
         18 : 0                  ------
          9   0                  28 : 6
          1   6                  ------
         -------
Content  28 : 6
         -------
```

To find the superficial Content.

The Perimeter of the greater Base is 72, and the Perimeter of the lesser Base is 48; add both the Perimeters together, the Sum will be 120; the half of which is 60; which being multiplied by 18 Feet, the Product will be 1080; which being divided by 12, the Quotient is 90 Feet; to which add the two Bases 2.25 Feet, and 1 Foot, the Sum will be 93.25 Feet, the whole superficial Content.

18

18	12	18 Height.
4	4	60
72	48	12)1080(
48		90 Feet.
		2.25 the greater Baſe.
2)120		1 the leſſer Baſe.
60		93.25 Sum.

Again, let A B C D be the *Fruſtrum* of a triangular Pyramid, each Side of the greater

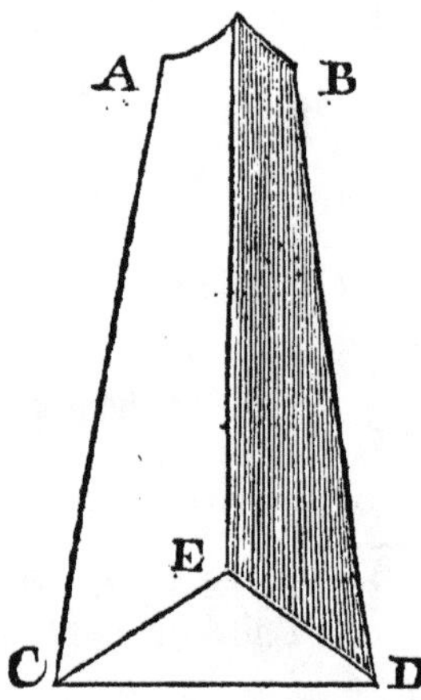

Baſe 25 Inches, and each Side of the leſſer Baſe 9 Inches, and the Length 15 Feet; what is the ſolid Content?

By the ſecond Rule, multiply 25 by 9, and the Product will be 225; and the Difference between 25 and 9 is 16; which being ſquared, makes 256, a Part of which is 85.333; which being added to 225, the Sum is 310.333; and this being multiply'd by 433, the Product will be 134.374, &c. which is the Area of a mean Baſe; and that multiply'd by 15 Feet, the Length, the Product will be 2015.610; which being divided by 144, the Quotient will be 13 99 Feet, the Solidity.

Or thus, by the latter Part of the firſt Rule: Find the Area of the greater Baſe, which will be 270.625, and the Area of the leſſer Baſe will be 35.073; theſe two Areas being multiply'd together, the Product will be 9491.639625, the ſquare Root of which is 97.425; to which add the two Areas, and the Sum will be 403.123; which multiply'd by a third Part of the Length 5, the Product will be 2015.615; and that divided by 144, the Quotient is 13.99 Feet, as before.

See the Operation of both.

```
        25    25
         9     9
       ----  ----
Product 225   16 Diff.
       ----   16
             ----
              96
             16
             ----
          3)256 the Square.
           --------
           85.333 a Third.
          225.   Add.
           --------
          310.333
             .433 tabular Number.
           --------
           930999
          930999
        1241332
        ----------
        134.374189 Mean Area.
                15 Length.
        ----------
        671.870945
       1343.74189
       ------------
  144)2015.61|2835 (13.99 Feet.
      ------
       575
       1436
        1401
       -----
        105
       -----
```

```
   25        9          433
   25        9           81
  ----      ----       ------
  125       81 Sq.      433
  50        ----       3464
  ----                 ------
  625 Square.          35.073
  433                  ------
  ----
 1875                 270.625
1875                   35.073
----                  --------
2500                   811875
-------               1894375
270.625 Area.        1353125
-------              811875
                    --------------
                     9491.630625
                     81       (97.0425
                    ------
                 187)1391
                     1309
                    ------
                1944)8263
                     7776
                    ------
               19482)48706
                     38964
                    -------
             194845)974225
                    974225
                    -------
                    ......
                    -------
```

270.625 greater Area.
97.425 the mean Proportional.
35.073 the leſſer Area.

403.123 the Triple of a mean Area.
5 a third Part of the Height.

144) 2015.615(13.99 Feet, the Solidity.
575
1436
1401

105

In finding the Area of the triangular Baſe, you multiply by .433, becauſe that is the Area of the equilateral Triangle, when the Side of it is 1, according to the Table of the Areas or Multipliers for finding the Areas of Polygons. See the Article POLYGONS.

Multiply the Square of the Side by the tabular Number, and the Product will be the Area of the Polygon.

To find the ſuperficial Content.

The Perimeter of the greater Baſe is 75, and the Perimeter of the leſſer Baſe is 27; the Sum of both is 102, and the Half is 51; which being multiply'd by 15 Feet, the Product will be 765; which being divided by 12, the Quotient will be 63.75; to which add the Sum of the two Baſes 2.12 Feet, and the Sum will be 65.87 Feet, the whole ſuperficial Content.

Note, That 51 ſhould have been multiplied by the ſlant Height; but the Difference it would make is but .06 of a Foot, which is inconſiderable.

Fruſtum of a Cone, is that Part of a Cone which remains when the top End is cut off by a plain Parallel to the Baſe.

To find the ſolid Content of it, are the ſame, in effect, as for the Fruſtum of a Pyramid.

RULE I.

To the Rectangle of the Diameters of the two Baſes add the Squares of the ſaid Diameters, and multiply the Sum by .7854, the Product will be the Triple of the mean Area; which multiplied by one Third of the perpendicular Height, that Product will be the ſolid Content.

Or thus:

Multiply the Area's of the greater and leſſer Baſes together, and out of the Product extract the ſquare Root; then add the ſquare Root and two Area's together, and multiply the Sum by one Third of the perpendicular Height, and the Product will be the ſolid Content.

RULE

RULE II.

To the Rectangle of the greater and lesser Diameters add $\frac{1}{3}$ Part of the Square of their Difference, and multiply the Sum by .7854, the Product will be a mean Area; which multiplied by the perpendicular Height, the Product will be the Solidity.

Example. Let ABCD be the *Frustum of a Cone*, whose greater Diameter CD is 18 Inches, and the lesser Diameter AB is 9 Inches, and the Length 14.25 Feet, which is the solid Content.

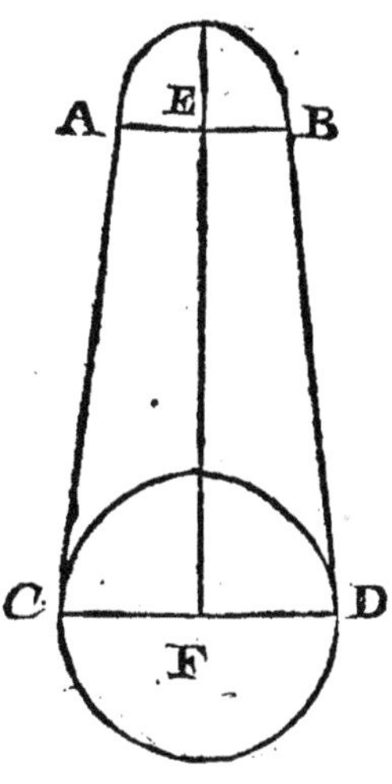

Multiply 18 by 9, and the Product will be 162; and the Difference between 18 and 9 is 9, whose Square is 81, a third Part of which is 27; which add to 162, and the Sum will be 189: This being multiply'd by .7854, the Product will be 148.44; which being divided by 144, the Quotient will be 1.03 Feet, the Area of the mean Base; which multiply 14.25 Feet, the Height, the Product is 14.6775 Feet, the solid Content.

Or thus, by the first Rule.

The Square of 18 (the greater Diameter) is 324, and the Square of 9 (the lesser Diameter) is 81, and the Rectangle or Product of 18 by 9 is 162; the Sum of these three is 567, which multiply'd by .7854, the Product is 445.3218; which divided by 144, the Quotient is 3.09 Feet, the triple Area of a mean Base: This multiply'd by 4.75, a Third of the Height, the Product will be 14.6775 Feet, the Solidity the same as before.

See the Operation.

```
     18       18 From        .7854
      9        9 Subt.         189
    ---       --             -----
    162        9 Rem.        70686
Add  27        9            62832
    ---       --            7854
Sum 189    3) 81 Square.    -----
           ---          144) 143.44|06 (1.03.
           27 a Third.       144
                             ---
                              444
                              432
                              ---
                               12
```

```
Height       14.25 Feet.
Area Base     1.03 Feet.
             -----
              4279
             1425
             -----
Solid Content 14.6775 Feet.
```

```
324 the Square of 18.
162 the Rectangle.
 81 the Square of 9.
---
567 the triple Square of a mean Diameter.
```

```
       .7854           3.09
         567           4.75
      ------          -----
       54978           1545
      74124           2163
     39270           1236
     -------         -------
144)445.32|18   Solid. 14.6775
     ------
      1332
      ----
        36
```

To find the superficial Content.

You will find the Circumference of the greater Base to be 56.5488, and of the lesser Base 28 2744; the Sum of both is 84.8232, the half Sum is 42.4116; which multiply'd by 14.25 Feet, and the Product is 604.36, &c. which divided by 12, the Quotient is 50.36 Feet, the Curve-Surface; to which add the Sum of the two Bases 2.21 Feet, the Sum is 52.57 Feet, the whole superficial Content.

To measure the *Frustum* of a rectangled Pyramid, called a *Prismoid*, whose Bases are parallel to one another, but disproportional.

The RULE.

To the greatest Length add half the lesser Length, and multiply

tiply the ſame by the Breadth of the greater Baſe, and reſerve the Product.

Then to the leſſer add half the greater Length, and multiply the Sum by the Breadth of the leſſer Baſe; and add this Product to the other Product reſerved, and multiply that Sum by a third Part of the Height, and the Product will be the ſolid Content.

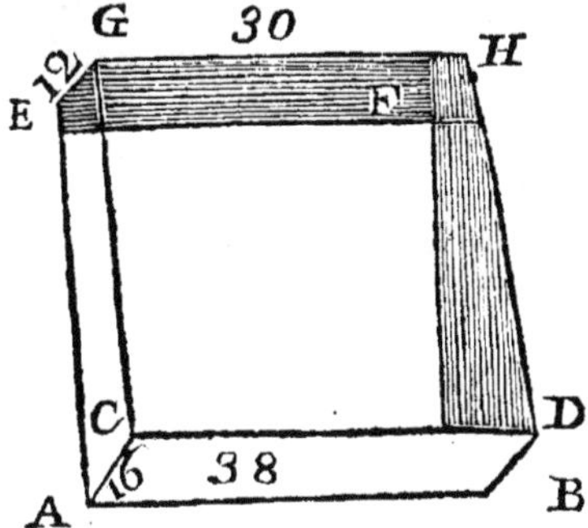

Example. Let ABCDEFGH be a Priſmoid given, the Length of the greater Baſe AB 38 Inches, and its Breadth AC 16 Inches; and the Length of the leſſer Baſe EF is 30 Inches, and its Breadth 12 Inches, and the Height 6 Feet: What is the ſolid Content?

To the greater Length AB 38, add half the leſſer Length EF 15, the Sum will be 35; which being multiply'd by 16, the greater Breadth, the Product will be 848; which reſerve.

Again, to EF 30 add half AB 19, and the Sum will be 49; which multiply by 12, the leſſer Breadth EG, and the Product will be 588: To which add 848, the reſerv'd Product, and the Sum will be 1436; which being multiply'd by 2, a third Part of the Height, the Product will be 2872; this Product divide by 114, and the Quotient will be 19.14 Feet, the ſolid Content.

38 = AB.
15 = ½ EF.
———
53
16 = AC.
———
318
53
———
848
588
———
1436
2 = a third Part of the Height.
———
2872

30 = EF.
19 = ½ AB.
———
49
12 = EG.
———
588
———

144) 2872 (19.94 Feet, the Content.
1432
———
1360
640
———
64
———

To prove this Rule: Let it be ſuppoſed the Solid cut into Pieces, ſo as to make it capable of being meaſured by the foregoing Rules, thus: Let ABCD repreſent the greater Baſe, and EFGH the leſſer Baſe; and let the Solid be ſuppoſed to be cut off through by the Lines *ac*, *bd*, and *ef*, *gh*, from the Top to the Bottom; ſo will there be a Parallelopipedon, having its Baſes equal to the leſſer Baſe EFGH, and its Height, 6 Feet, equal to the Height of the Solid: Multiply 30 (the Length of the Baſe) by 12 (the Breadth thereof) and the Product is 360; which multiply by the Height 6 Feet, and the Product is 2160.

Then there are two Wedge-like Pieces, whoſe Baſes are *ab* EF, and GH, *cd*; if theſe two Pieces are laid together, the thick End of one to the thin End of the other, they will compoſe a rectangled Parallelopipedon; which, to meaſure, multiply the Length of the Baſe 30 by its Breadth 2, and the Product will be 60; which multiply by 6, the Height, the Product is 360.

Then there are two other Wedges like Pieces, whoſe Baſes are *e* E, *g* G, and *f* F, *h* H; theſe two laid together, will compoſe a rectangled Parallelopipedon: To meaſure this, multiply the Length of the Baſe 12 by the Breadth 4, the Product is 48; which multiply by 6, the Height, the Product is 288.

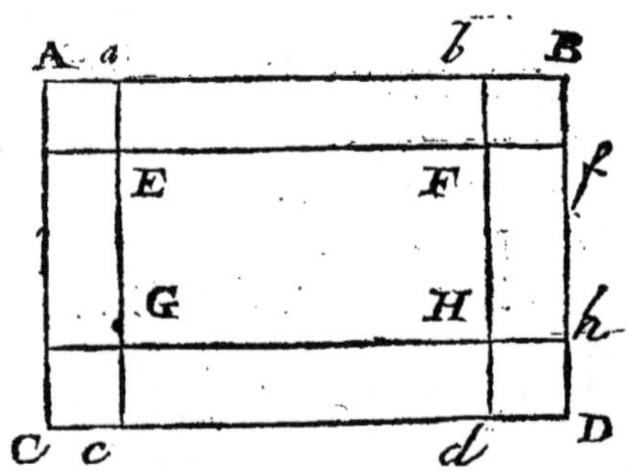

And laſtly, there are four rectangled Pyramids at each Corner; which, to meaſure, multiply the Length of one of the Baſes 4 by its Breadth 2, the Product is 8; which multiply'd by 2, (a third Part of the Height,) the Product is 16; and that multiply'd by 4, becauſe there are four of them, the Product is 64; then add all theſe together, and the Sum is 2872; and divide by 144, the Quotient is 19.94 Feet, the ſame as before, which ſhews the Rule to be true.

See the Operation:

12	30	12	4
30	2	4	2
360	60	48	8
6	6	6	2
2160	360	288	16
360			4
288			64
64			

144) 2872 (19.94 Feet the whole Content.

To find the ſuperficial Content.

Half the Perimeter of the greater Baſe is 54, and half the Perimeter of the leſſer Baſe is 42; which being added together, the Sum is 96; which being multiply'd by 6 (the Height) the Product will be 576: Then divide this Product by 12, and the Quotient is 48 Feet; to which add the Sum of the two Baſes 6 72 Feet, and the Sum will be 54.72 Feet, the whole ſuperficial Content.

FUNNELS *of Chimneys*. The *Funnel* is the Shaft, or ſmalleſt Part from the Waſte, where 'tis gathered into its leaſt Dimenſions.

Palladio directs, That the *Funnels of Chimneys* be carried through the Roof, three, four, or five Feet at the leaſt, that they may carry the Smoke clear from the Houſe into the Air.

He adviſes alſo, that Care be taken as to the Width of them; for that if they be too wide, the Wind will drive back the Smoke into the Room; and if they be too narrow, the Smoke will not be able to make its Way.

Therefore Chamber Chimneys muſt not be made narrower than ten or eleven Inches, nor broader than fifteen; which is the ordinary Depth of the *Funnels* of great Kitchen Chimneys, whoſe Breadth is four or five Feet within the Work, from the Place where the Breſt ends, to the Top of the *Funnel*.

Now the ſaid Breaſt reaches from the Mantle-Tree to the Ceiling or Pitch of the Arch, always diminiſhing within the Work, till you come to the Meaſures of Depth and Breadth before mentioned; and from thence to the End of the *Funnel*, it muſt be carried up as even as it poſſibly can be; for if there be a Failure in this, the Smoke happens to be offenſive.

FURRING, in Architecture, is the making good the Rafters Feet in the Cornice.

Thus, when Rafters are cut with a Knee, theſe *Furrings* are Pieces which go ſtraight along with the Rafter, from the Top of the Knee to the Cornice.

Alſo when Rafters are rotten, or ſunk hollow in the Middle, there are Pieces cut thickeſt in the Middle, and tapering towards each End, which are nailed upon them, to make them ſtraight. Such Pieces are called *Furs*, and the Putting them on, *Furring the Rafters.*

FUSAROLE, in Architecture, is a Moulding or Ornament placed immediately under the Echinus in the *Doric*, *Ionic*, and *Compoſite* Capitals.

The *Fuſarole* is a round Member carv'd in Manner of a Collar or Chaplet with oval Beads. The *Fuſarole* ſhould always anſwer exactly under the Eye of the Volute, in the *Ionic* Capital.

FUST, in Architecture, is the Shaft of a Column, or that Part comprehended between the Baſe and the Capital. Alſo called the *Naked.*

The *Fuſt* is that cylindrical Part which makes, as it were, the Body or Trunk of the Column, excluſive of the Head and Feet.

The Word is *French*, and literally ſignifies a *Caſk.* But ſome derive it from the *Latin Fuſtis*, a *Club.*

G A

GABLE-END *of a Houſe*, is the upright triangular End from the Cornice or Eaves, to the Top of its Roof.

To meaſure a Gable-End.

Multiply the Breadth at Bottom by half of the Perpendicular, or Line from the Angle of the Top to the Middle of the Bottom, or multiply half the former by the whole of the latter, and the Product will give the Content in ſuch Meaſures as the Dimenſions were taken in.

GAIN, the Bevelling Shoulder of the Joiſts, or other Stuff.

'Tis alſo uſed for the lapping of the End of the Joiſts, &c. upon a Trimmer or Girder; and then the Thickneſs of the Shoulder is cut into the Trimmer, alſo bevelling upwards, that it may juſt receive the *Gain*, and ſo the Joiſt and Trimmer lie even and level with their Surface.

This Way of working is uſed in Floors and Hearths.

GALLERY, in Architecture, is a covered Place in a Houſe, much longer than broad, and which is uſually on the Wings of a Building, ſerving to walk in.

Gallery is alſo a little Iſle or Walk, ſerving as a common Paſſage to ſeveral Rooms, placed in a Line or Row.

Their Length (according to *Palladio*) ought to be at leaſt five Times their Breadth. They may be ſix, ſeven, or eight Times their Breadth, but muſt not exceed.

GARD MANGER, a Storehouſe or Room to ſet Meat in.

GATE,

GATE, a large Door leading or giving Entrance into a City, Town, Caſtle, Palace, or other conſiderable Building; or a Place for Paſſage of Perſons, or Horſes, Coaches, or Waggons, &c.

As to their Proportion: The principal *Gates* for Entrance, through which Coaches and Waggons are to paſs, ought never to be leſs than ſeven Foot in Breadth, nor more than twelve Foot; which laſt Dimenſion is fit for large Buildings.

As to the Height of a *Gate*, it ought to be one and a half of the Breadth and ſomething more.

But as for common *Gates* in Inns, under which Waggons go loaded with Hay and Straw, &c. the Height of them may be twice their Breadth.

Of the Price of ſome Sorts of Gates.

As to the Price of *Gates*, it is various, according to the Sorts of *Gates*; which again will differ according to the Dimenſions and Workmanſhip. I ſhall at preſent mention only *Palliſadoe* and *Pold-Gates*.

Of Palliſadoe Gates.

Mr. *Wing* ſays, in *Rutlandſhire*, if the *Gates* be ſix or ſeven Feet high, and the Workman find Timber and Workmanſhip, they are worth about 9 or 10*s.* *per* lineal Yard; but if he find only Workmanſhip, then it is worth but 6 or 7*s.* *per* Yard.

If they are *Semi-Palliſadoe*, with kneeling Rails at the Top, handſomely moulded on both Sides, and ſquare Palliſadoes, raiſed Pannels, and Biſection-Mouldings on both Sides, the *Gates* about eight Feet high, and the Poſts a Foot ſquare, opened in the Front, or revailed with a Moulding ſtruck in it on both Sides the Revail, a Baſe and Capital laid on the Poſts, and the Heads cut into one of the *Platonick* Bodies; as ſuppoſe an *Icoſaedron*, and the Poſts were about ten or eleven Feet above Ground, then the Workmanſhip is worth 12 or 13*s.* a Yard lineal; but if the Workmen find Timber, it will be worth upwards of 20*s.* a Yard lineal; in ſuch *Gates*, to find all Iron-Work, Painting, &c. it would be worth above 30*s.* a Yard lineal.

Of Pold Gates.

Theſe are ſuch as are ſet up in Fences, for ſhutting up the Paſſages into Fields, and other Incloſures.

Theſe are of two Sorts, either of ſawed, or cleft Timber. For the making one of ſawn Timber, ſetting it up, and its Poſts, the Price in different Places, is from 3*s.* 6*d.* to 5*s.* But if the Carpenter pay for the Sawing, then the Price is from 5*s.* to 6*s.* 6*d.* Such a *Gate*, Timber and Work, is worth from 7*s.* to 10*s.* according to the Goodneſs; but with Poſts, from 12*s.* to 15*s.* but *Gate* and Iron-Work, from 10*s.* to 13*s.* and Poſts from 15*s.* to 18*s.*

Cleft Pold-Gates, Cleaving, Making, and Hanging, from 4*s.* to 5*s.* and ſo proportionably for all Timber, Iron, and Poſts. The Reaſon why the Prices are diffe-

rent,

rent, is, because they are according to the Customs of different Places.

GAVEL is used for what is more usually call'd the *Gable*.

GENERATED } in Mathe-
GENITED } maticks, is us'd to signify whatever is produced either in Arithmetick, by the Multiplication, Division, or Extraction of Roots; or in Geometry, by the finding out the Contents, Areas, and Sides, or of extream and mean Proportionals, without Arithmetical Addition, and Subtraction.

GENERATING LINE, or *Figure*, in Geometry, is that which, by its Motion or Revolution, produces any other Figure, plane or solid: Thus a Right Line, mov'd any way parallel to itself, *generates* a Parallelogram; round a Point, in the same Plane, with one End fasten'd in that Point, it *generates* a Circle; one entire Revolution of a Circle in the same Plane, *generates* the Cycloid; the Revolution of a Semicircle round its Diameter, *generates* a Sphere, *&c.*

GENESIS, in Geometry, is the Formation of any Plane or solid Figure by the Motion of some Line or Surface; which Line or Surface is always call'd the *Describent*; and that Line, according to which the Motion is made, is called the *Dirigent*.

GEOCENTRICK is apply'd to any Thing which has the Earth for its Centre.

GEODÆSIA, Surveying, or the Art of measuring Land.

GEODETICAL *Numbers*, are such Numbers as are considered according to those vulgar Names or Denominations, by which Money, Weights, Measures, *&c.* are generally known, or particularly divided by the Laws and Customs of several Nations.

GEOGRAPHY is the Science that teaches and explains the Properties of the Earth, and the Parts thereof that depend upon Quantity.

GEOMETRICAL, of or pertaining to Geometry.

Geometrical Plane. See PLANE.

Geometrical Place. See LOCUS.

Geometrical Solution of a Problem, is when the Thing is solved according to the Rules of Geometry, and by such Lines as are truly *Geometrical*, and agreeable to the Nature of the Problem.

GEOMETRY originally signify'd the Art of Measuring the Earth; but it is now the Science of whatever is extended, so far as it is, such, that is of Lines, Superficies, and Solids.

It is very probable, that Geometry had it first Rise in *Egypt*, where the *Nile* annually overflowing the Country, and covering it with Mud, obliged Men to distinguish their Lands one from another, by the Consideration of their Figure; and to be able also to measure the Quantity of it, and to know how to plot it, and to lay it out again in its just Dimensions, Figure, and Proportion. After which, 'tis likely, a farther Contemplation of those Draughts and Figures, help them to discover many excellent and wonderful Properties belonging to them; which Spe-

culation

culation continually was improving, and is still to this very Day.

But the *Geometry* of the Antients was contained within very narrow Bounds, in Comparison of the Modern, as well as their other Mathematical Speculations; for it only extended to Right Lines and Curves of the first Kind or Order: Whereas new Lines of infinite Orders, are received into the Modern *Geometry*; which Orders are defin'd by Equations, involving the Ordinates, and Absciffes of Curves.

Sir *Isaac Newton* was the first Person who gave any tolerable Account of the Nature of Curves above Conick Sections.

Geometry is divided into Speculative, and Practical: The former treats of the Properties of Lines and Figures, such as *Euclid's Elements*, and *Apollonius's Conicks:* And the latter shews how to apply these Speculations to Use in Life.

GILDING *with Gold or Silver.* Whatsoever you would *gild*, must be first drawn with Gold Size, according to the true Proportion of what you would have *gilt*, whether Figure, Letter, or whatever it be.

When the Gold Size has been thus laid on, it must stand till it is dry enough to *gild*, which is to be known by touching it with the End of your Finger; for if your Finger stick a little to it, and yet the Gold Size come not off, then it is dry enough; but if the Colour come off on your Finger, then it is not dry enough, and must be let alone a little longer; for if you should then lay your Gold on, it would so drown it, that it would have no Lustre; But on the other Hand, if your Size should be so dry, as not, as it were, to adhere a little to your Finger, then it is too dry, and the Gold will not take, for which there is no Remedy but new Sizing: Therefore you must watch the very Nick of Time, when it is neither too wet nor too dry, both Extreams being unfit for laying the Gold on it.

When your Size is ready for *Gilding* take your Book of Leaf-Gold, and open a Leaf of it; take it out with your Cane Plyers, and lay it on your *Gilding Cushion*; and if it lie not smooth, blow on it with your Breath, but very gently, which will make it lie flat and plane; then with a Knife of Cane, or for want of that a common Pocket-Knife, (that hath a smooth and sharp Edge, being wiped very dry on your Sleeve, that the Gold stick not to it,) cut your Leaf-Gold into such Pieces or Forms as you judge most suitable to your Work.

When you have thus cut a Leaf of Gold into proper Forms, then take your *Gilding Pallet*, and breathe upon it to make it dampish, that the Gold may stick to it: With this Tool take your Gold up, (by clapping it down on the several Pieces you had before cut into Forms,) and transfer it to your Size, upon which clap it down as dextrously as you can, and the Gold will leave the Pallet, and stick on the Size; which you must afterwards press down smooth with

 a Bunch

a Bunch of Cotton, or the Bottom of a Hare's Foot; and thus you must do Piece by Piece, till you have covered all your Size with Gold; and after it is fully dry'd, then with your Hare's Foot brush off all the loose Gold, and the *Gilding* will remain fair and beautiful.

If the Work to be *gilded* be very large, open your Book of Leaf-Gold, and lay the Leaf down on your Work without cutting it to Pieces, and so do Leaf by Leaf, till you have covered quite over what you intend to *gild:* And if some particular Places do miss, take up with a small Bunch of Cotton a Piece of Leaf-Gold cut to a fit Size, and clap it on, that the Work may be intirely covered: And if the Gold be to be laid in the Hollows of carv'd Work, you must take it up on the Point of a Camel's Hair Pencil, and convey it in, and with the said Pencil dab it down till it lie close and smooth.

Note, That after your *Gilding* is thus finished, you may, if you please, diaper or flourish on it with thin burnt Umber, whatsoever shall be suitable to your Design. The Umber must be tempered but thin, so that the Gold may appear through it.

Note further, That a Book of Gold contains 24 Leaves, each Leaf being three Inches square, the Price of each Book is two Shillings at the Gold-Beaters; one Book will cover 216 square Inches of Work, for so many square Inches are conden Bottom about six Inches square, and then well stuffed out with Cotton or Wooll very hard, plain, and flattish: Upon this *Gilding Cushion* the Gold Leaves are to be laid when you would cut them into such Scantlings as will best fit the Work you design to *gild*.

Gilding Knife, a Slip of the hollow *Spanish* Cane, cut up to a smooth and sharp Edge with a good Penknife. This Cane Knife is accounted the best, because, if well made, it will not only be very sharp, but also cut the Gold Leaf more naturally than any other; for a Steel Knife, though it cut very well, yet the Gold will stick to it, and so give you much Trouble to part the Leaf from it, except you are careful to keep the Edge very dry, by continually wiping it with a clean dry Cloth.

Gilding Pallet, is a flat Piece of Wood, about three Inches long, and an Inch broad, upon which is to be glued a Piece of fine Woollen Cloth, of the same Length and Breadth. Upon this Pallat do but breath with your Breath, that the Cloth may be made a little moist by it, and then if you clap it down gently upon the Gold that is cut out on the Cushion, it will stick to the Pallat, and may thence be conveyed to the Work you are to *gild* and lay down on it.

GIRDING BEAMS, are used by some Architects for the same as Girders.

GIRDERS, in Architecture, are some of the largest Pieces of Timber in a Floor, the Ends of which are usually fastened into Summers and Brest-Summers; and Joists are framed in at one End to the *Girders*.

The Scantlings and Size of *Girders* and Summers, upon the Rebuilding of *London*, after a Consultation of experienced Workmen, were reduced into an Act by the Parliament, and are thus set down, as fit for all Fabricks, great and small.

	Feet.		Feet.		Feet.		Feet.
Girders and *Summers*, must be in Length	10	to	15	and in Breadth	11	in Depth	8
	15		18		13		9
	18		21		14		10
	21		24		16		12
	24		26		17		14

How they are to be laid in the Brickwork.

No *Girder* or *Summer* ought to be less than ten Inches into the Wall, and their Ends must be laid in Loam.

That *Girders* and *Summer* be of good hearty Oak, as free from Knots as may be; because that will be the least subject to break, and may with more Safety be relied on in this cross and transverse Work.

In as much as there is a Moisture in Timber, a certain ingenious modern Builder advises,

 that

that all bearing Timber have allow'd it a moderate Camber or Roundness; for till that Moisture is in some sort dry'd out, the said Timber will sag with its own Weight; and that chiefly is the Reason why *Girders* are truss'd.

But here you must observe, that *Girders* are best truss'd when they are first sawn out; for by its drying or sinking, it lightens the Trusses in them yet more.

It is also to be observ'd, that all Beams or Ties be cut or forced in framing to a Camber or Roundness, such as an Inch in the Length of eighteen Feet; and that principal Rafters be also cut, or forced up to a Camber or Roundness, as before. The Reason of this is, all Trusses, though ever so well fram'd, by the Shrinking of the Timber, and Weight of the Covering, will sag, and sometimes so much, as to offend the Eye of the Beholder: So that by this Preparation, your Truss may ever appear well.

You should also observe, that all Case Bays, either in Floors or Roofs, do not exceed twelve Feet, if possible; that is, do not let your Joists in Floors, your Purlins in Roofs, *&c.* exceed twelve Feet in their Length or Bearing, but rather let their Bearing be eight, nine, or ten Feet.

Also in Bridging-Floors, do not place your binding or strong Joists above four or five Feet apart; nor let your Bridgings or common Joists be above twelve Inches apart, that is, between one Joist and another.

It should also be observ'd never to make double Tenants or Tenons for bearing Uses, such as Binding Joists, common Joists or Purlins; for in the first Place, it weakens very much whatever you frame it into; and, in the second Place, it is a Rarity to have a Draught in both Tenons, that is, to draw your Joint close by the Pin; for the said Pin, by passing through both Tenons, if there is a Draught in each, must bend it so much, that except the Pin be as tough as Wire, it must needs break in driving, and consequently do more Hurt than Good.

GIRT. See FILLET.

GIVEN is a Word often used in the Mathematicks, and signifies something which is suppos'd to be known.

Thus if a Magnitude be known, or we can find another equal to it, we say, it is a *Given Magnitude*, or that such a Thing is *given* in Magnitude.

If the Position of any Thing be supposed as known, then say, *Given in Position.*

Thus if a Circle be actually described upon any Plane, they say, its Centre is *given in Position*; its Circumference is *given in Magnitude*, and the Circles is *given both in Position and Magnitude.*

But a Circle may be *given in Magnitude* only; as when only its Diameter is *given*, and the Circle not actually described.

If the Kind or Species of any Figure be *given*, they say, *Given in Species*; if the Ratio between any two Quantities is known, they are said to be *given in Proportion.*

GLACIS, in Building, *&c.* is an easy insensible Slope or Declivity.

GLASS,

GLASS, a Diaphanous or Transparent Body, made by Art of Sand and Nitre, as *Pliny* says. It is also made of white glistering Flints mixed with *Sal Alkali*, or the Salt of the Herb Glasswort; or Salt of Fern-Ashes, for common *Glass*, some say.

M. *Blancourt* says, the *Venetians* use white Flints, and also a rich Sand, and likewise a sort of white Marble. He likewise adds, that all white transparent Stones, which will not burn to Lime, are fit to make; and that all Stones that are fit to strike fire, are capable to be made into *Glass*.

A certain learned and curious Author gives us the following Characters or Properties of *Glass*, whereby it is distinguish'd from all other Bodies, *viz.*

1. That it is an artificial Concrete of Salt, Sand, or Stones.

2. That it is *fusible* by a strong Fire.

3. That when fused, it is tenacious and coherent.

4. That it does not waste or consume in the Fire.

5. That when melted, it cleaves to Iron.

6. That when it is red-hot, it is ductile, and capable of being fashion'd into any Form; but not malleable; and capable of being blown into a Hollowness, which no Mineral is.

7. That it is frangible when thin, without annealing.

8. That it is friable when cold.

9. That it is always Diaphanous, whether hot or cold.

10. That it is flexible and elastick.

11. That it is dissoluble by Cold and Moisture.

12. That it is only capable of being graven or cut with a Diamond and Emery.

13. That it receives any Colour or Dye, both externally and internally.

14. That it is not dissoluble by Aqua Fortis, Aqua Regia, or Mercury.

15. Neither acid Juices, nor any other Matter, extract either Colour or Taste, nor any other Quality from it.

16. It admits of Polishing.

17. That it neither loses of Weight nor Substance, by the longest and most frequent Use.

18. That it gives Fusion to other Metals, and softens them.

19. That it is the most pliable Thing in the World; and that it best retains the Fashion given it.

20. That it is not capable of being calcin'd.

21. That an open *Glass* filled with Water in the Summer-Time will gather Drops of Water on the Outside, so far as the Water on the Inside reaches; and a Man's Breath blown upon it will manifestly moisten it.

22. Little *Glass Balls*, filled with Water, Mercury, or other Liquor, and thrown into the Fire; as also Drops of green *Glass* broken, fly asunder with a loud Noise.

23. That neither Wine, Beer, nor any other Liquor, will make it musty, or change its Colour, nor rust it.

24. That it may be cemented, as Stones and Metals.

25. That a *Drinking-Glass* partly filled with Water, and rubbed on the Brim with a wet Finger,

Finger, yields musical Notes, higher, or lower, as the *Glass* is more or less full; and makes the Liquor frisk and leap.

The *Sorts of* Glass.

There are various Sorts of *Glass*, which are made use of in the World; but at present, I shall only speak of those Sorts of *Glass* which Glaziers commonly use here in *England*; which are these following, *viz.* *Crown Glass*, which is of two Sorts; 1. *Lambeth* and *Ratcliff*. 2. *French* or *Normandy Glass*. 3 *German Glass* of two Sorts, White, and Green. 4. *Dutch Glass*. 5. *Newcastle Glass*. 6. *Staffordshire Glass*. 7. *Bristol Glass*. 8. *Looking Glass*. 9. *Jealous Glass*. Of which Sorts, I shall treat succinctly in their Order.

Crown Glass is of two Sorts, *Ratcliff* and *Lambeth Crown Glass*.

That Sort of *Crown Glass* that goes by the Name of *Ratclif Crown Glass*, is the best and clearest Sort of *Crown Glass*; which Sort was at first made at the *Bear-Garden*, on the Bank-side, in *Southwark*, in the Year 1691. which was published in the *Gazette*, and commended, as follows, and called *Crown Window-Glass*, much exceeding *French Glass* in all its Qualifications.

But the Maker of this Sort of *Crown Glass* being now removed to *Ratcliff*, it therefore now bears the Name of *Ratcliff Crown Glass*, as it did at first of the *Bear-Garden Crown Glass*.

This Sort of *Crown Glass* is of a light Sky-blue Colour, which may be very distinctly seen if it be laid on a Piece of a white Paper.

It has been reported, that an *English Glass-Maker* went over to *France* on Purpose to learn the *French* Way of making *Glass*; which he having attained to, came over again into *England*, and set up Making of *Crown Glass*, and in the Performance, outstripp'd the *French* his Teachers, as *Englishmen* usually do.

There are twenty-four Tables of this *Glass* to the Case, the Tables being of a circular Form about three Foot, six, seven or eight Inches in Diameter, and consequently each Table will be in Area about nine or ten Feet, and the Case betwixt two hundred and twenty, and two hundred and forty.

This *Glass* is brought from *Ratcliff* in such kind of Frames as *Newcastle Glass* is brought up to Town in, only the *Newcastle Glass* is brought on Shipboard; and this *Ratcliff Glass* upon Staves by two Men.

1. This *Glass* called *Ratcliff Crown Glass*, has been sold for about 9*d.* a Foot in *London*, cut into Squares, and when wrought in Lead, and set up, for about 18*d.* a Foot.

2. *Lambeth Crown Glass* takes its Name also from the Place where it is made. It is of a darker Colour than *Ratcliff Crown Glass*, and more inclining to the Green.

This Sort is sold for 8*d.* a Foot cut into Squares; and being wrought and set up into Windows with Lead, its Price is said to be worth about 16*d.* a Foot.

French

French Glaſs, alſo called *Normandy Glaſs*, becauſe it was formerly made at *Cherburg* in *Normandy*; and alſo *Lorrain Glaſs*, becauſe made there. Now it is made wholly in the Nine Glaſs-Works; five whereof are in the Foreſt of *Lyons*; four in the County of *Eu*; the leaſt at *Beaumont*, near *Rouen*. They alſo make *Glaſs* at *Nevers* in *Orleans*, and likewiſe at *St. Gobin*, near *La Fere* in *Picardy*; but from which of theſe Places, any *French Glaſs* comes, that is uſed in *England*, is uncertain.

It is a thinner and more tranſparent *Glaſs* than our *Newcaſtle Glaſs*, and when laid on a Piece of white Paper, it appears of a dirtyiſh green Colour.

It uſed to be of a middle Price, betwixt *Crown* and *Newcaſtle Glaſs*, which has been ſold for 12 . a Foor, wrought in Lead, and ſet up.

Of this *Glaſs*, there is but twenty-five Tables to the Caſe.

German Glaſs. Of this there are two Sorts, White and Green.

The *White German Glaſs* is of a whitiſh Colour, and free from thoſe Spots and Blemiſhes which our *Newcaſtle Glaſs* is ſubject to; but it has commonly ſome fine or ſmall carved Lines or Streak'd Lines, as the *Newcaſtle Glaſs* hath.

Green German Glaſs. This, beſides its greeniſh Colour, is ſubject to have thoſe fine Lines or Streaks which the White is; but both the *Green* and the *White German* are ſtraighter, and not ſo warp'd as the *Newcaſtle Glaſs* is. Both theſe Sorts of *Glaſs* are brought over from *Germany*, and yet is generally as cheap as *Newcaſtle Glaſs*.

Dutch Glaſs does not differ much from *Newcaſtle Glaſs* in its Colour and Price. 'Tis frequently much warp'd like that, and the Tables are but ſmall.

Newcaſtle Glaſs, is that which is moſtly uſed in *England*. 'Tis of an Aſh Colour, and ſubject to Specks, Streaks, and other Blemiſhes, and beſides, is frequently warp'd and crooked.

Mr. *Leybourn* ſays there are forty-five Tables to the Caſe, each Table containing five ſuperficial Feet; and conſequently a Caſe of forty-five Tables to the Caſe will contain two hundred and twenty-five Feet; tho' ſome ſay, there are but thirty-five Tables, and ſix Feet in each Table; which amount but to two hundred and ten Feet.

Mr. *Leybourn* ſays, that a Caſe of forty-five Tables, five Feet to a Table, equal to two hundred and twenty-five Feet, weighs about two hundred Weight, and conſequently nine Feet will weigh about eight Pound.

As to the Price of *Newcaſtle Glaſs*, it is uncertain: For when Coals are plenty, then *Glaſs* is cheap; and when Coals are dear at *London*, then *Newcaſtle Glaſs* is ſo likewiſe; not that they want Coals at *Newcaſtle*, but becauſe they have no other Conveyance for it to *London*: So that at ſome Times it has been at 30*s*. a Caſe, and at other Times 40*s*. But ſome ſay, that the moſt common Price is 34 *s*. the Caſe.

Some ſay, 'tis worth 6 or 7*s*. to cut a Caſe of this *Glaſs* into Quarries Diamond Faſhion (with Halfs, and Quarters, and Three Quarters of Quarries, as the *Glaſs* falls out;) and others again have ſaid

said they would do it for half the Money.

Newcastle Glass, cut into large Squares, are sold from 22*s.* to 25*s. per* hundred Feet, according to their Size; and small Squares from 19*s.* to 22*s. per* hundred Feet; and Quarries of *Newcastle Glass* for about 16*s. per* hundred Feet.

Glazing done with this *Newcastle Glass*, with Quarries, Banding, Soldering, and Pinning the Casements, being included, the usual Price in *London* is 5*d.* or 6*d. per* Foot. But in several Parts of the Country, they have 6*d. per* Foot, and will be paid for pinning the Casements beside.

Glazing, in some Places of *England*, as in *Rutland*, and other Parts towards the North, is done with Quarries of *Newcastle Glass* at 4*d.* ½ or 5*d. per* Foot; and Squares wrought into Lead, and set up, for 6*d. per* Foor.

But in *Sussex*, *Kent*, and the South Parts of *England*, they will not work so cheap; because the *Glass* costs them something dearer. They usually reckon 7*d per* Foor for Glazing with Squares of *Newcastle Glass*, and they will be paid for pinning the Casements besides.

Staffordshire Glass, is a sort of *Glass* that is seldom used but in that and the neighbouring Counties.

Bristol Glass is so called, because it is made at the City of *Bristol*; but very little of it comes to *London*, by reason they have not the Convenience of sending it by Sea, as they have from *Newcastle* by Coal Ships; though this is as cheap, and better than *Newcastle Glass*.

Looking Glass. As to *Looking Glass* Plates, they are made either at the *Old Bear-Garden* in *Southwark*, or at *Vaux-Hall*, near *Lambeth*.

I am not certain, whether this Sort of *Glass* be not made with the Sort of Sand which Dr. *Grew* mentions in his *Musæum Regalis Societatis*, p. 346.

Fine Sand, says he, from a Sand-Pit near *Bromley* in *Kent*, of which is made the clearest and best *English Glass*: It consists of some Grains as clear as Chrystal; which being mixed with others obscure, give a whitish Ash Colour to the whole Mass.

Some have a Way of examining which is the whitest and clearest *Glass*; which is as follows:

They take it up close by one Edge, betwixt the Edges of the Middle and Forefinger; and then looking against the cut or broken Edge, the Eyes being thus skreen'd by the Edges of the two Fingers, they say, 'tis easy by this Method to discern which is the whitest and clearest *Glass*.

These *Looking Glass* Plates are ground smooth and flat, and polish'd. They are sometimes used in Sashes, or Sash-Windows. But 'tis a dear sort of *Glass*; for they ask 4*s.* a Foot for such Squares, and if they are large, 'tis much more.

Looking Glasses, being foil'd, being in Vogue for Ornaments over Chimneys in Parlours, &c. I shall say something briefly concerning them.

Sir *William Petty* tells us, that the Value of *Looking Glass* Plates consists in a duplicate Proportion

portion of their Sides to their Squares.

Because you may not be left quite in the Dark as to this Matter, I shall give you the Price which I have known set down upon two Sizes of *Looking Glasses*, *viz.* one of five Feet long, and twelve Inches broad, in a Frame, to place over a Chimney, 40*s.* some of ten and eight Inches, in Walnut-Tree Frames, 4*l.* apiece, if they have the Diamond Cut; but if not, this Size is about 6*s.* apiece cheaper.

Jealous Glass is a sort of wrinkled *Glass*, of such a Quality, that what is done on the other Side of it cannot be distinctly seen; but yet it admits the Light to pass through it.

It is made of the same Materials that *Looking Glass* Plates are. It is cast in a Mould, and is composed all over its Surface with a Multitude of oblong circular Figures, (which are concave,) somewhat resembling Weavers Shuttles on the one Side; but the other consists of Figures a little convex; and this last Side is the Side they cut it on, when the Squares are too big for the present Use. It being very difficult to cut it on the concave Side.

Some Sorts of this *Jealous Glass* have a Convexity tiling in the middle of the Concavity; so that one Side or Surface of it doth much resemble the Boats made by Boys by folding of Paper, only in this *Glass*, the Concavities and Convexities are more obtuse and blunt.

But there are various Sorts of this *Glass*, in respect either to the Form or Size of the Figures of which this *Glass* is composed: Some of it having Shuttle-like Figures much larger than others; and some of it with the Points of Shuttles (as it were) very curved; and sometimes these Figures are in a perpendicular Position to one Edge of the Square, and other some are oblique to it.

This *Glass* is usually sold at about 18*d.* *per* Square, each Square being about twelve or fourteen Inches broad, and fifteen or sixteen Inches long.

The Reason why these Plates are so dear, is said to be, because the *Looking Glass Plate Makers* don't care to make them 'till their Pots of Metal are almost out, and that they are most at Leisure; for they say it wastes the Metal too much for their Profit.

This Sort of *Glass* is commonly used in and about *London*, to put into the lower Lights of Sash-Windows, &c. where the Windows are low next the Streets, to hinder People who pass by from seeing what is done in the Room: It is also sometimes set in Lead for the same Purpose.

This Sort of *Glass* must needs prevent People's seeing through it; because the Rays or Species of a visible Object, are by Reason of such a Variety of Refractions, (caused by the Inequality of the Surface of the *Glass*,) broken and confused, when they arrive at the Retina or Fund of the Eye.

The

The Method of Working or Blowing of WINDOW *or* TABLE GLASS.

The Method of making *Crown Window Glaſs*, as now practiſed in *England*, as has been before hinted, was borrow'd from the *French*, by an *Engliſh Glaſs Maker*, who went over to work in *France*, on purpoſe to penetrate into the Secret; which when he had attained to, he came back, and ſet up a *Glaſs Work*, wherein he far excelled the *French*, who were his Teachers.

This *Glaſs* is blown much after the Manner of *Looking Glaſs*, as follows:

The Furnace, Melting-Pots, Materials and Fire, are the ſame for *Window* or *Table Glaſs*, as for *Round Glaſs*; and the Difference in the Operation only commences after the Operator has dipped his Blowing-Iron a fourth Time in the melted Metal.

The *Glaſs* then being in this State, they blow it; but inſtead of rounding or forming it into a Punch, the particular Motion the Workman gives it in the directing and managing the Wind, and the Way of rolling ir on the Iron, make it extend in Length two or three Feet, and form a Cylinder; which at firſt, is but two Inches in Diameter; but which, by being again committed to the Fire, and taken out, and blown afreſh, becomes of the Extent requir'd for the Table of *Glaſs* to be formed; but with this Circumſtance, that the Side which is faſtened to the Iron, goes gradually diminiſhing, and ends in a kind of Cone or Pyramid.

In order to render the two Ends nearly of the ſame Diameter, after adding a little *Glaſs* to that oppoſite to the Iron, they draw it out with a Pair of Iron Pincers: After which, they cut off the ſame End with a little Water, and carrying the Cylinder back to the *Bocca*, they likewiſe incide or cut it with Water in two other Places, one, eight or ten Inches from the Iron, and the other the whole Length.

The *Glaſs Cylinder* being thus abridged of both its Extremities, is next heated on a kind of Eartrhen Table ſomewhat raiſed in the Middle, in order to promote its Opening at the Place incided or cut longitudinally or lengthways.

The Workman here makes uſe of an Iron, with which he alternately lowers and raiſes the two Sides or Halves of the Cylinder, which now begin to open and unfold like a Sheet of Paper, and at length grow perfectly flat.

The Table of *Glaſs* is now in its laſt Perfection, and needs nothing farther but to be heated over again.

They take it out, and lay it on a Table of Copper, from whence, after it has cool'd, and come to its Conſiſtence, they carry it on Forks, to the Tower of the Furnace, where it is left for twenty-four Hours to anneal.

The Number of Tables annealed at a Time, which ſometimes amount to a hundred, with the perpendicular Situation they are ſet in, occaſion'd, antiently, that thoſe which were firſt ſet in, ſuſtaining in ſome meaſure the Preſſure of all the laſt, were bent;

bent; and thus rendred inconvenient for Use.

But this Inconvenience has been since remedied, by separating them into Tens with an Iron Shiver; which diminishing the Weight by dividing it, keeps the Tables as flat and even as they were put in.

Glazier's Work, or Glazing.

Glazing is a manual Art, whereby the Pieces of Glass (by the Means of Lead) are so fitted and compacted together by *straight* or *curved* Lines, that it serves as well for the intended Use, (in a manner,) as if it were one entire Piece; nay, in some Respects, far better and cheaper, *viz.* in case of breaking, *&c.*

These two Heads of *Straight* or *Curved*, will admit of several Subdivisions. And,

First, of *Straight*; which contain a square Work, whose Angles are Right ones, as almost all Window-Lights are in Timber Window-Frames; and so likewise are the Squares, if *glaz'd* with such of which the Lights are composed.

Secondly, *Miter*, or such as make an Angle of 45 Degrees; this but seldom happens in this Profession, unless it be in some Piece of Fret-Work.

Thirdly, *Bevel*: This is the most common, especially in the Country; and ordinary Houses, for most such are *glaz'd* with Quarries, which is Bevel-Work; so likewise is a great deal of Fret, and all Snip Work.

Curved Work consists either of Circles, Ovals, or some distorted Arches.

Circles and Ovals are commonly used for Lights at some particular Place in a Building, as in a Pediment over a Door, or the like, in the Middle of a Front, *&c.*

Of Glazier's Draughts.

The most ingenious *Glaziers*, both in the City and Country, work by Design, (and not by Guess,) they making a Draught of all their Windows on Paper, in which they set down the Dimensions of each Light, both of Height and Breadth; and the Number of Squares, both in Breadth and Height in each Light, and also the Number of Lights in each Window, after the following Manner.

1			2		
3 6 0	3 6 0	3 6 0	4 0 0	4 0 0	4 0 0
6/4	C	6/4	7/4	C	7/4
2 1 0	2 1 0	2 1 0	2 1 0	2 1 0	2 1 0

	3		4	5		6
4 50	4 50	4 50	4 50	4 25	3 75	3 75
$\frac{8}{3}$	C	$\frac{8}{3}$	C $\frac{8}{3}$	$\frac{7}{3}$	C	$\frac{6}{3}$
1 50	1 50	1 50	1 50	12 5	1 75	1 25

N.B. Here are six distinct Windows, *viz.* the two upper ones are three-light Windows; and of the four lower ones, there is one of three Lights, two single Lights, and one double one.

N.B. A Number standing at the Top (of the oblong Figure in the Scheme above) is the Height of the Light; that at the Bottom the Breadth, and that Number in the Middle the upper one for the Number of Squares in Height, and the lower one for the Number in Breadth.

N.B. also, That the first and second Windows (which are three-light Windows,) have their Dimensions set down in Feet, and duodecimal Parts of Feet; *e.g.* In the first Window you have this Number 3 6 0 at the Top, which signifies the Height of the Light to be 3 Feet and 6 duodecimal Parts of a Foot; in the Middle there is $\frac{6}{4}$, which signifies there is 6 Squares in Height and 4 in Breadth, (equal to 24 in the whole Light; and below there stands 2 1 0, which signifies 2 Feet, and one duodecimal Part of a Foor. In the second or middle Light, there is a C, set to shew that there must be a Casement in that Light; and consequently that the upper Squares and lower ones must be cut somewhat shorter, (because of the Frame of the Casement,) and the Side Squares must be cut something narrower, and the four Corner-ones both shorter and narrower.

Now by such a Draught a *London Glazier*, when his Country Customers send to him for such a certain Parcel of Glass, he knows immediately how to cut it to fit his Work, and the *Country Glazier* knows how to work up his Glass by it; so that it shall fit each Window, though he be fifty Miles distant from it, as well as if he were by it.

The *London* Glass-Cutters commonly mark (with a Letter or Figure over them) all the Windows that are of one Size, and write the same Mark on a Piece of Paper, which is put in among that Parcel of Squares which belong to those Lights that are all of one Size. This Piece of Paper is so put in, that the Character is visible above the Edges of the Squares; by which distinguishing Character, the *Country Glazier* readily knows which Squares to take for any Window.

I shall

I ſhall add, as to this Article of *Draughts*, that thoſe *Glaziers* who underſtand decimal Arithmetick, ſet down their Dimenſions in Decimals, which better ſuits the *London Glaſs-Cutters*, becauſe they have their Rules centeſimally divided for that Purpoſe.

For that Reaſon I have here ſet down the Dimenſions of the four lower Windows in Feet, and centeſimal Parts: As for Example; in the third Window, at the Top, you have theſe Numbets 4 50, which ſignify that the Height of the Light is 4 Feet, and 50 centeſimal Parts; and at the Bottom there are theſe Numhers 1 50, which ſignify 1 Foot 50 centeſimal Parts; and ſo of the reſt.

Of Meaſuring Glazier's Work.

I ſhall *firſt* conſider the Cuſtoms uſed among them, (for Cuſtom is to be the greateſt Guide in all manner of Meaſures;) and *ſecondly*, the taking the Dimenſions, and computing the Quantity.

Note 1. That in *Glazing*, when Windows have a ſemicircular Top, (or any other curved Forms,) the Cuſtom is to take the full Height as if they were ſquare.

2. That all Windows conſiſting of intire Circles, or Ovals, or any other curved Form, the Dimenſions are taken the two longeſt Ways at Right Angles one to another, and from theſe Dimenſions the Areas are found as if they were ſquare.

3. That all Crochet Windows in Stone-Work, are all meaſured by their full Dimenſions in Height and Breadth, as if they were ſquare, and not curved.

4. That there is very good Reaſon for all theſe Cuſtoms, if we conſider,

Firſt, The Trouble in taking Dimenſions to make them by.

Secondly, The Waſte of Glaſs in working it to theſe Forms. And,

Thirdly, The extraordinary Time expended in ſetting it up, more than in that of ſquare Lights.

Of the taking Dimenſions.

Glaziers generally take them to Parts of Inches, and compute to the Nicety of a Fraction of an Inch, which may be done ſeveral Ways; four of which are practiſed by ſome Surveyors and Workmen; which are, *firſt*, by Vulgar Fractions; *ſecondly*, by Croſs Multiplication of Feet, Inches, and Parts; *thirdly*, by Duodecimals; and, *fourthly*, by Decimals.

But becauſe *Glaziers* uſually take Dimenſions to the Parts of an Inch, the beſt and readieſt Way to compute the Areas, is to take the Dimenſions with a Sliding Rule, ſuch as is generally uſed by *Glaziers*, which Rule is divided centeſimally; the Dimenſions being thus taken, and ſet down, are multiplied one into the other, as eaſily in Vulgar Arithmetick, as whole Numbers are.

As for the Manner of computing the Quantity, see CROSS MULTIPLICATION.

Of the Price of divers Sorts of Glaziers Work.

1. *Glazing with Squares:* For the Price of *French*, *German*, *Dutch*, and *English* Crown Glass, wrought in Lead, and set up, see before.

As to the Price of Square-Work, the Master finding Glass, and the *Glazier* Lead, Solder, and Workmanship, it is valued at about two Pence half-penny *per* Foot; but the *Country Glaziers* will be paid three Pence a Casement for pinning of them, (which is the putting of Leaden Pins through the Iron Frame, and soldering them, to fix the Glass to the Frame,) *viz.* Casements of four Foot and a half long, and so in Proportion, if they find Lead and Solder for it.

But for working up Squares, and setting up, and finding nothing but Workmanship, it is worth about one Penny, or three Half-pence *per* Foot.

2. *Of Glazing with Quarries*; which is for the most Part done with *Newcastle* Glass. See for the Price of new Work and Materials, what is said before in the Article *Newcastle Glass*.

But if the *Glazier* find only Lead, Solder, and Workmanship, it is worth about three Pence *per* Foot: But if they find nothing but Work, then three Half-pence or two Pence is a sufficient Price.

For taking down Quarry Glass, scowering and soldering it anew, and banding and setting up again, the usual Price is three Half-pence *per* Foot.

But in Churches, where they say they have usually more for Banding, &c. the Price is two Pence *per* Foot; and so likewise for taking down, scowering, soldering, banding, and setting up again of old-fashion'd Work, composed of Pieces of Glass of different Sizes and Figures, the Price is two Pence *per* Foot.

Mr. *Leybourn* tells us, that in *London* they generally use that Size of Quarries called 12 *s.* which he describes as follows: Quarries are for the most Part six Inches in Length, from one Acute Angle to the other; and in Breadth, from Obtuse Angle to Obtuse Angle, four Inches; so that each Quarry contains twelve superficial Inches; which Sort is that they call *Long Quarries*. See QUARRIES.

N. B. There are several Appellations given to the various Dimensions, &c. of Quarries, *viz.*

1. The Range, which is a Perpendicular let fall from one of the Obtuse Angles to the opposite Side.

2. And the Length is the longest Diagonal, from one Acute Angle to the other.

3. The Breadth is the shortest Diagonal, which is drawn between the two Obtuse Angles; as for the Sides, and Area of the Quarry, that is very well known to all.

You will find in the Word Quarries, that there have been, or still are twelve Sorts of Quarries, from whence arise divers Propositions

Propositions of great Use to *Glaziers*. As,

1. To find any of the five fore-cited Dimensions, as Range, Side, Length, Breadth, and Area of any of the Sort of Quarries.

2. To find the Area of any Sort of Quarries.

3. Having any of the Dimensions given, *viz.* Range, Side, Breadth, or Length, to find the Name or Denomination of the Size, *viz.* whether 8*s.* 10*s.* 12*s.* &c.

4. Having the Area of a Quarry given, to find of what Sort or Size it is.

5. To find whether a Window be *glazed* with those they call square Quarries, or long ones; for it is to be observed, that there are six Sorts of Sizes of square Quarries, and six Sizes of long Quarries; which make 12 Sorts in all.

Glazier's Work is measured by the Foot square; so that the Length and Breadth of a Pane of Glass in Feet, being multiply'd into each other, produceth the Content.

It is to be noted, That *Glaziers* usually take their Dimensions to a quarter of an Inch; and in multiplying Feet, Inches, and Parts, the Inch is divided into 12 Parts, as the Foot is, and each Part is divided into 12, &c.

Example I. If a Pane of Glass be four Feet eight Inches, and three Quarters long; and one Foot, four Inches, and one Quarter broad, how many Feet of Glass does it contain?

The Decimal of { 8 Inches $\frac{3}{4}$ / 4 Inches $\frac{1}{4}$ } is { .729 / 354 }

F.	I.	P.		
4	8	9		
1	4	3		
4	8	9		
1	6	11	0	
	1 :	2 :	2 :	3

```
  4.729
  1354
-------
  18916
 23645
14187
4729
-------
6.403066
```

Facit 6 Feet 4 Inches.

By Scale and Compasses.

Extend the Compasses from 1 to 1.354, and that Extent will reach from 4.729, to 6.4 Feet, the Content.

Example II. If there be eight Panes of Glaſs, each four Feet ſeven Inches three Quarters long, and one Foot five Inches and one Quarter broad, how many Feet of Glaſs are contain'd in the ſaid eight Panes?

The Decimal of $\left\{\begin{matrix} 7 \text{ Inches } \frac{3}{4} \\ 5 \text{ Inches } \frac{1}{4} \end{matrix}\right\}$ is $\left\{\begin{matrix} .646 \\ .437 \end{matrix}\right.$

F.	I.	P.		
4	7	9		
1	5	3		
4	7	9		
1	11	2	9	
	1	1	11	3
6 :	8 :	1 :	8 :	3
				8
53 :	5 :	1 :	6 :	0

```
    .464
    .437
   -----
   33522
  13938
 18584
 4646
--------
6.676302
       8
--------
53.410416
```

Facit 53 Feet 5 Inches.

By Scale and Compaſſes.

Extend the Compaſſes from 1 to 1.437, and that Extent will reach from 4646 to 6.676; then extend the Compaſſes from 1 to 8, and that Extent will reach from 6.676 to 53.4, the Content.

Example III. If there are ſixteen Panes of Glaſs, each four Feet five Inches and a half long, and one Foot four Inches and three Quarters broad, how many Feet of Glaſs is contained in them?

F.	I.	P.		
4 :	5 :	6		
1 :	4 :	9		
4 :	5 :	6		
1 :	5 :	10 :	0	
	3 :	4 :	1 :	6
6 :	2 :	8 :	1 :	6
				4
24 :	10 :	8 :	6 :	0
				4
99 :	6 :	10 :	0 :	0

```
   4.458
   1.395
  ------
   22290
  40122
 13374
 4458
--------
6.218910
       4
--------
24.87564
       4
--------
99.50256
```

Facit 99 Feet 6 Inches.

It

It may be obſerved, that, inſtead of multiplying by 16, I have multiplied by 4 twice, becauſe four Times 4 is 16

By Scale and Compaſſes.

Extend the Compaſſes from 1 to 1.395, and that Extent will reach from 4.458 to 6 219, then extend the Compaſſes from 1 to 16, and that Extent will reach from 6.219 to 99 Feet, the Content.

It muſt be obſerved, that when Windows have Half Rounds at the Top, they are to be meaſured at their full Height, as if they were ſquare. In like manner Round or Oval Windows are meaſured at the full Length and Breadth of their Diameters.

So alſo are Crocket Windows in Stone-Work meaſured by their full Squares.

The Reaſon is, that the Trouble of taking their Dimenſions to work by, the Waſte of Glaſs in working, and the Time ſpent in ſetting them up, is far more than the Value of the Glaſs.

GLUE. To make the beſt *Glue* for *gluing* the Joints of Deal Boards.

Set a Quart of Water on the Fire; then put in about half a Pound of good *Glue*, and boil them gently together over a ſoft Fire, till the *Glue* be entirely diſſolved, and of a due Conſiſtence; for if it be too thin, the Wood will ſo drink it up, that there will not remain a Body ſufficient to bind the Parts together: On the contrary, if it be too thick, it will not give way for the Joint to ſhut cloſe enough to be ſtrongly joined; for though it is *Glue* that makes the Joints ſtick, yet where there is ſo much of it, that the Joint cannot cloſe exactly, it will never hold firm.

When *Glue* is uſed, it muſt be made thoroughly hot; for *Glue* never takes firm hold of the Wood, when it is not thoroughly hot.

And ſee that the Joints to be *glued* have not been touch'd with Oil or Greaſe; for if ſo, the *Glue* will never take faſt hold.

The Joints of the Boards being ſhot true, and the *Glue* hot, ſet both the Faces of the Joint cloſe together, and both turn'd upwards; then dip a Bruſh in the *Glue*, and beſmear the Faces of the Joints as quick as poſſible, and clap the two Faces of the Joint together, and ſlide or rub them long Ways one upon another two or three Times, to ſettle them cloſe, and ſo let them ſtand till they are dry and firm.

GOLDEN RULE, in Arithmetick, a Rule or Praxis of great Uſe and Extent in the Art of Numbers, whereby we find a fourth Proportional to three Quantities given. It is alſo called the *Rule of Three*, and *Rule of Proportion.*

GORGE, in Architecture, is a ſort of Concave Moulding, wider, but not ſo deep as a Scotia, chiefly uſed in Frames, Chambranles, &c.

Gorge of a Chimney, is the Part between the Chambranle and the Crowning of the Mantle. Of this there are divers Forms, ſtraight, perpendicular, in Form of a Bell, &c.

Gorge is alſo ſometimes uſed for a Moulding, which is concave on the upper Part, and convex at Bottom, more properly called *Gula* and *Cymatium*.

Gorge is alſo uſed for the Neck of a Column, which is more properly called *Collarin* and *Gorgerine*.

GOTHIC *Architecture* is that which deviates from the Proportions, Characters, *&c.* of the Antique.

The *Gothic Architecture* is frequently very ſolid, heavy, and maſſive; and ſometimes, on the contrary, exceedingly light, delicate, and rich The Abundance of little, whimſical, wild, and chimerical Ornaments, are its moſt uſual Characters. The Profiles of this are generally very incorrect.

Authors diſtinguiſh *Gothick* Architecture into two Kinds, *viz. Antient* and *Modern*.

The *Antient* is that which the *Goths* brought with them out of the North into *Germany* in the Fifth Century. The Edifices built in this Manner were exceeding maſſive, heavy, and coarſe; from *Germany* it has been introduced into other Countries.

Thoſe of the *Modern Gothic* run into the other Extream, being light, delicate, and rich to Exceſs; witneſs *Weſtminſter-Abbey*, the Cathedral at *Litchfield*, the Croſs at *Coventry*, &c.

The laſt Kind continued long in uſe, eſpecially in *Italy*, *viz.* from the Thirteenth Century, to the Reſtoration of the Antique Building in the Sixteenth. All the antient Cathedrals are of this Kind.

It is not to be doubted, but that the Inventors of the *Gothic* Architecture thought they had far ſurpaſſed the *Greek* Architects. A *Greek* Building has not one Ornament, but what adds a Beauty to the Whole.

The Parts neceſſary to ſuſtain or ſhelter it, as the Columns, Cornices, *&c.* derive all their Beauty from their Proportions: Every Thing is ſimple, meaſur'd, and reſtrain'd to the Uſe it is intended for.

No daring out-of-the-way Strokes, nothing quaint to impoſe on the Eye. The Proportions are ſo juſt, that nothing appears very grand of itſelf, although the Whole is grand.

On the contrary, in the *Gothic* Architecture, we ſee hugh Vaults raiſed on ſlender Pillars, which one would expect every Minute to tumble down, though they will ſtand for many Ages. Every Thing is cramm'd with Windows, Roſes, Croſſes, Figures, *&c.*

Gothic Column is any round Pillar in a *Gothic* Building, either too thick, or too ſmall for its Height.

There are ſome of them found twenty Diameters in Height, without either Diminution or Swelling.

To draw a Gothic Arch *by the Interſection of Right Lines.*

Firſt, Draw the Baſe Line *a b*, and divide it in the Middle at *f*, and ſet up the Height of the Arch from *f* to *e*; then draw the Lines *a c* and *b d* perpendicular to *a b*, and equal to half the Height of the Arch *f e*, and draw the Lines *c e* and *e d*: Then divide *a c* and *b d bd* into any Number of equal Parts; alſo *ce*, *de* into the ſame Number of equal Parts, and draw Right Lines to every correſpondent Diviſion, as from 1 to 1, from 2 to 2, and ſo on, and the Interſection of thoſe Lines create the Arch *aeb*, which was to be done. See Plate, Fig. 1.

How to draw the Gothic Arch Reverſe *by the Interſection of Right Lines.*

Firſt, Draw the Baſe Line *ab*, divide it in the Middle at *e*; then ſet up twice the Height of that you deſign the Arch ſhall riſe from *e* to *c*, and draw the Lines *ac* and *cb*, and divide each of them into any Number of equal Parts, and draw Right-Lines from Diviſion to Diviſion; then will thoſe Lines create the Arch *adb*, which was requir'd. See Plate, Fig. 2.

To draw the Gothic Arch Reverſe *another Way.*

Draw the Baſe Line *ab*, then draw *cd* parallel to *ab*, and diſtant ſo much as the Arch is deſign'd to riſe; the Line *cd* is equal to half the Line *ab*, the Middle of which is at *e*: Then draw the Lines *ac* and *bd*; then divide *ac* and *bd* into any Number of equal Parts, and alſo *ce* and *ed*, and draw the Lines as taught before; which will make the Arch *aeb*. See Plate, Fig. 3.

N. B. The Line *cd* may be either longer, or ſhorter, as you have a Mind to ſhape the Arch; and this Arch, and the preceding Arch, are convenient for the gathering of large Chimneys.

To draw the Gothic Arch *ramping.*

Firſt draw the level Line *ga*, and divide it in the Middle at *f*; then at Pleaſure erect a Perpendicular at *g* towards *d*; alſo from *f* towards *e*, and from *a* towards *c*; then draw the Ramp Line *ab*, and ſet up the Height of the Arch from *a* to *e*, alſo the Lines *ac* and *bd*, and draw the Lines *ce* and *de*; then divide the Lines *ac* and *ce* into any Number of equal Parts; alſo the Lines *be* and *de*, and draw the Right Lines, as before taught, which will deſcribe the Arch *aeb*; which was the Thing to be done. See Plate, Fig. 4.

GOUGE, an Inſtrument uſed by divers Artificers, being a Sort of round hollow Chiſſel, uſed in cutting Holes, Channels, Grooves, &c. in Wood, Stone, &c.

GRADATION, in Architecture, ſignifies a Place by which we go up by Steps, particularly an Aſcent from the Cloiſter to the Choir in ſome Churches. Alſo an artful Diſpoſition of ſeveral Parts, as it were, by Steps or Degrees after the Manner of an Amphitheatre; ſo that thoſe which are placed before do no Diſſervice, but are rather ſervice-

Gradation, in Painting, is uſed to ſignify an inſenſible Change of Colour, by the Diminution of the Teints and Shades.

GRANARY, a Place for laying up or ſtoring Corn in, particularly

ticularly, for keeping a confiderable Time.

Sir *Henry Wooton* advises to make it look towards the North, as much as may be, because that Quarter is the coolest and most temperate.

Mr. *Worlidge* observes, that the best *Granaries* are built of Brick, with Quarters of Timber wrought in the Inside, to which the Boards may be nailed; with which the Inside of the *Granary* must be lin'd so close to the Bricks, that there may not be any Room left for Vermine to shelter themselves. There may be many Stories one above another, which should be near the one to the other; because the shallower the Corn lies, it is the better and more easily turn'd.

Some have had two *Granaries* one above the other, and have filled the upper with Wheat, or other Corn

The upper one having a small Hole in the Floor, by which the Corn fell down into the lower one, like the Sand in an Hour-Glass, which, when it was all come down into the lower *Granary*, it was then carried up again into the upper one; and by this Means was kept continually in Motion, which is a good Preservative for the Corn.

A large *Granary* full of square Wooden Pipes may keep Corn from heating.

GRANGE, an antient Term for a Barn, wherein to lay up and thrash Corn. The Word is sometimes also used in a more extensive Sense for the whole Farm, with all the Appendages, as Barns, Stables, Stalls, and other necessary Places for Husbandry.

GRATICULATION, a Term used by some for the dividing a Draught or Design into Squares, in order to the reducing it thereby.

GRAVITATION is the Exercise of Gravity, or it is the Pressure that a Body, by the Force of its Gravity, exerts on another Body under it.

It is one of the Laws of Nature discovered by Sir *Isaac Newton*, and now received by most Philosophers, that every Particle of Matter in Nature *gravitates* towards every other Particle; which Law is the Hinge whereon the whole *Newtonian Philosophy* turns.

All Bodies are mutually heavy, or *gravitate* mutually toward each other; and this Gravity is proportional to the Quantity of Matter; and at unequal Distances, it is inversly as the Square of the Distance.

What is called by us *Gravitation*, with respect to the *gravitating* Body, is called *Attraction* with respect to the Body *gravitated*.

GRAVITY, in Mechanicks, is the Conatus or Tendency, or that Force by which Bodies are carry'd or tend towards the Centre of the Earth.

That Part of Mechanicks which considers the Motion of Bodies arising from *Gravity*, is peculiarly called *Staticks*. See STATICKS.

Gravity is distinguish'd into *Absolute* and *Relative*.

Absolute Gravity, is that wherewith a Body descends freely through another resisting Medium; or it is the whole Force, by

Plate VIII.

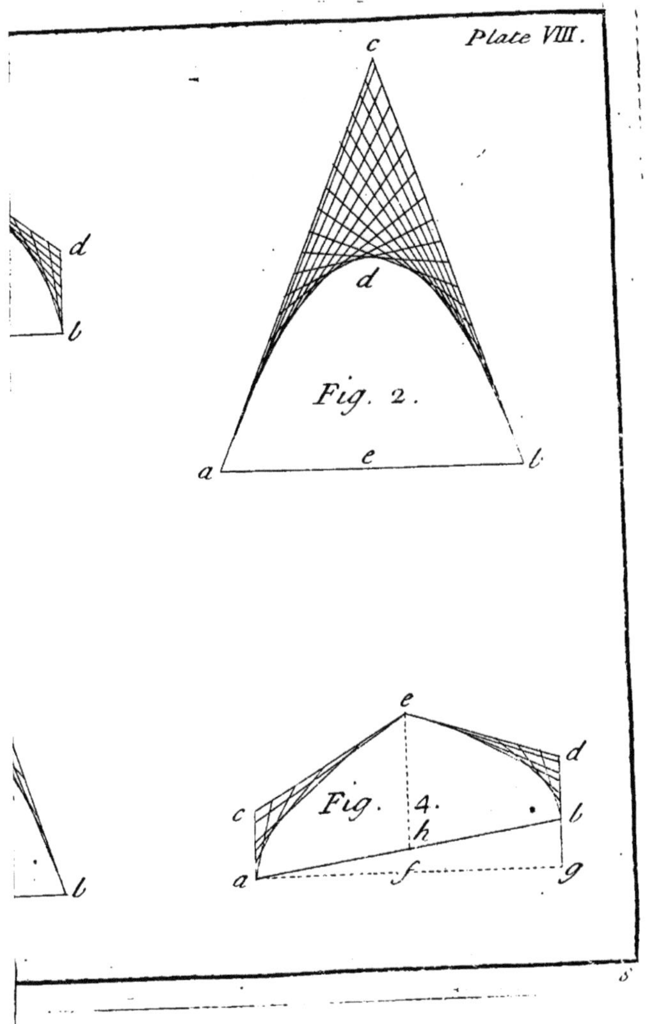

Fig. 2.

Fig. 4.

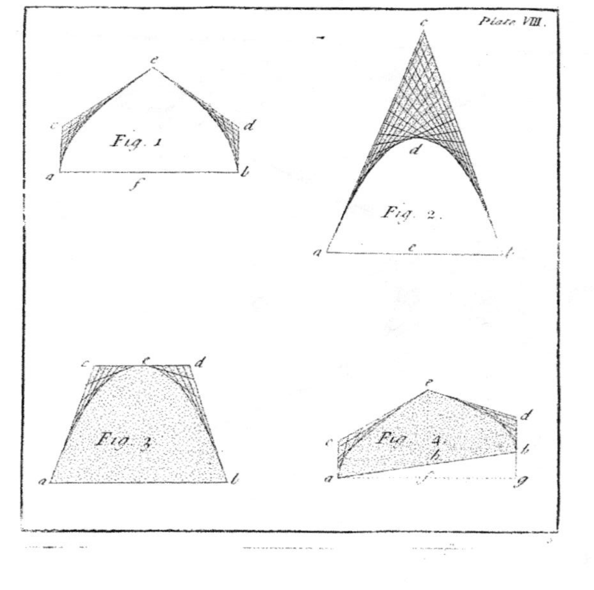
Plate VIII.
Fig. 1
Fig. 2.
Fig. 3
Fig. 4

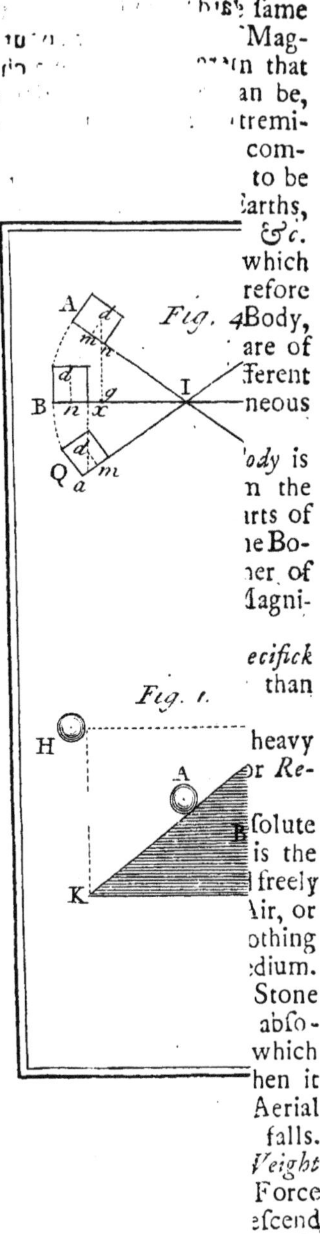

g elſe,
e Me-
an in-
cliu'd

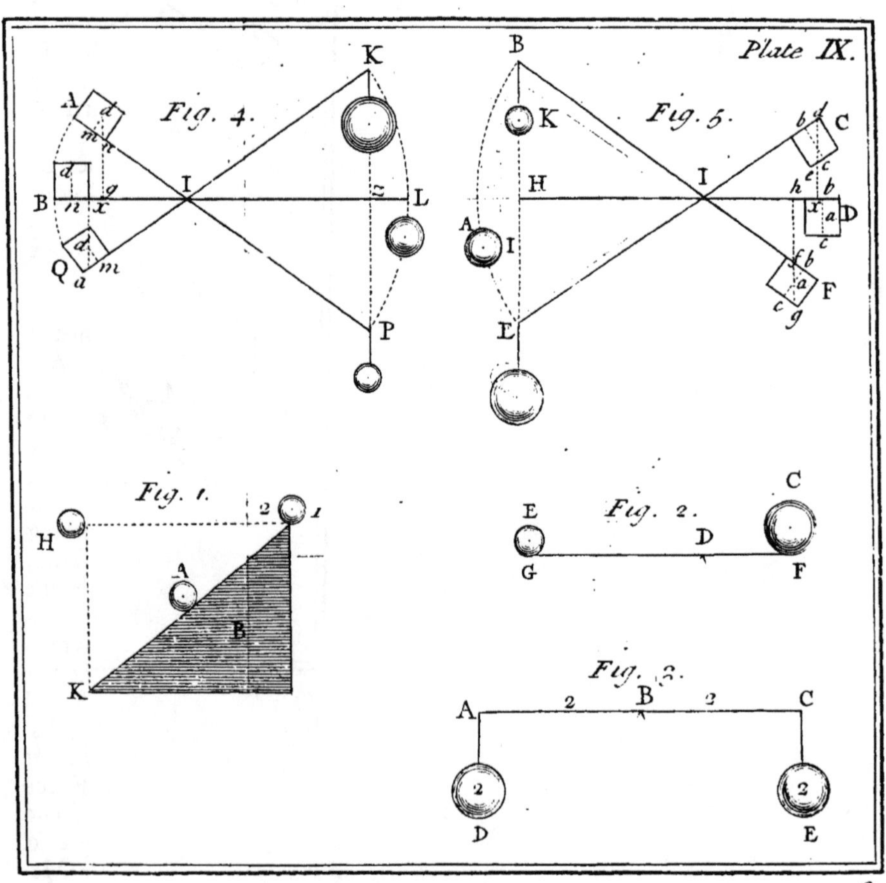
Plate IX.
Fig. 4.
Fig. 5.
Fig. 1.
Fig. 2.
Fig. 3.

by which any Body tends towards the Centre of the Earth.

Relative Gravity is that wherewith a Body defcends after having fpent part of its Weight in overcoming fome Refiftance.

Such is that wherewith a Body defcends along an inclin'd Plane, where fome Part is employed in overcoming the Refiftance or Friction of the Plane.

Accelerate Gravity is the Force of *Gravity*, confidered as growing greater, the nearer it is to the attracting Body or Point.

Gravity or *Weight*, is the Heavinefs of Matter, and is the natural Inclination which is in heavy Bodies to move downwards when they are not fuftain'd or held up, and fall towards the Centre of the Earth.

And for this Reafon, the Centre of the Earth is called, *The Centre of Heavy Bodies.*

The Centre of *Gravity* of a heavy Body, is a Point by which a Body being fufpended, all its Parts which are about that Point, will balance one another, and oppofitely hinder one another from falling, whereby the Body will remain in any given Pofition.

Whence it is plain, that a liquid Body cannot of itfelf have any Centre of *Gravity*, becaufe its Parts are not fixed to one another, but are in a continual Motion, as Water, Wine, Beer, &c.

The Centres of *Gravity* and *Magnitude* cannot be in the fame Point, but in a Body which is Homogeneous.

A Homogeneous Body is one whofe Matter is uniform, and every where of the fame Weight about its Centre of Magnitude, which is a Point in that Body as far diftant as can be, and equal from all its Extremities, which the Matter that compofes our Earth is known to be otherwife: Some being of Earths, Metals, Minerals, Water, &c. the fpecifick *Gravities* of which are very different; and therefore the Earth, or any other Body, whofe Parts or Matters are of different Weights in different Parts, are called Heterogeneous Bodies.

Specifick Gravity of a Body is that which proceeds from the natural Denfity of the Parts of its Matter, which makes one Body weigh more than another of the fame Dimenfions or Magnitude.

As for Example: The *Specifick Gravity* of Gold is greater than that of Lead, &c.

The *Specifick Gravity* of heavy Bodies is either *Abfolute* or *Relative.*

Firft, Abfolute. The abfolute Weight of a heavy Body is the Force which it has to defcend freely into a fluid Medium, as in Air, or Water, when it touches nothing elfe but the Parts of that Medium. Thus the Weight of a Stone which in the Air is called abfolute, from its own Force which it has to defcend freely, when it touches nothing but the Aerial Particles through which it falls.

Secondly, The *Relative Weight* of a heavy Body, is the Force which fuch a Body has to defcend when it touches fomething elfe, more than the Parts of the Medium, as when it bears on an inclin'd

Plane, as A on B, or on the End of a Lever, as the Body E on the Lever F G, where it often happens that the Body in Question becomes a Counterpoise to a greater and heavier Body, as the Body C, as it is nearer or farther from the Centre of Motion D, on which the Lever moves. See Plate, Fig. 1 and 2.

This Counterpoise of Bodies is called *Equilibrium*.

Now 'tis plain, that the Body H, (in the first Figure,) which is suppos'd equal to the Body A, in falling to K, must fall with greater Force than the Body A; because that the Body H has no other Resistance than that of the Air: But the Body A has the Resistance of the inclin'd Plane K I, and Air also.

Therefore it is evident, that the *Absolute Weight* of any Body is greater than the *Relative Weight*.

Note, That the Centre of Motion in a Lever, is that Point whereon it rests, and moves; as D, in a Balance, Fig. 2. is that Point which it hangs by, as the Point B of the Balance A C, Fig. 3 and in a heavy Body, it is that Point by which a Body is held, and about which it may be mov'd, as the Points or Centres of the Circles, square and equilateral DEF.

As to the Powers by which all Bodies are mov'd by the following Engines, and the Application to Practice, see POWER.

Centre of GRAVITY.

THEOREM I. *See Plate, Fig.* 4.

If a Power, as L, whose Line of Direction is perpendicular to to the Lever L I B, equipoise the Cube D M, whose Centre of *Gravity d* is above the Lever, its Power will be increased, as the said Cube is raised above A, and decreased, as the said Cube is let lower to Q, *&c.*

The Cause of this is very plain; for by the Body's Change of Place, the Line of Direction of both Power and Body are remov'd, whereby their Distances from the Fulcrum are not in the same Ratio as before.

First, Suppose the Cube *dm* on the Lever B L to be remov'd to Q, then 'tis plain that its natural Line of Descent or Direction *dm* will become *da*; wherefore the Bearing of the whole Weight is at *a*, which is as much further from the Fulcrum L, as the Line *am*.

Secondly, Suppose the Cube *dm* on the Lever B L to be raised as far above B to A, as it was before depressed from B to Q, then will its Line of natural Direction or Descent *dm* become *dn*.

Thirdly, Since the Cube at A is at the same Distance from B as Q is from B, therefore the other Ends K and P will also be equidistant from L; and therefore if the Right Line K P be drawn,

drawn, it will cut the Lever BL at Right Angles in Z.

Wherefore the Diſtance IZ is = the Diſtance of the Powers in both the Levers KQ and AP.

But ſince that by continuing the Lines of Direction da and dn up to the Lever BL, they meet at unequal Diſtances from the Fulcrum in x and g; therefore the Cube at A, whoſe Diſtance g I is the leaſt from the Fulcrum, I requires leſs Weight than the Cube at Q, whoſe Diſtance x I is greater.

Fourthly, Now ſeeing that a leſſer Power is required at P than at K, 'tis evident, that the ſame Power will equipoiſe a greater Weight; and therefore the Power L, in raiſing the Cube to A, is increaſed; and, on the the contrary, as aforeſaid, the Power L, in letting down the Cube to Q, is decreaſed Q E D.

This is the firſt Variety of the Effects of raiſing Bodies.

I ſhall next proceed to the ſecond, wherein, by having the Centre of *Gravity* placed below the Lever, the Power is decreas'd when the Body is raiſed out of a horizontal Poſition, and increaſed when let fall below the horizontal Poſition.

This being the contrary to the preceding, may, perhaps, at firſt Sight, appear impoſſible. But, obſerve. See Plate, Fig. 5.

Firſt, That all Things being equal, as before, the Diſtance of the Powers in the Levers BF and EC are equal to HI; but the Diſtances of the Weights F and C are unequal; that of the Lever BF being at b, and that of the Lever EC being at x.

Now ſeeing that the Diſtance of the raiſed Weight, which is the Point x, is farther from the Fulcrum I, than the Diſtance of the let fallen Body F, which is the Point b, and both the Diſtances of Power equal to one another, 'tis plain, that the Power at K will be increaſed, and that at H decreas'd. Q E D.

Now from the preceding Rules, and thoſe Remarks on the Manner of placing Bodies on a Lever, it is impoſſible, but that at all Times you may eaſily raiſe any heavy Body with the leaſt Power, and in the leaſt Time; which is the chief Work of Mechanicks, and are very uſeful to be known by Architects; for a good Architect ought to be a good Mechanick.

GREEN-HOUSE, or *Conſervatory*; a Place built for depoſiting exotick Plants, and ſuch as will not bear the Winter's Cold in our Climate.

Theſe Sorts of Houſes, as they are commonly built, ſerve more for Ornament than Uſe: Their Situation towards the South Sun is the only Thing that ſeems to be regarded towards the Health of the Plants they are to ſhelter.

It is rare to find one among them that will keep a Plant well in the Winter, either by reaſon of their Situation in moiſt Places, their Want of Glaſſes enough in the Front, and the Diſproportion of the Room within them; and ſometimes, where it happens that a *Green-Houſe* has been conſidered in theſe Points, all is confounded by the Flues under it, which convey the Heat from the Stoves.

Beſides

Besides what is commonly called a *Green-House*, it has been customary to provide Glass Cases of several Kinds, and Stoves, for the Preservation of Plants brought from different Countries.

But an ingenious Author has found them to be so many unnecessary Expences; and that a good *Green-House*, well contrived, will do all that is required for the Welfare of any Plant in the Winter; and that it may be so ordered, as to shelter at one Time Orange-Trees, Plants from the *Cape of Good Hope*, *Virginia*, *Carolina*, and indeed such as grow within ten Degrees of the Line.

The same Author says, that when he was first acquainted with Aloes, *Indian* Figs, and such like Plants, he confesses he thought they could never have Heat enough, and that he destroy'd many by that too common Notion; he could hardly venture them out of the Hot-Beds in the most extreme Heats of Summer; and that in the Winter, they were half-roasted with subterraneous Fires he made under the Glass Cases where they stood.

A good *Green-House* ought to be situated on the driest Ground, to be as free from Damps as possible; substantial Provision ought likewise to be made for keeping out the Cold, and yet upon Occasion to let in Air freely; but chiefly to contrive that the Front of the House be so disposed, that nothing may obstruct the Passage of the Sun's Rays, in the Winter, into the House.

It is generally allowed, that the South Aspect is the best for a *Green-House*, as it will in that Exposure receive the Sun for the greatest Part of the Day; but in case that cannot be had with Conveniency, the South-West Aspect is next to be coveted; and it would be pleasant, as well as beneficial to Plants, if the Conservatory was always joined to the Dwelling-House.

Nothing can be more agreeable in Winter, than to have a View from a Parlour or Study through Ranges of Orange-Trees, and curious Plants of foreign Countries, blossoming and bearing Fruit, when the Gardens without Doors are, as it were, in a State of Death; and to walk among those Curiosities of Nature, as in the most temperate Climate, without any Sense of the Frost, or pinching Cold that reigns abroad; and besides, there is this Conveniency in joining the Conservatory to the House, that in cold Weather you may go into it, without letting in the cold Air, or blighting Winds from abroad.

Thus much as to the Situation: The next Thing to be considered is the Proportion of the Building, and that chiefly in Relation to the Height and Breadth of the Room; which a certain Author directs, that for the better Admission of the Sun's Rays to pass all over the House, the Breadth of it be no more than the Height from the Floor to the Ceiling, which may be from ten to eighteen Feet.

That the Walls towards the North and the East, be of a good Thickness, and the Front towards

wards the South be all of Glaſs, except a low Wall about a Foot high from the Ground; that there be no Piers of Brick-Work or Timber in the glaz'd Part, for they caſt more Shade into the Houſe, in Proportion to their Bigneſs, than it can receive Light through the Glaſs; whereas, every one who underſtands exotick Plants will allow, that they ſhould have all the Advantages of the Sun's Rays in Winter, that they poſſibly can receive: And for this End, he is of Opinion, that it would be proper, in the colder Parts of *England*, to build the Front of a *Green-Houſe* in a Sweep, or in the Form of a Semicircle, which would then receive the Rays of the Sun from the Time of its Riſing, till its Setting.

That the Glaſs in the Front, whether it be in Saſhes or Caſements, be ſo contrived, that it may either be made to ſlide quite below or above the Frames, or to be taken away, as Occaſion ſhall offer, to give Air to the Plants, which for about a Fortnight or three Weeks after they are ſet into the Houſe, and as long before the Plants come abroad, ſhould be quite open Night and Day, if the Froſts or blighting Winds are not abroad.

Some have practiſed, with good Succeſs, to lay the Windows of their *Green-Houſes* ſloping about ten Inches; but he is of Opinion, that they will do as well upright.

He adviſes, that the Door be in the Middle of the Front, and at leaſt four Foot wide, to admit large Plants; that it be glaz'd, to which ſtrong Shutters ſhould be added, at leaſt an Inch thick, which in the Winter Time ſhould be ſhut every Night, for fear of Froſt; and alſo in extraordinary cold Weather, when violent Winds blow right againſt the Houſe.

That for the better Security of the Plants from Cold, a Place for the laying up the Gardiner's Tools be built at the *Green-Houſe*, and over it a Fruitery or Seed-Room, or in the Lien of the latter, the Room may be filled with dry Straw.

The beſt Pavement for a *Green-Houſe*, he ſays, is that made with ſquare Tiles, which quickly ſucks up Wet, and never ſweat, as Marble, or ſuch Kinds of hard Stone uſually do; and that for lining of the Walls, nothing is preferrable to *Dutch* glaz'd Tiles, which are ſoon warmed with the Sun, and reflect a great Heat into the Houſe.

That in the Diſpoſition of the Shelves in the *Green-Houſe*, one Third of the Floor be allowed for them to ſtand upon, one Third from the firſt Shelf to the Windows, and as much from the laſt Shelf to the Back of the Houſe; ſo that a Perſon may walk round the Plants, which being placed in the middle Line of the Houſe, are ſafe from the extreme Cold, which is generally nearer the Walls or Glaſſes.

The Chimney for warming the Air, he directs to be built between the Windows and the firſt Shelf at one End of the Houſe, about a Foot above the Floor, which will riſe afterwards, and ſpread itſelf over the Whole.

But

But the ingenious Mr. *Philip Miller* has given us a more accurate Design of a *Green-House*, which he describes as follows:

As to the Length of the House, he says, that should be proportioned to the Number of Plants it is to contain, or the Fancy of the Owner; but as to the Depth, that should never be more than sixteen Feet in the Clear, and the Length of the Windows should be at least equal to the Depth of the House; and if they are something longer, it will be still the better.

These Windows should be carried up quite to the Ceiling, that there may be no Room for dead Air in the upper Part of the House, and they ought to come down within about ten Inches or a Foot of the Floor; their Breadth should be proportioned to the Length of the House, which in a small *Green-House* may be four Feet broad, but in a large one they should be six Feet.

The Piers between these Windows should be as narrow as possible they may be, to support the Building, for which Reason he chuses to have them either of Stone or solid Oak; for it they are built with fine rubbed Bricks, they are generally so soft, that the Piers will require to be made thicker than can be allow'd, otherwise the Building will be in danger of falling in a short Time, especially it any Rooms be built over the *Green-House*; which would be of great Use in keeping out the Frost in very hard Winters.

If the Piers are made with Stone, he directs that they be twenty Inches broad in Front, and sloped off backwards to about ten Inches broad; whereby the Rays of the Sun will nor be taken off, or obstructed by the Corners of the Piers; which it would be, if they were square.

And if the Piers are made of solid Oak, eighteen Inches square he accounts strong enough to support the Building; and also sloped off, after the Manner before directed as to those of Stone.

A Tool-House may also be erected at the Back of the Building, which may also serve for many other Purposes, and will also be extremely useful, by preventing Frost from entering that Way; so that the Wall between these need not be more than two Bricks in Thickness; whereas if it were quite exposed, behind it ought to be two Brick and a half, or three Bricks in Thickness.

And thus also, if you have a Mind to make a handsome Building, and to have a noble Room over the *Green-House*, you may make the Room to come over the Tool-House, and carry up the Stair-Case in the Back, so as not to be seen in the *Green-House*: And by this Means you have a Room twenty or twenty-two Feet in Width, and of a proportionable Length.

And under this Stair-Case there may be a private Door into the *Green-House*, at which the Gardiner may enter in hard frosty Weather, when it will not be safe to open any of the Glasses in the Front.

The Floor of the *Green-House* may be laid with Marble, Stone, or or broad Tiles, according as the Owner pleaſes, and muſt be raiſed two Feet above the Level of the Ground on which the Houſe is ſituate, which will be ſufficient, if the Soil be dry; but if moiſt and ſpringy, and thereby ſubject to Damps, it will be neceſſary to raiſe it at leaſt three Feet above the Surface.

He adviſes alſo, to make a Flue of about ten Inches in Width, and two Feet in Depth, under the Floor, about two Feet from the Front; which Flue is to be carried the whole Length of the Houſe, which may be returned along the back Part, and be carried up in proper Funnels adjoining to the Tool-Houſe, by which the Smoak may paſs off.

The Fire-Place may be contriv'd at one End of the Houſe, and the Door at which the Fuel is put in, as alſo the Aſh-Grate, may be contrived to open into the Tool-Houſe; ſo as to be quite hid from the Sight, and be in the dry; and the Fuel may be laid in the ſame Place, and ſo will always be at Hand for Uſe.

He alſo adviſes to have good ſtrong Shutters to the Windows in the Front of the *Green-Houſe*, hung on Hinges to fold back, ſo that they may fall back quite cloſe to the Piers, ſo as not to obſtruct the Rays of the Sun.

Theſe Shutters may be an Inch thick, or a little more, made to join ſo cloſe, as to be able to keep out our common Froſts; and when the Weather is ſo cold as to endanger the freezing in the Houſe, it is but making a Fire in the Flue, and that will effectually paevent it.

The back Part of the Houſe ſhould be plaiſtered with Mortar, and white-waſh'd; or if lined with Wainſcot, ſhould be painted white, as ſhould the Cieling, and alſo every Part withinſide the Houſe; for White reflects the Rays of Light in a much greater Quantity, than any other Colour; and is of very great Service to Plants, eſpecially in the Winter Seaſon, when the Houſe is pretty much cloſ'd, ſo that but a ſmall Share of Light is admitted through the Windows: For he ſays, he has obſerved, that at ſuch Times, where a *Green-Houſe* has been painted black, or any dark Colour, the Plants have caſt moſt of their Leaves.

He adds, that to avoid the Inconvenience which attends the placing Plants of very different Natures in the ſame Houſe, it will be very proper to have two Wings added to the main *Green-Houſe*, which will greatly add to the Beauty of the Building, and alſo collect a greater Share of Heat.

The *Green-Houſe*, according to his Plan, is placed exactly fronting the South, and one of the Wings faces the South-Eaſt, and the other the South-Weſt; ſo that from the Time of the Sun's firſt Appearance upon any Part of the Building, until it goes off at Night, it is conſtantly reflected from one Part to the other, and the cold Winds are alſo kept off from the Front of the main *Green-Houſe* hereby.

And in the Area of this Place you may ſo contrive, as to place many

many of the moſt tender exotick Plants, which will bear to be expoſed in the Summer Seaſon; and in the Spring, before the Weather will permit you to ſet out the Plants, the Beds and Borders of this Area may be full of Anemonies, Ranunculus's, early Tulips, *&c.*

In the Centre of this Area may be contrived a ſmall Baſon for Water, which will be very convenient for watering Plants, and will alſo very much add to to the Beauty of the Place; beſides, the Water being thus ſituated, will be ſoftened by the Heat which will be reflected from the Glaſſes upon it, whereby it will be render'd much better than raw cold Water for tender Plants.

The two Wings of the Building ſhould be ſo contrived, as to be fit for placing Plants of different Degrees of Hardneſs, which muſt be effected by the Situation and Extent of the Fire-Place, and the Manner of conducting the Flues. For which, ſee the Articles STOVES.

The Wing facing the South-East ſhould always be preferred for the warmeſt *Stove*, its Situation being ſuch, as that the Sun, upon his firſt Appearance in the Morning, ſhines directly upon the Glaſſes; which is of great Service in warming the Air of the Houſe, and adding Life to the Plants, after having been ſhut up during the long Nights in the Winter Seaſon.

Theſe Wings may be allow'd ſixty Feet in Length, and may be divided in the Middle by Partitions of Glaſs, with Glaſs Doors to paſs from one to the other. And the Fire-Place may be ſo ordered as to warm both Diviſions, by placing an Iron Regulator in the Flue, ſo that the Smoak may paſs through the Flues of which Part ſoever you pleaſe.

By this Contrivance you may keep ſuch Plants as require the ſame Degree of Heat in one Part of the Houſe, and thoſe which will thrive in a much leſs Warmth in the other Part; but this will be more fully explained under the Article STOVE.

The other Wing of the Houſe facing the South-Weſt, may alſo be divided in the ſame Manner, and Flues carried through both Parts, which may be uſed according to the Seaſons, or the particular Sorts of Plants which are placed therein.

So that by this Diſpoſition here will be four Diviſions in the Wings, each of which may be kept up to a different Degree of Heat; which, together with the *Green-Houſe*, will be ſufficient to entertain Plants from all the ſeveral Countries of the World.

And without having theſe ſeveral Degrees of Warmth, it will be impoſſible to preſerve the various Kinds of Plants from the ſeveral Parts of *Africa* and *America*, which are every Year introduced into the *Engliſh* Gardens.

For when Plants from very different Climates are placed in the ſame *Green-Houſe*, ſome periſh for want of Heat, while others are deſtroy'd by having too much of it; and this is often the Caſe in ſuch *Green-Houſes*, where there are large Collections of Plants.

To

To GRIND *Colours in Oil:* Let the *Grinding-Stone* be placed about the Height of your Middle, let it ſtand firm and faſt, ſo that it joggle not up and down; then take a ſmall Quantity of the Colour you intend to *grind*, (two Spoonfuls is enough,) for the leſs is *ground* at a Time, the eaſier, and finer, will the Colour be *ground*.

Lay theſe two Spoonfuls of the Colour in the Middle of the Stone, and put a little Linſeed Oil to it, (but take Care not to put too much at firſt;) then mix it together a little with the Muller, and turn the Muller five or ſix Times about; and if you find there be not Oil enough, put a little more, and *grind* it till it come to the Conſiſtence of an Ointment, or appears free from any Sort of Lumps, and ſmooth as the moſt curious Sort of Butter; for it *grinds* much better and ſooner when it is ſtiffiſh, than when it is ſo thin as to run about the Stone; and in *grinding*, you muſt often bring the Colour that has ſpread together into the Middle of the Stone with a Piece of Lanthorn Horn.

And in *Grinding* hold your Muller down as hard as you can, and alſo move it with ſuch a Slight, as to gather the Colour under it; and that no Knots or Grittineſs remains, and that it is become as fine as Butter itſelf.

When it is *ground* enough, cleanſe it off the Stone with the Horn into a Gallipot or Pan, and lay on more Colour, and proceed as before, till you have *ground* what Quantity you want.

If you *grind* a conſiderable Quantity, to be uſed not till ſome Time after, put it into Bladders, tie it up cloſe, and hang it up.

Thoſe who care not for the Trouble to *grind* the Colours, may have all Manner of Colours ready *ground* at Colour Shops.

How to order Colours for working, after they have been ground.

When you uſe Colours, you muſt add more Oil to them, but not ſo much as to make them ſo thin, that they will let the Ground be ſeen through them, or run about; and if your Colour be as ſtiff as it ought to be, once doing will be more than twice doing with thin Colour.

Painters make uſe of a common Fraud and Deceit, when they agree to do Work by the Yard at a common Price, to be coloured three Times over. In painting with ſuch thin Colours that at three Times doing over, it is not ſo ſubſtantial as one Time would be, if the Colour had a thick and ſubſtantial Body.

Three Times colouring with ſubſtantial and well-bodied Colour, will laſt ten Times as long as that which has been ſo ſlightly coloured.

The Priming Colour indeed ought to be very thin, that it may have Oil enough to penetrate into the Wood, which tends much to its Preſervation; but the ſecond muſt be thicker than the firſt.

To find the Angle of a regular GROIN, *or* MITER BRACKET *of a Cove.*

First, Draw the Line of Projection A B, also the Mitre Line D B; then will A D be at Right Angles with A B; then from A strike the Quadrant C B, and draw the Line A C, which will be at Right Angles with A B; divide A B any how, either into equal or unequal Parts, and from those Divisions raise Perpendiculars from the Line A B to touch the Arch B C: Also continue these Right Lines to cut the Miter Line B D, as you see by the dotted Lines, which will divide the Line D B into the same Number of Parts, and in Proportion with the Line A B; then from those Divisions made by the dotted Lines on the Line D B, raise Perpendiculars at Pleasure, and take the Line A C in your Compasses, and set it up the first Line, as from D to E; then take the Line 1, 1, on the Quadrant, and set it up the Line 1, 1; also from 2, 2, to 2, 2, from 3, 3, to 3, 3, and from 4, 4, to 4, 4, and so on, till the Points are laid down; into which Points you must strike Nails; then bend a thin Lath round them, and by its Edge strike the Arch for the *Groin* or *Mitre Bracket* E B.

N. B. This is work'd in the same Manner, let the given Arch B C be what it will, or let the Line B D be true Mitre, or irregular.

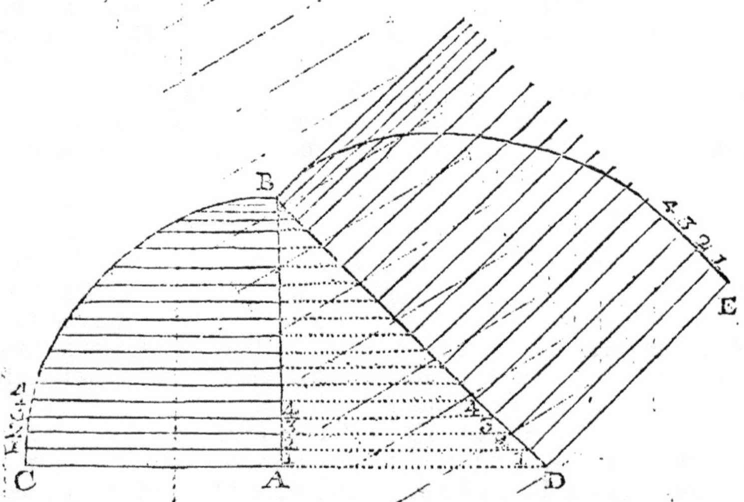

To make Centres for regular or irregular Groins, *so that the Mitres shall be true.*

This Figure represents an irregular *Groin*, because *a b* is longer than *a c*.

Let the Arch *c f d* be given, and let the Curve be what you will.

What must be the Curve of *a k c*, so that when the two different Centres are set in their Places, their Mitre or Angle shall

Plate X.

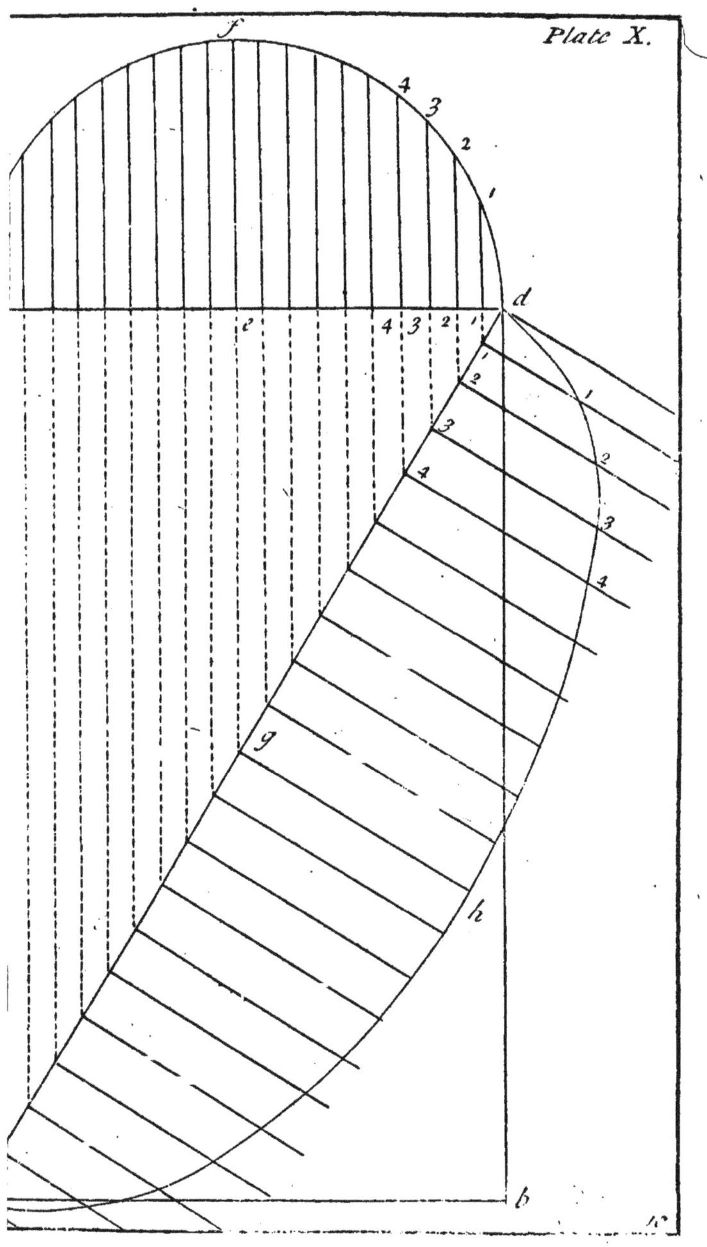

Plate X.

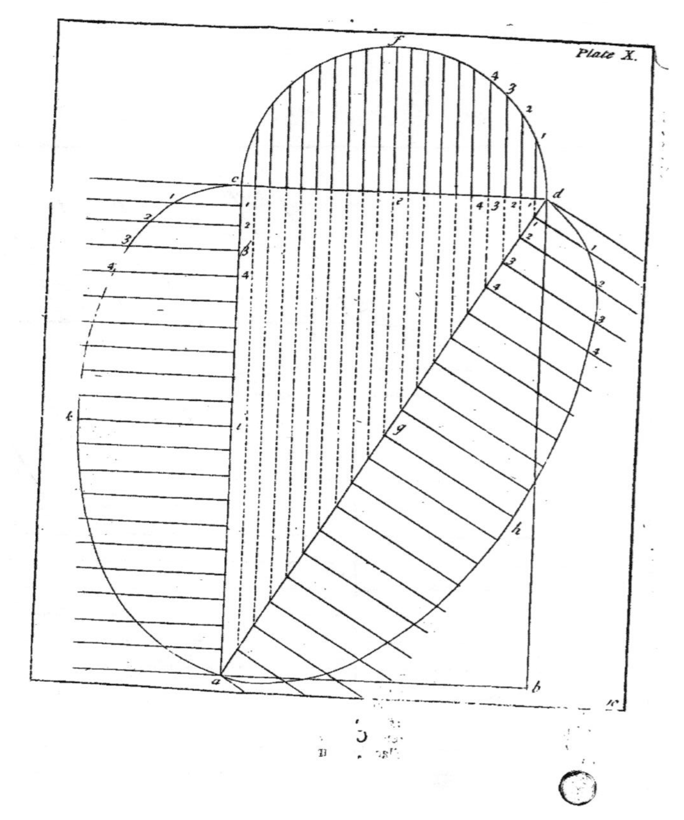

ſhall be perpendicular over the Angle Line *ad*?

Or if the *Groin* be of Wood for the Ceiling of a Room, then you muſt find the Arch *dha*, which muſt be your Hip to faſten the other Ribs to.

Firſt, Deſcribe the Figure *abcd*, and draw the Line *ad*; then ſtrike the Arch *cfd* from the Point *e*, and divide the Line *cd* any how; from which raiſe Perpendiculars to touch the Arch *cfd*, and continue thoſe Perpendiculars to the Line *ad*, as by the dotted Lines which divides the Line *ad* into the ſame Number of Parts, and in Proportion to *cd*; then from thoſe Diviſions created by the dotted Lines, raiſe Perpendiculars at Pleaſure from the Line *ad*: Having done this, lay your ſquare or parallel Ruler at Right Angles with the Line *ac*; and from the Diviſions on the Line *ad*, draw Perpendiculars from the Line *ac* of a Length at Pleaſure, which will divide the Line *ac* into the ſame Number of Parts, and in Proportion with thoſe on the Lines *cd* and *ad*; then you muſt take the Line 1, 1, on the Arch *cfd*, and ſet it on the firſt Perpendiculars on the Lines *ad* and *ac*, as from 1 to 1; and ſo that every Line marked with the ſame Figures are of an equal Length, as 1, 1, to 1, 1, = 2, 2, to 2, 2, 3, 3 = 3, 3, 4, 4 = 4, 4, alſo *gh* and *ik* to *ef*; and ſo of the other Lines, as they follow.

And when the Points 1, 2, 3, 4, and all the reſt are found by the foregoing Method, you muſt ſtrike a Nail in every one of them, and bending a thin Lath round them, draw the Arches *akc* and *ahd*, and thoſe you find will anſwer the Purpoſe deſigned. See the Plate.

How to prepare the Boarding for the Covering of the Centres of any Kind of Groins, *and to cut them to their right Lengths and Bevels, before the Centres are ſet in their Poſition.*

Let ABCD repreſent the Plan of an irregular *Groin*; draw the Curves of the two different Centres BFC and CDG, by the Rule laid down in the foregoing Figure for making of Centres of *Groins*, &c. in the preceding Figure.

Then continue the Line BC both Ways from B to I and from C to K, ſo that IK be equal in Length to the Girt of the Curve BFC; and draw IH and KH, which are each equal in Length to the Girt of the Curves at the *groining* or *mitering* of the two different Centres; and draw the Lines IO and KP perpendicular to IK.

Then will the Plane HIOPK be equal in Quantity to the Back of the Centre BFC.

To find the Bevels of the Boards, lay them all down together, juſt as many as will fit between the Lines IO and CP, letting their Ends reach over the Lines IH and HK.

Then ſtrike a Line on their Ends, as from I to H, and from H to K, which is the true Bevel of every Board that covers the Centre BFC; and for thoſe on the Centre CDG, do in the ſame Manner. As for Example:

 Produce

Produce CD, from C to L, and from D to M, so that LM is equal in Length to the Girt of the Curve CGD; and draw the Lines LH and MH; also the Lines LQ and MR perpendicular to LM.

Then will the Plane HLQRM be equal in Quantity with the Back of the Centre CGD, and consequently the covering Boards must be equal; therefore, as on the Plane HLOPK, lay the Boards on the Plane HLQRM, letting their Ends reach over the Lines LH and HM, on which strike a Line from L to H, and from M to H, which will give the true Bevels for the Boards on the Centre CDG; which was to be done. See the Figure.

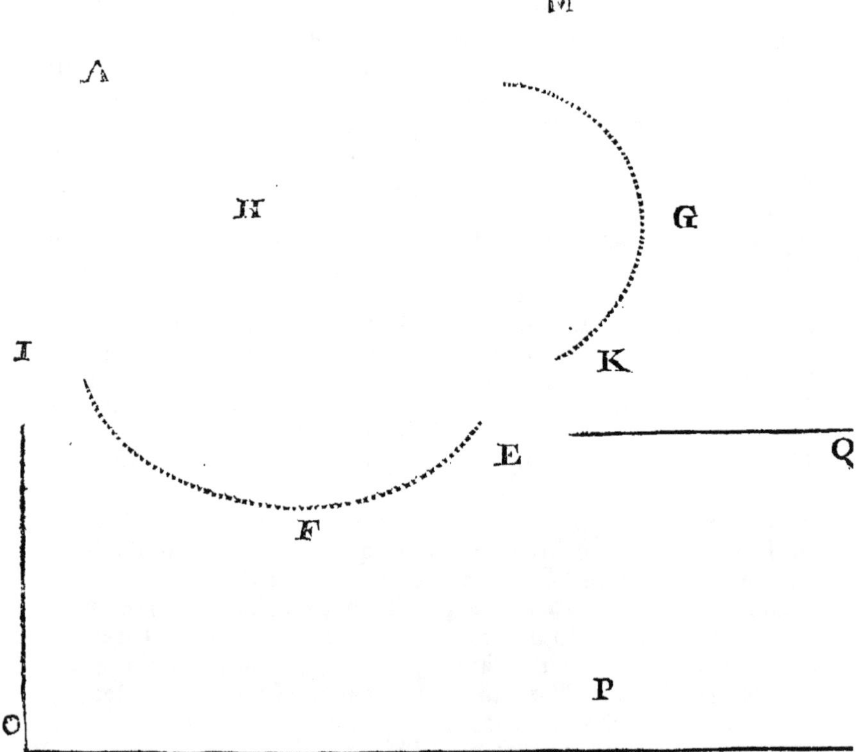

GROTESQUE, GROTESK, GROTESCO } A wild, whimsical Figure of a Painter or Carver, containing something whimsical, ridiculous, extravagant, and even monstrous in it.

The Word is also particularly apply'd to a Work or Composition in Sculpture or Painting in the *Grotesque* Manner or Taste; consisting either of Things that are merely imaginary, and have no Existence in Nature, or of Things

Things turned and diſtorted out of the Way of Nature, ſo as to raiſe Surpriſe and Ridicule.

Grotesk Work is the ſame with what is ſometimes called *Antique*. The Name is ſaid to have taken its Riſe hence, that Figures of this Kind were in antient Times much uſed in adorning the Grotto's, wherein the Tombs of eminent Perſons or Families were incloſed; ſuch as that of *Ovid*, whoſe Grotto was diſcovered near *Rome* not ſixty Years ſince.

GROTESQUES } GROTESKS } are uſed particularly to ſignify thoſe fanciful Ornaments of Animals, interſpers'd among Foliages, Fruit, &c. as thoſe painted by *Raphael Urbin* in the *Vatican*, and thoſe carved by *Michael Angelo*, in the Ceilings of the Portico of the Capitol.

Theſe Kind of Compartiments are called by *Vitruvius Harpagenituli*.

GROTTO } GROTTA } is a large, deep Cavern or Den in a Mountain or Rock.

Grotto is alſo uſed for a ſmall artificial Edifice made in a Garden, in Imitation of a natural *Grotto*.

The Outſides of theſe *Grotto's* are uſually adorned with Ruſtick Architecture, and their Inſide Shell-Work; and alſo furniſhed with various *Jet d'Eaux*, or Fountains, &c.

GROUND to build on. See Foundation.

GROUND, in Painting, is uſed to ſignify the Surface upon which the Figures and other Objects are raiſed and repreſented.

The *Ground* is properly underſtood of ſuch Parts of the Piece, as have nothing painted on them; but retain the original Colours upon which the other Colours are applied to make the Repreſentations.

A Building is ſaid to ſerve as a *Ground* to a Figure, when the Figure is painted on the Building.

GROUND GUTS. See Alder.

GROUNDSELL, or PLATE. See Sell.

GROUND-PLAT, or PLOT, a Piece of *Ground* on which a Building is to be erected.

GROVE } GROOVE } in Joinery, &c. a Term uſed to ſignify the Channel that is made by their Plough in the Edge of a Moulding, or Stile, or Rail, &c. to put their Pannels in, in Wainſcotting.

GROUP, in Painting, Sculpture, &c. is a Term uſed to expreſs the Aſſemblage or Knot of two or more Figures of Men, Beaſts, Trees, Fruits, or the like, which have ſome apparent Relation to each other.

A *Group of Columns*, in Architecture, is uſed when we ſpeak of three or four Columns joined together on the ſame Pedeſtal; but when there are but two, the Word *Couple* is uſed, and not a *Group of Columns*.

GRY, a Meaſure containing one Tenth of a Line.

A Line is one Tenth of a Digit, and a Digit one Tenth of a Foot, and a Philoſophical Foot one Third of a Pendulum, whoſe

Diadromes or Vibrations, in the Latitude of forty-five Degrees, are each equal to one Second of Time, or one Sixtieth of a Minute.

GUEULE, in Architecture. See GULA.

GULE, GULA, GOLA, in Architecture, a wavy Member, the Contour of which resembles the Letter S, which the *Greeks* call *Cymatium*, *q. d.* a little Wave; and our Architects an *Ogee*.

This Member is of two Kinds: *Recta* and *Inversa*.

The first and principal has its Cavities or Hollows above, and Convexities below. This always makes the Top of the Corona of the Cornice, jetting over the Drip of the Cornice, like a Wave ready to fall.

It is called *Gula Recta*, and by the *French*, *Doucine*.

It is sometimes called absolutely the *Entablature*, as being the first or uppermost Member of it.

The second, or *Gula Inversa*, is exactly the Reverse of the former, the Cavity or Hollowness of it being at the Bottom; so that with respect to the former, it appears inverted. This is used in the Architrave, and sometimes in the Cornice, along with the former, only separated by a Reglet.

Some derive the Term *Gula* from the Resemblance these Members bear to the *Gula*, or Throat of a Man: Others from *Gueles*, a Term in Heraldry, as supposing the Moulding form'd from the antient Manner of wearing their Garments, which consisted of Slips, or Swathes, alternately, Furr and Stuff of various Colours; the Intervals between which were *Gules*, or *Guales*.

GUNTER's LINE, is a Line of Numbers which is upon the ordinary Two-Feet, or Eighteen-Inch Rules, commonly used by Carpenters, Joiners, &c.

This Line being the Scale recommended to in those Operations in this Book, that are wrought with Scale and Compasses, I shall give some Directions for the Use of it, as follows:

If the Number you would find on the Line, consists only of Unites, then the Figures upon the Line represent the Number sought. Thus if the Number be 1, 2, 3, &c. then 1, 2, 3, &c. upon the Line, represents the Number sought. But if the Number consists of two Figures, that is, of Units and Tens, then the Figure upon the Rule stands for Tens, and the larger Divisions stand for Units: Thus, if 34 were to be found upon the Line, the Figure 3 upon the Line is 30, and 4, of the large Divisions, (counted forward,) is the Point representing 34; and if 340 were to be found, it will be at the same Point upon the Line; and if 304 were to be found, then the 3 upon the Line is 300, and 4, of the smaller Divisions, (counted forward,) is the Point representing 304. If the Number consists of four Places, or Thousands, then the Figure upon the Line stands for Thousands, and and the larger Divisions are Hundreds,

dreds, the leſſer Diviſions are Tens, and the Tenth Parts of thoſe leſſer Diviſions are Units.

Thus, if 2735 were to be found, then the 2 is 2000; and the 7 larger Diviſions (counted forward) is 700 more; and 3 of the leſſer Diviſions is 30 more; and half of one of the leſſer Diviſions is 5 more, which is the Point repreſenting 2735. You muſt remember, that between each Figure upon the Line there are 10 Parts, which are called the larger Diviſions; and each of thoſe larger Diviſions are ſubdivided (or ſuppoſed ſo to be) into 10 other Parts, which are called the ſmaller Diviſions; and each of thoſe Parts ſuppoſed to be ſubdivided again into 10 other Parts, &c. You muſt alſo remember, that if one in the Middle of the Line, ſtands only for 1, then 1 at the upper End will be 10, and 1 at the lower End will only be $\frac{1}{10}$; but if 1 at the lower End ſignifies 1, then 1 in the Middle ſtands for 10, and 1 at the upper End is 100, &c.

There is one Thing more which I would have my Reader to underſtand; and that is, how to find all ſuch proportional Numbers made uſe of in the Proportions about a Circle, and of a Cylinder, and in other Places; which Thing may be of good Uſe, to know how to correct a Number which may happen to be falſe printed, or to enlarge any Number to more decimal Places, for more Exactneſs; for though it is mentioned what ſuch Numbers are, yet it has not been ſhewn how to find them; which a Learner may be a little at a Nonplus to do; though they are eaſily found by the Rules there laid down. I ſhall therefore give two or three Examples, in this Place, of finding ſuch Numbers, which may enable my Reader to find out the reſt.

And, firſt, let it be required to find the Area of a Circle, whoſe Diameter is an Unit.

By the Proportion of *Van Culen*, if the Diameter be 1, the Circumference will be 3.1415926, &c. whereof 3.1416 is ſufficient in moſt Caſes. Then the Rule teaches to multiply half the Circumference by half the Diameter, and the Product is the Area, that is, multiply 1.5708 by .5, (*viz.* half 3.1416 by half 1) and the Product is .7854, which is the Area of the Circle whoſe Diameter is 1.

Again, if the Area be required, when the Circumference is 1, firſt find what the Diameter will be, thus, as 3.7416 : 1 : : ſo is 1 to .318309, which is the Diameter when the Circumference is one. Then multiply half .318309 by half 1, that is .159154 by .5, and the Product is .079577, which is the Area of a Circle whoſe Circumference is 1.

If the Area be given to find the Side of the Square equal, you need but extract the ſquare Root of the Area given, and it is done. So the ſquare Root of .7853 is .88.62, which is the Side of a Square equal when the Diameter is 1. And if you extract the ſquare Root of .079577, it will be .2821, which is the Side of the Square equal to the Circle whoſe Circumference is 1.

If the Side of a Square within a Circle be required, if you square the Semidiameter, and double that Square, and out of that Sum extract the square Root, that shall be the Side of the Square which may be inscribed in that Circle; so if the Diameter of the Circle be 1, then the Half is .5, which squared, is .25, and this doubled, is .5, whose square Root is .7071, the Side of the Square inscrib'd.

Again, if the Diameter of a Globe be 1, to find the Solidity. it is demonstrated, that the Globe is $\frac{2}{3}$ of a Cylinder of the same Diameter and Altitude. Thus, if the Cylinder's Diameter be 1, and its Altitude or Length be also 1, find the Solidity thereof, and take $\frac{2}{3}$ of it, and that will be the Solidity of the Globe required. Now if the Diameter be 1, the Area of the Circle, or Base of the Cylinder, is .7854, (as is above shewn;) which multiply'd by 1, the Altitude of the Cylinder, and the Product is also .7854, the Solidity of the Cylinder; $\frac{2}{3}$ whereof is .5236, which is the Solidity of the Globe, whose Diameter is 1.

GUTTÆ, in Architecture, are Ornaments in the Form of little Cones, used in the Platfond of the *Doric* Cornice, or on the Architrave underneath the Triglyphs, representing a sort of Drops or Bells, and usually six in Number.

GUTTERS, in Architecture, are a kind of Valleys in the Roofs of Buildings, serving to receive and drain off the Rain Waters.

These *Gutters* are of two Kinds in respect to their Position; for they are either such as come something near a Parallelism with the Horizon, or such as incline towards a vertical Position to the Horizon.

The first Kind of *Gutters* may be called *Parallel Gutters*, and may be distinguish'd into three Sorts, which are covered with Lead: For,

First, Either it is a *Gutter* between two Roofs, which stand parallel to each other, being made upon the Feet of the Rafters of two Roofs, which meet together. Or,

Secondly, A *Gutter*, where a Building has a Cantaliver or Modillion Cornice, which projects one Foot and a half, or two Feet (according to the Design of the Building) beyond the Walls; then the Roof is set with the Feet of the Rafters no farther out than the Wall, but rather within it; so that the Joists of the upper Floor lie out beyond the Walls, and also beyond the Feet of the Rafters, which is yet cover'd with Lead.

The third Sort of these *Parallel Gutters* are in flat Roofs, which are usually called *Platforms*; where are also *Gutters* for the Water that run from the Platform to descend to, which is from thence convey'd off from the Building, either by Spouts or Pipes.

Secondly, *Vertical Gutters* are such as are made by two Roofs meeting at Right Angles one to another, or (which is the same Thing) made by the End of one Roof

Roof joining to the Side of another

As for Example: If a Building be in the Form of a *Roman* L, it is then common to have one *Gutter* on the Inside of the L; but if the Building be in the Form of a T, it has two *Gutters*; but if in the Form of an H, it has four.

These *Gutters* also are of two Sorts, *viz.* either of Lead or Tile: All which shall be treated of in Order.

Of the Laying of Parallel Lead Gutters.

In speaking to this Head, I shall first give a necessary Caution, which is, *viz.* first to take care that the *Gutter-Boards, &c.* lie not too near Parallel with the Horizon; but in such a Position, that there may be a good Current, (as the Workmen phrase it,) for if it be laid too near a Level, the Water will be very subject to stand in Plashes, if the *Gutter* chances to stick a little in the Middle, *&c.* which some *Gutters* are apt to do: But this is according as they are posited in the Building.

Some *Gutters* have a Layer of Sand for the Lead to lie upon; but there are two Reasons that may render this Method not approveable.

First, Because some Sorts of Sand does very much corrode and decay the Timber that lies near it.

Secondly, That when a *Gutter* is laid on Sand, but a very little squatting, *viz.* by jumping upon it with the Heels of one's Shoes, will make Dents in it, and in those Dents the Water will stand; and this will be a Means of decaying the Lead the sooner.

In laying of Leads for *Gutters upon Boards*, 'tis common for Plumbers, to solder them, when they are so long, that a Sheet of Lead will not reach. To do this, they usually cut a Channel cross the *Gutter-Boards* at the End of the Sheet where the Soldering is to be, and to beat down the Ends of both the Sheets (that are to meet there) into the Channel; which, when it is done, there will remain a little Cavity, which is filled up by the Solder level with the rest, when it has been soldered.

The Lead which is usually laid in *Gutters* is that which weighs about eight or nine Pound to the Foot. See LEAD.

Of Vertical Gutters.

These *Gutters* are made either of Lead, or Tile. As to those made with Lead, I shall forbear saying any Thing, because they are almost the same in Effect as the Parallel ones. But, that unless the Builder will be at the Charge, they need not be altogether so thick for these Vertical ones, as for the Parallel ones: For these Vertical ones will last as long, if laid with Lead of about six or seven Pound to the Foot, as Parallel ones with Lead of eight, nine, or ten Pound to the Foot.

Gutters laid with Tiles, are also of two Kinds: Those made of of *Concave* or *Gutter-Tiles*, and *Plain Tiles*. Of which I ſhall omit ſpeaking here, but recommend to the Article GUTTER-TILES.

Plain Tile-Gutters are alſo diſtinguiſh'd into two Sorts, *viz.*

I. *Plain Tile-Gutters*, (properly ſo called.) And,
II. *Point-Gutters*. Of both which I ſhall treat in their Order.

First, *Of* Plain Tile-Gutters, (*properly ſo called.*)

In theſe *Plain Tile-Gutters*, there is a *Gutter-Board* laid, which raiſes them from pointing to an Angle. And in laying on the Tiles, the Workman begins at one Side of the *Gutter*, and ſo works acroſs, as if it were plain Work, and then brings the next Row of Tiles back again; ſo that he works forth and back, or to and fro, from Right to Left.

So that *Gutters* which are laid after this Manner, are not angular, but of a kind of diſtorted curvilineal Form; by which Means they are not ſo ſubject to be furr'd up with the Mortar which waſhes out of the adjacent Tiles.

II. *Of* Three-Point Gutters.

Theſe are the ſecond Sort of *Gutters*, which are laid with plain Tiles: In laying of which they begin and lay one Tile on one Part of the Roof, (it is no Matter which Part firſt) and lay one Corner of the Tile juſt in the Middle of the *Gutter*; and then they lay another on the other Part of the Roof, with its Corner juſt in the Middle of the *Gutter*, alſo that the Corner of the ſecond Tile is contingent with the firſt; and then lay another Tile in the *Gutter*, with its Corner, as it were, betwixt the other two, and to them.

When they have done thus, they proceed in the Work, and lay a Tile on each Part of the Roof, as before, and another betwixt them in the *Gutter*, proceeding in their Work in this Manner, till they have finiſh'd the *Gutter*. And this is what is called a *Three-Point Gutter*: For three Points, or Angles of Tiles, always come together, *viz.* one Angle of three diſtinct Rules, which makes it very uniform and handſome.

Here you are to take notice, that only three Inches ſquare of the middle Tile is viſible (if the Gage be ſeven Inches,) the reſt of that Tile being covered with the next Row of Tiles above it.

But notwithſtanding theſe *Gutters* are very handſome, and if well done, ſecure alſo; yet if they let the Water into the Houſe (by reaſon of ſome Stoppage, or broken Tile in the *Gutter*,) they are very troubleſome to mend.

Of Meaſuring Gutters *or* Valleys.

There are uſually different Cuſtoms in different Parts of the Kingdom, as to the meaſuring of *Gutters* or Valleys in Tiling: For in ſome Places, they but ſeldom,

ſeldom, if ever, allow any Thing for the *Gutters*; but include them in the reſt of the Roof at Flat and Half. And ſome ſay, at *London*, they very ſeldom meaſure the *Gutters*, but only as they are Part of the Roof; ſo they are included in the Flat and Half-Meaſure.

Some Workmen at *Tunbridge-Wells* never demand any other, but only as it is included in the Plain Meaſure; which is an Area found by Multiplication of twice the Length of the Rafters by the Length of the Building; or, which is the ſame Thing (when it is three quarters pitch,) the Flat and Half-Flat.

In laying of *Gutters* with concave Tiles, the Workmen in ſome Pars of *Suſſex* and *Kent*, have brought up a Cuſtom of being allowed ſo many Feet more than the Plain Meaſure, as there are *Gutter-Tiles*, (and alſo including *Corner-Tiles*, *Ridge-Tiles*, and *Dorman-Tiles*,) in the whole Roof.

At ſome other Places, they claim ſo many Feet more to be added to the Plain Meaſure, as the *Gutters* (and alſo Corners) are in Length, including *Gutters* at the Sides of Dormans and Lutherns, if there be any Dorman-Tiles uſed.

In ſome Places, the Workmen inſiſt upon a Cuſtom of having double Meaſure allowed for *Plain-Tile* (eſpecially, *Three-Point*) *Gutters*, *e. g.* if there were but one *Gutter* in a Roof, and this *Gutter* fifteen Feet long, then their Cuſtom is to have thirty Feet more than the Area of the Roof amounts to; and this Allowance ſome Workmen claim in both Sorts of *Gutters with Plain Tiles*.

Either of theſe *Plain-Tile Gutters* are cheaper to the Maſter-Builder, than Concave ones; becauſe Plain Tiles are cheaper than *Gutter-Tiles*, they being in many Places not above one fourth Part of the Price.

And beſides, if the Workmen be allow'd ſo many Feet more than the Area of the Roof, as there are *Gutter-Tiles* that will be one half as much more as the Double Meaſure; for if it be gaged ſo ſlight as eight Inches, then in a *Gutter* of fifteen Feet long, there would be forty-five Tiles, which will be reckon'd forty-five Feet; whereas at Double Meaſure, it amounted but to thirty Feet.

There is another Way of computing Double Meaſure; for the Account of which, I ſhall refer you to the Article SLATING.

GUTTERING, in Carpentry, is uſually done by the Lineal Foor, which is by ſome valued at *London*, for Materials and Workmanſhip, at 1 *s*.

GUTTER-TILES are whilſt they are flat and plain, (before they are bent fit for the Uſe they are intended,) ſeemingly at a Diſtance, a kind of Triangle, with one convex Side. But although they ſeem to be ſo at a Diſtance, they are not in Reality ſo; for they are of a quadrangular Form, conſiſting of two ſtreight Sides, of about ten, or ten Inches and a half long, (for

(for ſo much they ought to be,) and of two circular Sides, the one convex, the other concave; the convex Side is about fourteen Inches, and the concave one about two Inches. This is their Form as to their Edges or Sides. I ſhall next deſcribe the Form of them, in reſpect of the Plane; at the little End they are bent circular, and ſo likewiſe at the convex great End, at firſt like a Corner Tile; but then they bend the Corners of the great End back again; ſo that if a Perſon look againſt the Edge of the broad End, it conſiſts of a circular Line betwixt two ſtreight ones, like the upper Part of the Character of the Sign *Libra* ♎: This, you muſt underſtand, is when you hold the concave Side of the Tile downwards.

Theſe Tiles are laid with their broad Ends and hollow Sides upwards.

As to the Weight of Gutter-Tiles.

Theſe *Tiles*, whoſe Dimenſions were 10 Inches on the ſtreight Edges, 14 Inches on the great convex Edge, when preſs'd down flat, as they were in the Mould, and two Inches at the concave Edge, and about $\frac{5}{8}$ Inch thick; 100 of them weigh about 321 or 322 Pounds, and conſequently 1000 would weigh about 3210 or 3220 *lb.* which is near 29 *C.* Weight, and conſequently 682 would be a Ton Weight.

As to their Price.

Mr. *Leybourn* ſays, that they are ſold at *London* at 1 *d* $\frac{1}{2}$ or 2 *d. per Tile*, or between 10 and 15 *s. per* 100. In ſome Places their conſtant Price is 1 *d.* $\frac{1}{2}$ *per* Piece, or 12 *s. per* 100.

H A

HAIR, with Plaiſterers, is Bullocks *Hair*, &c. which is uſed in white Mortar; a certain Quantity of which is put to a certain Quantity of Lime. See LIME and MORTAR.

As to the Price: This varies, according to the Plenty or Scarcity of it in *London.*

In ſome Places in *Kent* it has been ſold for ſeven Pence *per* Buſhel; and in *Suſſex* for ten Pence and twelve Pence; ſo that a Horſe-Load, which is ſixty Buſhels, may be from thirty Shillings to three Pounds, more or leſs.

HALF-ROUND. See CAPITAL.

HALL, in Architecture, is a large Room at the Entrance of a fine Houſe, Palace, or the like.

Vitruvius mentions three Sorts of *Halls:* The *Tetraſtyle*, which has four Columns, ſupporting the Plafond or Ceiling; the *Corinthian*, which has Columns all around let into the Wall, and is vaulted over; and the *Egyptian*, which had a *Periſtyle* of inſolated *Corinthian* Columns, bearing a ſecond Order with a Ceiling.

The *Hall* is properly the firſt and fineſt Partition or Member of an Apartment; and in the Houſes

Houſes of Miniſters of State, publick Magiſtrates, &c. is that wherein they diſpatch Buſineſs, and give Audience.

In very magnificent Buildings, where the *Hall* is larger and loftier than ordinary, and placed in the Middle of the Houſe, it is called a *Saloon.*

Of their Dimenſions.

A certain noted *French* Architect directs, that the Length of a *Hall* be at leaſt twice and a quarter its Breadth, and that in great Buildings you may allow it three Times the Breadth; which laſt Length, he ſays, will be the moſt beautiful and convenient.

As to the Height of *Halls*, it may be two Thirds of the Breadth, or ſixteen or eighteen Feet in noble Buildings.

In large and ſtately Buildings, the *Halls*, and other Rooms of the firſt Story, may be arched; by which Means they will be rendered much handſomer, and leſs ſubject to Fire.

The Height is to be found by dividing the Breadth of the *Hall* into ſix Parts, and five of thoſe ſhall limit the Height of the Room, from the Floor to the under Side of the Key of the Arch.

HALLS and ANTICHAMBERS, and other Rooms of the firſt Story that are arched, which will be much more handſome, and leſs ſubject to Fire; their Height may be adjuſted by dividing the Breadth into ſix Parts, and taking five of them for the Height from the Floor to the Superficies, unto the Bottom of the Key of the Arch.

As for Example: Let the Figure be twenty-four Feet in Breadth, more or leſs, and be divided into ſix equal Parts; take five of them, which will be twenty Feet in Height from the Floor unto the Bottom of the Key of the Arch. See the Figure.

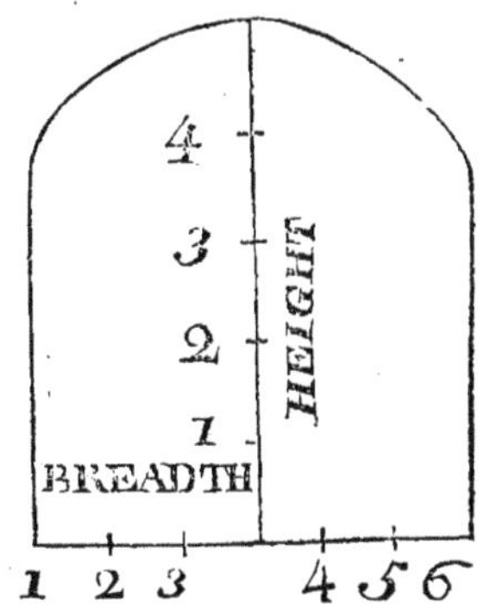

And if you would have it higher, you muſt divide the ſaid Breadth into eight Parts, and take ſeven of them for the Height, which will make twenty-one Feet.

And if you divide the ſame Breadth into twelve Parts, take eleven of them, which will make the Height twenty-two Feet from the Floor to the Bottom of the Key of the Arch.

The Chambers of the ſecond Story muſt be a ſixth Part leſs in Height, than the Chambers below.

As if the firſt Story be twenty Feet in Height, divide the twenty Feet into ſix equal Parts, and take five of them, which will make the ſecond Story ſixteen Feet

Feet eight Inches from the Foor to the Joist.

Again, if the first Story be twenty one Feet in Height, divide the twenty-one Feet into six equal Parts; take five of them, and they will make the second Story seventeen Feet six Inches from the Height of the Floor to the Joist.

If the first Story be twenty-two Feet in Height, divide the twenty-two Feet into six equal Parts, and take five of them, which will make the second eighteen Feet four Inches in Height.

HAMMER HARDENING is mostly used on Iron and Steel Plates for Saws, *&c.*

To HANG OVER. See to BATTER.

HANSE. See ARCHES.

HEAD, in Architecture, *&c.* is an Ornament of carved Work or Sculpture, frequently serving as the Key of an Arch, or Plat-Band on other Occasions.

These sort of *Heads* usually represent some of the Heathen Divinities, Virtues, Seasons, Ages; with their Attributes, as a Thunderbolt for *Jupiter*, a Diadem for *Juno*; a Trident for *Neptune*, a Crown of Ears of Corn for *Ceres*, a Helmet for *Mars*, a Caduceus for *Mercury*, *&c.*

The *Heads* of Beasts are also used in Places suitable, as an Horse's *Head* for an Equerry, a Deer's or Boar's for a Park or Forest, a Dog's for a Kennel, a Bullock's or Sheep's for a Shambles or Market-House.

In the Metopes and Friezes, and other Parts of certain *Antique Doric* Temples, we see Representations of Bullocks or Rams' *Heads* flead, as a Symbol of the Sacrifices offered there.

HEADS, with Bricklayers, a Term used to signify half a Tile in Length; but to the full Breadth of a Tile: These they used to lay at the Eaves of a Roof.

HEADING *Architrave*. See ARCHITRAVE.

HEALING, in Architecture, signifies the covering the Roof of a Building, either with Lead, Tiles, Slate, Horsham Stone, Shingles, Reeds, Straw, *&c.*

HEARTH *Stones*. See FIRE-STONES.

HEAT, in Smithery. See Iron.

HEEL, in Carpentry, an inverted Ogee.

HEIGHT is the third Dimension of a Body, considered with respect to its Elevation above the Ground.

Altimetria is an Art or Science, which teaches the Measuring of all *Height*, both accessible and inaccessible.

The Instruments chiefly used in taking *Heights*, are the Quadrant, and the Geometrical Quadrant.

HELICOID *Parabola*, or the *Parabolick Spiral*, is a Curve arising upon a Supposition of the Axis of the common *Apollonian Parabola's*, being bent round into the Periphery of a Circle.

The *Helicoid Parabola* then is a Line passing through the Extremities of the Ordinate, which now converse towards the Centre of the said Circle.

HELIOSCOPE, in Opticks, a Sort of Tellescope peculiarly fitted for viewing and observing the

the Sun, as his Spots, Eclipses, *&c.*

HELIX, the Word is *Greek*, and literally signifies a Wreath, or Winding.

Helix, in Architecture, is the Caulicoles, or little Volutes under the Flower of the *Corinthian* Capital, called also *Urilla*.

Helix, in Geometry, is a Spiral Line; but some Authors in Architecture make a Difference between *Helix* and *Spiral*

A Stair-Case, according to *Daviler*, is an *Helix*, or is *helical*, when the Stairs or Steps wind round a cylindrical Newel; whereas the Spiral winds round a Cone, and is continually approaching nearer and nearer its Axis.

HEMISPHERE, in Geometry, is one half of a Globe or Sphere, when divided into two, by a Plane passing through its Centre.

HEMISPHEROIDAL, in Geometry, approaching near to the Figure of a *Hemisphere*, but is not justly so.

HEPTAGON, in Geometry, a Figure consisting of seven Sides and seven Angles; which, if the Sides be all equal, is called a regular *Heptagon*.

HEPTAGONAL *Numbers*, are a Sort of Polygonal Numbers, wherein the Difference of the Terms of the corresponding arithmetical Progression is five.

One Property, among others of these Numbers, is, that if they be multiply'd by 40, and 9 be added to the Product, the Sum is a square Number.

HEXAEDRON } in Geometry,
HEXAHEDRON } is one of the five regular Bodies properly called a *Cube*.

HEXAGON, in Geometry, a Figure of six Sides, and as many Angles. If these Sides and Angles be equal, it is called a *regular Hexagon*.

The Side of a *Hexagon* is demonstrated to be equal to the Radius of a Circle circumscribed about the same.

Hence a *regular Hexagon* is inscribed in a Circle, by setting the Radius off six Times upon the Periphery.

HEXASTYLE, in the antient Architecture, a Building, having six Columns in Front.

HINGES, in Building, are those necessary Irons by Means of which Doors, Lids, Folds of Tables, *&c.* whether of Houses or other Buildings, make their Motion, whether of opening or shutting, or folding, *&c.*

The Sorts or Kinds are many; as *Beds*, *Box*, *Buts*, *Casement*, *Lancashire* and *Smooth-filed*; *Casting*, *Chest-black Lancashire*, *Smooth-filed Coach*, *Desk*, *Dove-Tails*, *Esses*, *Folding*, *Garnets*, *Dozen-Ware long*, *Dozen-Ware short*, *Weighty long*, *Weighty short*, *Lambs-Heads*, *Port-Side Lancashire*, *Side Smooth-filed*, and *Smooth-filed*, *Side with Squares*, *Side with rising Joints*, *Lancashire* and *Smooth-filed Stall*; *Trunk* of sundry Sorts; *Screw*, *Scuttle*, *Shutter*, *Lancashire Joints*, *Lancashire Dozen-Ware with Hooks*, *Dozen-Ware long*, *Dozen-Ware short*, *Weighty long*, *Weighty short*.

The Price of ſome of theſe *Hinges* are as follows:

Bed-Hinges, from 5 *s.* to 7 *s.* *per* Dozen.

Box-Hinges, from 1 *s.* to 4 *s.* *per* Doz.

Small Braſs ones, from 2 *s.* to 2 *s.* 6 *d.* *per* Doz.

Dove-Tails, from 1 *s.* to 4 *s.* *per* Doz.

Hooks and *Hinges*, &c. *per lb.* from 3 *d.* ¼ to 4 *d.*

Side-Hinges, from 3 *s.* to 16 *s.* *per* Doz.

With a Square, from 20 *s.* to 36 *s.* *per* Doz.

Screw-Hinges, from 30 *s.* to 48 *s.* *per* Doz.

HIPS, in Carpentry, are thoſe Pieces of Timber which are placed at the Corner of a Roof.

The *Hips* are much longer than the Rafters, by reaſon of their oblique Poſition; and are planted not with a right or ſquare Angle, but a very oblique one; and conſequently are not, or at leaſt ought not to be ſquare at any Angle, as Rafters are not at all, but level at every one of them; and which is yet more, as Rafters have but four Planes, theſe commonly have five. They are commonly, by Country Workmen, called *Corners*, and ſome call them *principal Rafters*, and others *Sleepers*.

The Truth is, *Hips* and *Sleepers* are much the ſame, only the *Sleepers* lie in the Valleys, (and join at the Top with the *Hips*;) but thoſe Surfaces or Planes which make the Back of the *Hip*, are the under Sides of the *Sleeper*.

The Backs of a *Hip* are thoſe two Superficies or Planes on the Outſide of the *Hip*, which lie parallel, both in reſpect of the Length and Breadth with the Superficies of the adjoining Side, and End of the Roof.

Hip-Mould is by ſome uſed for the Back of the *Hip*: But others underſtand it to mean the Prototype, or Pattern, commonly made of a thin Piece of Wainſcot, by which the Back and Sides of the *Hip* are ſet out.

I ſhall here give you the Method of finding the Length and Backs of *Hips*, &c. in ſquare Frames, and alſo of the Rafters, Diagonals, Half Diagonal, and Perpendicular, as follows:

	Feet.			
As 20	15 : 00	:: Breadth of the Houſe :		ten of the Rafter.
	18 : 00			ten of the *Hip*.
	11 : 18			Perpendicular.
	28 : 20			Diagonal.
	16 : 63			neareſt Diſtance.

Hip Angles	at Foot	38—22	Rafter Angles	at Top	41—50
	at Top	51—28		at Foot	48—10
	at Back	116—12			

The Angles are always the ſame in all ſquare Frames that are true Pitch.

Hip-Roof, with Carpenters, called alſo *Italian Roof*, is a *Roof* which has neither *Gable-Head*, *Shread-Head*, nor *Jerkin-Head*; (by which is meant ſuch Heads as are borh *Gable* and *Hip* at the ſame End :) For 'tis a Gable or Upright as high as the Collar Beam; and then there are two ſhort *Hips*, which ſhut up with their Tops to the Tops of a Pair of Rafters, which Country Carpenters call *Singlars*.

For a *Hip-Roof* hath Rafters as long, and with the Angles at the Foot, *&c.* at the Ends of Buildings, as it has at the Sides; and the Feet of the Rafters on the Ends of ſuch Buildings as have *Hip-Roofs*, ſtand on the ſame Plane, *viz.* parallel with the Horizon, and at the ſame Height from the Foundation with Rafters on the Sides of the *Roof*.

Theſe are the *Hip-Roofs* that are by ſome called *Italian Roofs*.

The Method of Meaſuring Hip-Roofs.

If they are three quarters, or true Pitch, as it is commonly called, then it is only multiplying the Length of the Building by the Breadth, and adding half as much more to the Area found; or elſe multiplying the Length by the Breadth and half, or the Breadth by the Length and half: Either of theſe three Ways will produce the Flat and half, which is equal to the Content of the *Roof* in plain Meaſure, if nothing be allowed for *Hips* and *Valleys*; but if the *Roof* have no Cornice, but the Rafters have Feet, then they muſt be added; and alſo the Eaves-Board, in a Bill of Meaſurement.

Or you may meaſure ſuch a Roof, by multiplying the Length of it by the Length of the Rafter, and it will give the Half Content; or elſe by multiplying the Length of the Building by twice the Length of the Rafter; which will give the whole Content.

How to find the Curve-Lengths and Backs of the Hip, *either bevel or ſquare of* Roofs *in general, obſerve the following Examples.* See Plate XII.

The Figure I. is an *Hexagon Plan*, and an *Ogee Rafter*.

Firſt, Draw the Plan *abcdef*, alſo draw the Line *b h*; then divide the Line *ab* in the Middle at *i*, and draw the Line *ih*; then will *bh* be the Baſe of the *Hip*, and *ih* the Baſe of the Rafter: From *h* draw a Line to *k*, perpendicular to *ih*, and equal in Length to the Perpendicular of the Rafter; alſo from *h* draw a Line to *g*, perpendicular to *bh*, and equal to *hk*; then draw the moulding Part of the Rafter *ik* in what Form you pleaſe: Having ſo done, divide the Line *ih* any how; from which Diviſions raiſe perpendicular Lines to touch the Curve Line *ik*; continue thoſe Lines, to touch the Line *bh*, as the dotted Lines in the Example ſhew, which will divide the Line *bh* into the ſame Number of Parts and Proportion with the Line *ih*; then from thoſe Diviſions raiſe perpendicular Lines at Pleaſure, and take the perpendicular Line 1, 1, on the Line *ih*, to the Curve of the Rafter *ik* in your Compaſſes, and ſet it up the correſpondent perpendicular Line on the Line *bh*, as 1, 1; alſo the Line 2, 2; and 3, 3; and ſo of all the reſt: And in each of theſe Points ſtick a Nail, and bend a thin Lath round, to touch them all at once; then on the Edge of it draw the Curve of the *Hip g h*; which was to be done.

This Fig. II. *repreſents the* Hip bg *in* Fig. I. *and* 1, 2, 3, 4, *at the Point* e, *repreſents the Sole of the Foot of the* Hip, *before the Back is work'd.*

Firſt, Draw the Lines on the *Hip*, at any convenient Diſtance, parallel to the Foot of the Baſe *ac*; then draw the Sole of the Foot of the *Hip*, as 1, 2, 3, 4, at the Point *e*, of the preceding Figure, Number I. and take in your Compaſſes the Diſtance between the Point 1 to the Line *ef*, or from 2 to the Line *ed*, and ſet it from the Back of the *Hip ab*, on thoſe parallel Lines which you ſee marked by Dots; then ſtrike a Nail into each of theſe Dots or Points, and bend a thin Lath to touch them all at once, and on the Edge of it ſtrike a Curve Line; then draw a middle Line down the Back of the *Hip*, and between that Line and the Curve, which is created by theſe Dots, hew off the ſuperfluous Wood, which will make the true Back of the *Hip*; and ſo of all other *Roofs*, in what Form ſoever: But only you muſt obſerve, if your Plan is bevel, as one End of 5, to ſet the Superfluity of the Sole of the *Hip* at the Point *e*, which is from 3 to the Line *cb*, and from 4 to the Line *cd*, on their proper Sides of the *Hip*, becauſe one Side will be wider than the other, which is the Caſe on the Back of all bevel *Hips*.

The Plan *abcdef*, in Fig. III. is a *Hexagon*, the ſame as Fig. I. and the Lines *bh*, *gh*, *hk*, and *hi* in the one, is equal to *bh*, *ih*, *hg*, and *hk* in the other; ſo alſo are the Soles of the Feet of the *Hips* 1, 2, 3, 4, at the two Points *e*; and there is no other Difference than the Curves of the Rafters, and, of conſequence, needs no other Explanation; and ſo likewiſe of the two *Hips*, Fig. II. and Fig. IV. the two laſt Figures being laid down only for Variety ſake.

How to find the Length and Bevel, and the Mould for a Hip, *either bevel or ſquare, whether it be above Pitch, or under Pitch.* Plate XIII.

Firſt, Draw the Plan *a b c d*, and let one End be bevel, as *bc*, and the other End ſquare, as *ad*, which divide in the Middle by the dotted Line *tv*.

Then draw the Line *ef* parallel to *ad*, and diſtant as far as *at* or *td*, and draw the Lines *ia* and *id*.

Then take the Line *ad* in your Compaſſes, and ſet it on the Line *ab*, as *km*, and draw the Lines *kl* and *ml*, to repreſent the Pitch of the Rafters, and let fall the Perpendicular *ln*; which take in your Compaſſes, and ſet it from *i* to *g*, and from *i* to *h*, in a ſtraight Line with *id* and *ia*, and draw the Lines *ga* and *hd*.

Then

: c), and from 4 to
l, on their proper
. Hip, because one
wider than the other,
Case on the Back of
r.

abcdef, in Fig. III.
; the same as Fig. I.
s bb, gh, hk, and
ne, is equal to bh,
kk in the other; so
Soles of the Feet of
3, 4, at the two
d there is no other
han the Curves of
nd, of consequence,
er Explanation; and
of the two Hips,
Fig. IV. the two
being laid down on-
y side.

the Length and Be-
he Mould for a Hip,
ther square, whether
e Pitch, or under
ate XIII.

w the Plan abcd,
End be bevel, as bc,
r End square, as ad,
: in the Middle by
ne tr.
w the Line ef pa-
and distant as far as
d draw the Lines

e the Line ad in
..s, and set it on
, as km, and draw
l and ml, to repre-
of the Rafters, and
Perpendicular ln;
n your Compasses,
m i to g, and from
aight Line with id
draw the Lines ga

Then

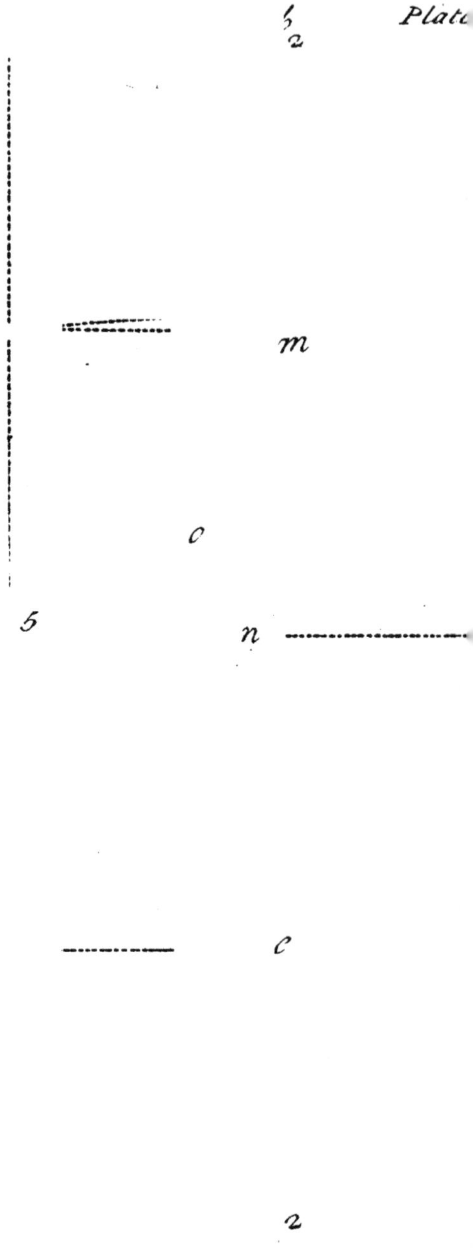

Plate

m

o

5 n

c

2

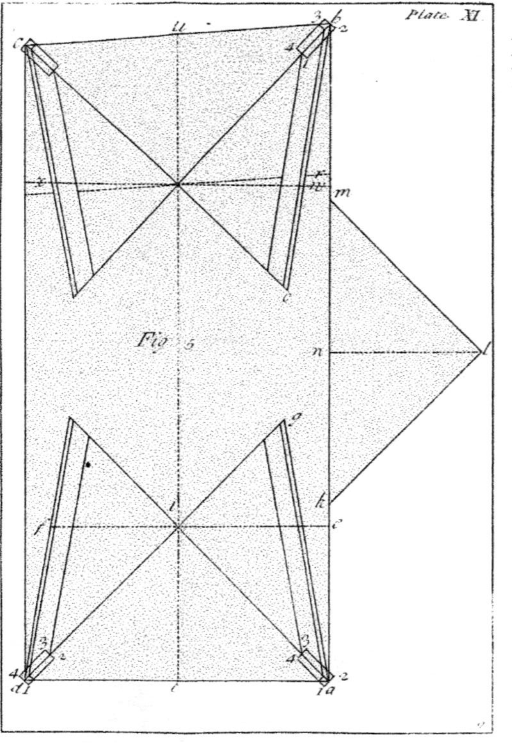
Plate XI
Fig. 5

Then draw the Line *r s* parallel to *bc*, and distant so much as *ei* or *it*.

Then draw the Line *w x* through the Point *p*, and parallel to *ef*, and draw the Lines *bp* and *cp*.

Then take the Perpendicular *ln*, and set it from *p* to *o*, and from *p* to *q*, in Right Lines with *bp* and *cp*, and draw the Lines *bo* and *cq*.

Then is *bo*, *cq*, *dh*, and *ag*, the Length of the four *Hips*.

And if the Lines *po* and *pq* were raised up, they would meet perpendicular over the Point *p*; so would the Lines *ig* and *ih* meet over the Point *i*; then draw the Breadth of the *Hip* parallel to the Line *ag*.

To find the Back of the Hip.

Set its Splay or Foot in its Position on its proper Place; the Sole of which is represented by its proper Figures 1, 2, 3, 4, in this Example; and strike it by on the under Side or Sole of the Hip to the Line *ab* and *ad*, which will shew you how much of it hangs over the Plates *ad* and *ab*, as from 1 to the Line *ad*, and from 1 to the Line *ab* at the Angle *a*, and from the Extremity of those Strokes, strike a Line on each Side the Hip parallel to the Back or Line *ag*.

Then strike another Line on the Middle of the Back, and those three Lines give its true Bevels, to answer both Sides of the Roof, whether it be bevel or square, by hewing off the superfluous Wood between Line and Line.

N. B. If you do not approve of setting the *Hip* up in its Position, you may find its Back by drawing a Line on the Angle of your Plate, as *ai*; also a middle Line on the Sole of the Foot of the *Hip*; and take the Bevels from the Line *ia* and *b*, and *ia* and *d*, and set them on the Foot or Underside of the Splay, which will be of the same Effect, as setting it up in its Position.

To find the Length of the Hip, *and Mould for its Back, another Way.* Plate XIV.

Draw the Plan *acdb*, and divide it in the Middle by *bn*, and draw the Line *iq* parallel to *ab*, and equal to *an* or *nb*; also the Line *zy* parallel to *cd*, and distant equal to *ai*; then draw the Line *gx* through the Point *w*, parallel to *iq*; and draw the Lines *cd*, *dw*, *af*, and *bf*.

Having done this, take the Width of the Span *ab*, and set it on the Line *ac* any where, as *eg*, and draw the Lines for the Pitch of the Rafter *ef* and *gf*; also the Perpendicular Line *fh*, which take in your Compasses, and set it from *w* to *t*, and from *w* to *u*, in a straight Line with *cw* and *dw*; also from *f* to *k*, and from *f* to *r*, in a straight Line with *fa* and *fb*; and draw the Lines for the Length of the *Hips* *ct*, *du*, *ak*, and *br*.

To draw the Lines for the Mould for the Back of the Hip.

Lay a straight Rule from *i* to *n*, and make the Point *m* from *n* to *q*, and make the Point *o*; alſo from *z* to *b* make the Point *a*, and from *b* to *y* make the Point *c*; then take the Compaſſes, and ſet one Foot in the Point *m*, and move the other Foot, till it touches the neareſt Place of the *Hip-Line ak*; which move about and make the Point *l*, and draw the Lines *il* and *ln*, then is *iln* the Mould for the Back of the *Hip ak*; alſo *npq* for the *Hip br*.

The ſame Method is uſed for the *Bevel-Hips ydb*, as is for the Back of the *Hip du*, and *bez* for the *Hip ct*; which was to be done. See Plate XIV.

HIP-TILES. See CORNER-TILES.

HOLLOW, in Architecture, a concave Moulding about a Quadrant of a Circle, by ſome called *Caſement*, by others *Abacus*.

HOLLY. The Timber of *Holly* is the whiteſt of all hard Wood, and therefore uſed by the Inlayers. It is alſo fit for all ſturdy Uſes; the Mill-Wright, Turner, and Engraver, prefer it to all others. It makes the beſt Handles and Stocks for Tools, Flails, Carters Whips, Bowls, Shivers, and Pins for Blocks. It is excellent for Door-Bars, *&c.*

HOMOLOGOUS, in Geometry, is apply'd to the Sides of ſimilar Figures, which are ſaid to be *Homologous*, or in Proportion the one to the other. Thus the Baſe of one Triangle is *Homologous* to the Baſe of another ſimilar Triangle, ſo in ſimilar Triangles, the Sides oppoſite to equal Angles, are ſaid to be *Homologous*.

HOOKS, in Building, are a neceſſary Sort of Utenſils which are uſeful for ſeveral Purpoſes. They are of various Sorts, ſome of Iron, and others of Braſs: Some of the Names of which are as follows:

1. *Armour-Hooks*. Theſe are generally of Braſs, and are to lay up Arms upon, as Guns, Muſkets, Halberts, Half-Pikes, Pikes, Javelins, *&c.*

2. *Caſement-Hooks*.

3 *Chimney-Hooks*, which are made both of Braſs and Iron, and of different Faſhions. Their Uſe is to ſet the Tongs, Fire-Shovel, *&c.* againſt. Theſe are ſold from 2*s.* to 2*s* 6*d.* a Pair; the Iron ones from 1*s.* to 1*s.* 6*d.* a Pair.

4. *Curtain-Hooks*.

5. *Hooks for Doors, Gates*, &c. Theſe are from 3*d.* ½ to 4*d.* a Pound.

6. *Double Line-Hooks*, large and ſmall.

7. *Single Line-Hooks*, large and ſmall.

8. *Tenter-Hooks*, of various Sorts, *viz.* 2*d.* 3*d.* 4*d.* 6*d.* 10*d.* 20*d.* and 40*d.*

HORSHAM STONE, is a Kind of thin broad Slate, of a greyiſh Colour, formerly much uſed, eſpecially in *Suſſex*, to heal or cover Churches and Chancels, great Houſes, *&c.*

It is called *Horſham-Stone* in that County, becauſe it is chiefly brought from the Town called *Horſham*.

This Sort of Stone, or rather Slate, is laid of different Sizes,

viz.

[illegible] Building, are a [illegible] Utensils which [illegible] several Purposes. [illegible] various Sorts, some [illegible] others of Brass; [illegible] Names of which [illegible]

H[illegible]ks. These are [illegible] Brass, and are to [illegible], as Guns, Mus- [illegible] Half-Pikes, Pikes,

-Hooks.

H[illegible]ks, which are [illegible] Brass and Iron, [illegible] Fashions. Their [illegible] Tongs, Fire- [illegible]. These are [illegible] 2*s.* 6*d.* a Pair; [illegible] 1*s.* to 1*s.* 6*d.*

H[illegible]ks.

[illegible] *D[illegible]rs, Gates,* &c. [illegible] 3*d.* ½ to 4*d.* a

Line-Hooks, large

Line-Hooks, large

H[illegible]ks, of various [illegible] 3*d.* 4*d.* 6*d.* [illegible]

[illegible] STONE, is a [illegible] Slate, of a [illegible] formerly much [illegible] in *Sussex,* to heal [illegible] and Chancels, [illegible]

H[illegible]-Stone in [illegible] it is chiefly [illegible] Town called

[illegible] Stone, or rather [illegible] different Sizes, *viz.*

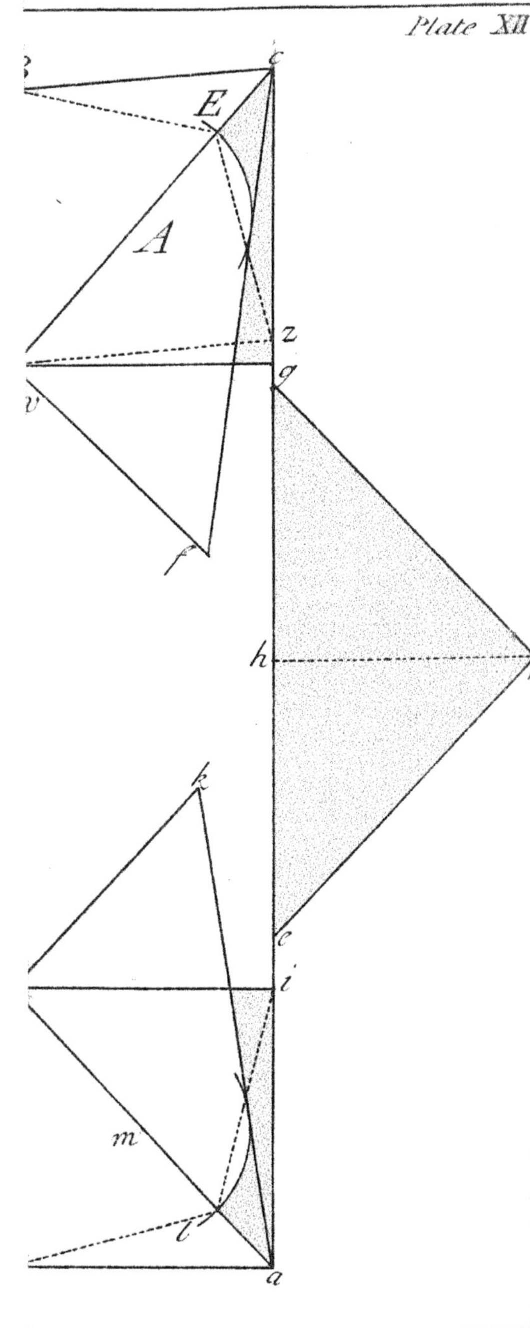

Plate XII.

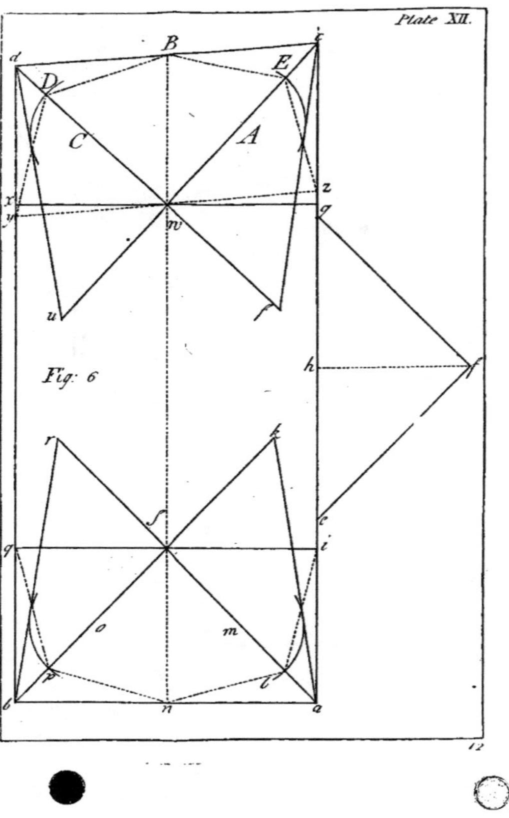

Fig: 6

12

viz. from eight or nine Inches to twenty-four Inches, or more, in Length and Breadth, *&c.* It is commonly from half an Inch, to an Inch thick.

Of the Price.

The Value of theſe Stones is according to the Diſtance from the Quarry, *viz.* from 10*s.* to 20*s.* the Load. Some of them have been laid down for 17*s.* or 18*s.* at eighteen or twenty Miles Diſtance from the Quarry.

A Load of theſe will cover about three Fourths of a Square.

Of the Price of Laying.

The Price of laying a Square, and Pointing (which is ſtriking Mortar under the lower Ends) in new Work, is about 5 or 6*s.*

But to rip it from old, and new lay and point it, is worth not leſs than 6 or 7*s. per* Square.

Of the Weight of this Sort of Healing.

A Square of this kind of Covering has been found to weigh about thirty-three or thirty-four hundred Weight; whereas a Square of Tiling does not weigh above ſixteen or ſeventeen hundred Weight, or not above eighteen hundred Weight, though it be gaged at ſix Inches, and the Tiles not exceeding the Length of ten Inches.

Of the Properties of this Sort of Covering.

It will appear by what has been already ſaid, that this Covering is dearer than Tiling; for the Charge of a Square of Tiling is from 23*s.* to 30*s.* or, as ſome ſay, from 24*s.* to 28*s.* a Square; whereas a Square of Covering with *Horſham-Stone* will be worth from 32*s.* to 38*s.*

And, beſides, for this Sort of Covering, the Timber for the Roof ought to be conſiderably ſtouter and ſtronger; becauſe a Square of this Sort of Stone is almoſt as heavy again as a Square of Tiling.

But then theſe Sorts of Stones are choſen as fitter for Churches, and other ſtrong Buildings; becauſe they are far more durable than Tiles, they being for the moſt part very hard, ſo that no Weather will hurt them, as it will Tiles

HOUSE, a Habitation or Place built with Conveniencies for dwelling in; or it is a Building wherein to ſhelter a Man's Perſon and Goods from the Inclemencies of the Weather, and the Injuries of ill-diſpoſed Perſons.

In treating on this Article HOUSE, I ſhall do theſe four Things:

I. Diſcourſe concerning the Situation of a *Country-Houſe.*

II. Of the Ground-Work of *Houſes.*

III. Concerning Building in *London.*

IV. Of Party Walls.

A *Country House*, or *Pleasure House*, is one built for a Person to enjoy and divert himself occasionally in.

This is the *Villa* of the antient *Romans*; and what in *Spain* and *Portugal* they call *Quinta*; in *Provence*, *Cassine*; in some other Parts of *France*, *Closerie*; and in *Italy*, *Vigna*.

Of the Site of a Country-House.

It is a Thing principally to be aim'd at in the Site or Situation of a *Country House* or *Seat*, that it have Wood and Water near it, they being principal Accommodations to a Rural Seat. If it cannot be conveniently built among Trees, yet there are but few Places where Trees may not be speedily raised about it.

It is far better to have a *House* defended by Trees than Hills; for Trees yield a cooling, refreshing, sweet, and healthy Air, and Shade during the Heat of Summer, and very much break the cold Winds and Tempests from every Coast in the Winter.

The Hills, according as they are situated, defend only from some certain Winds; and if they are on the North Side of the House, as they defend from the cold Air in the Winter, so they also deprive you of the cool refreshing Breezes, which are commonly blown from thence in the Summer.

And if Hills be situated on the South Side, it also then proves very inconvenient.

Besides, they yield not the Pleasures and Contentments, nor the Varieties of Delights to the ingenions Rustick, as the tall plumps of Trees, and pleasant Groves do.

Yet Hills which are cloath'd with Coppices, or otherwise improv'd, are pleasant Objects of Sight, if they stand not too near a *House*.

A *House* should not be too low seated, since this would cause you to lose the Conveniency of Cellars: But if you cannot avoid building on low Grounds, set the first Floor above the Ground the higher, to supply what you want to sink in your Cellar in the Ground; for in such low and moist Grounds, it conduces much to the Dryness and Healthiness of the Air to have Cellars under the House, so that the Floors be good, and ceiled underneath.

Mr. *Worlidge* says, that Houses built too high in Places obvious to the Winds, and not defended by Hills or Trees, require more Materials to build them, and also more Reparations to maintain them, and are not so commodious to the Inhabitants, as the lower-built Houses; which may be built at a much easier Rate, and also as compleat and beautiful as the other.

Of the Ground-Work of Houses.

In *Buildings* or *Houses* not above two Stories with the Ground Room, and not exceeding twenty Feet to the Raison-

Place,

Place, and upon a good Foundation, the Length of two Bricks, or eighteen Inches, for the Heading Courſe, will be ſufficient for the Ground-Work of any common Structure; and ſix or ſeven Courſes above the Earth to a Water-Table; where the Thickneſs of the Walls are abated or taken in on either Side the Thickneſs of a Brick, namely, two Inches and a quarter.

But for large and high *Houſes* or *Buildings* of three, four, or five Stories with the Garrets, the Walls of ſuch Edifices ought to be, from the Foundation to the firſt Water-Table, three Heading Courſes of Bricks, or twenty-eight Inches at leaſt; and at every Story a Water-Table or Taking-in, on the Inſide, for the Summers, Girders, and Joiſts to reſt upon, laid into the Middle, or one quarter of the Wall at leaſt for the better Bond.

But as for the innermoſt or Partition-Wall, half a Brick will be of a ſufficient Thickneſs; and for the upper Stories, a Nine-Inch (or Brick-Length) Wall will ſuffice.

The Parts, Proportions, &c. of the *Houſes* in *London* are regulated by a Statute made for rebuilding the City after the Fire, what here follows, is ſo much of the Act as relates to the Bricklayers Work, the Heights and Number of Stories, and Thickneſs of Walls, of the four ſeveral Rates of Houſes, which is as follows:

And be it farther enacted, That the *Houſes* of the firſt and leaſt Sort of Building, fronting by Streets or Lanes, ſhall be two Stories high, beſides Cellars and Garrets; that the Cellars thereof be ſix Feet and a half high, if the Springs of Water hinder not, and the firſt Story being nine Feet high from the Floor to the Ceiling, and the ſecond Story as much: That all the Walls in Front and Rear, as high as the firſt Story, be of the full Thickneſs of the Length of two Bricks; and thence upwards to the Garrets, of the Thickneſs of one Brick and a half; and that the Thickneſs of Garret-Walls on the back Part be left to the Diſcretion of the Builder, ſo that the ſame be not leſs than one Brick Length; and that the Thickneſs of the Party-Wall in the Garret be of the Thickneſs of the Length of one Brick at leaſt.

And be it farther enacted, That the *Houſes* of the ſecond Sort of Building, fronting Streets, and Lanes of Note, and the River of *Thames*, ſhall conſiſt of three Stories high, beſides Cellars and Garrets; that the Cellars thereof be ſix Feet and a half high, (if the Springs hinder not,) that the firſt Story contain full ten Feet in Height from the Floor to the Ceiling; the ſecond full ten Feet; the third nine Feet; that all the ſaid Walls in Front and Rear, as high as the firſt Story, be two Bricks and a half thick; and from thence upward, to the Garret-Floor, of one Brick and an half thick; and the Thickneſs of the Garret-Walls on the back Part be left to the Diſcretion of the Builder, ſo that the ſame be not leſs than one Brick thick:

And alſo that the Thickneſs of the Party-Walls between every *Houſe* of this ſecond and larger Sort of Building, be two Bricks thick, as high as the firſt Story; and thence upwards to the Garrets of the Thickneſs of one Brick and a half. Alſo that the *Houſes* of the third Sort of Building fronting the high and principal Streets, ſhall conſiſt of four Stories high, beſides Cellars and Garrets, as aforeſaid; that the firſt Story contain full ten Feet in Height from the Floor to the Ceiling; the ſecond ten Feet and a half; and the third nine Feet; the fourth eight Feet and a half. That all the ſaid Walls in Front and Rear as high as the firſt Story, be two Bricks and a half in Thickneſs; and from thence upwards to the Garret-Floor of the Thickneſs of one Brick and an half: That the Thickneſs of the Garret-Walls on the back Part be left to the Diſcretion of the Builder, ſo as the ſame be not leſs than one Brick.

And alſo that the Party-Walls between every *Houſe* of this third and larger Sort of Building be two Bricks thick as high as the firſt Floor, and thence upwards to the Garret-Floor, the Thickneſs of a Brick and half.

And be it further enacted, That in all *Houſes* of the fourth Sort of Building, being *Manſion Houſes*, and of the greateſt Bigneſs, not fronting upon any of the Streets or Lanes, as aforeſaid, the Number of Stories, and the Height thereof ſhall be left to the Diſcretion of the Builder, ſo as he exceeds not five Stories.

The ſame Act alſo enjoins, That no Timber be laid within twelve Inches of the Chimney Jaumbs; and that all Joiſts on the Back of any Chimney be laid with a Trimmer at ſix Inches diſtant from the Back: Alſo that no Timber be laid within the Funnel of any Chimney, upon Penalty to the Workman for every Default 10*s* and 10*s*. every Week it continues unreform'd.

Thus far the Act.

Note further, When you lay any Timber or Brick-Work as Taſſels (or Torſels) for Mantle-Trees to lie on, or Lintels over Windows, or Templets under Girders, or any other Timbers, they muſt be laid in Loam, which is a great Preſerver of Timber; whereas Mortar eats and corrodes it. Likewiſe the Joiſts Ends and Girders, which lie in Walls, muſt be loam'd all over, to preſerve them from the Corroding of the Mortar.

Some Workmen pitch the Ends of Timber that lie in Walls, to preſerve them from the Mortar.

Concerning Party-Walls.

In treating of theſe, I will preſent the Reader with two different Methods of valuing ſuch Walls, according to two different Surveyors, *viz.* Mr. *Leybourn* and Mr. *Philips*.

And first, according to Mr. *Leybourn*.

He says, forasmuch as the Buildings of *London* join one upon another, and almost every several *House* hath a distinct Proprietor, the Parliament hath decreed, That the Wall dividing the Proprietors Ground shall be built at the equal Charge of both the Owners: It will not therefore be unnecessary to shew how these Party-Walls are to be valued.

How all Brick Works, whether one, two, three, four, or any other Number of Bricks Lengths in Thickness, are all to be reduced to the Thickness of a Brick and a half.

It hath been observed, (saith he,) that about 4500 of Bricks, a Hundred and quarter of Lime, two Loads and a half of Sand, at 3*s. per* Load, will compleatly raise one Rod of Brick-Work of a Brick and a half in Thickness.

	l.	*s.*	*d.*
Now 4500 Bricks, at 16*s. per* 1000, is ———	3	12	0
A Hundred and quarter of Lime, at 10*s. per* Hund.	0	12	6
Two Loads and a half of Sand, at 3*s. per* Load ——	0	7	6
In all ——	4	12	6

And thus much will a Rod of Party-Wall (the Materials only reduced to a Brick and a half thick) amount to, at the former supposed Rates; to which may be added, for Workmanship, 1*l.* 8*s.* which added to 4*l.* 12*s.* will make 6*l.*

So that for every Rod of Party-Wall, they allow 3*l.* a-piece. Whence, if a Party-Wall be measured, and the Measure, when reduced to a Brick and half, be found to contain 16 Rods, that 16 Rods multiply'd by 3*l.* will give 48*l.* and so much is the one Proprietor to allow the other.

But here you are to note by the Way, That although this Rule here delivered be general, yet the Price of the Party-Wall will be more or less, according as Materials shall be cheaper or dearer; for sometimes a Rod or Wall of Brick-Work, of a Brick and a half thick, will cost but 5*l.* 10*s.* and then each Proprietor must pay but 2*l.* 15*s. per* Rod.

Thus far Mr. *Leybourn*. I shall next add Mr. *Philips*'s Way.

Now (says he) having the Dimensions, both in Length and Height, of the Cellar, and all other Stories in the *House*, then the following Tables will shew (according to the Thickness of the Wall) how many Bricks your Neighbour is to pay for towards his Party-Wall.

For which Purpose, the ensuing Tables will serve very well; for those Walls, according to the Act of Parliament for that Purpose, are to be made Part of them two Bricks thick, Part of them one Brick and half thick, and Part of them one Brick thick.

Now knowing the Number of Bricks which go to the making of the Wall, you may easily compute the Charge of the Mortar and Workmanship thereof, and from thence find the whole Charge;

Charge; which you will find (says he) about 30 *s.* for every 1000 of Bricks.

This Computation of Mr. *Philips*'s being made when Bricks were about 18 or 20*s. per* 1000, makes his Price too great; which if they be less, may not amount to but about 25 or 26*s. per* 1000.

He proceeds to an Example; as suppose a *House* of the third Rate, the Party-Wall of which being 30 Feet long, and you would know how many Bricks are to be paid for towards this Party-Wall.

First, Measure the Cellar, where the Party-Wall is to be two Bricks thick, the Length of which is 30 Feet, and the Depth 7 Feet; find this Length in the first Column, and the Depth in the Top of the Table; and in the Square of Meeting in the Table for one Brick thick, you will find 2314 Bricks are to be paid for.

Then proceed to the first Story, which will be likewise 30 Feet long, and 10 Feet high, and also two Bricks thick, the same Table shews the Allowance for this, which is 3306.

The second Story also is 30 Feet long, and 10 ½ high; but the Party-Wall is to be but a Brick and a half thick, the Half whereof is three Fourths of a Brick, yields for 30 Feet long, and 10 Feet high, 2479.

And for half a Foot more in Height 124.

The third Story is 30 Feet long, and 9 Feet high, being likewise a Brick and half thick; and for this the Table shews the Half to be paid for to be 2231.

The fourth Story is 30 Feet long, and 8 Feet and a half high, for the 8 Feet the Table shews 1983: And for the half Foot 124.

All which added together, make 12559, which are to be paid for the half of the Party-Wall; which, at 26 *s. per* Thousand, comes to 16 *l.* 6 *s.* 6 *d.*

Thus you may see what any Party-Wall comes to, though your Neighbour's *House* joins never so little or much to yours, as readily as you can by measuring by the Rod.

And whereas the Floors of the several Stories add somewhat to the Height, you may add something for them, according as you find them in Thickness.

Lastly, For the Garrets; the Walls of which being but one Brick thick, you may take half the Number in the Table of one Brick's Thickness, and add to the rest of the Account.

All the Difference that can be between Neighbours herein, will be about the Price of Bricks, and the Lime and Workmanship; but if Neighbours build together, they will easily determine it: But if they do not, yet the first Builder is sufficiently provided by his Workmen to rectify his Charge, and by Act of Parliament is allowed full Satisfaction, with Interest from the Time of Building.

By a Statute made in 22 *Car.* II. *cap.* 11. *it is enacted*, That no Builders shall lay Foundations, until that proper Surveyors (appointed by the Lord Mayor of the City of *London*, Aldermen, and Common-Council,) have viewed

viewed the ſame, and ſeen the Party-Walls and Piers equally ſet out.

But before ſuch Survey is taken, the Builders ſhall go to the Chamberlain, and enter their Names, and the Places where their Buildings are to be erected; and at the ſame Time pay 6*s.* 8*d.* taking an Acquittance for the ſame: And upon the Builders exhibiting the ſaid Receipt unto the proper Surveyors, or any of them, they ſhall ſurvey and ſet out the Foundation within three Days after ſuch Requeſt: And in default of Payment, the Chamberlain may ſue for it before the Mayor and Aldermen.

As to Party-Walls: The better to prevent Fire from having a free Paſſage from *Houſe* to *Houſe*, it is enacted by *Stat.* 19 *Car.* II. That between every two *Houſes* there ſhall be one Party-Wall of Brick or Stone, and of ſuch Thickneſs as hereafter mentioned.

And to prevent Diſputes between Landlord and Landlord, in reſpect to the Expences thereof, *it is hereby enacted,* That there ſhall be Patty-Walls and Party-Piers, ſet out equally on each Builder'sGround; and whoever firſt builds his *Houſe*, ſhall be obliged to leave a convenient Toothing in the Extremes of his Front and Rear Walls, that when his Neighbour, or Neighbours, is, or are diſpoſed to build up his or their *Houſe*, or *Houſes*, the Walls of them may be incorporated, and firmly bound together.

Nor ſhall the ſecond Perſon build againſt the ſaid Party-Walls, or on their own contiguous Grounds, until they have paid the firſt Builder the Moiety of the Charge of ſuch Party-Walls, with Intereſt at 6 *per Cent.* from the Beginning of firſt building: And provided that any Differences ariſe concerning the Value of ſuch Walls, they ſhall be referred to the Alderman of the Ward and his Deputy; and where one of them is a Party, or where they cannot compoſe ſuch Difference, the Lord Mayor and Court of Aldermen ſhall.

But by an Act made in the 7th Year of Queen *Anne*, intitled, *An Act for the better preventing of Miſchiefs that happen by Fires*, it is enacted, That the firſt Builder ſhall be paid by the Owner of the next *Houſe*, after the Rate of 5 *l. per* Rod, as ſoon as he ſhall have built the ſaid Party-Wall.

And in Conſideration that divers new *Houſes* have been, and may be erected ſingly on new Foundations, within the Limits of the Cities of *London* and *Weſtminſter*, or other Pariſhes or Places compriz'd within the Bills of Mortality, there was an Act made in the 11th Year of King *George* I. intitled, *An Act for the better regulating of Buildings*; which ſtrictly forbids all ſecond Builder or Builders, whomſoever, to make uſe of, or take the Benefit of ſuch Party-Wall and Fence-Wall ſo firſt built, at the Expence of the firſt Builder; nor ſhall any ſuch ſecond Builder or Builders, his, her, or their Executors, Adminiſtrators, or Aſſigns, on any Account whatſoever, lay any Wood, or Timber, or cut any Hole for Cup-Boards,

Preſſes, &c. in ſuch Party-Wall, under the Penalty of forfeiting the Sum of 50 *l.*

The Thickneſs of Patty-Walls, by 19 *Car.* II. were appointed to conſiſt of one Brick and half in the Cellars; and Stories above Ground, the Garrets excepted, which were to be of one Brick, or nine Inches Thickneſs only.

But by the Acts made in the 6th and 7th of Queen *Anne*, *it is enacted*, That from and after the firſt of *May* 1708, all and every *Houſe* and *Houſes*, that ſhall be built or erected upon any Foundations, either new or old, with the above Limits, ſhall have Party-Walls between *Houſe* and *Houſe*, wholly of Stone or Brick, and of the Thickneſs of two Bricks Length at leaſt in the Cellar and Ground Stories, and one Brick and a half, or 13 Inches upwards, from thence quite through all the remaining Stories, unto 18 Inches above the Roof.

And to prevent the ill Conſequences that may ariſe from Wood or Timber laid in Party-Walls, which may communicate Fire from one *Houſe* into the next, *it is enacted by the aforeſaid Act, of the 11th of King George* I. That it ſhall not be lawful to make or have in any Party-Wall of any *Houſe*, which after the 24th of *June* 1725. ſhall be erected or built within the preceding Boundaries or Limits, any Door-Caſe, Window, Lentil, Breſt-Summer, or Story-Poſts or Plates whatſoever, unleſs where two or more *Houſes* are joined or laid together, and ſo uſed as one ſingle *Houſe*; and that to be no longer than during the Time of ſuch Uſage, upon Pain of Penalty, that the Owner of every ſuch *Houſe*, for every ſuch Offence, ſhall forfeit the Sum of 50 *l.*

And in conſideration that Party-Walls built upon old Foundations may decay, and become dangerous, and needful to be rebuilt; and whereas Differenees have, and may again ariſe between the two Landlords, concerning the Expences of taking down the ſame, ſhoring up the Floors, and rebuilding them again; *it is therefore by the aforeſaid Act enacted*, That from and after the 24th Day of *June*, 1725. all and every Perſon and Perſons, inhabiting in any Place or Places, in and about the Cities of *London* and *Weſtminſter*, or any other Place or Places compriz'd within the Weekly Bills of Mortality, or within the Pariſhes of *St. Mary le Bone* and *Paddington*, or within the Pariſhes of *Chelſea* and *St. Pancras*, who ſhall build, or cauſe to be built, any *Houſe* or *Houſes*, upon any Foundation, old or new, and who ſhall find it abſolutely neceſſary to take down any decay'd Party-Wall between ſuch *Houſe* and the next adjoining *Houſe*, ſhall give Notice thereof in Wriring to the Owner or Occupier of ſuch adjoining *Houſe*, full three Months before ſuch Party-Wall ſhall be begun to be pulled down, to the Intent that the ſame may be viewed by four able Workmen, within the Space of one Month next after the Service of ſuch Notice; which four Workmen are to be equally appointed by both Patties, that is, each

each Perſon to appoint two of them, or more, if required, when they both do agree thereto.

But in caſe that the Landlord or Occupier of the next adjoining *Houſe*, will not agree to the rebuilding of ſuch Party-Wall, or Walls, or is uncapable of paying the immediate Moiety thereto, and ſhall neglect to nominate and appoint, within three Weeks next after the Service of Notice, as aforeſaid, ſuch Workmen, that then the other of the ſaid Parties ſhall nominate or appoint four or more able Workmen, who ſhall view the Party-Wall required to be taken down and rebuilt; which Workmen, or the major Part of them, ſhall certify in Writing under their Hands to the Juſtices of the Peace, in the next General or Quarter Seſſions of the Peace, holden for the City or County where ſuch Party-Wall is ſituated and being, and that ſuch Party-Wall is ruinous, and needful to be rebuilt, *&c.*

And provided that any Perſon or Perſons whomſoever, ſhall think him, her, or themſelves injured by ſuch Certificate, the ſaid Juſtices ſhall ſummon before them one or more of the ſaid Workmen, or other Perſon or Perſons whom they ſhall think fit, and ſhall examine the Matter upon Oath, and their Determination ſhall be final and concluſive to all Parties, without any Appeal from the ſame.

But it is to be obſerved, that a Copy of the Workman's Certificate muſt be delivered to the Occupier or Owner of ſuch next adjoining *Houſe*, or left there, within three Days after ſuch Certificate ſhall be made to the Juſtices, as aforeſaid; and if there ſhall be no Appeal from the ſame within three Months after, in every ſuch Caſe, if ſuch Landlord or Occupier ſhall refuſe or neglect to ſhore up and ſupport his, her, or their *Houſes*, within ſix Days after the Expiration of the ſaid three Months Notice, that then the firſt Builder or Builders, with his or their Workmen, (giving Notice as aforeſaid,) may lawfully enter into ſuch *Houſe* or *Houſes* (at all ſeaſonable Times,) with Workmen and Materials, and therewith ſhore up and ſupport the ſame; the Expence whereof ſhall be paid by the Landlord or Occupier; as alſo the half Expence of the Party-Wall built by the firſt Builder, after the Rate of 5 *l.* *per* Rod, for every Rod of Work contained therein.

And when the firſt Builder ſhall have built the ſaid Party-Wall, he ſhall leave at ſuch next *Houſe* with the Landlord or Occupier a true Meaſurement of the Quantity of Brick-Work contained therein, within ten Days after ſuch Party-Wall ſhall be ſo built and compleated; of which one half Moiety, at the Rate aforeſaid, as alſo the Expence of ſhoring and ſupporting, ſhall be paid by the Landlord or Landlords thereof, or their Tenants or Occupiers, who are hereby empowered to pay and deduct the ſame out of the next Rent that ſhall become due.

And provided, That Neglect or Refuſal of the Money ſo due be made, and remain unpaid for the Space of twenty-one Days after Demand thereof; then it ſhall

ſhall and may be lawful to and for ſuch firſt Builder or Builders, his, her, or their Executors and Adminiſtrators, to ſue ſuch Landlord or Landlords for ſuch Sums ſo proportionably due, by Action of Debt, or on the Caſe, Bill, Plaint, or Information, in any Court of Record at *Weſtminſter*, &c.

And here note, That the Law here delivered relating to the rebuilding of decay'd Party-Walls, of either Brick or Stone, the ſame is to be underſtood and obſerved of old *Houſes*, where inſtead of having one Party-Wall between them, as this Act directs, have two Timber Walls or Partitions, one belonging to each *Houſe*, and ſeparate from one another; therefore be it underſtood on all Sides, That whoſoever, for the Safety of his or their *Houſes*, will pull down his own Wooden Walls or Partitions, and inſtead thereof build a Party-Wall of Brick or Stone, he or they are alſo empowered to pull down the next Wooden Wall or Partition of the next adjoining *Houſe* or *Houſes*, (if the Landlord will not agree thereto,) and proceed in every Step, as before delivered for the rebuilding of decay'd Party-Walls of Brick or Stone.

Which new-built Wall muſt be placed equally on both Premiſes, that is to ſay, half the Thickneſs of the Foundation on one Landlord's Land, and the other half on the other; and that all Settings-off in the Foundation be equally the ſame on both Sides, as directed in the Beginning thereof.

The ſeveral Rates of *Houſes*, or *Buildings*, appointed after the Fire in 1666. were four.

Firſt, Thoſe of Allies, By-Lanes, *&c.* were termed Buildings of the firſt Rate, and were ordained to conſiſt but of two Stories, excluſive of the Cellars and Garrers, whoſe reſpective Heights were ſettled as follows, *viz.* the Height of the Cellar is ſix Feet and a half, the Height of the firſt and ſecond Stories each nine Feet, and the Height of the Garrets at Pleaſure.

The Scantlings appointed for the Timber of theſe Buildings, are as follows:

Summers or Girders, whoſe Lengths are not to exceed 15 Feet, muſt conſiſt of 12 Inches in Breadth, and 8 Inches in Depth or Thickneſs; and Wall-Plates 7 Inches by 5 Inches.

Principal Rafters, under 15 Feet, to be 8 Inches by 6 Inches at their Feet, and 5 Inches by 6 Inches at their Top. Single Rafters to be 4 Inches by 3 Inches; and Joiſts, whoſe Lengths are more than 10 Feet, muſt be 7 Inches deep, and 3 Inches in Breadth; excepting thoſe for the Garret Floors, which muſt be 3 Inches by 6 Inches.

And here obſerve, Stat. 22. *Car.* II. That no Joiſts or Rafters be laid at greater Diſtance from one another, than 12 Inches, and no Quarters at greater than 14 Inches.

Secondly, *Houſes* of the ſecond Rate are ſuch as front Streets and Lanes of Note, conſiſting of three Stories in Height, excluſive

clusive of the Cellars and Garrets.

The Height of the Cellars must be 6 Feet and a half, (if Springs will allow it;) the Height of the first and second Stories 10 Feet each, the Height of the third Story 9 Feet, and the Height of the Garrets at Pleasure.

The Scantlings appointed for the Timber of these Buildings, are as follows:

First, for the Floors.

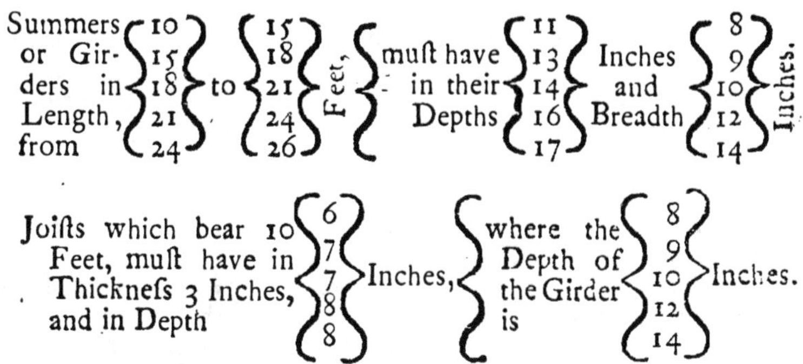

Summers or Girders in Length, from		to		Feet, must have in their Depths	Inches and Breadth	Inches.
	10		15	11		8
	15		18	13		9
	18		21	14		10
	21		24	16		12
	24		26	17		14

Joists which bear 10 Feet, must have in Thickness 3 Inches, and in Depth	Inches, where the Depth of the Girder is	Inches.
6		8
7		9
7		10
8		12
8		14

Binding Joists, with their Trimming Joists, 5 Inches in Breadth, their Depth equal to their own Floors.

Wall-Plates, or Raising-Pieces and Beams	Inches, and	Inches
10		6
8		6
7		5

Lintels of Oak in the	Story,		and		Inches.
first, second,		8		6	
and third		5		4	

Secondly, for the Roof.

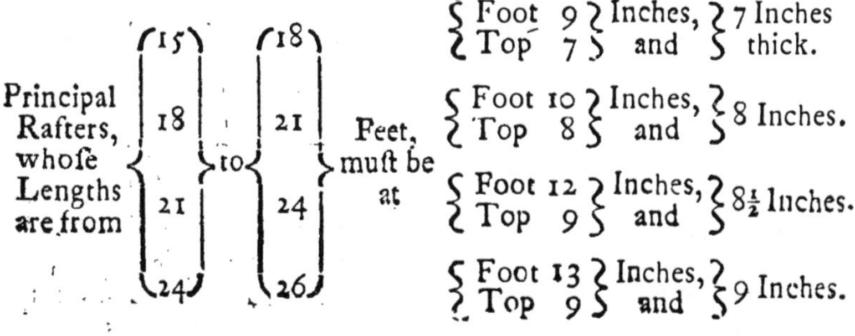

Principal Rafters, whose Lengths are from	to	Feet, must be at			
15	18	Foot 9 / Top 7	Inches, and	7 Inches thick.
18	21	Foot 10 / Top 8	Inches, and	8 Inches.
21	24	Foot 12 / Top 9	Inches, and	$8\frac{1}{2}$ Inches.
24	26	Foot 13 / Top 9	Inches, and	9 Inches.

Purlins,

Purlins, whose Lengths are from	15 / 18	to	18 / 21	Feet, must have in their Squares	9 / 12 Inc. by	8 / 9 Inc.
Single Rafters, whose Lengths do not exceed	9 / 6			Feet, must have in their Squares	5 / 4 Inches by	4 / 3½ Inch.

Thirdly, Buildings of the third Rate, are such as front the most principal Streets of Trade, as *Cheapside*, *Fleet-Street*, the *Strand*, *&c.* consisting of four Stories in Height, exclusive of the Cellars and Garrets.

The Height of the Cellars are as in the last preceding, the Height of the first Story 10 Feet, the second 10 Feet and a half, the third 9 Feet, the fourth 8 Feet and a half, and the Garrets at Pleasure.

The Scantlings of Timber appointed for this third Rate of *Houses*, are the same of those of the second.

The fourth Rate of *Houses* being such as are appointed for Persons of extraordinary Quality, situate in magnificent Squares, *&c.* may have the Height of their Stories and Scantlings of their Timber at Pleasure; but they must not exceed four Stories in Height, exclusive of the Cellars and Garrets.

And here it is to be noted, That the Height of the first Floor over the Cellars, in *Houses* of the second and third Rates, shall not be more than 18 Inches above the Pavement of the Street, nor less than 6 Inches, with a circular Step without the Building.

Scantlings of Stone appointed for the first, second, and third Rates of Buildings.

First Rate.

	Inches.		Inches.
Corner Piers	18	by	18
Middle or Single Piers	14		12
Double Piers between *House* and *House*	14		18
Door Jaumbs and Heads	12		18

Second and Third Rates.

	Feet.		Inches.
Corner Piers	2	by	6
Middle or Single Piers	1		6
Double Piers between *House* and *House*	2		18
Door Jaumbs and Heads	14 Inches by 10.		

As to Materials: And first of Quartering.

		Feet.					
Single	Quarters, whose	8	must	$3\frac{1}{2}$	and in	$1\frac{3}{4}$	Inches in
Double	Lengths are	8	have	4	Breadth	$3\frac{1}{2}$	Thickness

Secondly, of Laths.

Laths, whose	5	Feet, must have one	$\frac{1}{4}$	of an Inch in
Lengths are	4	Inch in Breadth, and	$\frac{1}{2}$	Thickness.

As to the Front and Rear Walls.

By the Stat. 19 of *Car.* II. *Houses* of the first Rate shall have their Cellar Walls in Front and Rear of two Bricks in Thickness, the first and second Stories of one Brick and a half, and the Garrets of one Brick only.

Houses of the second Rate shall have their Cellar Walls in Front and Rear two Bricks and a half in Thickness, the first and second Stories two Bricks, the third Story one Brick and a half, and the Garrets one Brick only.

Houses of the third Rate shall have their Cellar Walls in Front and Rear three Bricks thick, in the first Story two Bricks and a half, in the second, third, and fourth Stories one Brick and a half, and in the Gartets one Brick only.

Houses of the fourth Rate, being chiefly for Noblemen, *&c.* have their Thickness left to the Discretion of the Architect.

By Stat. 7 of Queen *Anne*, no Modilion or Cornice of Wood or Timber should hereafter be made, or suffered to be made, or suffered to be fixed under the Eaves of any *House*, or against any Front or Rear Wall thereof; but the Front and Rear Walls of every *House* and *Houses*, shall be built intirely of Brick or Stone, (the Windows and Doors excepted,) to be carried two Feet and a half high above the Garret Floor, and coped or covered with Stone or Brick.

Also by Stat. 7. of Queen *Anne*, it is enacted, That all Jaumbs and Backs of Chimneys, which shali or may be built, shall consist of one Brick in Thickness at the least, from the Cellars to the Roof; that all the Insides of such Chimneys shall be four Inches and a half in Breadth; that all Funnels shall be plaistered or pargetted within, from the Bottom to the Top; that all Chimneys be turned or arched with a Trimmer under the Hearths with Brick, the Ground Floor excepted; and that no Timber shall be nearer than five Inches to to any Chimney, Funnel, or Fire-Place; that all Mantles between the Jaumbs be arched with Brick or Stone, and no Wood or Wainscot shall be placed or affixed to the Front of any Jaumb or Mantle-Tree of any Chimney, nearer than five Inches from the Inside thereof.

That all Stoves, Boilers, Coppers, and Ovens, shall not be nearer than nine Inches, at the least, to the adjoining *House*; and no Timber or Wood to be nearer than five Inches to any Fire-Place or Flue.

But by Stat. 22 *Car.* II. *it is Enacted*, That no Timber be laid within twelve Inches of the Foreside of Chimney Jaumbs; and that all Joists on the Back of every Chimney be laid with a Trimmer of six Inches Distance therefrom; and that no Timber be laid within the Funnel of any Chimney, on Penalty to the Workman for every Default 10*s.* and 10*s.* more every Week it remains unreformed

A TABLE for one Brick in Thickneſs, or the Half of Two Bricks.

The Height of the Walls, in Feet.

Foot long.	Half Brick.	I Brick.	II Bricks.	III Bricks.	IV Bricks.	V Bricks.
1	5	11	22	33	44	55
2	11	22	44	66	88	110
3	16	33	66	99	132	165
4	22	44	88	132	176	220
5	27	55	110	165	220	275
6	33	66	132	199	264	331
7	39	77	154	231	309	386
8	44	88	176	264	353	441
9	50	99	198	298	397	496
10	55	110	220	331	441	551
11	61	121	244	364	485	606
12	66	132	264	397	529	661
13	72	143	286	431	573	716
14	77	154	309	462	617	771
15	83	165	331	496	661	826
16	88	176	355	529	705	882
17	94	187	375	562	749	937
18	99	198	397	595	793	992
19	105	209	419	628	837	1047
20	110	220	441	661	882	1102
21	116	231	463	694	926	1157
22	121	242	485	726	970	1212
23	127	253	507	760	1014	1267
24	132	264	529	793	1058	1322
25	138	275	551	826	1102	1377
26	143	286	573	860	1146	1432
27	154	309	617	926	1234	1543
28	165	331	661	992	1322	1653
29	220	441	881	1322	1763	2204
30	275	551	1102	1652	2204	2755

A TABLE for one Brick in Thickneſs, or the Half of two Bricks.

The Height of the Walls in Feet.

Foot long.	VI Bricks.	VII Bricks.	VIII Bricks.	IX Bricks.	X Bricks.
1	60	77	85	99	110
2	132	154	176	198	220
3	198	231	264	298	331
4	264	309	353	397	441
5	331	386	441	496	351
6	397	463	529	595	661
7	463	540	617	694	771
8	529	617	705	793	882
9	595	694	793	893	992
10	661	771	882	992	1102
11	727	848	970	1091	1212
12	793	926	1058	1190	1322
13	859	1003	1146	1289	1433
14	926	1080	1234	1388	1543
15	992	1157	1322	1488	1653
16	1085	1234	1410	1587	1763
17	1124	1311	1499	1686	1873
18	1190	1388	1587	1787	1983
19	1256	1466	1675	1884	2094
20	1322	1543	1763	1983	2204
21	1388	1620	1851	2083	2314
22	1455	1697	1939	2182	2424
23	1520	1774	2028	2281	2534
24	1587	1851	2116	2380	2645
25	1653	1928	2204	2479	2755
26	1719	2006	2292	2578	2865
27	1857	2160	2468	2777	3085
28	1983	2314	2645	2975	3306
29	2645	3085	3526	3967	4408
30	3306	3857	4408	4959	5510

A TABLE for three-Quarters of a Brick thick, being the Half of a Brick and a half.

The Height of the Walls in Feet.

Foot long.	Half a Brick.	I Brick.	II Bricks.	III Bricks.	IV Bricks.	V Bricks.
1	4	8	17	25	32	41
2	8	17	33	50	66	83
3	12	25	50	74	99	124
4	17	33	66	99	132	165
5	21	41	83	124	165	207
6	25	50	99	149	198	248
7	29	58	116	174	231	289
8	33	66	132	198	264	331
9	37	74	149	223	298	372
10	41	83	165	238	331	413
11	45	91	182	273	364	455
12	50	99	198	298	397	496
13	54	107	215	322	430	537
14	58	116	231	347	463	578
15	62	124	248	372	496	620
16	66	132	264	397	529	661
17	70	140	281	421	562	702
18	74	149	298	446	595	744
19	79	157	314	471	628	785
20	83	165	331	496	661	826
21	87	174	347	521	694	868
22	91	182	369	545	727	909
23	95	190	380	570	760	950
24	99	198	397	595	793	992
25	103	206	413	620	826	1033
26	107	215	430	645	860	1074
27	116	231	463	694	926	1157
28	124	248	496	744	992	1240
29	165	331	661	992	1322	1653
30	207	413	826	1240	1653	2066

A TABLE for three Quarters of a Brick thick, being the Half of a Brick and half.

The Height of the Walls in Feet.

Foot long.	VI Bricks.	VII Bricks.	VIII Bricks.	IX Bricks.	X Bricks.
1	50	58	66	74	83
2	99	116	132	149	165
3	149	174	198	223	248
4	198	231	264	298	331
5	248	289	331	372	413
6	298	347	397	446	496
7	347	405	463	521	579
8	394	463	529	595	661
9	446	521	595	660	744
10	496	579	661	744	826
11	545	636	727	818	900
12	595	691	793	893	992
13	645	752	860	976	1074
14	694	810	926	1041	1157
15	743	868	992	1117	1240
16	793	926	1058	1189	1322
17	843	983	1124	1264	1405
18	893	1041	1190	1339	1488
19	942	1099	1256	1413	1570
20	992	1157	1322	1488	1653
21	1041	1215	1388	1562	1736
22	1091	1273	1455	1636	1818
23	1140	1331	1521	1711	1901
24	1190	1388	1587	1785	1983
25	1240	1446	1652	1860	2066
26	1290	1504	1709	1934	2149
27	1338	1620	1851	2083	2314
28	1488	1736	1983	2231	2479
29	1983	2324	2625	2975	3306
30	2479	2893	3306	3719	4132

HOUSING, with Bricklayers, a Term which they use when a Tile or Brick is warp'd, or cast crooked or hollow in burning; then they say, *such a Tile or Brick is housing*. Tiles are apt to be *housing* or *hollow* on the Struck-Side (*i. e.* that which was uppermost in the Mould,) and Bricks on the contrary Side.

Some have made this Observation, That Tiles are always smoothest when burnt on the Struck-Side, by reason the Sand sticks to the Underside, which they strow on the Stock of the Mould, to prevent the Earth sticking to it.

HYDRAULICKS, so called of Ὕδωρ Water, and αὐλὸς, *Gr.* a Pipe or Flute; because, at the first Invention of Organs, being unacquainted with the Method of applying Bellows to blow them, they made use of a Cataract or Fall of Water, to make a Wind, and sound them.

The Organs, says *Vitruvius*, were played by the Help of two Suckers, which were pull'd up or let down in the Body of the Pump; which Suckers press'd the Air with Violence into a Funnel revers'd in a Copper Coffer, half full of Water, and pressed the Water, and constrained it so to ascend round about within the Coffer; which operated so, that its Weight, in making it reenter into the Funnel, push'd the Air into the Pipes, and made them play, producing the same Effects which the Bellows did.

Hydraulicks is that Part of the Science of *Staticks*, which considers the Motion of Fluids, and particularly Water, with the Application thereof in artificial Water-Works.

To *Hydraulicks* belong not only the conducting and raising of Water, with the constructing of Engines for those Purposes, but also the Laws of the Motion of Fluid Bodies.

Hydraulicks, therefore, comprehends the Art of conducting Water into Pipes, Canals, Drains, &c. Also the raising it, with the several Engines employ'd for that Purpose; as *Siphons*, *Pumps*, *Syringes*, *Fountains*, *Jets d'Eau's*, *Fire-Engines*, *Mills*, &c.

HYDROSTATICKS, [of Ὕδωρ Water, and στατική, *Gr. Staticks*] a Science that explains the *Equilibrium* of Fluids, or the Gravitation of Fluids at rest: Upon the Removal of that *Equilibrium*, Motion ensues; and here *Hydraulicks* commence.

Hydraulicks therefore suppose *Hydrostaticks*; and the Generality of Writers, from the immediate Relation between these two, join them together, and call them both either *Hydraulicks*, or *Hydrostaticks*.

But Mr. *Harris*, in his *Lexicon Technicum*, blames Mr. *Ozanam* for mixing and confounding *Hydrostaticks* and *Hydraulicks* the one with the other; since by the first is explained the natural *Equilibrium* or Motion of Water and other Fluids; and by the latter, the Force of mechanical Engines for the forcing it up to great Heights.

HYPÆTHRON } in antient
HYPÆTHROS } Architecture, a kind of Temple open at the Top.

Vitruvius ſays it is an open Building or Portico, ſuch as ſome antient Temples were, which had no Roof or Covering, as the Temple of *Jupiter Olympius*, built by *Coſſatius*, a *Roman* Architect at *Athens*.

HYPERBOLA, in Geometry, is one of the two Lines formed by the Section of a Cone.

The *Hyperbola* ariſes when the Plane that cuts the Cone is not parallel to one of the Sides, as it is on the *Parabola*; but diverges from it outwards, not inwards.

HYPERBOLIFORM *Figures*, are ſuch Curves as approach in their Properties to the Nature of the *Hyperbola*, called alſo *Hyperboloids*.

HYPERTHYRON, in the antient Architecture, is a ſort of Table uſed after the Manner of a Frieze over the Jaumbs of *Dorick* Doors and Gates, and the Lentils of Windows. It lies immediately under the Corona; and our Workmen uſually call it the *King-Piece*.

The END of the First Volume.

The following are ADDITIONS and CORRECTIONS communicated to the Compiler of this Work after the Sheets were printed off; therefore not being willing to omit any Thing that may be of Service to the Publick, but to make this Work as Compleat as possible, we have inserted them here by Way of

SUPPLEMENT.

IN the Article ASHLERING, instead of 4 *d.* to 6 *d.* read from 18 *d.* to 2 *s.* *per* Square.

In BALUSTER, for 3 *d.* *per* Yard, read 3 *s.* *per* Yard; tho' the Prices are various, according to the Goodness of Workmanship.

In the Article BARNS, have no Dependance on the Prices set down.

In the Article BATTEN *Doors*, as to their Price, have no Dependance; for no Price can be set on them, without knowing the Dimensions.

BATTER is a Term used by Workmen, to signify that a Wall, Piece of Timber, or the like, doth not stand upright, but leans into the Building; if it leans from the Building, they say it *over-hangs*.

BAULKS } Are small young
BALKS } Fir-Trees, the slender Tops being cut off, and hew'd up, brought from *Norway*.

Load

Load: BAULKS, are large Pieces of Fir-Timber, which are brought from *Norway*, from one Load to four or five in one Piece.

In the Article BEAM FILLING, add 3*d*. or 4*d*. *per* Foot has been given for Workmanship, where it has been troublesome; as in a Country Church, or where they have been obliged to scaffold, or use long Ladders.

BOND, a Term among Workmen, chiefly Bricklayers, who say, *Make good Bond*; by which they mean so to dispose the Bricks or Stones, that the Joints may not be immediately over others.

First, a Bricklayer lays a Stretcher, or Brick long-ways in the Building, beginning at the Corner, and so on all Stretchers in that Course; then upon that he lays next a Header, beginning at the same Corner; next to that a Closer, which is Part of a Brick, about two Inches; which, with the Header already laid, is about six Inches and a half with the Mortar between them, then there is left about two Inches and a half for *Bond*, as they call it, which will cause the Middle of the next Header to lie over the Joint of the two Stretchers of the under Course; and so they lay Headers all along in the same Wall, which they call *Flemish Bond*.

Or first they lay a Header, then a Closer, next a Stretcher, then a Header, next a Stretcher, and so on to the End of the Wall; then on the next Course a Stretcher to begin with, which will span over the under Header and Closer, and cover about two Inches of the under Stretcher; then next a Header, and so on to the End of the Wall, which they call *English Bond*.

To the Article BOULDER-WALLS add, Some Workmen lay Laths in the Wall angleways, and then cross them somewhat like a Net, every two or three Feet in Height, which prevents it falling down in moist and rainy Weather.

In the Article BRICK-WORK, instead of *thinner*, read *thicker*.

Also to that Paragraph, the *Base of the Gable*, in the same Article, add, or the Base of the Gable being 24 Feet, take three Fourths of that, which is the Length of the Rafter, which is 18 Feet; three Fourths of that, which is 13 Feet 6 Inches, is the Perpendicular nearly.

So few Writers having said any Thing of Timber-Bridges, a late Author having presented us with the following Plan of one, I have here inserted it. This he explains as follows:

Let A* be the Plan of the Bridge, supposed to extend any Length, not exceeding one hundred Feet, nor twenty-four Feet in Breadth; let B* be the Side or Upright of the same; and let C* be the Section of the same by a larger Scale.

The better to conceive the Particulars, he directs to observe in A*, that *aaaa* are the Butment or Support to each Shore; and let *bb* be the tying Beams, which are

are halved into the Posts; also let *cc* be the bearing Beams; and let *dddd* be the binding Joists, which are let into the bearing Beams, as in the Plate C* D* at T; also let *eeee* be the Plan of the several King-Posts.

And in B observe, that *ff* is the Top of the Water at its common Level, and let *gg* be the Butments or Support to each Shore; also let *hh* be the tying Beams, as halved into the Posts; let *ii* be the Plate for the Braces *ll* to rest on, which support the Posts *kk*; so do the Braces *mm* discharge the whole Weight; and let *nn* be Struts to help the Strength, as by butting against each Brace; let *ooo* be the top Place or Rail, and *pp* a Plank weathered to throw the Water off.

N.B. The additional Beams &c &c do add prodigiously to its Strength.

And in C*, which is the Section by a larger Scale, let *qq* be the Posts, and *rr* the bearing Beam, framed therein, and let *ss* be the binding Joists: Also let *tt* be the top Rail, being wider than the rest, to preserve the Joints the better; and let *uu* be the Plank weathered to throw the Water off; yet better, as at *ww*.

It is necessary to let the tying Beam into the Posts a small Matter, because the Plank *xx* bears on it, as well as on the binding Joists; let *yy* be Straps of Iron bolted through the Posts, in order to strengthen the same; the lower Bolt goes through the said Strap, and comes under the bearing-Beam, and which, with the Joggle *zz*, preserves a good Bearing for the Beam, which ought to be truss'd, as shewn in the Plate B; and &c &c is the Gravel and Paving.

To preserve the Timber the better, let the Truss B* be boarded on each Side.

In the Article CANT, for *turn it about*, read *turn it over*.

In the Article CEILING, as to the Price, add, This is to be understood of the Journeyman's Price from the Master; and also the Price, with Materials, is for common Camp Work; for Work which is done very well is worth twice that Sum.

To the Article CHIMNEYS add this, which is a more easy and natural Method.

Let the Stack of *Chimneys* to be measured, be as in the Plate.

First, prepare the Measuring-Book, by ruling it into ten Columns; the first for Remarks, the second for so many Times over as you are to measure Things of the same Dimensions. As for Example: If you have two Hearths in a Stack of *Chimneys* of the same Dimension, and on one Floor, you need set but one down in your Book, and say twice over, that is, put down 2 in the second Column, and put down double the Product under that Word, fifth Column; if the first Dimension was to be done twice over, you set down 2 in the next Column on the

the ſecond Column over-againſt 6 *F.* 6 *I.* and make the Product 9 Feet 8 Inches.

The third Column is for the Dimenſions; the fourth for ſo many Bricks as the Wall is thick; the fifth for the Product of the Dimenſions, when multiplied together; the ſixth for the Products of the Deductions; the ſeventh for reducing the Products into 1½ Brick thick, as the eighth is for one Brick; and the other two for reducing the Deductions to their Thickneſs.

If you are to reduce the firſt Dimenſions, whoſe Product is 4 10, and 5 Bricks thick, multiply it by 5, which is 24 2 in one Brick. The next Dimenſion is 6 6 by 1 1, whoſe Product is 7 Feet (not regarding the odd 6 Parts) and 4 Bricks thick; put 7 twice down in the ſeventh Column, which is a Three-Brick Wall, and 1 down in the eighth Column for one Brick more, which is the four Bricks Thickneſs.

If you are to reduce the Product of 13 1, which is a Deduction, and 2½ Bricks thick, put firſt 13 1 in the ninth Column, and 13 1 in the tenth.

If you are to reduce the Product of 9 2, F. 1⅓ Brick thick, put 9 2 in the eighth Column, and ⅓ of 9 2 under it in the ſame Column, for that will be 12 2 in one Brick in Thickneſs, and ſomething more; but thoſe odd Parts are ſeldom regarded in Practice; it gives the Turn of the Scale to the Maſter, and amounts to a very ſmall Matter at the End.

When all this is done, you add up the four laſt Columns; the firſt, or ſeventh Column is 331 3 reduced to 1½ Brick; the next 216 4 of 1 Brick thick, then you deduct the third or ninth Column of 1½ Brick from 331 3, there remains 312 8; then deduct the laſt Column 61 8 from the eighth, 216 4, there remains 154 8; which multiply by 2, and divide by 3, which brings 103 the Thickneſs of 1½ Brick, which add to 312 8, is in all 415 8, the Standard Thickneſs of a Brick and half Wall; which divide by 272 (the ¼ being not regarded in this Work) there will be 1 Rod 143 Feet, which divide by 68, will bring it to Quarters and 7 Feet remaining; which is in all 1 Rod ½ and 7 Feet, as will appear by the Table.

Remarks,

Remarks.	Times over.	Dimensions.	Bricks thick.	Product.	Product deducted.	Reduced to $1\frac{1}{2}$ Brick.	Reduced to one Brick.	Deductions reduced to $1\frac{1}{2}$ Brick.	Deductions reduced to one Brick.
		F. *I.*		*F.* *I.*			*F.* *I.*		
Fronting of Chimney		6 6 9	5	4 10			24 2		
Foundation above setting off		6 6 1 1	4	7 0		7 7	7		
Deduct		3 6 1 7	$2\frac{1}{2}$		5 6			5 6	5 6
First Story		9 7 5 10	4	55 11		55 11 55 11	55 11		
Deduct Chimney		3 6 3 9	$2\frac{1}{2}$		13 1			13 1	13 1
Second Story		8 7 5 6	4	47 2		47 2 47 2	47 2		
Deduct Chimney		3 9 3 0	$2\frac{1}{4}$		11 3				11 3 11 3 2 9
Part of lower Funnel		8 7 1 10	1	15 9			15 9		
Third Story		7 6 4 3	$3\frac{1}{2}$	31 10		31 10 31 10	15 11		
Deduct Chimney		3 3 2 9	2		8 11				8 11 8 11
Part of lower Funnel		5 0 1 10	$1\frac{1}{2}$	9 2		9 2			
Part of the other		5 0 1 10	$1\frac{1}{3}$	9 2			9 2 3 0		
Shaft		9 0 4 3	$2\frac{1}{2}$	38 3		38 3	38 3		
						331 3	216 4	18 7	61 8
						18 7	61 8		
						312 8 103	154 8 2		
					Standard Feet	415 8	308 1 4		
						3)	309 4	(103	

272) 415 (1 Rods.
68) 143
136 (2 Quarters.

7 Feet.

In all $1\frac{1}{2}$ Rod and 7 Feet, at 5 *l.* 15 *s.* *per* Rod, } 8 15 $5\frac{1}{2}$

To bring this into Money, look on the other Side.

At five Pounds fifteen Shillings *per* Rod, what is the Amount of one Rod and a half and seven Feet? First,

One Rod is	5	15	00
Half a Rod is	2	17	06
Seven Feet is as below	0	02	$11\frac{1}{2}$
Whole Amount	8	15	$05\frac{1}{2}$

For the 7 Feet bring the Pounds into Shillings, which is 115; which multiply by 12, to bring it into Pence, which is 1380 Pence, the Pence in five Pounds fifteen Shillings: Then say,

If 272 Feet be worth 1380 Pence, what is 7 Feet?

```
              7
         272) 9660 (35 Pence.
              816
              ----
              1500
              1360
Divided by 68) 140 (2 Farthings.
              136
              ---
                4
```

Which is the Quarter of 272, which will bring it into Farthings; as to what remains, is but $\frac{4}{68}$ of a Farthing, and not worth regarding.

A *Chimney*'s Proportion, according to some Moderns.

Chimneys in	Breadth.	Height.	Depth.
Kitchen	6, 8, or 10	$4\frac{1}{2}$, 5, or 6	$2\frac{1}{2}$, or 3
Halls	4, 5, or 6	4, or $4\frac{1}{2}$	2, or 2-3
Chambers	3--6, to 4	3--9 to 4	22 Inches.
Studies and Wardrobes	2--6 to 3	3-6 to 3-9	18 Inches.

To

To the Article CLINKERS add, a sort of small yellow Bricks which are brought from *Holland*.

To the *Method of drawing a* COLUMN, add this engraven *Column* following, which is by far more preferable.

In the Article CONTENT, for *forty-three* read *forty*.

In the Article COPING, instead of 4 *d. a Foot*, read *that it will cost in* London 2 s. *a Foot running Measure*.

In the Article CORNICE, in the Paragraph beginning thus, Mr. *Wing* says, &c. add, Mr. *Wing*'s Free-Stone *Cornice* is very cheap: He should have told us where it was done for that Money, for it will not suit *London*.

The Price of Free Stone or *Bath* Stone *Cornice* about *London*, of about 18 or 20 Inches thick, is worth about 10 or 11 *s. per* Foot running Measure of *Portland* Sone, 15 or 16 *ditto*.

Wooden *Cornice* is worth about 10 *d. per* Foot superficial Measure.

DENTILS. *Vitruvius* is said to prescribe the Breadth of each *Dentil* or Tooth to be its Height, and the Indenture, or Interval between each two, he directs to be two Thirds of the Breadth.

As *Dentils* are much in use among the modern Workmen, so are they fit to be recommended as a very pretty and beautiful Ornament, if perform'd well.

Lest the young Workman should fall into any Mistake, by what is said above from so great an Author as *Vitruvius*, I shall make the following Observation, and give you the Words of Mr. *Evelyn*, in his Account of *Architects* and *Architecture*, page 35 and 36.

I do not remember any of the Architects who have wrote since *Vitruvius*, that gives these Dimensions to *Dentils*. They may appear in large and massive Buildings very grand, according to *Vitruvius*; but the Proportions which are given them by other Masters suit best with our modern Buildings: And what *Vitruvius* means by Modilions, representing Purlins, *Dentils*, and the Ends of Rafters, I know not, for I never understood *Dentils* to represent or signify any Thing more, than a Row or Gang of Teeth.

Dentils (saith Mr. *Evelyn*) are the Teeth (a Member of the Cornice) immediately above the Cymatium of the Frieze, by some named also *Asseri*, from their square Form; I say, in the *Corinthian* and *Ionic*, &c. for in the *Doric* Order they were not antiently admitted, or rather properly, according to the Opinion of our Master, (viz. *Vitruvius*,) though we must needs acknowledge to have found them in the most authentical Pieces extant. As for their Dimensions, they kept to no certain Rule, but made them sometimes thicker, sometimes thinner, square or long, and more in Number; commonly the Spaces less by an half, sometimes by a third Part, than the Teeth, which were themselves twice as high as their Breadth, and frequently (especially

cially in more polite Orders,) beginning with the Cone of a Pine pendent at the very Point over the angular Column. *Lomatius* is yet more precise in this Particular, and gives them as much Height as the middle Fascia of the Architrave Projecture, equal (somewhat too much,) Front, twice the Breadth of their Height, and a third Part less than their Breadth, for Vacuity. The *Dentils* have sometimes a small Regula, and now and then more than one, as usually in the *Ionica*, where it has likewise an Ovolo or Echinus for the Bedding of the Corona; but if enrich'd, and that two of them encounter, one should be simple and plain, as where it happens to be inserted beneath it. Next to this superior Echinus are the Modilions; but instead of them *Dentils* are thought to have been first instituted, and for that Reason superfluously join'd where Mutules are; and therefore, where we find Tænia under Modilions, it is not properly divided into Teeth; nor is it rashly to be imitated, though we have some great Examples to countenance it. That of the *Pantheon* may safely guide us herein, where it is left plain for this very Cause, and that the Reason of the Thing does not in Truth allow it. However, it must be acknowledged, nothing has been more grosly abused; even amongst our most renowned Masters.

In the Article DIAL, and in the Paragraph, The best Wood for this Purpose is the *clearest Oak, and the reddest Fir, if it be not turpentiny*, read *the clearest Wainscot, and yellow Fir, clear of dead Turpentine Knots.*

In the Article DOORS, and Paragraph beginning thus, In small Buildings, &c. add, the Moderns seldom exceed three Feet for the Front *Doors* of small Buildings, and the Chambers from two Feet four to two Feet ten, and two Diameters and one Third in Height.

In the Article FACIA, and to the Paragraph beginning thus, The Price of Fascia's is, &c. add, the Price of Brick Fascia's, with Materials, is one Shilling two Pence *per* Foot superficial Measure, Moulding on the same is one Shilling and ten Pence *per* Foot superficial Measure.

In the Article FEATHER-EDG'D, instead of *Side* read *Edge.*

To the Article FENCING, add, these Prices of *Fencing* here mentioned, may be what poor labouring Men may have in the *Weald* of *Sussex*, but are not fit for the rest of the Kingdom.

Paling with three Rails and Pales, is worth in some Parts of *Kent* fifteen or sixteen Shillings *per* Rod, if done well, finding all Materials. And *Paling* with two Rails and Pales, is worth thirteen or fourteen Shillings *per* Rod; Posts and Rails cross a Field is worth four Shillings *per* Rod, finding all Materials.

In the Article FLOORS, and in Paragraph The Price, &c. after *eleven Shillings*, add the Word *Workmanship.*

To the Article FLOORING, to the Paragraph beginning thus, Mr. *Leybourn* says, *&c.* add,

The Price of Boarding *Floors* in and about *London*, is as follows:

Boarding with whole yellow Deals, with folding Joints, from twenty-two to twenty-four Shillings *per* Square.
Ditto, Sap-fisted, that is, the Sap cut all out, two Pounds *per* Square.
Common straight Joint Boarding, thirty-five Shillings *per* Square.
Second best *ditto*, forty-six Shillings *per* Square.
Dowell'd, fifty-six Shillings *per* Square.
Chan *Floors*, i. e. Boards without Knots, and dowell'd, from five to seven or eight Pounds *per* Square.

In the Article FRAMING add, Mr. *Leybourn* certainly means that he fells, hews, and finds the Timber into that Price. Instead of twelve Shillings *per* Ton, it should be forty Shillings *per* Ton; add that to the rest, and it will be about a *Sussex* Price for Oak.

If you would estimate the Value of a Square of *Framing* for a Barn after the *Sussex* Method, for some Place near *London*, it is thus:

	l.	s.	d.
To twenty-five Feet of *Mardo* Fir Timber, at thirty Shillings *per* Load	0	15	0
To sawing *ditto*	0	03	0
To *Framing*	0	05	0
To Weather-Boards, sixteen at nine Pence *per* Piece, and Coin Boards	0	13	0
To Work and Nails	0	05	0
	2	01	0

This is above three Times the *Sussex* Price; but it is easy to see where he failed in his Estimate: He has under-valued the Timber much, as to Price; has accounted for sawing the Boards, but says nothing about the Timber they are to be sawed out of; the Sawing too little by one Third, for he supposeth rough Timber: In short, it is all Blunders.

But if the Workman would make a true Estimate of a Barn, the Scantlings must be all ascertained and fixed, the Dimenſions of the Barn given, as the Length, Width, and Height, if the Planks be Oak, and the reſt Fir.

Firſt, cube the Plates, that is, meaſure how many ſolid Feet there is in the ſaid Plates; then cube the Fir, that is, find the ſolid Content in Feet in the whole Carcaſe in the large and ſmall Timbers; then find how many Squares of Weather-Boarding, and Squares of Thatching, with what Locks, Hinges, &c. is wanted, according to your Agreement; then make a fair Bill of it all, as if it were already done, as under: This will take up ſome Time, (for I would adviſe the Workman to draw it all up on Paper, with the ſeveral Scantlings, which will make it very eaſy to compute, and he will avoid Miſtakes,) but it will anſwer his End, for he will be ſure of his Gain before-hand, and not work by Gueſs, as is the Way among moſt Workmen that are not acquainted with Figures. And here I cannot but condemn the Method uſed by the ordinary Workmen in *London*, of meaſuring the Area or Ground Plat of the Building only, and making their Eſtimate from thence, as erroneous.

For let the Plan be ten Feet ſquare, as for Inſtance, a Summer-Houſe, the Walls of that Building will be forty Feet about, and the Area one Square; then admit another Building of twenty Feet ſquare, which is four ſquare on the Plane or Area, the Wall of this Building will be 80 Feet about; ſo the building of one Square hath Walls half the Quantity of that of four ſquare: But this is but one Error among a great many that accrue in this Method of eſtimating; and I would adviſe the young Workman to have no Regard to this lazy and idle Method, leſt he pay too dear for it in the End.

The Manner of the Bill for an Eſtimate.

	l.	*s.*	*d.*
To ſawed cub'd Oak in Plates, 40 Feet at 3*s.*	06	00	0
To ſawed cub'd Fir in Carcaſe, 571 Feet at 20*d.*	47	11	8
To 28 Squares 75 Ft. broad Weather-boarding, at 18*s.*	25	17	6
To 18 Squares Pantiling, at 18*s.*	16	04	0
To 160 Feet Underpinning, at 6 *d. per* Foot	04	00	0
To Hinges, Locks and Staples, &c.	00	14	0
	100	07	0

The above Prices are the *London* Prices for Work and Stuff.

FRONT

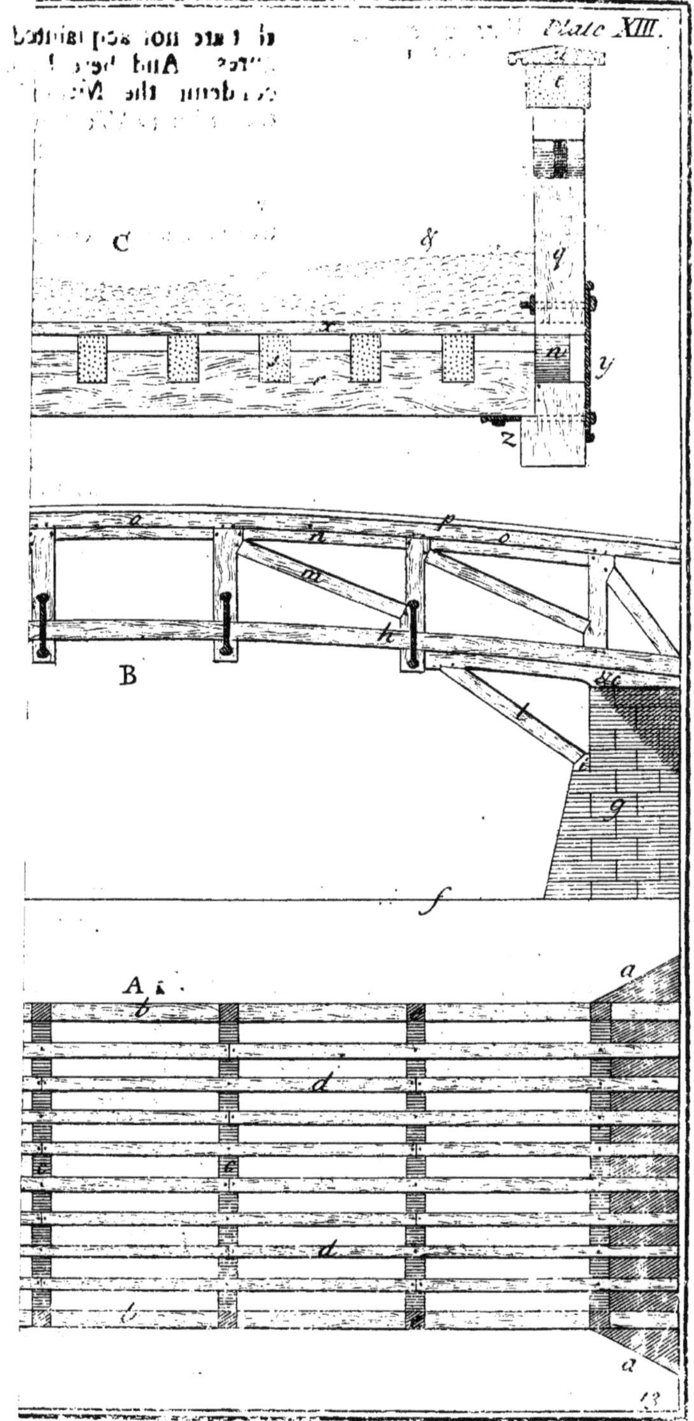

Plate XIII.
C
B
A

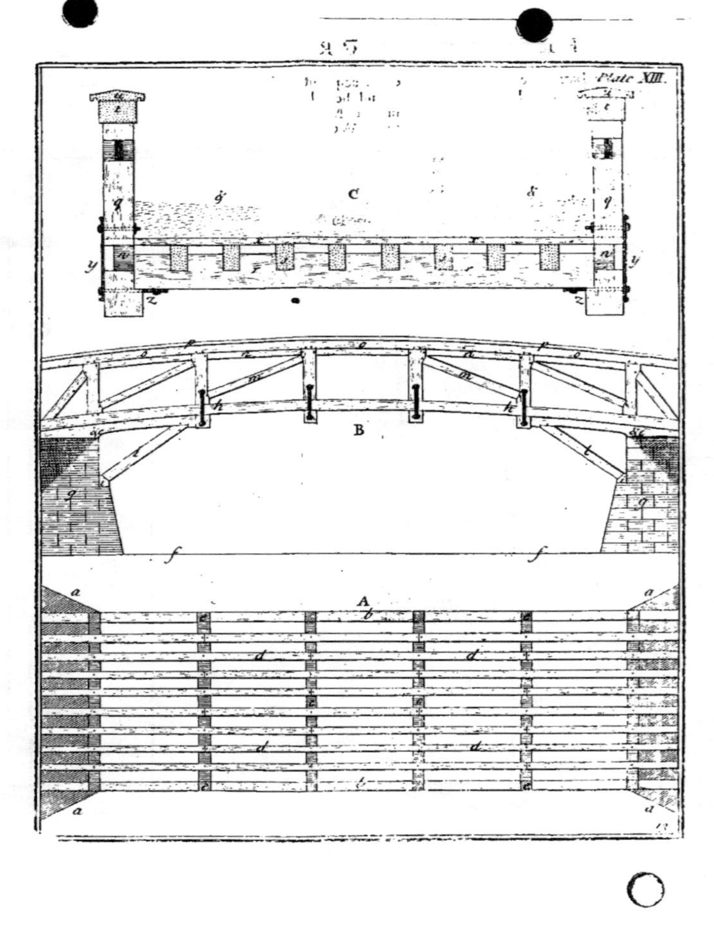
Plate XIII.
A
B
C

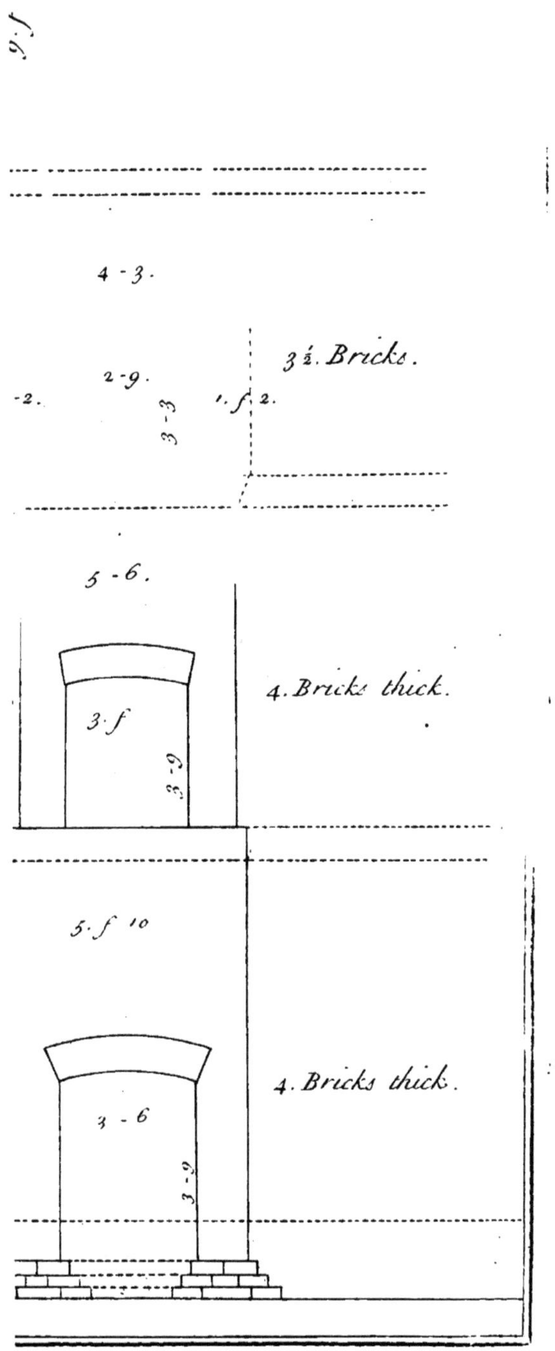

9.f

4 - 3.

3½. Bricks.

2 - 9.

-2.

3 - 3

1. f. 2.

5 - 6.

4. Bricks thick.

3.f

3 - 9

5.f 10

4. Bricks thick.

3 - 6

3 - 9

Plate XIV.

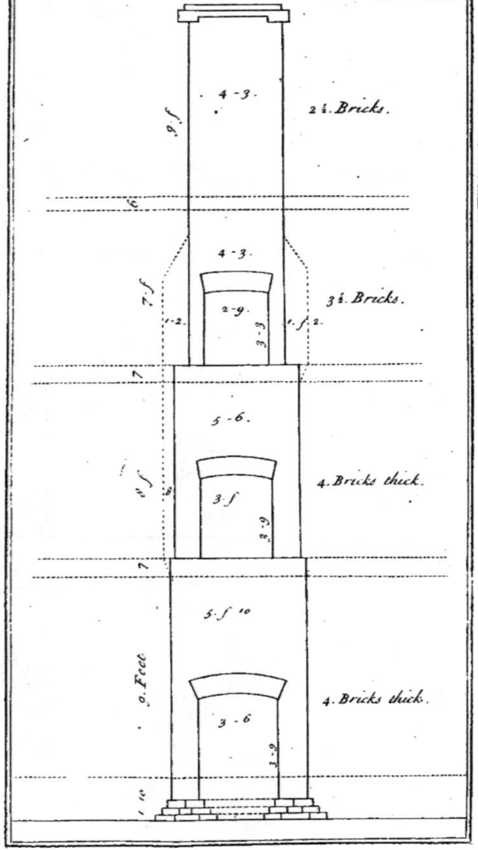

2½. Bricks.
3½. Bricks.
4. Bricks thick.
4. Bricks thick.
4 - 3.
4 - 3.
2 - 9.
3 - 3
1 - 2.
1. 5 2.
5 - 6.
3. 5
3 - 9
5. 5 10
3 - 6
3 - 9
9. 5
6
7. 5
7
7
9. Feet
1 10

Plate XV.

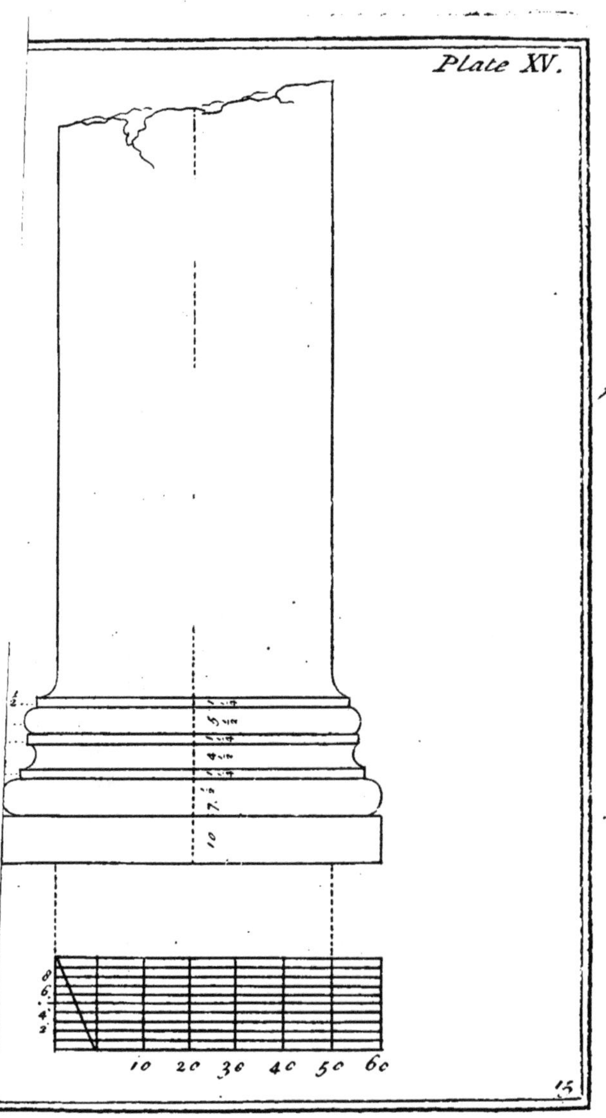

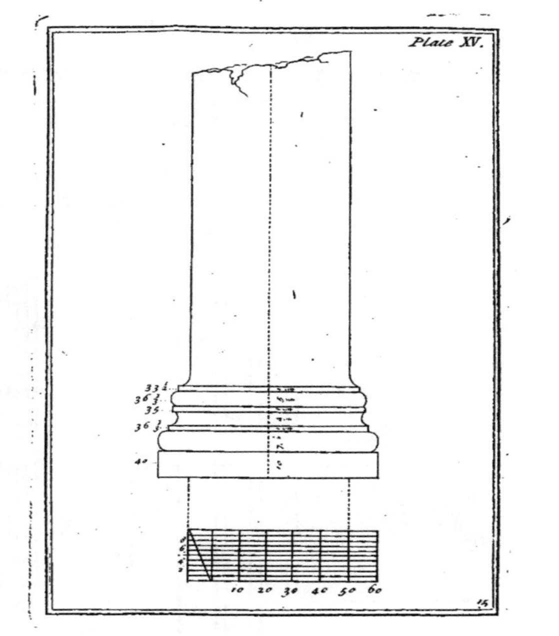
Plate XV.
33 1/2
36 2/3
35
36 2/3
40
10 20 30 40 50 60
15

FRONT. To this Article add, There is no certain Price for this Work by the Rod, but it is always done by the Foot, the Price more or leſs, according to the Goodneſs and Variety of the Workmanſhip.

GUTTERS. To this Article add, *Gutters* ſhould never have leſs than a quarter of an Inch to a Foot for Drip, and the Soldering croſs the *Gutter* is always to be avoided, and the Length of the Lead, from Fall to Fall, ſhould never exceed 14 Feet.

In the Article HOUSE, to the Paragraph beginning thus, Some Workmen pitch the Ends of Timber, &c. add, The beſt Way to preſerve the Ends of Timber in the Walls, is to let them have Air, and nothing to touch them.

In the Article about LATHS, inſtead of four Inches in Breadth muſt have half an Inch in Thickneſs, read, a quarter of an Inch.

N.B. We hope the Reader will make all proper Allowances in Prices, when better Work or Materials ſhall exceed the Scheme of theſe Computations.

BOOKS *Printed for* A. Bettesworth *and* C. Hitch, *in* Pater-noster-Row; *and* S. Austen, *in* St. Paul's Church-Yard.

A Book of Psalmody, containing Chanting Tunes for *Venite Exultemus, Te Deum Laudamus, Benedicite, Jubilate Deo, Magnificat, Nunc Dimittis*, and the Reading Psalms; with 18 Anthems, and Variety of Psalm Tunes. In Four Parts. The Seventh Edition, Corrected and Enlarged by JAMES GREEN. Price stitch'd 2 *s.* being the compleatest Book extant.

CAMPANOLOGIA: Or, The Art of Ringing. Improved and Compleated. The Second Edition. Price 1 *s.* 6 *d.*

A New ENGLISH DISPENSATORY. In Four Parts. Containing, I. A more accurate Account of the Simple Medicines, than any hitherto extant. II. The Officinal Compositions, according to the last Alterations of the College at *London:* To which are added, The Emendations of the *Edenburgh Dispensatory*; and many other Compositions, taken from the Practice of our Hospitals, and the most Celebrated Authors. III. Extemporaneous Prescriptions, taken from the best Authors, and the most Eminent Physicians now in Practice. IV. A Rational Account of the Operation of Medicines. To which are added, The Quantities of the middle Syllables of the *Latin* Names, express'd by long and short Marks. So that this DISPENSATORY answers at the same Time the Purpose of a *Prosodia Pharmaceutica.* By JAMES ALLEYNE, M. D. Price 6 *s.*

The ANTIQUITIES of LONDON and WESTMINSTER; being an Account of whatsoever is Antient, Curious, or Remarkable, as to Palaces, Towers, Castles, Walls, Gates, Bridges, Monasteries, Priories, Sanctuaries, Nunneries, Religious Houses, Cathedrals, Churches, Chapels, Colleges, Inns of Court, Hospitals, Schools, and other Magnificent Buildings; as Exchanges, Halls, Crosses, Markets, Goals, and all publick Edifices; also Rivers, Brooks, Bourns, Springs, &c. and many other Curious Matters in Antiquity; whereby will plainly appear the Difference between the Antient and Present State of those two famous Cities.

The Second Edition. Price 2 *s.*

Lightning Source UK Ltd.
Milton Keynes UK
UKHW02f2129090818
327015UK00011B/690/P

9 781332 956623